THE LIBRARY
ST. MARY'S COLLEGE OF MARYLAND
ST. MARY'S CITY, MARYLAND  20686

R.V. Fisher · H.-U. Schmincke

# Pyroclastic Rocks

With 339 Figures

Springer-Verlag
Berlin Heidelberg New York Tokyo 1984

Professor RICHARD V. FISHER
Department of Geological Sciences
University of California
Santa Barbara, CA 93106, USA

Professor HANS-ULRICH SCHMINCKE
Institut für Mineralogie
Ruhr-Universität Bochum
4630 Bochum, FRG

ISBN 3-540-12756-9 Springer-Verlag Berlin Heidelberg New York Tokyo
ISBN 0-387-12756-9 Springer-Verlag New York Heidelberg Berlin Tokyo

Library of Congress Cataloging in Publication Data. Fisher, Richard V. (Richard Virgil), 1928–. Pyroclastic rocks. 1. Volcanic ash, tuff, etc. I. Schmincke, Hans-Ulrich. II. Title. QE461.F55. 1984. 552'.23. 83-20042.

This work is subject to copyright. All rights are reserved, whether the whole or part of the material is concerned, specifically those of translation, reprinting, re-use of illustrations, broadcasting, reproduction by photocopying machine or similar means, and storage in data banks. Under § 54 of the German Copyright Law, where copies are made for other than private use, a fee is payable to "Verwertungsgesellschaft Wort", Munich.

© by Springer-Verlag Berlin Heidelberg 1984
Printed in Germany

The use of registered names, trademarks, etc. in this publication does not imply, even in the absence of a specific statement, that such names are exempt from the relevant protective laws and regulations and therefore free for general use.

Media conversion, printing and binding: Brühlsche Universitätsdruckerei, Giessen.
2132/3130-54321

*For Beverly and Irma,*
*Susan, Dianne, Dick and Peter*
*Anna, Polly, Max and Paco*

# Preface

> "What is volcanic ash? Volcanic ash is not ash at all. It is pulverized rock. A one-inch layer of dry ash weighs 10 pounds per square foot as it lands. It often contains small pieces of light, expanded lava called pumice or cinders. Fresh volcanic ash may be harsh, acid, gritty, glassy, smelly, and thoroughly unpleasant."
>
> *Leaflet by the U.S. Federal Emergency Management Agency published for the public following the May 1980 eruption of Mount St. Helens.*

Perhaps the most challenging aspect of writing this book has been the explosive increase in publications on "ash", or "tephra", to use the more general term, which has forced us to revise many of the chapters, several of them more than once. We started to write this book some 10 years ago because we recognized a need for a summary on pyroclastic rocks, their physical aspects and stratigraphic relationships. Our intention was to provide a guide through the bewildering array of volcaniclastic rock types that are so characteristic of volcanic complexes, volcanic fields, and volcaniclastic materials in sedimentary basins.

We need not emphasize that – as with many other rock types – the past is the key to the present. The very nature of large explosive volcanic eruptions makes direct observations of processes in the vent, in the eruption column, or in live nuées ardentes impossible, and so our understanding of eruptive and transport processes must come largely from detailed stratigraphic and sedimentological studies. On the other hand, students of volcanoes and volcanic rocks in general are blessed with several tens of active volcanoes, many of which erupt each year. Well-studied historic eruptions like those of Mt. Pelée, Asama, Kilauea, and Mount St. Helens have provided quantum jumps in our understanding of the mechanisms of volcanic eruptions. Process-oriented studies in which the present is the key to the past are increasing in number and are accompanied by theoretical modeling and experimental studies. This type of research, in part dependent on sophisticated instrumentation, will certainly increase in the future. In our present account, we can only provide a glimpse of this work, since we are mostly concerned with empirical aspects of pyroclastic rocks.

We divide this book into three main parts. Part I (Chaps. 2–4) is a brief account of volcanoes, volcanic rocks, physical properties of magmas and clastic and eruptive processes. It provides some background for understanding the origin of deposits discussed in Part II (Chaps. 5–11, pyroclastic deposits) (particles, fallout, pyroclastic flow, hydroclastic and submarine pyroclastic rocks, and lahars). This part also contains a chapter on alteration of volcanic glass, the dominant constituent of tephra (Chap. 12). Part III (Chaps. 13 and 14) is concerned with stratigraphic relationships and some tectonic settings of pyroclastic rocks.

To keep a book on such a broad topic at reasonable length, we treat only highlights of each subject, and have heavily referenced each chapter to lead the interested reader deeper into the subject. We concentrate on literature written in English, and invite critical comments and suggestions about important papers we have missed. The bibliography is put at the end of the text to eliminate duplication and to allow its use for author information independently of the text. Vigorous research presently being done on many aspects of pyroclastic processes and rocks assures that parts of this book will be out of date at the moment of printing, but one of our purposes is to point out gaps in our present knowledge and thereby provide an organization that will stimulate further research.

*Pyroclastic Rocks* is somewhat of a misnomer for the title of this book. Strictly speaking, "pyroclastic" refers to clastic materials ejected from volcanic vents, whereas we include discussions of rocks ranging from lahars remobilized from loose surficial volcanic fragments to pillow breccias formed by fragmentation of quickly erupted submarine pillow lavas. We largely exclude igneous and clastic materials formed beneath the earth's surface, such as brecciated intrusions, and only briefly discuss epiclastic material originating from weathered and eroded pre-existing volcanic rocks.

We owe a great debt to many colleagues and students for stimulating discussions, ideas, and unpublished data. Aaron C. Waters and Donald A. Swanson have been inspiring and critical friends for more than two decades and have thoroughly criticized most of the chapters. It is a pleasure to acknowledge the helpful comments on individual chapters by several colleagues. We are particularly grateful to D. Crandell, D. Johnston, M. Mazzoni, K. Nakamura, F. Seifert, F. Spera, T. A. Steven, R. E. Wilcox, L. Wilson, J. V. Wright, S. Yokoyama, and R. Zielinski. Much of our field and laboratory work over the years was financed by several research organizations. For continuing support we would like to thank especially the National Science Foundation and the Deutsche Forschungsgemeinschaft. Parts of the book were written with support from the Volkswagenwerk Foundation to H-U.S. during a year's leave (1976–1977) in Santa Barbara, and from Alexander von Humboldt Foundation to R.V.F. (Senior Scientist Award) during a year's leave in Bochum (1980–1981).

Almut Fischer cheerfully enlarged many more photographs than were finally selected and Horst Bauszus and David Crouch carefully drafted the illustrations. Meryl Wieder, Evelyn Gordon, and Diane Mondragon spent long hours accurately typing first and revised early drafts. The agonies of shaping the final manuscript was done at the end by Ellie Dzuro. Our thanks to Dieter Hohm, Wolfgang Engel, and Claudia Grössl of Springer-Verlag for their patience and help in seeing the book to press.

R. V. FISHER
H.-U. SCHMINCKE

# Contents

*Chapter 1* **Introduction**

The Sedimentary Record: Volcanic Contribution . . . . . . 3
Chemical Composition of Tephra . . . . . . . . . . . . . 6
Types of Pyroclastic Accumulations . . . . . . . . . . . 8

*Chapter 2* **Volcanoes, Volcanic Rocks and Magma Chambers**

Tectonic Setting of Volcanoes . . . . . . . . . . . . . . . 11
    Divergent Margins . . . . . . . . . . . . . . . . . . . 12
    Convergent Margins . . . . . . . . . . . . . . . . . . 13
    Intraplate Volcanoes . . . . . . . . . . . . . . . . . 14
Form of Volcanoes . . . . . . . . . . . . . . . . . . . . 15
Origin of Magmas and Classification of Volcanic Rocks . . 16
    Basaltic Magmas . . . . . . . . . . . . . . . . . . . 17
    Basalts . . . . . . . . . . . . . . . . . . . . . . . 18
    Andesite-Suite Magmas . . . . . . . . . . . . . . . . 20
    Andesites . . . . . . . . . . . . . . . . . . . . . . 20
    Differentiated Magmas . . . . . . . . . . . . . . . . 21
Magma Chambers . . . . . . . . . . . . . . . . . . . . . 22
    Volumes of Magma Chambers . . . . . . . . . . . . . . 22
    Zonation in Magma Chambers . . . . . . . . . . . . . . 23
    Large Calcalkalic Systems . . . . . . . . . . . . . . 27
    Small Highly Alkalic Systems . . . . . . . . . . . . . 30
    Oceanic Rhyolite-Basalt Systems . . . . . . . . . . . 33
    Very Small Mafic-Intermediate Magma Systems . . . . . 33

*Chapter 3* **Magmatic Volatiles and Rheology**

Volatiles . . . . . . . . . . . . . . . . . . . . . . . . 35
    Methods of Determining Volatiles . . . . . . . . . . . 36
    Composition and Amount of Volatiles . . . . . . . . . 37
        Water 38 – Carbon Dioxide 43 – Sulfur 47
    Volatile Distribution in Magma Columns . . . . . . . . 48
Rheology . . . . . . . . . . . . . . . . . . . . . . . . 51
    Viscosity . . . . . . . . . . . . . . . . . . . . . . 52
Vesiculation . . . . . . . . . . . . . . . . . . . . . . 55

## Chapter 4  Explosive Volcanic Eruptions

Types of Volcanic Activity . . . . . . . . . . . . . . . . . 59
Types of Eruptions . . . . . . . . . . . . . . . . . . . . . 60
Pyroclastic Eruptive Systems . . . . . . . . . . . . . . . . 61
    Eruption Columns . . . . . . . . . . . . . . . . . . . . 61
    Plinian Eruptions and Eruption Columns . . . . . . . 63
    Hawaiian and Strombolian Eruptions . . . . . . . . 69
        Observations 69 – Eruptive Mechanisms 69
Hydroclastic Eruptive Processes . . . . . . . . . . . . . 74
    Formation of Glassy Pyroclasts . . . . . . . . . . . 75
    Granulation . . . . . . . . . . . . . . . . . . . . . . 77
    Inhibition of Vesiculation . . . . . . . . . . . . . . . 77
    Transfer of Heat . . . . . . . . . . . . . . . . . . . 78
    Steam Eruptions . . . . . . . . . . . . . . . . . . . . 78
    Magma-Water Mixing and Fuel-Coolant Interaction . . 80
    Phreatic and Phreatomagmatic (Vulcanian) Eruptions . 82
Volcanic Energy . . . . . . . . . . . . . . . . . . . . . . 85

## Chapter 5  Pyroclastic Fragments and Deposits

General Components . . . . . . . . . . . . . . . . . . . . 89
Vitric Particles and Pyrogenic Crystals . . . . . . . . . . 96
    Glass Shards . . . . . . . . . . . . . . . . . . . . . 96
    Shards Formed by Vesiculation . . . . . . . . . . . 96
    Pumice . . . . . . . . . . . . . . . . . . . . . . . . 103
    Pyrogenic Minerals . . . . . . . . . . . . . . . . . 103
    Surface Textures of Minerals and Shards . . . . . . 105
Structure . . . . . . . . . . . . . . . . . . . . . . . . . 107
    Pyroclastic Bed or Stratum . . . . . . . . . . . . . 107
    Graded Bedding . . . . . . . . . . . . . . . . . . . 109
    Cross Bedding . . . . . . . . . . . . . . . . . . . . 110
    Massive Beds . . . . . . . . . . . . . . . . . . . . 114
    Alignment and Orientation Bedding . . . . . . . . 114
    Penecontemporaneous Deformation Structures . . . 115
Texture . . . . . . . . . . . . . . . . . . . . . . . . . . 116
    Grain Size and Size Distribution . . . . . . . . . . 116
    Distribution Curves . . . . . . . . . . . . . . . . . 118
    Shape and Roundness . . . . . . . . . . . . . . . . 122
    Fabric . . . . . . . . . . . . . . . . . . . . . . . . 123

## Chapter 6  Subaerial Fallout Tephra

Components of Subaerial Fallout . . . . . . . . . . . . 128
Areal Distribution . . . . . . . . . . . . . . . . . . . . 132
    Distribution and Thickness . . . . . . . . . . . . . 132
    Volume . . . . . . . . . . . . . . . . . . . . . . . 139

Structures . . . . . . . . . . . . . . . . . . . . . . . . 141
    Bedding . . . . . . . . . . . . . . . . . . . . . . . . 141
    Mantle Bedding. . . . . . . . . . . . . . . . . . . . . 144
    Graded Bedding. . . . . . . . . . . . . . . . . . . . . 148

Fabric . . . . . . . . . . . . . . . . . . . . . . . . . . 148
Size Parameters . . . . . . . . . . . . . . . . . . . . . 150
    Maximum Size of Components . . . . . . . . . . 150
    Median Diameter . . . . . . . . . . . . . . . . . . . 151
    Grain-Size Distribution (Sorting) . . . . . . . . . . 153

Eolian Fractionation . . . . . . . . . . . . . . . . . . . 156

## Chapter 7 Submarine Fallout Tephra from Subaerial Eruptions

Chemical Composition . . . . . . . . . . . . . . . . . . 163
Structures of Submarine and Lacustrine Ash Layers . . . . 166
Areal Thickness Distribution and Volume . . . . . . . . 168
Grain Size and Sorting . . . . . . . . . . . . . . . . . . 173
Regional Distribution and Tephrochronology . . . . . . . 176
    Source . . . . . . . . . . . . . . . . . . . . . . . . . 176
    Correlation and Age . . . . . . . . . . . . . . . . . . 177
    Pacific Region. . . . . . . . . . . . . . . . . . . . . . 179
    Atlantic Region . . . . . . . . . . . . . . . . . . . . . 182

## Chapter 8 Pyroclastic Flow Deposits

Historic Development of Concepts . . . . . . . . . . . . 187
The Deposits. . . . . . . . . . . . . . . . . . . . . . . . 192
    Volume . . . . . . . . . . . . . . . . . . . . . . . . 192
    Relationship to Topography . . . . . . . . . . . . . 193
    Flow Units and Cooling Units . . . . . . . . . . . . 195
    Components . . . . . . . . . . . . . . . . . . . . . . 197
    Primary Structures in Unwelded Deposits . . . . . . . 198
        Internal Layering 198 – Gas-Escape Structures 200
    Emplacement Facies . . . . . . . . . . . . . . . . . . 203
    Texture . . . . . . . . . . . . . . . . . . . . . . . . 206
        Pyroclastic Flow Deposits 207 – Pyroclastic Surge
        Deposits 208 – Segregation of Crystals and Lithics 208
    Chemical Composition . . . . . . . . . . . . . . . . . 209
    Temperature Effects . . . . . . . . . . . . . . . . . . 210
        Measured Temperatures 210 – Inferred Temperatures
        211 – Welding and Compaction 213 – Structures
        Related to Temperature and Viscosity 215
    Thermoremanent Magnetism . . . . . . . . . . . . . 218

Classification and Nomenclature . . . . . . . . . . . . . 218

The Flows . . . . . . . . . . . . . . . . . . . . . 222
   Origin . . . . . . . . . . . . . . . . . . . . . 222
   Transport and Mobility . . . . . . . . . . . . 225
   Tufolavas, Froth Flows, Foam Lavas and Globule Flows 227
Ignimbrite Vents: Speculation . . . . . . . . . . . 230

## Chapter 9 Deposits of Hydroclastic Eruptions

Definition of Terms . . . . . . . . . . . . . . . . 231
Components of Hydroclastic Deposits . . . . . . . . . 234
   Grain Size Distribution . . . . . . . . . . . . . . 234
   Characteristics of Essential Components . . . . . . . 235
   Accretionary Lapilli . . . . . . . . . . . . . . . 238
   Accidental Clasts . . . . . . . . . . . . . . . . 239
      Maximum Size of Fragments Related to Energetics 239
      Ultramafic Xenoliths 241
Structures . . . . . . . . . . . . . . . . . . . . 242
   Penecontemporaneous Soft Sediment Deformation . . . 242
   Vesicles (Gas Bubbles) . . . . . . . . . . . . . . 242
   Bedding Sags . . . . . . . . . . . . . . . . . . 245
   Mudcracks . . . . . . . . . . . . . . . . . . . 246
Base Surge Deposits . . . . . . . . . . . . . . . . 247
   Bed Forms from Base Surges . . . . . . . . . . . 249
      Sandwave Beds 250 – Plane-Parallel Beds 254 –
      Massive Beds 254
   Bed Form Facies . . . . . . . . . . . . . . . . 254
   U-Shaped Channels . . . . . . . . . . . . . . . 256
Maar Volcanoes . . . . . . . . . . . . . . . . . . 257
   Classification . . . . . . . . . . . . . . . . . . 257
   Origin . . . . . . . . . . . . . . . . . . . . . 258
   Dimensions . . . . . . . . . . . . . . . . . . . 260
      Areal Extent and Geometry 260 – Volume 261
   Chemical Composition . . . . . . . . . . . . . . 262
Littoral Cones . . . . . . . . . . . . . . . . . . . 263
   Deposits . . . . . . . . . . . . . . . . . . . . 263
   Origin . . . . . . . . . . . . . . . . . . . . . 264
Peperites . . . . . . . . . . . . . . . . . . . . . 264

## Chapter 10 Submarine Volcaniclastic Rocks

Deep Water Stage . . . . . . . . . . . . . . . . . 265
   Pillow Breccias . . . . . . . . . . . . . . . . . 267
   Fine-grained Hyaloclastites . . . . . . . . . . . . 270
Shoaling Submarine Volcano . . . . . . . . . . . . 274
Transition Submarine – Subaerial . . . . . . . . . . 275
Volcaniclastic Aprons . . . . . . . . . . . . . . . 276

Silicic Submarine Eruptions . . . . . . . . . . . . . . . 279
Subaqueous Pyroclastic Flows . . . . . . . . . . . . . 281
    Terminology . . . . . . . . . . . . . . . . . . . . 285

Nonwelded Deposits . . . . . . . . . . . . . . . . . . 285
    Environment of Deposition . . . . . . . . . . . . . 285
    Components . . . . . . . . . . . . . . . . . . . . 285
    Grain Size, Sorting and Fabric . . . . . . . . . . . 287
    Bedding and Grading . . . . . . . . . . . . . . . 287
        The Massive Lower Division 289 – Upper Division 290
    Relationship to Eruptions and Eruptive Centers . . . . 292

Welded Deposits . . . . . . . . . . . . . . . . . . . . 293
    Discussion . . . . . . . . . . . . . . . . . . . . . 294

## *Chapter 11* Lahars

Debris Flows as Fluids . . . . . . . . . . . . . . . . . 298
Distribution and Thickness . . . . . . . . . . . . . . . 299
Surface of Lahars. . . . . . . . . . . . . . . . . . . . 301
Basal Contact of Lahars . . . . . . . . . . . . . . . . 302
Components of Lahars . . . . . . . . . . . . . . . . . 303
Grain-Size Distribution . . . . . . . . . . . . . . . . . 303
Vesicles . . . . . . . . . . . . . . . . . . . . . . . . 306
Grading . . . . . . . . . . . . . . . . . . . . . . . . 307
Fabric . . . . . . . . . . . . . . . . . . . . . . . . . 308
Comparison of Lahars with Other Kinds of Coarse-Grained
    Deposits . . . . . . . . . . . . . . . . . . . . . . 309
Origin. . . . . . . . . . . . . . . . . . . . . . . . . 309

## *Chapter 12* Alteration of Volcanic Glass

Diagenesis. . . . . . . . . . . . . . . . . . . . . . . 312
Alteration of Basaltic Glass . . . . . . . . . . . . . . 314
    Palagonite . . . . . . . . . . . . . . . . . . . . . 314
        Physical Properties 315 – Textural Changes 315 –
        Mineralogical Changes 317 – Zeolites 318 –
        Chemical Changes 320
    Process of Palagonite Formation . . . . . . . . . . . 323
    Rate of Palagonitization . . . . . . . . . . . . . . 326

Alteration of Silicic Glass . . . . . . . . . . . . . . . 327
    Hydration and Ion Exchange . . . . . . . . . . . . 327
    Advanced Stages of Alteration . . . . . . . . . . . 329
    Saline Alkaline Lake Environment . . . . . . . . . . 330
    Marine Environment . . . . . . . . . . . . . . . . 333
    Bentonites and Tonsteins . . . . . . . . . . . . . . 336
    Burial Diagenesis and Metamorphism . . . . . . . . 340

## Chapter 13 Stratigraphic Problems of Pyroclastic Rocks

Relation of Volcanic Activity to Rock Stratigraphy . . . . 347
    Volcanic Activity Units . . . . . . . . . . . . . . . . 347
    Eruption Unit . . . . . . . . . . . . . . . . . . . . . 349
    Stratigraphic Problems in Young Volcanic Terranes . . 350
    Stratigraphic Nomenclature in Older Volcanic Terranes . 350

Tephrochronology . . . . . . . . . . . . . . . . . . . . 352
Volcanic Facies . . . . . . . . . . . . . . . . . . . . . . 356
    Facies Based upon Position Relative to Source . . . . 356
        Near-Source Facies 358 – Intermediate-Source Facies
        359 – Distant-Source Facies 359 – Caldera Facies 359
    Facies Based upon Environment of Deposition . . . . 361
    Facies Based upon Primary Composition . . . . . . . 361
        Compositional Facies 362 – Petrofacies 365
    Diagenetic Rock Facies 365

Stratigraphic Examples . . . . . . . . . . . . . . . . . 367
    Oshima Volcano, Japan . . . . . . . . . . . . . . . 368
    San Juan Volcanic Field, USA . . . . . . . . . . . 371
    Archean Greenstone-Belt Volcanoes, Canada . . . . . 378

## Chapter 14 Pyroclastic Rocks and Tectonic Environment

Convergent Margins, Magmatic Arcs, and Sedimentation . 383
    The Trench . . . . . . . . . . . . . . . . . . . . . . 386
    Fore-Arc and Back-Arc Basins . . . . . . . . . . . 391
    The Cordilleran System . . . . . . . . . . . . . . . 392
        Western North America: Paleozoic Rocks 392 –
        Southern South America: Upper Mesozoic Flysch 396 –
        Cenozoic Tectonism and Volcanism: Western North
        America 398
    Oceanic Island Arc Settings . . . . . . . . . . . . . 400
        Volcaniclastic Rocks and Facies; Cenozoic 400 –
        Lau Basin and Tonga Arc 405 – Lesser Antilles Arc 405

The Pre-Cambrian . . . . . . . . . . . . . . . . . . . 408

*References* . . . . . . . . . . . . . . . . . . . . . . . 410
*Subject Index* . . . . . . . . . . . . . . . . . . . . . 449
*Locality Index* . . . . . . . . . . . . . . . . . . . . . 465

## Chapter 1  Introduction

Volcanic eruptions are spectacular natural events that have piqued man's curiosity since prehistoric times. On the one hand they can be of great benefit to man, but on the other can cause great harm and thereby provide a major impetus to their study. Broader aspects make use of volcanoes as windows into the interior of the earth. Magma, molten rock, provides one of the major clues to the earth's origin and to the evolution of its mantle and crust. Our planet's hydrosphere and atmosphere – and thus the origin and evolution of life – owe their origin to degassing of the earth, a process largely accomplished by volcanic eruption. Periods of especially intense volcanic activity can affect climate and thus the world's flora and fauna. Volcanic rocks are the source material from which many sedimentary and metamorphic rocks are derived.

The field of volcanology per se delves into the chemical and physical evolution of present-day volcanoes. Most volcanological work in the past has been done on the composition of gases, lavas, and minerals to decipher the origin and evolution of magmas and the causes of volcanism. Other studies, largely by geophysical methods, include monitoring the degree of inflation of a volcano and associated earthquakes in attempts to predict eruptions and to understand the nature of the "plumbing" beneath and within a volcano. Some volcanologists are interested in the different kinds of volcanic eruptions and their products; this knowledge is necessary for geologists who study older deposits in order to determine the kind of volcanic eruption that produced them.

Pyroclastic rocks are a major, but often neglected, product of volcanism, and their study in years past was mainly the concern of igneous petrography (e.g., Williams et al., 1982). Pyroclastic rocks, however, are being increasingly discussed in texts and journals on sedimentary rocks and sedimentology (Carozzi, 1972, 1976; Fairbridge and Bourgeois, 1978; Friedman and Sanders, 1978; Schmincke, 1974a; Pettijohn et al., 1972; Pettijohn, 1975; Tucker, 1981; Allen, 1982). Recent textbooks on volcanology that discuss various aspects of pyroclastic rocks are by Rittmann (1962), Macdonald (1972), Francis (1976), Williams and McBirney (1979) and Decker and Decker (1981). Simkin et al. (1981) have contributed a directory and gazetteer of the world's volcanoes including a chronology of volcanism during the last 10,000 years, and Self and Sparks (1981) have edited an important book devoted exclusively to pyroclastics. "Pyroclastic geology" is therefore an interdisciplinary subject that includes parts of the fields of volcanology, igneous petrology, geochemistry, sedimentology, sedimentary petrology, stratigraphy, and paleontology. It involves the study of active volcanoes as well as ancient volcanic sequences.

Petrologic principles and petrographic and geochemical techniques are applied to nearly all studies in pyroclastic geology. Determination of magma composition and differentiation processes is usually attempted by chemically analyzing the rocks and individual minerals or glass shards collected from pyroclastic deposits. Also, pyroclastic materials, a dominant part of many volcanoes, can give information about the interaction of water and magma, changing magmatic volatile content and explosivity during eruptions. In the simplest case, many eruptions begin with explosive ejection of fragments and end with outpourings of lava. Long quiescent periods following effusive activity may once again be followed by explosive activity as the result of volatile build-up, buoyant rise of magma and interaction with groundwater or influx of hotter magma into cooler more differentiated magma. Renewed activity may bring to the surface products of a different composition than before, indicating petrologic changes at depth that can best be determined from a comprehensive study of the pyroclastic products and lavas. In addition to studies of magmatic problems, petrographic and mineralogical techniques that involve grain size studies as well as thin-section analyses of rocks are important in correlating ash layers between different areas.

Determining the depositional sequence and three-dimensional geometry of pyroclastic rocks – their stratigraphic relationships – is the first step in nearly all studies of both ancient and modern pyroclastic deposits. In older rock sequences, standard techniques of physical stratigraphy or basin analysis are used; recognition of primary pyroclastic debris in ancient rocks provides evidence of concurrent volcanism. In late Cenozoic volcanic terranes, the lateral chemical and mineralogical composition, and sedimentological attributes of pyroclastic layers may be studied for clues about the source volcano, mechanics of transport and deposition, as well as environments of deposition, that are frequently reflected in diagenetic alteration. From depositional patterns, size variations and mineralogical changes of pyroclastic layers, a geologist may infer ancient wind directions, height of eruption columns and energy of an eruption, and many other aspects of eruptive behavior, transport mechanisms, and depositional environments.

Studies of eruptive phases and consequent deposits of the 1980/81 eruptions of Mount St. Helens, Washington, illustrate the above approaches for monitoring volcanoes and the study of volcanic products. These are presented in a collection of papers published by the U.S.Geological Survey (Lipman and Mullineaux, 1981), which also include accounts of the social and economic effects, hazards assessments and eruption prediction. Of special importance for this book is the integration of eyewitness accounts and extensive photographic coverage of the May 18 and other 1980 eruptive phases of the volcano along with studies of the deposits which began immediately following and sometimes during an eruptive phase. We have incorporated some of these findings within relevant chapters.

Physically viewed, the origin of most pyroclastic fragments is a consequence of rapid energy changes whereby magma moves and is violently erupted from a deep higher temperature and higher pressure environment to the earth's surface. Pyroclastic fragments such as pumice and glass shards derived this way are unique because they form almost instantly, as distinct from slower fragmentation of pre-existing rocks by weathering and erosion. Unlike the relatively slow crystallization process in stationary magma bodies underground, extrusion to the surface is too rapid for crystals to grow in the liquid except in trace amounts. Thus, one common product of surface eruption is metastable glass.

It is important to find out whether volcanic fragments in a clastic rock are recycled by weathering and erosion or are newly formed and deposited contemporaneously with volcanic action. Such determinations may be difficult, however, because fragments derived from the erosion of rocks consisting of lava flows or tuff frequently become mixed with fresh pyroclastic debris and with nonvolcanic material. This mixing can take place in any physiographic environment – terrestrial or marine. Fragmentation also may occur by steam explosions and thermal shock when magma comes in contact with ground water, surface water, ice, or wet sediments.

## The Sedimentary Record: Volcanic Contribution

The great wealth of data from the Deep Sea Drilling Project, recent investigations of island arcs, and studies of the ancient "geosynclinal" assemblages and greenstone belts that occur in all shield areas of the earth reveal the nearly ubiquitous presence of volcanic rocks as flows, pyroclastic layers, and epiclastic volcanic material intermixed with nonvolcanic debris within sedimentary basins throughout geologic time.

Volumetric estimates for some volcanic fields and for several individual eruptions indicate that the volcanic contribution to the world's supply of sediment is high. However, estimates of the relative percentages of common sedimentary rocks within today's geologic record (Tables 1-1, 1-2) usually do not specify, and in many instances do not recognize, the amount of volcanically derived sediments within marine rock sequences, nor the enormous volumes of both young and ancient pyroclastic rocks on continents. For example, large volumes of Permian, Ordovician, Cretaceous and Tertiary volcanic ash occur within the United States (Ross, 1955). Estimates range as high as $2.1 \times 10^4$ km$^3$ of material for some individual fields. In the western United States alone, more than $5 \times 10^4$ km$^3$ of debris, largely as pyroclastic flows, was distributed over much of Nevada, Arizona, and western Utah from early Oligocene to late Pliocene time (Mackin, 1960). Yet this is only a small part of the total volcanic record of that time. Volume estimates of two large

**Table 1-1.** Percentages of common sedimentary rocks estimated from measurements. [After Garrels and Mackenzie (1971, Table 8.2, p. 206) and Pettijohn (1975, Table 2.2, p. 21)]

|  | Leith and Mead (1915) | Schuchert (1931) | Kuenen (1941) | Krynine (1948) | Horn and Adams (1966) | | Ronov (1968) | |
|---|---|---|---|---|---|---|---|---|
|  |  |  |  |  | Continental shield | Mobile belt shelf | Platform | Geosyncline |
| Shale | 46 | 44 | 56 | 42 | 53 | 59 | 49 | 39 |
| Sandstone | 32 | 37 | 14 | 40 | 28 | 36 | 24 | 19 |
| Limestone | 22 | 19 | 29 | 18 | 19 | 5 | 21 | 16 |
| Volcaniclastics | n.d.[a] | n.d. | n.d. | n.d. | n.d. | n.d. | 6 | 25 |

[a] n.d. = not determined

**Table 1-2.** Percentages of common sedimentary rocks based upon geochemical calculations. [After Garrels and Mackenzie (1971, Table 8.2, p. 207)]

|  | Mead (1907) | Clarke (1924) | Holmes (1913) | Wickman (1954) | Garrels and Mackenzie (1971) |
|---|---|---|---|---|---|
| Shale | 82 | 80 | 70 | 83 | 74 |
| Sandstone | 12 | 15 | 16 | 8 | 11 |
| Limestone | 6 | 5 | 14 | 9 | 15 |
| Volcaniclastics | n.d.[a] | n.d. | n.d. | n.d. | n.d. |

[a] n.d. = not determined

**Table 1-3.** Approximate volumes of deposits from some well-known eruptions. (After Izett, 1981)

| Eruption or ash bed | Volume estimate ($km^3$) | Date |
|---|---|---|
| Toba (Sumatra, Indonesia) | 2000 | 75,000 B.P. |
| Bishop Tuff (California, U.S.A.) | 500 | 0.74 m.y. |
| Tambora (Indonesia) | 100–300 | 1815 A.D. |
| Mount Mazama (Oregon, U.S.A.) | 30 | 6500 B.P. |
| Santorini (Greece) | 30 | 1500 B.C. |
| Mount St. Helens (Washington, U.S.A.) | 1.0 | May 18, 1980 |

prehistoric recent eruptions are: Mount Mazama (Crater Lake), Oregon, 40 to 56 $km^3$, which includes pyroclastic flows and fallout tephra (Williams and Goles, 1968); and White River fallout ash, upper Yukon basin, 25 $km^3$ (Lerbekmo et al., 1975). Izett (1981) shows the volumes of deposits of several well-known eruptions (Table 1-3) and discusses the frequency of eruptions.

Sapper (1927) estimated that 328 $km^3$ of pyroclastic debris and 64 $km^3$ of lava were erupted between the years 1500 to 1914 A.D. This gives a pyroclastic:effusive ratio of 5:1 and an average annual volume of pyroclastics of about 0.8 $km^3$. Without considering changes in production rates, this totals $3.6 \times 10^9$ $km^3$ of pyroclastic debris added to the earth's surface over approximately 4.5 b.y. Using more recent information, and taking sea floor volcanism into account, Nakamura (1974) estimated a total annual volcanic (effusive and pyroclastic) production rate of $6 \sim 8$ $km^3$ (Table 1-4), with most of the material ($\sim 60\%$) being lava flows from ocean-ridge volcanoes along divergent (accreting) plate boundaries. Here the water pressure on the sea floor is generally too high for pyroclastic (*sensu stricto*) processes to occur (McBirney, 1963). Accordingly, the world pyroclastic:effusive ratio would decrease by an unknown amount. However, using arbitrary ratios of 1:1 or 0.25:1 would give $6.8 \sim 9 \times 10^9$ $km^3$ or $3.4 \sim 4.5 \times 10^9$ $km^3$ respectively of pyroclastics added to the earth's surface during 4.5 b.y. of earth history. If either Sapper's or Nakamura's estimates are used, the total pyroclastic volume is an order of magnitude greater than the volume estimates of present-day sedimentary rocks (Table 1-5). Although the figures for volcanic rocks take into account neither recycling by weathering, nor loss by metamorphism or remelting in subduction zones, nor

**Table 1-4.** Estimated annual production rate of volcanic material. (After Nakamura, 1974)

| Plate tectonic setting | Estimated volume (km$^3$) |
|---|---|
| Convergent boundaries (island arc volcanoes) | ~0.75 |
| Divergent boundaries (ocean-ridge volcanoes) | 4~6 |
| Intraplate, oceanic (seamounts, etc.) | ~1 |
| Intraplate, continental (flood basalts, etc.) | ~0.1 |
| Total | 6~8 |

**Table 1-5.** Estimated total volume of present-day sedimentary rocks. (Pettijohn, 1975, Table 2-1, p. 20)

| Reference | km$^3$ |
|---|---|
| Clarke (1924) | $3.7 \times 10^8$ |
| Goldschmidt (1933) | $3.0 \times 10^8$ |
| Kuenen (1941) | $13.0 \times 10^8$ |
| Wickman (1954) | $4.1 \pm 0.6 \times 10^8$ |
| Poldervaart (1955) | $6.3 \times 10^8$ |
| Horn and Adams (1966) | $10.8 \times 10^8$ |
| Ronov (1968) | $9.0 \times 10^8$ |
| Blatt (1970) | $4.8 \times 10^8$ |

changes in rates of production through time, the pyroclastic contribution to the world's budget of sedimentary rocks has been significant.

Ronov (1964, 1968) emphasizes changes in the relative volumes of sedimentary rocks since the earth's beginning (Fig. 1-1). Based upon measured sequences, sandstones in early Archean time are thought to be dominantly graywackes derived from the weathering and erosion of early crustal volcanic rocks, with steadily increasing amounts of more silicic differentiates to the present. The high proportion of mafic volcanic rocks and their derivative graywackes in the early to middle Precambrian rock record suggests that large quantities of Ca, Si, Fe and Mg derived from weathering of volcanic rocks were available for the development of carbonate rocks, chert and iron compounds.

Based largely on Ronov's (1964, 1968) work, Garrels and Mackenzie (1971) emphasize the importance of volcaniclastic sediments in the physical and chemical evolution and recycling of sedimentary rocks, including carbonates. They calculate estimated masses of limestone, shale, sandstone, and volcaniclastic sediments in Precambrian and post-Precambrian time and compare them with Ronov's (1968) estimates obtained by measurement (Table 1-6). Despite the paucity of data and the errors inherent in attempts to determine the amounts of common sediments through time, the similarities between calculations and measurements are surprising. The volumetric importance of volcaniclastic materials in the sedimentary record is clear, but their contribution is even greater than shown because an unknown amount of fine-grained tephra, commonly altered to montmorillonite and, with greater age, to illite, goes largely unnoticed in rocks classified as shale. Judging

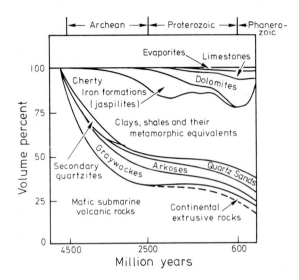

Fig. 1-1. Changes in relative volumes and evolution of sedimentary rocks since the earth's origin. (After Ronov, 1964)

by the frequency of volcanic eruptions and abundance of tephra expelled yearly (Simkin et al., 1981), the volcanic contribution to the world's sedimentary budget, even as a thin dust film, is high.

Lindström (1974) suggests that the non-calcareous component of Lower Paleozoic (Tremadocian) pelagic shales of northern Europe (Baltic Shield) may have been volcanically derived. The shales are black, carbonatic, and cover an area of 500,000 km$^2$. The average rate of sedimentation of the non-carbonate component of the shales is estimated at 0.1 mm/10$^3$ yrs. The slow sedimentation rate, uniformity of facies over a vast area and lack of current-induced sedimentary structures suggest a pelagic environment far removed from land during much of the early and middle Ordovician, yet the shales contain abundant iron. The ratios $SiO_2:Al_2O_3:(Fe_2O_3):MgO:TiO_2$ of about 3.0:1:0.66:0.17:0.035 in the shales can be explained by sedimentation of fine-grained basaltic ashes far from land even though volcanic particles have not been positively identified.

Table 1-6. Percentage estimates of common sedimentary rocks

|  | Shale | Volcaniclastic | Limestone | Sandstone |
|---|---|---|---|---|
| Precambrian[a] | 62 | 27 | 6 | 4 |
| Precambrian[b] | 48 | 27 | 14 | 11 |
| Post-Precambrian[b] | 54 | 27 | 11 | 8 |

[a] Measured (Ronov, 1968)
[b] Calculated (Garrels and Mackenzie, 1971)

## Chemical Composition of Tephra

The number of chemical analyses of whole-rock unaltered tephra is small compared to the more than 50,000 analyses published on lavas (Le Maitre, 1976). This is because of (1) the ease with which tephra is chemically changed during alteration, so that lava flows commonly yield more reliable information on original magma compositions, (2) the difficulty of recognizing epiclastic or accidental mixtures in tephra and (3) sorting of tephra components during transport. Chemical data (Table 2-2) are useful for the comparison of weathered rocks and soils with parent pyroclastic material (Mohr and Van Baren, 1954), in problems of diagenesis (Chap. 12) and in monitoring lateral compositional changes of tephra layers (Baak, 1949). The great importance of pyroclastic deposits in evaluating magma chamber processes is only now becoming appreciated (Chap. 2).

Pettijohn (1975, pp. 208–305) emphasized the usefulness of bulk chemical analysis of clastic rocks, two of the most cogent reasons being (1) to have standards for recognizing chemical analogs where metamorphism has obliterated evidence of the original parent rock, and (2) for studies of the mass balance in the overall evolution of the earth. Taylor and coworkers (e.g., Taylor and McLennan, 1981) have argued that Archean sedimentary rocks differ significantly from post-Archean sedimentary rocks in their REE patterns. The Archean rocks resemble island-arc volcanic rocks, indicating a much less evolved upper Archean crust, their salient features being: lower total REE and La/Yb ratios and a lack of an Eu-anomaly (Fig. 1-2). In some Archean provinces like the Yellowknife Supergroup (Canada), however, sedimentary rocks have higher La/Yb ratios than the Australian Archean rocks studied by Taylor owing to a higher proportion of felsic volcanics in their source areas, but they are still regarded as characteristically Archean (Jenner et al., 1981). This view of significant differences in REE patterns between Archean and post-Archean sedimentary rocks is challenged by Dymek et al. (1983) who hold that the "typical Archean pattern" merely reflects a nonrepresentative sampling of volcaniclastic graywackes, and that REE patterns of highly metamorphosed Archean sedimentary rocks believed to represent shelf sediments are indistinguishable from those of post-Archean sedimentary rocks. Obviously, there is a great need to study the geochemistry of both modern volcanic terranes and their associated volcaniclastic rocks.

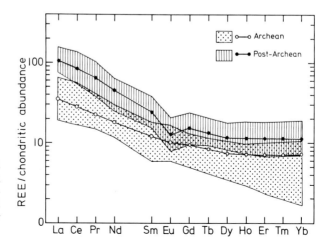

**Fig. 1-2.** Comparison of REE pattern of Australian Archean and post-Archean sedimentary rocks showing averages and ranges for each group. (After Taylor and McLennan, 1981)

# Types of Pyroclastic Accumulations

Pyroclastic rocks are composed of fragments that originate from volcanic eruptions or as a direct consequence of an eruption. Eruptions giving rise to pyroclastic fragments may be grouped into two general categories: (1) those caused by expansion of gases initially contained within the magma [i.e., pyroclastic (magmatic) eruptions], and (2) those caused by vaporization of external water in contact with hot magma or lava (hydroclastic eruptions). The initial driving force of rising magma columns is buoyancy and, at low pressures, internal gas expansion in explosive eruptions. The magma, however, may come in contact with water as it ascends and be blown to bits by expansion of steam. Thus, magmatic and hydroclastic processes may overlap. Whatever causes an eruption, or whether an eruption is subaerial or subaqueous, fragments are either (1) abruptly transported through the air or water and fall back to the surface to become *fallout deposits*, or else (2) become *pyroclastic flow* deposits from flows along the ground surface on land or beneath the sea.

Major kinds of subaerial deposits that originate from pyroclastic eruptions (Fig. 1-3) are fallout tephra and the deposits from pyroclastic surges, pyroclastic flows and lahars, all of which may occur from a single eruption as at Mount St. Helens on May 18, 1980 during a 12-h period (Lipman and Mullineaux, 1981) (Fig. 1-4). Fallout deposits (Chap. 6) come to rest after fragment transport by initial trajectory (large fragments), or after scattering by wind from turbulent eruption clouds. Fallout deposits also form from fine-grained particles that move turbulently upward from the tops of moving pyroclastic flows. Flows of hot pyroclastic material originate from collapsing eruption columns, by the low-pressure "boiling-over" from vents, and by the laterally directed eruptions frequently associated with domes and dome collapse. Pyroclastic flow deposits and their origin are discussed

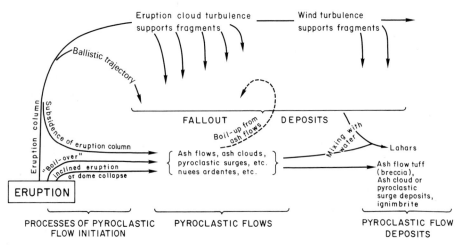

Fig. 1-3. Processes by which subaerial pyroclastic flow and fallout deposits originate

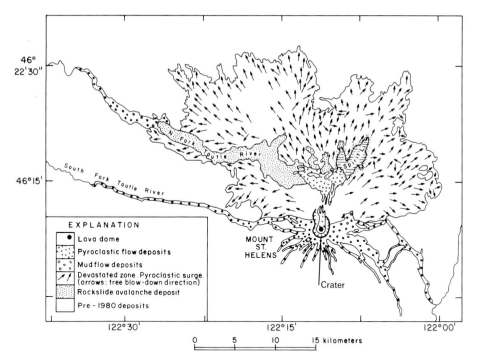

**Fig. 1-4.** Major kinds of volcaniclastic deposits, excluding fallout ash (see Chap. 5), from the May 18, 1980 eruption of Mount St. Helens, Washington. (After Lipman and Mullineaux, 1981, Plate 1)

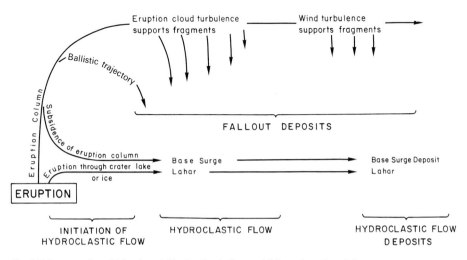

**Fig. 1-5.** Processes by which subaerial hydroclastic flow and fallout deposits originate

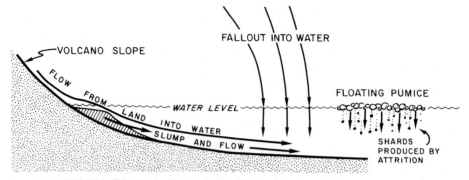

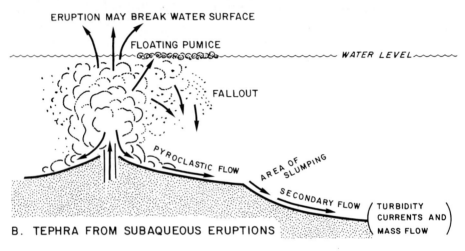

**Fig. 1-6 A, B.** Processes by which subaqueous pyroclastic flows and fallout originate

in Chapter 8. Two ways in which lahars (Chap. 11) can form are directly by eruptions through crater lakes, or from hot pyroclastic flows that mix with snow or water on the flanks of volcanoes. They may also originate not from an eruption, but from postdepositional failure of water-soaked pyroclastic and fragmented lava flow debris resting on steep slopes.

Subaerial deposits that originate from hydroclastic eruptions (Fig. 1-5) show characteristic differences in fragment types and in the textures and structures of the layers from pyroclastic eruptions in air. Hydroclastic deposits include fallout tephra (tuff, breccia, etc.) and flow deposits from base surges and lahars, discussed more fully in Chapters 9 and 11.

Subaqueous deposits (Fig. 1-6) result from two general processes: (1) those that originate from underwater eruptions and (2) others that originate from subaerial eruptions whose products enter water. Slumping, mass flow and turbidity current action of previously deposited debris are a third major underwater process category.

# *Chapter 2* Volcanoes, Volcanic Rocks and Magma Chambers

Volcanoes are hills, mounds or sheets of relatively localized igneous rock assemblages made up of pyroclastic rocks, lava flows, and intrusions in varying proportions. Volcanoes differ notably in their geometry, volume, and relative amounts of pyroclastic rocks and lava flows, with differences mostly dependent upon eruptive mechanisms and rates of extrusion. These in turn depend chiefly upon the magma composition. Chemical composition of magma is also responsible for, or can be correlated with, physical properties such as volatiles and viscosity, which govern to a large extent the nature of many pyroclastic eruptions.

In a broad way, magma types can be divided into mafic, intermediate and silicic varieties, magma being defined as melt $\pm$ crystals $\pm$ gas (dissolved or in bubbles). The dominant and most widespread supply of pyroclasts on land is from the intermediate and highly silicic/alkalic magmas.

Among the major new aspects in the study of pyroclastic rocks is the increasing recognition that these rocks harbor a wealth of petrologic information about their parent magmas. Foremost in this respect is the fundamental observation that many magma chambers are mineralogically and chemically zoned, and a topic of much current interest centers on the nature of processes that lead to such zonations.

## Tectonic Setting of Volcanoes

Most active and dormant volcanoes occur in belts that coincide with zones of earthquakes. The main volcanic and seismic belts define boundaries of large lithospheric plates. Lithospheric plates which range in thickness from less than 30 to 200 km are believed to move over the asthenosphere along a layer of lower density that may contain a small amount of magma. Major types of plate margins are: (1) divergent or constructive, (2) convergent or destructive, and (3) margins that slide laterally past one another (Figs. 2-1, 2-2). Of the active subaerial volcanoes associated with plate boundaries, about 80 percent occur within convergent and 15 percent in divergent settings (Fig. 2-1). About 5 percent occur away from plate margins in intraplate settings. Magmas generated along sites of divergence – mid-ocean rifts and rises and marginal basins – and within oceanic and continental plates are dominantly mafic; more silicic magmas are generated chiefly near or above sites of subduction (convergence) in regions of island arcs and active continental margins, or else within continental and, more rarely, oceanic plates.

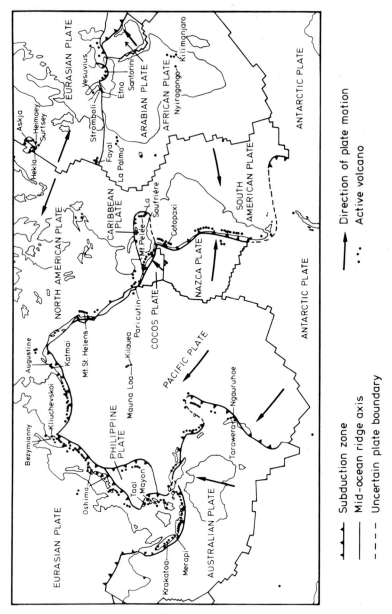

**Fig. 2-1.** Major plates, mid-ocean ridges and subduction zones. Names of active volcanoes shown are those mentioned in text. (After Macdonald, 1972; Press and Siever 1978; Simkin et al., 1981)

**Divergent Margins**

Divergent margins are sites of extension where oceanic lithospheric plates are generated and move away from one another. Important divergent plate margins are the present-day mid-oceanic rises and ridges such as the Mid-Atlantic Ridge, the East Pacific Rise and the Carlsberg Ridge in the Indian Ocean. They are charac-

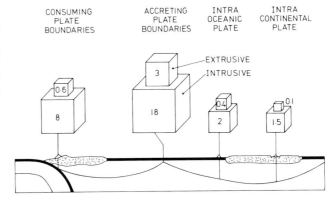

Fig. 2-2. Magma production rates (in km³ per year) at divergent and convergent plate boundaries and intraplate settings. (After Schmincke, 1982b; based on data by Nakamura, 1974; Fujii, 1975; Crisp, 1984)

terized by high heat flow and are sites where most of the earth's volcanic products are extruded. About 3 km³ of basalt is added annually to the earth's crust at oceanic ridges (Table 1-4, Chap. 1). It forms the upper part of layer 2 of the oceanic crust which is from 1 to 2 km thick. Volcanism at mid-ocean ridges occurs mostly at water depths generally exceeding the partial pressure of magmatic gases, except for minor $CO_2$-degassing. Consequently, most of the volcanic products are lavas.

Two major kinds of submarine volcanoes have been found. Steep-sided pillow volcanoes 50 to 200 m high and several tens to hundreds of meters wide have long been recognized along the Mid-Atlantic and East Pacific Ridges (Ballard and van Andel, 1977; Ballard and Moore, 1977; van Andel and Ballard, 1979). Sheet flows are also a significant component within rift zones (Lonsdale, 1977; Ballard et al., 1979). Primary pyroclastic material makes up a small part of such submarine volcanoes, although spalling of pillow rinds plus lava spatter can form fragmental deposits. Magmas are dominantly basaltic with low volatile content. Because of the increasing importance of exsolution of magmatic volatiles for disrupting a magma with decreasing water depth, the ratio of pyroclastic debris to lava flows is higher in shallow marine basins than in deep-sea deposits. Volcanism along mid-ocean ridges has only been studied in the last few years by direct observation from submersibles and deep-tow instruments. Numerous drillings by the Glomar Challenger in all major ocean basins have greatly enlarged our knowledge and understanding of clastic rocks formed in this environment. Studies of the extrusive rocks in ophiolites, which are uplifted oceanic crust formed at divergent plate margins, or above subduction zones, increasingly helps to understand submarine volcanic processes.

**Convergent Margins**

Convergent margins are sites where two plates move toward one another, collide, and one plate dives beneath the other. Opposing plates may be composed of either oceanic or continental lithosphere, or both. Volcanism is most common where oceanic lithosphere is subducted beneath continental lithosphere. Major deep-focus earthquake belts, island arcs and young fold mountains all are associated with convergent margins. The foci of intermediate and deep earthquakes that occur beneath the arcs or continental margins define dipping zones called Wadati-Benioff subduction zones. Volcanic belts above subduction zones are made up of great volumes of lava flows and especially pyroclastic debris ranging from basaltic through andesitic, rhyodacitic to rhyolitic in composition. Large volumes of volcanic debris from these volcanoes enter the ocean basins through the air or as pyroclastic flows, debris flows and slides which may generate turbidity currents that spread out over vast areas within adjacent oceanic basins. The most extensive island arcs, volcanic chains and major earthquake zones rim the Pacific; other belts occur in the Greater and Lesser Antilles and along the collision zone between the Eurasian, African and Arabian plates (Fig. 2-1). Two-thirds of the world's active subaerial volcanoes occur around the Pacific Ocean margin (Simkin et al., 1981).

**Intraplate Volcanoes**

Oceanic intraplate volcanoes include large volcanic edifices that emerge above water to form oceanic islands as well as tens of thousands of generally smaller submarine volcanoes (seamounts), which include flat-topped guyots (see Chap. 10). Studies by Batiza (1977, 1982) show that there may be 2 to $6 \times 10^4$ seamounts in the Pacific Ocean basin alone, representing between 5 and 25 percent of the oceanic volcanic layer. The total volume of extrusive material produced by these volcanoes is smaller than that of volcanoes built above convergent plate boundaries, but except for volcanoes that rise above sea level, little is known about their structure and lithology. Seamounts contain a higher percentage of volcaniclastic rocks than the mid-ocean ridge volcanoes because they are higher, have steeper slopes, and commonly are constructed from more volatile-rich alkalic magmas. Therefore many may build up to shallow water depths – or even islands – where exsolution of magmatic volatiles can lead to pyroclastic activity. Archipelagos such as the Hawaiian, Samoan and Galapagos Islands in the Pacific, and the Azores, Madeira, Cape Verde and Canary Islands in the Atlantic, are typical examples of oceanic intraplate volcanoes. The term intraplate is somewhat of a misnomer for many oceanic volcanic islands, some of which have formed on or close to a mid-ocean ridge (Iceland), others on or close to transform fractures (Azores) or near the boundary of oceanic and continental lithosphere (Canary Islands). Some archipelagos are arranged in a linear chain that may form above more or less stationary melting spots localized in the sublithospheric mantle. Most of the pyroclastic (and hydroclastic) rocks of volcanic islands probably occur within the upper submarine part of an island. The submarine volcano includes volcanic aprons that may extend outward for 100 km or more around a volcanic island.

Intraplate volcanism within continental plates occurs mostly within or near major rift zones and in uplifted areas. Although eruptive centers within continental plates are rather numerous, the total volume of extrusive material is small compared to that at divergent and convergent plate boundaries. Exceptions are regions of flood basalts, sites of some of the most voluminous land-based eruptions on earth. Basaltic cinder cones are perhaps the most characteristic subaerial volcano of intraplate settings, but there are numerous larger and less mafic central volcanoes that have produced a large spectrum of effusive and pyroclastic rocks.

# Form of Volcanoes

Until a few decades ago, the description of volcanic land forms was a major aim of many volcanologists. Such works as Sapper (1927) and Cotton (1944) have provided a widely used classification of different volcanic forms. Volcanological research has now become more quantitative, and also seeks to understand the origin and evolution of magmas, and the dynamics of volcanic eruptions.

Basically, volcanoes have three major geometric forms: cones, shields, and sheets. The cone can be symmetrical, as for many andesite volcanoes; it can be truncated by a central caldera; or it may be a short stubby cone with a large central crater, such as a tuff ring. In this simplified scheme even a broad but high shield volcano exemplified by the huge Hawaiian volcano Mauna Loa can be regarded as a broad cone. Viscosity, eruptive rates, duration of eruptive phases, and type of explosive mechanisms are the major factors responsible for these constructional forms: highly viscous lava (or viscous pyroclastic flows) will accumulate on the slopes and at the base of a cone even at high production rates, whereas highly fluid flows and voluminous pyroclastic flows move rapidly from a volcanic center onto low slopes, resulting in low shields.

Shield volcanoes also can be viewed as transitional to plains or plateaus of flood basalts. These are major accumulations of sheet-forming extrusive materials, some flows covering areas in excess of 100,000 km$^2$ with little change in thickness. Some evolved lavas of low viscosity, such as the trachytes and phonolites of East Africa, have also formed widespread sheets.

Of major interest and emphasis in this book are the sheetlike pyroclastic accumulations: pyroclastic flows and fallout tephra. Plateaus built of pyroclastic flow deposits resemble those consisting of flood lavas, but the overall dimensions of pyroclastic flow plateaus are mostly smaller and the thickness of individual cooling units is generally greater than for flood basalt flows. For many years, welded pyroclastic flows were thought to be lava flows, but their high silica content and therefore presumed high initial viscosity defied explanation of how they could be emplaced as widespread lava sheets. It has now been established beyond doubt that pyroclastic flows are transported as particulate flow systems rather than as lava flows (Chap. 8). Most appear to be derived from central vents or from ring fractures which encircle large calderas that formed synchronously with the eruption of large-volume pyroclastic flows.

Areally, the most extensive kind of volcanic sheet deposits are fallout tephra forming widespread blankets of pumice lapilli or ash. Individual lobate sheets commonly have elliptical forms in plan view due to unidirectional wind transport and may be many hundreds or even thousands of kilometers long. Most are thin, however, and do not have the volume of flood basalts or of large pyroclastic flows. Such large single sheets are the result of powerful explosions and can generally be traced to a source, commonly a caldera. They are especially well-preserved as ash layers in deep-sea sediments. On continents most parts of them are eroded, although remnants may be preserved in peat bogs, lake sediments, beneath pyroclastic flows and in other areas of quiet sediment accumulation.

## Origin of Magmas and Classification of Volcanic Rocks

For simplicity, magmas are divided here into two major groups: (1) parental magmas formed by partial melting of the upper mantle or the crust, and (2) derivative magmas formed by differentiation from parent magmas or by mixing of magmas. More complete treatments can be found in Wyllie (1971), Carmichael et al. (1974), Ringwood (1975), Yoder (1976), Gill (1981), Best (1982) and others. Volcanic rocks are most simply classified by chemical criteria (Fig. 2-3; Table 2-1).

Two sets of conditions for magma evolution must be distinguished: those acting at the source of the magma and those acting on the magma up to the time of surface eruption. Basaltic magmas are commonly regarded as parental to the silica-, alumina- and alkali-rich magmas, the most differentiated of which have compositions close to petrogeny's residua system (Tuttle and Bowen, 1958). Basically, however, any magma can be parental to a more differentiated one. Differentiation of oceanic basalts by crystal settling without contamination by continental rocks or marine sediments is perhaps the simplest system that can be studied. Differentiation processes may be greatly complicated by contamination along continental margins and there are many as yet unresolved problems about the origin of magmas.

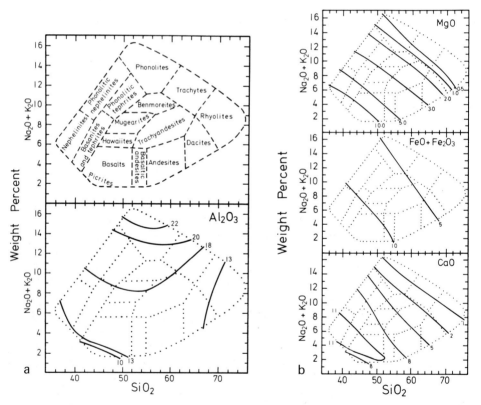

**Fig. 2-3 a, b.** Classification of major groups of volcanic rocks in alkali-silica variation diagrams showing weight percent concentrations of the oxides $Al_2O_3$, MgO, CaO and $FeO+Fe_2O_3$. (After Cox et al., 1979)

**Table 2-1.** Average chemical composition of some common volcanic rocks (Le Maitre, 1976)

|  | Nephelinite | Basanite | Hawaiite | Tephrite | Basalt | Tholeiite | Mugearite |
|---|---|---|---|---|---|---|---|
| $SiO_2$ | 42.43 | 45.46 | 48.65 | 48.73 | 50.06 | 50.72 | 52.28 |
| $TiO_2$ | 2.71 | 2.56 | 3.30 | 1.80 | 1.86 | 1.96 | 2.11 |
| $Al_2O_3$ | 14.90 | 14.89 | 16.32 | 17.31 | 15.99 | 14.98 | 16.98 |
| $Fe_2O_3$ | 5.78 | 4.14 | 4.92 | 4.21 | 3.92 | 3.51 | 5.17 |
| $FeO$ | 6.60 | 8.02 | 7.73 | 5.48 | 7.46 | 8.22 | 6.52 |
| $MgO$ | 6.76 | 8.93 | 5.15 | 4.80 | 6.96 | 7.38 | 3.52 |
| $CaO$ | 12.32 | 10.53 | 8.21 | 9.34 | 9.66 | 10.35 | 6.14 |
| $Na_2O$ | 4.97 | 3.58 | 4.15 | 3.77 | 2.97 | 2.44 | 4.87 |
| $K_2O$ | 3.53 | 1.88 | 1.58 | 4.58 | 1.12 | 0.45 | 2.46 |

|  | Andesite | Phonolite | Trachyte | Latite | Dacite | Rhyodacite | Rhyolite |
|---|---|---|---|---|---|---|---|
| $SiO_2$ | 56.86 | 57.49 | 62.61 | 62.80 | 66.36 | 67.52 | 74.00 |
| $TiO_2$ | 0.88 | 0.64 | 0.71 | 0.83 | 0.58 | 0.60 | 0.27 |
| $Al_2O_2$ | 17.22 | 19.47 | 17.26 | 16.37 | 16.12 | 15.53 | 13.53 |
| $Fe_2O_3$ | 3.29 | 2.87 | 3.07 | 3.34 | 2.39 | 2.46 | 1.47 |
| $FeO$ | 4.26 | 2.28 | 2.42 | 2.27 | 2.41 | 1.80 | 1.16 |
| $MgO$ | 3.40 | 1.12 | 0.95 | 2.25 | 1.74 | 1.68 | 0.41 |
| $CaO$ | 6.87 | 2.80 | 2.34 | 4.27 | 4.29 | 3.35 | 1.16 |
| $Na_2O$ | 3.54 | 7.98 | 5.57 | 3.88 | 3.89 | 3.90 | 3.62 |
| $K_2O$ | 1.67 | 5.38 | 5.08 | 3.98 | 2.22 | 3.16 | 4.38 |

## Basaltic Magmas

Depending upon the prevailing geothermal gradients, the earth's uppermost mantle consists of garnet peridotite at high pressures (above about 22 kb), changing to spinel peridotite at intermediate pressure (10 to 22 kb) and to plagioclase peridotite at low pressure (<10 kb). The most primitive kinds of magmas, the mafic basaltic magmas, are inferred to be generated by partial melting of such peridotite.

The variables that control the composition of basaltic magmas are complex. These include (1) composition – chemical and mineralogical composition of the source rock, type and relative abundance of volatiles, and (2) melting processes – degree of partial melting (as a function of pressure, temperature, and volatile content) and depth of magma generation. Basaltic magmas are called primary when they rise straight to the earth's surface along fissures and erupt with practically no chemical change during transit. They must be in equilibrium with peridotite and are characterized by high Mg/Fe ratios, Cr and Ni contents; many alkalic basalts also contain ultramafic inclusions of high pressure origin. However, magmas that can be called primary are rare. Most magmas cool, fractionate and mix to varying degrees en route to the surface (Fig. 2-4).

Experimental and chemical data suggest that tholeiitic basaltic magmas from mid-ocean ridges are derived at shallow mantle depths (<30 km) by partial melting (20 to 30 percent), whereas silica undersaturated alkalic basaltic and the rarer basanite and olivine nephelinite magmas may result from a smaller amount of par-

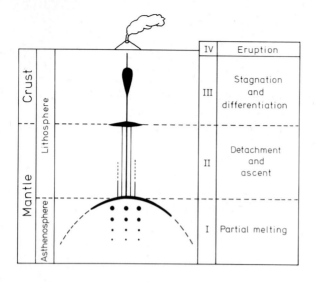

Fig. 2-4. Schematic model for four stages of an intraplate basaltic volcano and its roots. *I* Partial melting *II* Detachment of magma from melting zone and rise to the crust *III* Potential stagnation and differentiation at several crustal levels (very primitive magmas may not stagnate at this level) *IV* Eruptive processes at very high crystal levels and at the surface. (After Schmincke, 1982b)

tial melting (<5 to 10 percent) at higher pressure at mantle depths of about 80 to 150 km. Moreover, highly alkalic magmas appear to be generated at greater mantle depths in regions of low thermal gradients (many continental areas).

Degrees of partial melting (at any depth) are a function of pressure and of the volatile content of the system. Work during the last ten years has shown that the type and amount of volatiles greatly influence melting temperature, the degree of partial melting of mantle peridotite, and the chemical composition of magmas generated by partial melting. The main volatiles studied are $H_2O$ and $CO_2$. At mantle conditions they probably occur as ions dissolved in the interstitial melt, or else locked up in accessory hydrous minerals such as phlogopite, amphibole and carbonates (magnesite, dolomite) and fluid inclusions. Highly silica-undersaturated alkalic magmas such as nephelinites and kimberlites appear to be generated at high $CO_2/CO_2 + H_2O$ ratios.

Explosive eruptions characteristic of highly undersaturated mafic magmas probably stem from abundant volatiles in the magmas. The generally non-explosive extrusion of tholeiite basalt is probably due to initially low volatile content. Low volatile content could be caused by large degrees of partial melting or an initially low volatile content of the mantle source. Large degrees of melting cause dilution of the volatiles, which may be concentrated in the first melt because of early breakdown of hydrous phases.

## Basalts

Basalt is by far the most abundant volcanic rock, although the ratio of basaltic pyroclastic to non-pyroclastic material is less than with other volcanic rocks. In general, two major groups of basalt can be distinguished: tholeiites and alkali basalts.

*Tholeiite*, by (CIPW-norm) definition is hypersthene-normative, and may be either quartz- or olivine-normative. Tholeiites can be further subdivided according to tectonic setting and chemical and mineralogical composition.

*Ocean-Floor Tholeiites* (Mid-Ocean Ridge Basalts = MORB), the earth's most abundant volcanic rock type, make up most of the upper oceanic crust (layer 2), a layer about 2 km thick consisting of pillow and sheet lava flows and underlying dike swarms. Chemically, they are characterized by very low concentrations of incompatible elements such as K and P among the major elements and Rb, Zr, Nb, U among the trace elements. Olivine and calcic plagioclase are the dominant phenocrysts; clinopyroxene is rare. Rocks of intermediate and highly evolved composition are uncommon. Clastic rocks are chiefly hyaloclastites from spalled pillow and sheet flow surfaces and various types of pillow breccias.

*Ocean-island tholeiites* are represented by the voluminous shield-forming lavas in Hawaii and Iceland, are rare on other volcanic islands, but may be common in seamounts. They are richer than ocean floor tholeiites in volatiles, alkalis, and incompatible elements and may carry orthopyroxene or Ca-poor clinopyroxene. Central volcanic complexes with highly evolved rocks have developed in some areas such as Iceland. Pyroclastic rocks are mainly hyaloclastites made of vesicular shards formed in shallow water during the seamount stage of an oceanic island, and also form during the flow of lava into the sea and from lava fountaining.

*Island arc tholeiites* are less mafic and more siliceous than ocean floor tholeiites. They form differentiation series with Fe-enrichment through basaltic andesites, Fe-rich andesites to dacites in several island arcs and are thus represented by a broad range of clastic rocks. This magma suite may be common in Archean greenstone terranes forming shallow submarine to subaerial crustal complexes. Primary tephra and epiclastic debris derived from island arcs form a significant portion of pelagic sediments in adjacent trenches and basins.

*Continental tholeiites* are represented by large flood basalt provinces such as the Columbia River, Parana, Karroo and Deccan basalt fields. Most of these are more alkalic and richer in Ti and P than other tholeiites. Plagioclase and augite are the most common phenocrysts, with sparse olivine occurring in some suites. Pyroclastic rocks are rare except as agglutinates (spatter) along dikes or local vents, and as hyaloclastites and breccias where lavas flowed into, or were erupted under, water.

*Alkali basalts* are more silica-undersaturated than tholeiites, have up to about 5% normative nepheline (CIPW norm), and are enriched in incompatible elements and volatiles. They are common among the shield-building lavas of most oceanic islands and seamounts and locally occur along fracture zones in the oceanic crust. In number of eruptive centers, alkali basalts are probably the most common basaltic lava on continents, although they are far exceeded in volume by the tholeiitic flood basalts. Titanaugite is perhaps the most characteristic phenocryst, with olivine, plagioclase, and titanomagnetite in lesser abundance. Olivine may, in contrast to tholeiites, commonly occur in the groundmass. Kaersutitic amphibole, sphene and apatite may occur in less mafic varieties. In many volcanic islands and some intraplate alkali basalt volcanic fields, central volcanic complexes of more differentiated magmas have developed. The ratio of pyroclastics to lava in alkali basalt suites is relatively high in two settings: (a) Shallow submarine and early shield-building activity generate thick deposits of vesicular pyroclasts underlying subaerial shield-building series, (b) Cinder cones, dominated by scoria breccias and agglutinates, and maar volcano assemblages are more voluminuous than associated lava flows in many alkali basalt provinces.

*Basanites,* nephelinites, and leucitites are more silica-undersaturated and contain more alkalis and volatiles than alkali basalts. Clinopyroxene, olivine, and, in

K-rich varieties, phlogopite are the main phenocrysts. Nepheline or leucite instead of feldspar microlites occur in the highly undersaturated nephelinites and leucitites. The overall volume of basanites, nephelinites, and leucitites is much smaller than that of alkali basalts; they occur on many oceanic islands and as intraplate volcanoes on continents, especially in association with rift zones and uplifted areas.

*Kimberlites* are even more silica-undersaturated and alkalic than nephelinites and are characterized by an abundant and varied suite of xenoliths mostly of mantle origin, but some from crustal wall rock. Phenocrysts, commonly altered, include olivine, phlogopite, ilmenite and, with luck, diamond, indicative of a high pressure origin of these magmas of extreme composition. Kimberlites occur almost exclusively in pipes, but these diatremes generally grade into solid dikes with depth.

## Andesite-Suite Magmas

The vast majority of andesite-suite or calc-alkaline magmas occur above Benioff zones, strongly suggesting a genetic link with subduction zone processes. Because dehydration would be an expected result of the temperature rise in a subducting slab consisting of water-rich, partly metamorphosed oceanic crust with local overlying sediments and underlain by depleted mantle peridotite, major questions about water release and its possible effect in reducing the melting temperature of crust or mantle become important. However, opinions are sharply divided as to whether andesite magmas are formed directly by partial melting of peridotite (Kushiro, 1974) or by fractionation of hydrous basaltic magmas generated along or above Benioff zones (Ringwood, 1975). It is increasingly realized, however, that there is a large spectrum of andesite-type magmas that must have formed by a correspondingly large variety of petrogenetic processes. The present consensus is that most andesite magmas form by differentiation from basalt magmas modified by fluids from the subducted slabs and locally by partial crustal melts.

Clear evidence that some andesitic magmas are water-rich comes from experimental studies (Eggler, 1972), studies of glass inclusions (Anderson, 1974a, b), and by the explosive behavior of andesite volcanoes. The question remains, however, whether the explosive nature of many andesite eruptions is due mainly to high primary volatile content, high viscosity, or the interaction of the magma with external water. Moreover, it is not known how much of the primary water is enriched in andesite magmas at the site of generation by dehydration of oceanic crust or is gained by fractional crystallization of basaltic magma during storage. On the other hand, there is growing evidence from oxygen and hydrogen isotope studies that much meteoric water can be absorbed in hot-but-congealed magmatic bodies (Taylor, 1974) or even by still-fluid magma bodies (Friedman et al., 1974).

## Andesites

Andesites are the second most abundant volcanic rock type next to basalt, and are the most common type of lava erupted in present-day subaerial volcanoes. They generally construct large stratovolcanoes with abundant pyroclastic rocks. Chemically, andesites are much higher in silica (average 58%) and alumina, and are lower in magnesia than most basalts, but they are not particularly rich in alkalis. Moreover, they have high calcium/alkali ratios giving rise to the name calc-alkaline suite. Calc-alkaline suites are typically devoid of iron enrichment during differentiation. Andesites are highly porphyritic, the commonest phenocrysts being strongly and complexly zoned plagioclase that generally have very calcic cores and andesine rims. Diopsidic clinopyroxene, and hypersthene are present in smaller amounts. Hornblende and biotite also may occur, chiefly in the continental andesites. Olivine-bearing andesites are common among the "basaltic andesites" of the circum-Pacific chains of volcanoes. Common among the pyroclastic rocks are (1) pyroclastic flow deposits, especially those formed from collapsing domes, (2) lahars, and (3) various types of breccias. Except for lahars, most andesitic pyroclastic rocks are confined to the stratovolcanoes. Andesite volcanoes are a source of abundant epiclastic volcanic sediments, much of which is deposited in sedimentary basins next to continental margins and island arcs.

**Differentiated Magmas**

Highly silicic calc-alkalic magmas can be generated by differentiation from a more mafic parent magma as well as by crustal anatexis. Several lines of evidence suggest that the origins of some rhyolitic and andesitic magmas are closely related. Firstly, their tectonic settings are similar. Secondly, many andesitic-dacite-rhyolite suites show large chemical and mineralogical similarities. These data favor derivation by low pressure differentiation from mantle-derived basaltic magmas or from partial melting of mafic mantle-derived rocks. For some rhyolite suites, a good case has been made for their derivation by melting of crustal sedimentary rocks which have been chemically processed by weathering and sorted by transport, a process believed to generate most granitic magmas.

Among pyroclastic rocks, rhyolitic compositions are represented by voluminous ash flow and Plinian fallout sheets. Indeed, volume alone (100's of km$^3$) leads some to believe that ash flow sheets represent the cupolas of large magma chambers, with most of the magma crystallizing at depth to form large plutons.

Alkalic or peralkalic rhyolites, trachytes, and phonolites of oceanic islands, continental intraplate environments and some andesitic chains can be considered separately from the above because they are generally associated with much larger abundances of mafic rocks and occur in any type of tectonic setting. These magmas may have formed at low pressure in shallow magma chambers by differentiation processes such as crystal fractionation and diffusion. In these high level magma reservoirs, conditions are commonly appropriate for explosive pyroclastic eruptions to develop. The most important conditions are the saturation and oversaturation of volatiles caused by such factors as crystallization of anhydrous phases, absorption of water from country rocks, and pressure release. In shallow magma reservoirs, particularly where convection is governed by pipe-like geometric restrictions, chemically and mineralogically zoned magma bodies may develop. Because many pyroclastic eruptions emit large volumes of magma within a very short time, compositional zonations are well-developed in the deposits, particularly in ash flow and large Plinian accumulations. Such deposits are among the most important rocks for deciphering magmatic processes that occur at low pressures. We will discuss the petrologic importance of compositionally zoned pyroclastic deposits at the end of this chapter.

*Dacites-rhyodacites-calcalkalic rhyolites,* generally part of the andesite (or calcalkalic or orogenic) suite, are more abundant as pyroclastic rocks than as lava flows. Common phenocrysts in rhyodacite are andesine, OH-bearing ferromagnesian phases such as biotite and/or hornblende as well as pyroxene. Quartz and sanidine also occur but are more common in rhyolite, which also carries oligoclase but only minor amounts of ferromagnesian phenocrysts (biotite, amphibole, ferroaugite, hypersthene, and fayalite). Pyroclastic rocks of this suite include widespread ash flow and fallout sheets that are especially good stratigraphic markers in continental and marine deep-sea deposits. In eruptions of smaller volume, breccias may be common and coarse-grained pyroclastic debris may accumulate close to the vent.

*Peralkalic rhyolites, trachytes, and phonolite* are treated together here because they are usually associated with their parent basaltic magmas and occur most commonly in alkalic intraplate basaltic provinces. Mildly alkalic basalts give rise to peralkalic trachytes and rhyolites, characterized by phenocrysts of anorthoclase, ferroaugite or hedenbergite, richteritic amphibole, aenigmatite, and sometimes fayalite. With increasing alkalinity and silica undersaturation in the basalts, the derivative volcanic rocks are trachyte, containing phenocrysts of anorthoclase or sanidine, pyroxene, biotite and/or amphibole, and, depending on their silica saturation, small amounts of either quartz or feldspathoid. In phonolites, feldspathoids (particularly nepheline, leucite and, more rarely, nosean or hauyne) plus sanidine or anorthoclase are the dominant felsic phenocrysts, whereas sodic pyroxene, amphibole or biotite are the mafic phenocrysts. Sphene, apatite, and titanomagnetite are common accessories. Pyroclastic deposits, especially ash flows and fallout ashes, are very common in rocks with these compositions.

# Magma Chambers

Magma chambers are a fundamental concept in igneous petrology and volcanology and have become of more general interest in the field of geothermal energy and ore deposits. Lava flows erupted from silicic magma chambers are usually of small volume owing to their high viscosity. During pyroclastic eruptions that produce large Plinian fallout and pyroclastic flow deposits, very large volumes of magma are erupted practically instantaneously. These quenched samples from the upper part of magma columns enable the petrologist to study many magmatic processes more precisely than in plutons, where crystallization and alteration during cooling mask the primary evidence. Important petrologic aspects studied during the last 20 years based on ash flows and Plinian fallout deposits include: estimation of volumes and depths of near-surface silicic magma chambers; mineralogical and chemical zonation of magma chambers (Table 2-2); differentiation mechanisms; long-term thermal and chemical evolution of magma systems.

### Volumes of Magma Chambers

Smith (1979) has discussed the relationship between volumes of ash flow eruptions, size of calderas and magma chambers and zonation in magma columns (Figs. 2-5,

2-6, 2-7). Although field data to estimate volumes of proximal extra-caldera deposits may be fairly precise, large volumes of fine ash winnowed from the flows during transit may become airborne and transported much farther, while large calderas may contain significant intracaldera ponded tuff. Thus, volumes are generally minimum values. Smith (1979) considers a hierarchy of volumes of magma erupted (Figs. 2-5, 2-6), with the lower three orders of magnitude (0.001 to 1.0 km$^3$) characteristic of avalanche deposits from domes and backfall pyroclastic flows from small central vent volcanoes (Pelée, Mayon, St. Helens). Magma volumes of 1 to 100 km$^3$ are from larger stratovolcanoes and small ring structures (with caldera formation) such as Krakatau and Crater Lake. Volumes of $10^2$ to $10^3$ km$^3$ are associated with larger calderas such as Valles, Long Valley, and Yellowstone (Fig. 2.5). Pyroclastic flows of still larger volumes ($>10^3$ km$^3$) are erupted from very large calderas and volcanotectonic depressions such as Toba in Sumatra, or the San Juan Mountains in Colorado.

Silicic magma chambers are believed to be less than 10 km thick, with the larger chambers being more slab-like and the smaller ones more cylindrical. No more than 1/10 of the chamber volume may be erupted during any one pyroclastic eruption (Smith, 1979) (Figs. 2-6, 2-7).

## Zonation in Magma Chambers

Most pyroclastic flow deposits that are related to calderas studied in detail show systematic changes in mineralogical and chemical composition which reflect thermal, mineralogical, chemical and other gradients in magma chambers. Such

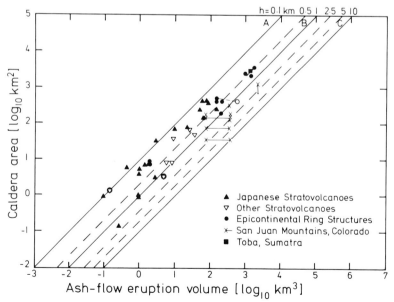

**Fig. 2-5.** Relationship between caldera area and volume of erupted ash flows. *Diagonal lines* indicate depth of drawdown (*h*) in magma chambers assuming vertical walls and a flat roof. (After Smith, 1979)

**Table 2-2.** Chemical analyses of tephra from fallout and flow deposits, mostly from chemically zoned eruptions

| | (1) BTE | (2) BTL | (3) K-42 | (4) K-45 | (5) K-123 | (6) TSB | (7) TSM | (8) TST | (9) P1R | (10) P1M | (11) P1B |
|---|---|---|---|---|---|---|---|---|---|---|---|
| I. Major elements (wt.-%) recalculated $H_2O$- and $CO_2$-free | ▼ | | | | | ▼ | | | ▼ | | |
| $SiO_2$ | 77.4 | 75.5 | 77.62 | 64.81 | 58.6 | 76.6 | 71.8 | 68.8 | 66.88 | 62.89 | 49.27 |
| $TiO_2$ | 0.07 | 0.21 | 0.17 | 0.71 | 0.71 | 0.10 | 0.31 | 0.41 | 0.82 | 1.69 | 4.07 |
| $Al_2O_3$ | 13.0 | 12.4 | 12.40 | 15.70 | 16.90 | 12.7 | 14.9 | 16.8 | 16.16 | 15.57 | 14.19 |
| $Fe_2O_3$ | – | – | 0.39 | 1.97 | 2.97 | 0.86 | 1.3 | 1.4 | 2.91 | 4.67 | 9.37 |
| FeO | 0.7* | 1.1* | 0.85 | 3.45 | 4.17 | 0.07 | 0.08 | 0.74 | 0.50 | 1.12 | 3.84 |
| MnO | 0.04 | 0.02 | 0.05 | 0.12 | 0.14 | 0.06 | 0.09 | 0.10 | 0.20 | 0.23 | 0.17 |
| MgO | 0.01 | 0.25 | 0.11 | 2.14 | 3.81 | 0.09 | 0.43 | 0.45 | 0.76 | 1.75 | 4.24 |
| CaO | 0.45 | 0.95 | 0.87 | 4.99 | 7.34 | 0.6 | 0.7 | 1.1 | 1.35 | 2.64 | 9.18 |
| $Na_2O$ | 3.9 | 3.35 | 4.25 | 4.21 | 3.55 | 3.4 | 4.2 | 4.3 | 7.22 | 6.44 | 3.49 |
| $K_2O$ | 4.8 | 5.55 | 3.2 | 1.74 | 1.29 | 5.1 | 6.0 | 6.0 | 2.99 | 2.54 | 1.16 |
| $P_2O_5$ | 0.01 | 0.06 | 0.03 | 0.14 | 0.13 | 0.02 | 0.04 | 0.1 | 0.16 | 0.43 | 1.01 |
| $H_2O$ | – | – | (2.66) | (2.52) | (0.88) | – | – | – | (0.30) | (1.64) | (0.60) |
| $CO_2$ | – | – | – | – | – | – | – | – | (0.05) | (0.07) | (0.09) |
| II. Trace elements (ppm) | | | | | | | | | | | |
| Cr | 3.0 | 2.5 | 2 | 8 | 27 | < 10 | < 10 | < 10 | < 10 | 6 | 29 |
| Co | 0.3 | 1.0 | – | – | 18 | – | – | – | 11 | 14 | 42 |
| Ni | < 3 | 3 | – | – | – | – | – | – | 6 | 9 | 22 |
| Cu | – | < 3 | – | – | – | – | – | – | 18 | 29 | 42 |
| Zn | 38 | 40 | 37 | 73 | 85 | – | – | – | 134 | 135 | 130 |
| Rb | 190 | 95 | 71 | 38 | 20 | – | – | – | 58 | 46 | 24 |
| Sr | < 10 | 110 | 40 | 237 | 327 | 30 | 70 | 500 | 328 | 463 | 875 |
| Y | 25 | 12 | 24 | 24 | 7 | 30 | 50 | 30 | 64 | 41 | 43 |
| Zr | 91.3 | 69.7 | 158 | 151 | 113 | 100 | 200 | 500 | 1022 | 738 | 308 |
| Nb | < 25 | 5 | – | – | – | 30 | 20 | 10 | 123 | 100 | 53 |
| Ba | < 10 | 465 | – | 600 | 410 | 70 | 200 | 3000 | 931 | 502 | 307 |
| Cl | – | – | 990 | 700 | 800 | – | – | – | – | – | – |
| S | – | – | 1375 | 150 | 140 | – | – | – | – | – | – |
| F | – | – | < 10 | 400 | 400 | – | – | – | – | – | – |

Arrow indicates top of zoned magma chamber
\* total Fe as FeO
– not determined or reported

|  | (12) STH | (13) FAB | (14) FAM | (15) FAT | (16) 1002 | (17) 1034 | (18) 1088 | (19) Oct. 14 | (20) Oct. 23 | (21) 507F |
|---|---|---|---|---|---|---|---|---|---|---|
| I. Major elements (wt.-%) recalculated $H_2O$- and $CO_2$-free | | | | | | | | | | |
| $SiO_2$ | 64.74 | 65.62 | 62.98 | 60.15 | 56.63 | 60.03 | 58.62 | 51.61 | 49.17 | 51.3 |
| $TiO_2$ | 0.61 | 0.37 | 0.77 | 1.31 | 0.12 | 0.28 | 0.97 | 0.92 | 0.91 | 1.85 |
| $Al_2O_3$ | 17.59 | 16.67 | 18.94 | 17.42 | 20.92 | 19.93 | 19.33 | 20.91 | 18.73 | 13.56 |
| $Fe_2O_3$ | 0.48 | 0.35 | 1.73 | 3.28 | 1.87 | 1.57 | 2.25 | – | – | 0.36 |
| FeO | 3.70 | 3.08 | 1.79 | 1.54 | 0.49 | 0.63 | 1.53 | 8.45* | 10.81* | 12.69 |
| MnO | 0.07 | 0.25 | 0.14 | 0.13 | 0.53 | 0.24 | 0.13 | – | – | 0.21 |
| MgO | 1.88 | 0.15 | 0.55 | 1.63 | 0.23 | 0.18 | 0.87 | 4.00 | 7.84 | 6.96 |
| CaO | 4.95 | 0.66 | 1.40 | 3.20 | 1.14 | 1.46 | 3.68 | 9.79 | 8.96 | 10.92 |
| $Na_2O$ | 4.49 | 7.20 | 5.51 | 5.30 | 12.18 | 8.98 | 5.22 | 3.57 | 2.95 | 2.32 |
| $K_2O$ | 1.34 | 5.62 | 6.07 | 5.78 | 5.83 | 6.68 | 7.16 | 0.75 | 0.62 | 0.07 |
| $P_2O_5$ | 0.14 | 0.03 | 0.13 | 0.25 | 0.03 | 0.05 | 0.24 | – | – | 0.16 |
| $H_2O$ | (0.45) | (2.83) | (3.93) | (1.23) | (2.10) | (2.10) | (0.98) | – | – | (0.25) |
| $CO_2$ | (0.06) | (0.41) | (0.12) | (0.07) | (0.04) | (0.02) | (0.02) | – | – | (0.06) |
| II. Trace elements (ppm) | | | | | | | | | | |
| Cr | 0 | 8 | 7 | 13 | 2.0 | 0 | 6.52 | 17 | 33 | 150 |
| Co | < 10 | 6 | 5 | 6 | 0.44 | 0.8 | 4.55 | 28 | 63 | 78 |
| Ni | < 10 | 13 | 7 | 13 | < 10 | 5 | 10 | 20 | 43 | 78 |
| Cu | 28 | 33 | 7 | 43 | 49 | 23 | 10 | 109 | 99 | 147 |
| Zn | 58 | 203 | 96 | 117 | 309 | 113 | 69 | 84 | 90 | 105 |
| Rb | 31 | 461 | 160 | 123 | 712 | 285 | 117 | 10.7 | 8.3 | < 11 |
| Sr | 480 | 10 | 106 | 372 | 5 | 185 | 1192 | 559 | 524 | 86 |
| Y | 13 | 118 | 43 | 34 | 43 | 21 | 27 | – | – | 46 |
| Zr | 128 | 1745 | 1022 | 592 | 2614 | 738 | 269 | 91.3 | 69.7 | 115 |
| Nb | 9 | 469 | 146 | 99 | 413 | 144 | 113 | – | – | < 11 |
| Ba | 324 | 46 | 236 | 634 | 359 | 419 | 673 | 425 | 300 | 63 |
| Cl | – | 4600 | 900 | 700 | 2600 | 3000 | 1300 | – | – | 100 |
| S | – | < 100 | < 100 | 200 | 260 | 420 | 1200 | – | – | 500 |
| F | – | 3300 | 1100 | 800 | 7000 | 2100 | 700 | – | – | – |

(1) and (2) from early and late phases of Bishop ash flow tuff (Hildreth, 1979); (3) early and (4) intermediate fallout, and (5) top of late ash flow, 1912 Katmai, Alaska (USA) eruption (Hildreth, 1983); (6) base to (8) top of zoned Tertiary Topopanah ash flow, Nevada (USA) (Lipman et al., 1966); (9) Miocene low silica basal rhyolitic through (10) mixed to (11) upper basaltic pyroclastic flow (Gran Canaria, Canary Islands) (Schmincke, 1969b and unpublished data); (12) gray dacite from May 18, 1980 blast deposit, Mount St. Helens, Washington (USA) (Schmincke, unpublished); (13) base, (14) middle and (15) top of Plinian trachytic Fogo A deposit, 1653 eruption, São Miguel (Azores) (Walker and Croasdale, 1971; chemical data from Schmincke, unpublished); (16) base, (17) middle and (18) top of Plinian late Quaternary phonolitic Laacher See Tephra (Eifel, Germany) (Wörner and Schmincke, 1984; (19) and (20) high-Al-basaltic fallout tephra, 1974 eruption of Fuego (Guatemala) (Rose et al., 1980; (21) tholeiitic deep sea tephra from the Galapagos Mounds area (Eastern Pacific) (Schmincke, 1983)

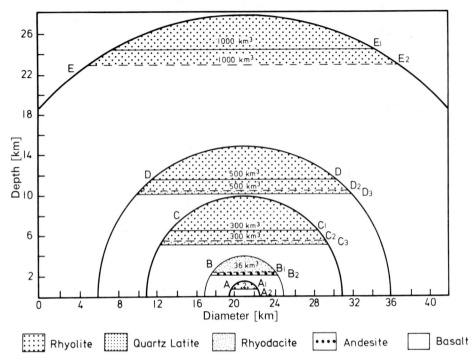

**Fig. 2-6.** Schematic cross sections of different types of zoned magma chambers. Diameters are those of associated calderas. Erupted magma is assumed to have occupied top of hemispheres. *A* Askja (1875), *B* Crater Lake, *C* Bandelier Tuff, *D* Timber Mountain, *E* Yellowstone. Subscripts denote different eruptive units. (After Smith, 1979)

zonations occur both in large calcalkalic sheets as well as in many small sheets of peralkalic composition (Figs. 2-6 to 2-16).

Salient features are summarized in two recent reviews of fundamental importance (Smith, 1979; Hildreth, 1981). Important conclusions of these essays include:

1. All pyroclastic eruptions exceeding 1 km$^3$ are compositionally zoned and many much smaller ones show pronounced zonations as well.
2. Erupted parts of magma chambers range from nearly uniform rhyolitic composition to strongly contrasting basalt-rhyolite compositions.
3. With depth, magma columns become hotter, more mafic chemically and more phenocryst-rich.
4. There are pre-eruptive gradients in T, $f_{O_2}$, major and trace element and isotopic composition, volatiles (H$_2$O, Cl, F) and in types, abundance and composition of phenocrysts.
5. A zoned magma column may be vertically layered with abrupt transitions between zoned subunits.
6. Wide compositional gaps are common at all levels of SiO$_2$-concentration and must have developed within a magmatic system.
7. Small-volume systems tend to show stronger compositional contrast than large-volume systems.

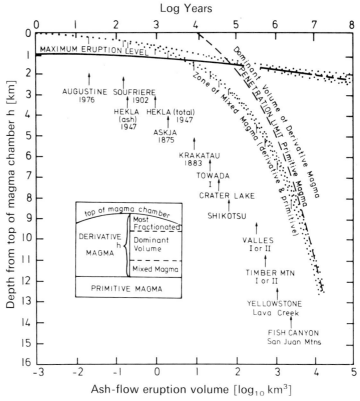

**Fig. 2-7.** Relationship between volume of ash flows, depth of drawdown from top of magma chambers, and time interval for development of zonation. The *lower dotted zone* indicates the transition zone beneath which the viscosity of magmas is too low to form ash flows while the *upper dotted zone* – representing an advanced stage of crystallization – separates the upper eruptible part of magma columns (<ca. 50% phenocrysts) from the lower part that is too viscous to erupt. Also shown are the maximum eruption levels, the penetration limit of primitive magma, and the zone of mixed magma. (After Smith, 1979)

8. Volatile-rich aphyric boundary layers or cupolas occur on top of more crystal-rich zoned magma columns.
9. Crystal fractionation is only one of several processes resulting in compositional gradients and is much less important than liquid-state thermodiffusion and liquid complexing in some high-silica and high-alkali systems.

Similar zonations also occur in large silicic and alkalic fallout deposits and within small single scoria cones of mafic to intermediate composition. Below, we briefly review four different kinds of zoned magma chambers, with data spanning the spectrum from large ash flow fields to single scoria cones.

**Large Calcalkalic Systems**

The classic examples of zoned magma chambers are large volume ash flow sheets of calcalkalic composition such as those in the Western US and Japan (Smith and

**Fig. 2-8.** Chemically and mineralogically zoned late Quaternary pyroclastic flow deposits at Crater Lake (Oregon). Note sharp transition between *white* rhyolitic *lower* and *dark* andesitic *upper half*

**Fig. 2-9.** Hand specimen of Miocene compositionally zoned strongly welded pyroclastic flow deposit. Lower rhyolitic half rich in anorthoclase phenocrysts overlain with a sharp boundary by mixed rock consisting of about 50% of rhyolite and 50% of chilled basalt (dark wispy lapilli). Composite flow, P 1, of about 300 km$^2$ areal extent from Gran Canaria (Canary Islands). (Schmincke, 1969b)

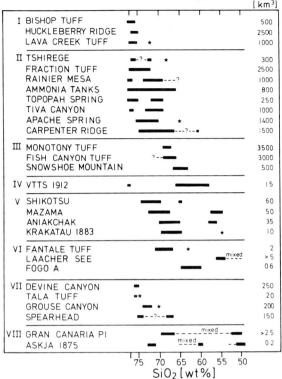

Fig. 2-10. Range in SiO$_2$-concentration and volumes of compositionally zoned pyroclastic flow deposits. Laacher See and Fogo A dominantly fallout. Stars represent postcaldera lavas. (After Hildreth, 1981)

Bailey, 1966; Lipman et al., 1966; Aramaki and Ui, 1966). Hildreth (1981) has divided these into five classes based on silica content, range of composition and volume, including some alkaline systems (Fig. 2-10). The ash flow sheet studied in most detail is the Bishop Tuff (Hildreth, 1979; 1981) (Figs. 2-12 to 2-14).

The 0.7-m.y.-old Bishop Tuff which erupted from Long Valley caldera (Bailey et al., 1976) is typical of many other high-silica calc-alkalic ash flow tuffs (Hildreth, 1979; 1981): the total deposit, including preceding Plinian fallout tuff, represents about 500 km$^3$ of magma. Phenocryst content increases from about 5 volume percent in the beginning to 30 volume percent at the end of the eruption. The mineral assemblage (quartz, sanidine, plagioclase, biotite, magnetite, ilmenite, apatite, zircon, allanite) changed about half way through the eruption by the addition of orthopyroxene, clinopyroxene, and pyrrhotite and, somewhat later, the disappearance of allanite. Temperatures calculated from the composition of coexisting magnetite and ilmenite ranged from about 720 °C for the first to about 780 °C for the last eruptive products. Major elements show significant changes. The early products are highly differentiated and later products are more mafic, although differences in SiO$_2$ are only about 2 percent absolute. The most drastic compositional zonation is shown by trace elements which may have enrichment or depletion factors of between 2 and 4, and up to 10 for some elements.

Compositional zonation cannot be explained by crystal fractionation, assimilation, or progressive melting of some source rocks (Hildreth, 1979, 1981). Instead, zonation may be the result of diffusion-convection-controlled differentiation

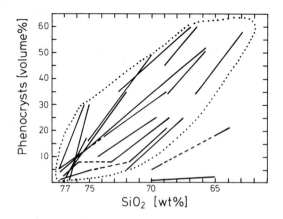

**Fig. 2-11.** Range of modal amount of phenocrysts in compositionally zoned pyroclastic flow deposits, phenocryst-rich compositions being erupted last. (After Hildreth, 1981)

mechanisms which develop along the temperature gradient as a result of the establishment of volatile gradients. These further lead to gradients in melt structure which, together with the other factors, appear to control the relative enrichment or depletion of elements in the non-convecting, highly differentiated roof portion of the magma chamber. Hildreth uses the term "thermogravitational process", which is a combination of Soret diffusion and boundary layer convection. Soret diffusion may in part be controlled by trace element complexing and structural gradients. The compositional gradients must have existed in the magma prior to crystallization. Michael (1983), however, argues that the compositional contrast between liquids in the Bishop Tuff magma chamber was originally caused by crystal-liquid fractionation. The more evolved liquid may have migrated toward the top of the magma column on account of its lower density.

**Small Highly Alkalic Systems**

Mineralogical and chemical zonation of major and trace elements is especially pronounced in the phonolitic fallout and minor pyroclastic flow deposits erupted

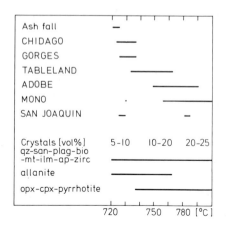

**Fig. 2-12.** Temperatures (based on Fe-Ti-oxide thermometry) and phenocryst contents for fall and flow units of Bishop Tuff. (After Hildreth, 1979)

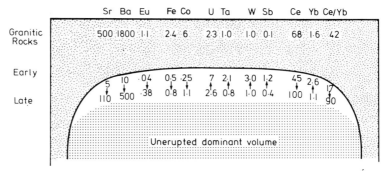

**Fig. 2-13.** Schematic cross sections through top part of Quaternary rhyolitic Bishop Tuff magma chamber showing concentration gradients in several trace elements (in ppm) and Fe (in %). *White zone* represents erupted top layer of magma beneath granitic roof rock. (After Hildreth, 1981)

11,000 years ago from the Laacher See volcano (Eifel, Germany) (Wörner and Schmincke, 1984). The tephra, composed of widespread fallout, ash flow, and base surge deposits, represents a magma volume of more than 5 km³. The composition ranges from early-erupted highly differentiated peralkalic, virtually crystal-free phonolite representing the cupola, to a less differentiated phenocryst-rich (about 50 volume percent) phonolite, erupted from a lower portion of a zoned crustal magma column (Figs. 2-15, 2-16). Major and trace element gradients define four distinct magma layers: The crystal-poor roof part (I) is strongly enriched in Na and incompatible (Zr, Th, U, Zn) and volatile elements (F, Cl); the main volume of the magma column (II) is characterized by moderately differentiated phonolite,

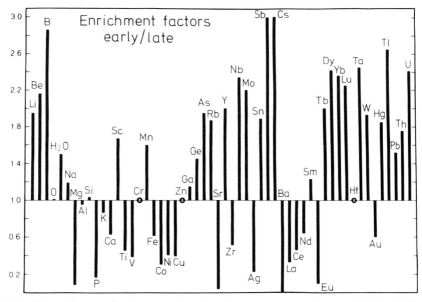

**Fig. 2-14.** Enrichment factors of major trace elements of early to late-erupted parts of Bishop Tuff. (After Hildreth, 1981)

31

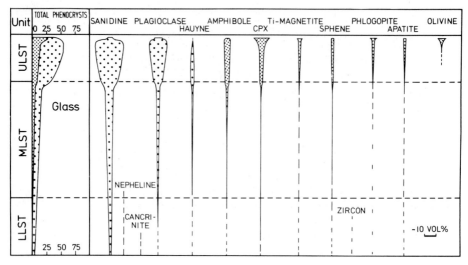

**Fig. 2-15.** Mineralogical zonation in late Quaternary Plinian phonolitic Laacher See Tephra sequence. Overall modal composition on left. *Dark stippling* = mafic, *light stippling* = felsic crystals. Volume-normalized stratigraphic subdivision: lower (L), middle (M) and upper (U) Laacher See Tephra on left. (Wörner and Schmincke, 1984)

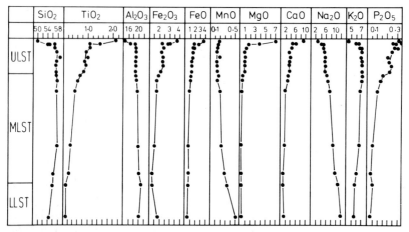

**Fig. 2-16.** Major element zonation of phonolitic Laacher See Tephra sequence. See Fig. 2-15. (Wörner and Schmincke, 1984)

phenocryst contents ranging from 5 to 15 volume percent. A sharp interface separates this from the lower, phenocryst-rich, mafic phonolite (III), relatively enriched in compatible elements (Sr, Sc, Cr, Co) and depleted in others (Zr, Th, U). The top of the erupted sequence (IV) is represented by phonolite-basanite mixed tephra as the final eruptive product.

The formation of the phonolite magma as well as the origin of the zonation is interpreted to result from crystal-liquid fractionation, with a primitive basanite being parental to the mafic phonolite. Convection-controlled diffusion processes significantly modified the trace element abundances within the entire Laacher See magma column and resulted in the volatile-rich uppermost magma layer. In contrast to the Bishop Tuff magma system discussed above, these late-stage liquid-liquid differentiation processes developed mainly *after* fractional crystallization; only the uppermost layer of the Laacher See magma chamber appears to be similar in origin to the roof portion of the Long Valley rhyolitic magma chamber, though much smaller in volume.

## Oceanic Rhyolite-Basalt Systems

The volume of highly differentiated (evolved), as compared to mafic magmas is small on oceanic islands, with Iceland and Gran Canaria (Fig. 2-9) being two notable exceptions. However, evolved oceanic magma systems have played a prominent role in petrogenetic discussions, following the observation by Bunsen (1851) on the co-existence of mafic and silicic magmas in Iceland and his idea on the formation of intermediate magmas by hybridization of these end members. Eruptions from such zoned columns represent much deeper drawdown and enable us to assess the role of mafic magma in generating derivative magmas and influencing eruptive processes in more detail than in large differentiated magma chambers that probably inhibit the upward penetration of basaltic magma to eruption levels. Tephra from such eruptions shows widespread mixing of mafic, intermediate, and silicic magmas, and makes reconstruction of petrogenetic processes especially speculative. A good example for this setting is Askja volcano.

The 1875 eruption of Askja volcano (Iceland) produced 0.2 km$^3$ of rhyolitic tephra, some pumice showing pre-eruptive mixing of rhyolitic with small amounts of icelandite and basaltic liquids (Sigurdsson and Sparks, 1978, 1981). The rhyolite contains less than 0.5 percent of crystals (plagioclase, clinopyroxene, hypersthene and magnetite). Mineral geothermometry, minimum composition of partially melted trondhjemite inclusions, and estimated gas exit velocities at the vent all indicate that the rhyolite magma had a water content close to 3 weight percent, much of which was apparently of meteoric derivation as indicated by low $\delta^{18}O$ values. The rhyolite magma is believed by Sigurdsson and Sparks to have formed by partial melting of older trondhjemite by basalt, accumulation of the silicic liquid at the top of the chamber and mixing with 7–14 percent basaltic magma, and development of a stratified rhyolitic magma column by fractional crystallization of the hybrid magma. Mixing of the rhyolite with new primitive basalt destroyed the stratification and may have triggered the 1875 Plinian eruption of Askja volcano.

## Very Small Mafic-Intermediate Magma Systems

Most literature of the past 15 years on zoned magma chambers has been on silicic magma systems, especially the voluminous calc-alkalic ash flow sheets. Basaltic scoria cones, although widespread in continental and oceanic intraplate tectonic settings are generally assumed to be homogeneous in composition. In many volcanic fields, however, there is a wide range in composition from very primitive to intermediate, clear evidence of differentiation. This is documented by chemical and mineralogical studies of stratigraphically controlled samples in well-exposed scoria cones in the Quaternary Eifel volcanic field.

The late Quaternary Rothenberg volcano, about 150 m high and 500 m wide, consists of extra- and intra-crater facies (Duda and Schmincke, 1978; Karakuzu et al., 1982). Seven phases of scoria, block tuff, tuff breccia, and agglutinates can be distinguished. The variety, abundance, and size of phenocryst phases vary systematically. Rocks from the first stage contain less than 10 percent total phenocrysts, dominated by phlogopite 200 µm in diameter and clinopyroxene, 500 µm in diameter. Phenocrysts increase in abundance, size and type – with olivine appearing in phase 5 and becoming increasingly abundant in the later eruptive products, rocks from the final phase containing 20 percent clinopyroxene (800 µm) plus olivine (300 µm) and 3 percent of phlogopite (700 µm). Simultaneously, the whole rock chemical composition changes from tephrite (stage 1) to basanite (stage 7). Fractionation calculations using both major and trace elements, the compositions of the observed phases, and the volumes of the different phases show that the tephrite can be generated by crystal fractionation from a basanite parent magma, with about 50 percent of the parent basanite magma not having been erupted.

# *Chapter 3* Magmatic Volatiles and Rheology

There are many questions about magmas that pertain to the origin of pyroclastic rocks. Why are some magmas quietly effusive, others violently explosive, and still others alternate between both characteristics? Why do some magmas build high-standing pyroclastic cones, some build small cinder cones, and still others enormous shield volcanoes or lava plateaus? Some of the most scenic volcanoes are towering cones composed primarily of pyroclastic debris knit together by a skeletal framework of lavas and dikes that help maintain the high-standing edifice. Is the commonly observed change from initial explosive to later more quiet eruptive activity within the same volcano caused by a decrease in the amount of dissolved volatiles, or is it caused by differences in the way rising magma interacts with its environment during its ascent? Which are the most important processes that cause magma to break up into particles?

Scrope (1862) recognized the importance of juvenile gases in propelling magma to the earth's surface and in causing explosive (pyroclastic) eruptions. Since then, most volcanologists have emphasized volcanic gases as the dominant agent that comminutes magma. Today we realize that the nature of explosive eruptions is governed by a variety of physical and chemical magma properties as well as their dynamic interplay. Especially important are (1) viscosity which depends on chemical composition, volatile content, temperature, and the amount of bubbles and crystals, (2) type and amount of magmatic volatiles as well as (3) rate of ascent and depth of vesiculation. Moreover, factors such as availability of external water also play an important role. Hydroclastic processes, whereby comminution results from the explosive interaction of water and magma, have been shown in the last few years to be much more common than traditionally thought.

## Volatiles

Magmas are dominantly silicate melts that contain variable amounts of crystals and volatiles. Under high pressure, the volatiles are dissolved, but under low pressure when a magma approaches the surface, the volatiles exsolve to form a free vapor (fluid) phase. Volatile components probably make up less than 5 weight percent of most magmas, but their molecular weights are lower than those of most of the silicates and metallic oxides. Thus, the mole fraction of water (W) $X_W^m$ is 0.5 in a water-albite melt (m) that contains only about 6.4 percent $H_2O$ by weight (Burnham, 1975a,b, 1979). A general account of the thermodynamic behavior of $H_2O$ and of other volatile species in albite melts is given by Burnham (1979), who extrapolates these models to natural magmas. There is little doubt that volatiles, whether dissolved or in a gaseous state, greatly influence the eruptive behavior of magmas.

In this section we concentrate on the transition of dissolved to exsolved volatiles, the concentration of volatiles in magmas, and resultant theories of vesiculation and ash formation. Mineralogical and petrological aspects of magmatic volatiles are reviewed by Carmichael et al. (1974), Burnham (1979), and others.

## Methods of Determining Volatiles

It is difficult to determine the amount and composition of volatiles in magmas. Volcanic rock analyses commonly report concentrations of $H_2O$ and $CO_2$, and, more rarely, S, F or Cl, but these concentrations are not representative of volatile concentrations in the magma because magma loses volatiles during eruption and crystallization. Moreover, unless the rocks are extremely fresh and glassy, volatile elements may have been gained or lost after cooling by alteration during weathering and diagenesis.

Another method of estimating the volatile contents of magma is to analyze the composition of volcanic gases directly at the vent, or from cooling intrusions, lava flows or pyroclastic flows (Shepherd, 1938; White and Waring, 1963; Lipman and Mullineaux, 1981). Direct collecting methods are generally hazardous and beset with formidable logistic, technical, and analytical problems, although much advance has recently been made by collecting or directly analyzing gases in eruption clouds (Rose, 1977; Harris and Sato et al., 1981).

But even the most successful direct analyses of volcanic gases may differ appreciably from the actual volatile content in the magma. Different volatile species in the magma have different solubilities, therefore less soluble gases are released more readily than the more soluble species, even if the less soluble may be less abundant than the more soluble gas.

There are several indirect approaches to the problem, providing us with the most realistic estimates so far available of the volatile contents of natural basaltic magmas. One is the study of glass and volatile inclusions trapped in phenocrysts (e.g. Roedder, 1972; Anderson, 1974a, b, 1975; Sommer, 1977; Delaney et al., 1978; Muenow et al., 1980). The other, pioneered by Moore (1965, 1970) (see also Moore and Schilling, 1973; Mathez, 1976), is the analysis of fresh glass and of gas and mineral phases within vesicles of glassy submarine lavas quenched at pressures higher than the vapor pressure of the volatile components. A third approach is to analyze rocks for less volatile elements, such as K and F, whose concentrations are assumed to be correlative with magmatic $H_2O$ content (Moore, 1970; Aoki et al., 1981).

Further information is provided by experimental solubility studies. For example, maximum or saturation solubilities have been determined in silicate magmas at different pressures and temperatures for $H_2O$ (e.g. Hamilton et al., 1964; Shaw, 1965; Kadik et al., 1972; Burnham, 1975a), for $CO_2$ (Brey and Green, 1975, 1976; Mysen, 1977; Spera and Bergman, 1980), and for the simultaneous solubility of both $H_2O$ and $CO_2$ (Brey and Green, 1977; Eggler, 1973; Eggler and Rosenhauer, 1978) (Fig. 3-1). The knowledge of saturation values establishes limits for estimating maximum magmatic gas components in the emitted total. Furthermore,

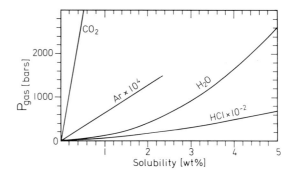

**Fig. 3-1.** Solubility of four gas species in mafic melts at different temperatures [$CO_2$: 1,200 °C; Ar: 1,500 °C (enstatite melt); $H_2O$: 1,100 °C; HCl: 1,200 °C]. (After Anderson, 1975)

experimental studies on the composition of coexisting phenocrysts (amphibole, olivine, biotite, magnetite, and feldspar), combined with determinations of the effect of water on lowering the liquidus temperatures of magmas, allow us to estimate the partial pressure of water in some types of magmas if temperature and oxygen fugacity are known (Eggler, 1972; Nicholls and Ringwood, 1973).

## Composition and Amount of Volatiles

Most workers agree that $H_2O$ (35 to 90 mol%), $CO_2$ (5 to 50 mol%) and $SO_2$ (2 to 30 mol%) are the most abundant volatiles in basaltic and andesitic magmas, whereas $H_2$, CO, COS, $H_2S$, $S_2$, $O_2$, HCL, $N_2$, HF, HB, HI, metal halogens, and noble gases are minor constituents (<2 mol%) (Anderson, 1975; Basaltic Volcanism Study Project, 1981; Gerlach, 1981). However, the type of molecular species of a gas component depends very much on temperature-dependent equilibria (Heald et al., 1963; Nordlie, 1971; Gerlach, 1982). For example, at equal oxygen fugacities, $SO_2$ is common at higher and $H_2S$ at lower temperatures. At constant T, $SO_2$ is more abundant at higher $f_{O_2}$ and $H_2S$ at lower $f_{O_2}$ (Fig. 3-2).

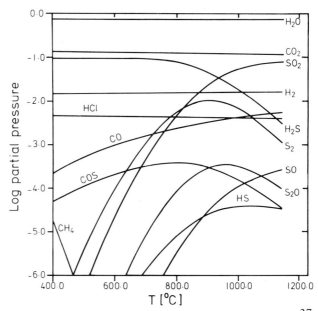

**Fig. 3-2.** Simulation of the composition of a volcanic gas at 1 bar total pressure as a function of temperature with $f_{O_2}$ buffered by cooling lava. (After Gerlach, 1982)

## Water

A long-standing problem is the source of the water vapor observed in many volcanic eruptions. Water vapor is the most abundant volatile species in most magmas, but how much is derived from the magma itself, how much is recycled meteoric water, and how much is meteoric water absorbed by the magma in the magma chamber from the country rock? Interaction of meteoric water and silicic magma bodies has been postulated based on oxygen-isotope studies (Lipman and Friedman, 1975) but how common such processes many occur is unknown.

The solubility of $H_2O$ in silicate melts increases strongly with pressure below about 10 kb and decreases slightly with temperature (Fig. 3-3). Water solubility increases from mafic to silicic compositions if considered on a weight percent basis, although this difference is largely due to the differences in liquidus temperatures (Hamilton et al., 1964). The compositional dependency also disappears when water is considered on an equimolar basis (Burnham, 1975a, 1979). It is commonly assumed that $H_2O$ is incorporated into the melts as $OH^-$ groups, acting as a network-modifier. It will thus break Si-O-Si bonds and depolymerize a melt. The detailed mechanism and thus the format of thermodynamic description is still a matter of debate (Holloway, 1981; Bottinga et al., 1981; Newton et al., 1981). Because the behavior of HF, HCl and $H_2S$ is similar to that of $H_2O$, their solubility is depressed compared to water at high fugacities of $H_2O$. Stolper (1982) has suggested that $H_2O$ occurs in silicate melts both as hydroxyl groups and as molecular water.

The experimental determinations of $H_2O$ solubilities only give us saturation values, but from other lines of evidence discussed below we can assume that most natural magmas are appreciably undersaturated at their depth of origin and become saturated only at very low pressures.

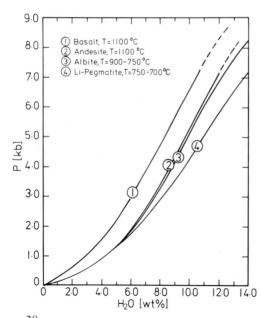

Fig. 3-3. Experimentally determined solubility of $H_2O$ in different aluminosilicate melts at various temperatures. (After Burnham and Jahns, 1962; Hamilton et al. 1964)

**Fig. 3-4.** Concentration of $P_2O_5$, F and Cl versus $H_2O^+$ of submarine quenched fresh basaltic glass from Juan de Fuca (MORB), Hawaii and Revillagigedo region. (After Moore, 1970)

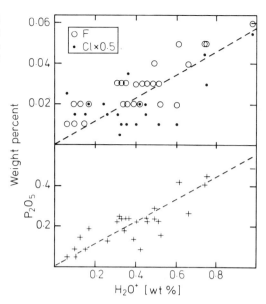

The best evidence for undersaturation, and indeed the most reliable measurements of water content of basaltic magmas, comes from analysis of deep sea basalt glass quenched at a total pressure greater than the partial pressure of water (Moore, 1965, 1970; Moore and Schilling, 1973; Delaney et al., 1978; Muenow et al., 1980; Garcia et al., 1979). Water contents ($H_2O$) are positively correlated with $K_2O$, $P_2O_5$, F and Cl (Fig. 3-4). In a more detailed study on a wide compositional spectrum of basalts, Aoki et al. (1981) noted a direct correlation of F with $K_2O$, both of which increase from tholeiitic rocks to basalts of highly potassic composition. The positive correlation of both elements with $H_2O$ was refined and allows a rough estimate of the primary $H_2O$ content of basaltic magmas (Fig. 3-5; Table 3-1). All three elements are believed to be derived mainly from the breakdown of phlogopite in the source rock because data for basalts other than MORB-tholeiites plot along phlogopite control lines (Aoki et al., 1981). Submarine tholeiites, however, are relatively enriched in F compared to $K_2O$ and may have gained their F, $K_2O$ and $H_2O$ from pargasite (Fig. 3-5) (Table 3-1) (Aoki et al., 1981).

The amount and diameter of vesicles in glassy rims of basalt lava erupted under water decrease with depth because of expansion of exsolved volatiles and more complete exsolution with decreasing pressure (Moore, 1965, 1970; Moore and Schilling, 1973). Vesicularity curves show a pronounced change in slope at depths which depend in part upon composition (Fig. 3-6). For the K-poor tholeiites, this change in slope occurs at about 500 m, for Kilauean tholeiites at about 800 m, and for alkali basalts at about 1800 m, representing the confining pressure at which volatiles become saturated in the melt and exsolve to form vesicles (Moore, 1970). Using these values for confining pressures and extrapolating the solubility data of Hamilton et al. (1964) to lower pressure, Moore argued that saturation values of water for K-poor tholeiite magmas are about 0.25 weight percent $H_2O$, for Hawaiian tholeiite magmas 0.5 weight percent, and 0.9 weight percent for alkali-rich

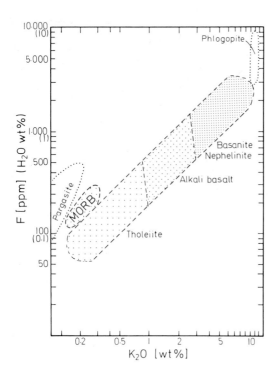

**Fig. 3-5.** Potassium versus F and ($H_2O$) concentrations in mafic volcanic rocks including mid-ocean ridge basalts (MORB) and fields of phlogopite and pargasites. (After Aoki et al., 1981)

basalt magmas – in agreement with water solubility values in experimental studies of silicate melts. Later analysis of the composition of gases within these vesicles showed, however, only $CO_2$ and crystallized sulfides along their inner walls, indicating that most of the vesicles were formed by exsolution of $CO_2$ and $H_2S(?)$ (Moore et al., 1977) (see below).

Vigorous vesiculation and loss of volatiles from the magma, however, do not necessarily occur at the saturation values (depths) inferred from vesicularity data. In a later study, Moore and Schilling (1973) found that actual loss of $H_2O$- and

**Table 3-1.** Estimated $H_2O$ content (wt.-%) in mafic magmas. (After Aoki et al., 1981)

|  | Range | Average |
|---|---|---|
| Continental and island arc |  |  |
| Quartz tholeiite | 0.14–0.20 | 0.15 |
| Tholeiite | 0.14–0.60 | 0.45 |
| Alkali basalt – nephelinite | 0.45–2.0 | 0.90 |
| perpotassic basalt | 2.0–8.0 |  |
| Ocean floor |  |  |
| Tholeiite | 0.15–0.70 | 0.30 |
| Oceanic island |  |  |
| Tholeiite | 0.15–0.70 | 0.50 |
| Alkali basalt – basanite | 0.50–1.5 | 1.0 |
| Nephelinite | 0.60–2.0 |  |

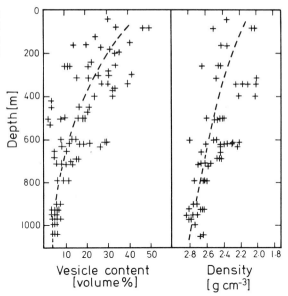

**Fig. 3-6.** Relationship between depth of eruption below sea level, modal abundance of vesicles and density in tholeiitic basalt. (After Moore and Schilling, 1973)

S-volatiles by vesiculation in ocean-floor tholeiites takes place only above about 200 m water depths (Fig. 3-7), although at greater depths they exsolve slowly from cooling pillow interiors. Moreover, escape of volatiles may not correlate linearly with decreasing water depth. At 43 m water depths (about 5 atmospheres), for example, half of the original sulfur is still present. Most of the gases may in effect be lost only in the upper 100 m or so for these tholeiitic basaltic compositions.

Aside from indicating the depths at which vesiculation occurs, Moore and Schilling's (1973) study indicates that actual explosive disintegration of tholeiite-basalt magma is not likely at water depths greater than about 100 m and may even be confined to depths of less than 50 m. For example, Reykjanes Ridge basalts have about 0.35 weight percent water ($H_2O^+$) (Fig. 3-7) and vesicularity ap-

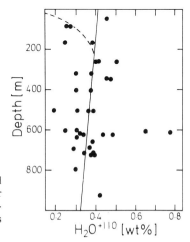

**Fig. 3-7.** Relationship between eruption depth below sea level of tholeiitic basalt and $H_2O^+$ concentrations in glassy outer 1 cm. *Dashed line* is assumed degassing trend at depths shallower than 200 m, assuming surface degassed lava contains 0.05 wt.% $H_2O^+$. (After Moore and Schilling, 1973)

proaches 50 volume percent at less than about 80 m (Moore and Schilling, 1973). Thus, disintegration of magma due to explosive degassing of magmatic volatiles probably occurs at shallower depths than theoretical studies suggest. McBirney (1963) postulated that explosive volcanic eruptions from expansion of magmatic volatiles are unlikely if gas/solid volume ratios are <1, and further estimated that explosive eruptions will not occur at water depths >500 m, for basaltic magma with 0.1 weight percent water.

The quantity of water lost from a basaltic magma during subaerial eruption is indicated by studies at Kilauea Volcano, Hawaii. Vesicle-poor Hawaiian submarine tholeiites contain about 0.45 weight percent water, whereas the highly vesicular tholeiitic basalt pumice and spatter from the August 1963 on-land-eruption of Kilauea Volcano contained an average of only 0.05 weight percent $H_2O$ (Friedman, 1967). Moreover, a clear relationship exists between the mode of eruption and the amount of water loss: basaltic pumice erupted from fountains only $\sim 40$ m high at Kilauea Volcano contained about 0.07 weight percent water, whereas lava erupted during vigorous activity from the more degassed high fountains ($\sim 500$ m) contained only 0.03 weight percent (Swanson and Fabbi, 1973). Basalts that form the shields of most oceanic islands and seamounts are generally moderately alkalic and probably have higher $H_2O$ and $CO_2$ contents. Thus, explosive activity of these magmas probably begins at somewhat greater water depth than that of tholeiite magmas.

Most andesite volcanoes are highly explosive. Sapper (1927), for example, estimated that more than 95% of the rocks in areas of andesitic volcanism are pyroclastic. It is commonly inferred that this high degree of explosivity is caused by the high $H_2O$ content of andesitic magmas in combination with their high viscosity. Because evidence is increasing that phreatomagmatic eruptions are common both in basaltic and andesitic volcanoes (McBirney, 1973; Schmincke, 1977), it is no longer permissible to automatically equate high degrees of explosivity with high magmatic volatile concentrations. Nevertheless, water contents in andesite are higher than in most basaltic magmas, as indicated by several lines of evidence.

Glass from andesitic pillow margins collected from the Mariana Arc contain about 1 weight percent $H_2O$, and glass/vapor inclusions in these rocks have high $H_2O/CO_2$ ratios compared to basalts (Garcia et al., 1979). Anderson (1974a, b; 1975) analyzed glass inclusions in phenocrysts of andesitic and basaltic rocks with the electron microprobe and assumed that the difference of the sum of the elements (to 100) could be accounted for by the $H_2O$ content of the glass. From such summation differences, he concluded that many andesitic magmas contain from 2–4 weight percent $H_2O$. Anderson suggested that the andesite magma at Paricutin, Mexico, had contained about 2 weight percent $H_2O$ prior to eruption.

Eggler (1972) and Sekine et al. (1979) compared observed phenocryst crystallization sequences with phase relationships determined experimentally in the same rocks. Because the thermal stability of plagioclase is depressed relative to pyroxene but $P_{H_2O}$ must be high enough to retain plagioclase as a liquidus phase, the water contents of these magmas can be bracketed. Thus, Eggler (1972) estimated a water content of about 2 weight percent for the Paricutin andesite magma based on the depression of the plagioclase liquidus, in agreement with Anderson's study of inclusions on the same rock (see above). Eggler (1972) further concluded that the observed phenocryst assemblage could have formed at a water saturation of <0.7 kb pressure and that plagioclase-bearing andesite magmas had $H_2O$ contents of $<3$ weight percent, a conclusion corroborated by the study of Sekine et al. (1979). Moreover, the eruption temperatures of about 1100 °C of the 1952 Paricutin andesite studied by Eggler (1972), and the silicic rocks investigated by Sekine et al. (1979) correspond to the experimentally determined liquidus temperature of the same rocks at $H_2O$-contents of 1–2 weight percent.

As mentioned in Chapter 2, hornblende andesites are common in continental provinces. A liquid that is crystallizing amphibole requires a minimum of about 3 weight percent $H_2O$ (Burnham, 1979) and a $H_2O$ content of 4 percent was inferred for a mafic hornblende-bearing andesite magma by Ritchey and Eggler (1978).

For fresh, nonhydrated rhyolitic glasses (obsidian), water contents are generally very low, mostly 0.2 to 0.4 weight percent $H_2O$ (Ross and Smith, 1955; Ross, 1964; Friedman et al., 1963). These values are obviously minimal because the amounts of volatiles lost during crystallization of magmas are usually made assuming saturation, i.e., $P_{H_2O} \sim P_{total}$. When an estimate of the lithostatic load can be made, the saturation concentrations of $H_2O$ in the magma at this pressure can be inferred from experimental solubility data. Wood and Carmichael (1973), by independently determining $P_{total}$ for three rhyolites, showed that the phenocrysts precipitated at $P_{H_2O}$ that may equal $P_{total}$ only during the last stage of crystallization. Ewart et al. (1975) inferred $H_2O$-concentrations of 7–8 weight percent for Quaternary rhyolitic magmas in New Zealand.

The relationship between the water content of a rhyolitic magma and its explosivity during eruption is quite uncertain. Only a small amount of pressure release at depth will cause $H_2O$-saturated rhyolitic magmas to crystallize if nucleation and growth of phases is fast compared to extrusion rates. On the other hand, rhyolitic lavas (domes) are commonly assumed to be derived from hot and dry magmas (Winkler, 1962; Harris et al., 1970).

In some volcanic provinces where effusive and pyroclastic rocks are interlayered, there is an apparent relationship between mode of eruption and inferred water content of magmas as based upon amount and type of phenocrysts. For example, pyroclastic flow deposits in the Miocene rhyolite-trachyte series and the Pliocene basanite-tephrite series on Gran Canaria are porphyritic and carry hydroxyl-bearing phenocrysts (amphibole with apatite inclusions and mica), while interbedded lavas of similar bulk chemical composition are less phyric and lack amphibole or mica (Schmincke, 1969a, 1976; Brey and Schmincke, 1980). This suggests that episodic volatile accumulation in the magma columns may have been responsible for crystallization of hydroxyl-bearing phenocrysts as well as the explosive nature of eruptions.

Carbon Dioxide

$CO_2$ has been known for some time to be the second most abundant magmatic gas species. Only recently, however, has its importance for determining the composition of mafic magmas in the source region and for governing the vapor phase at elevated pressure become recognized through experimental work on simple systems, and whole rock compositions as well as analytical work on glass and fluid inclusions, quenched glasses and unruptured vesicles of quenched magmas. Simultaneously, there has been a re-evaluation of the importance of $CO_2$ in causing highly explosive eruptions of mafic silica-undersaturated magmas, the extreme case being kimberlite magmas.

Maar volcanoes and their pyroclastic deposits were traditionally interpreted as caused by gas explosions from $CO_2$-rich alkalic mafic magmas (Noll, 1967; Ollier, 1967; Ringwood, 1975). However, all maars studied in detail to date show overwhelming evidence for a phreatomagmatic eruptive mechanism, that is, interaction of magma and meteoric water (Chap. 9). The empirical evidence assembled by volcanologists suggesting a phreatomagmatic origin of maars and possibly kimberlite breccias is not fully appreciated by experimental petrologists. On the other hand, the relatively large amounts of $CO_2$ in mafic alkalic magmas and especially its low solubility compared to $H_2O$ must undoubtedly influence the eruptive behavior of these magmas in a major way, but the exact role of $CO_2$ in influencing viscosity, density, and the rate and mechanisms of ascent and eruption are poorly understood.

Experimental evidence reviewed by Wyllie (1979) clearly shows that high $CO_2$ concentrations and high $CO_2/H_2O$ ratios during partial melting of source peridotites lead to alkalic and silica-undersaturated magma compositions (basanites, nephelinites), with kimberlite magmas apparently being very $CO_2$-rich. The association of some of these magmas with carbonatites, the occurrence of primary groundmass carbonate in rocks such as melilite nephelinites, and the predominance of $CO_2$ in the gas of the most alkalic (melilite nephelinite) active volcano studied in detail (Nyiragono) (see review by Gerlach, 1981), all show that in these magmas, $CO_2$ is a major gas species at low pressure as well.

The solubility of $CO_2$ is strongly dependent on pressure and varies widely in magmas of different composition owing to differences in melt structure (Fig. 3-8). The structural role of $CO_2$ in silicate melts has recently been evaluated by Mysen and Virgo (1980). It may either form $CO^{-2}$ anions and, thereby, lead to a polymerization of the melt or, through the formation of neutral metal-carbonate complexes, depolymerize a melt. The direction the system goes depends on the structural role of the modifier cation forming the carbonate complex. The degree of polymerization thus largely determines the solubility of $CO_2$. Solubility is low in silica-rich melts but is relatively high in the more strongly depolymerized mafic melts. For the same reason, even small amounts of $H_2O$ in the magma resulting in some depolymerization, favor solubility of $CO_2$ (Eggler, 1973). At pressures above about 10 kb, $CO_2$ may enter aluminosilicate melts in molecular form (Brey and Green, 1975; Brey, 1976; Burnham, 1979).

Carbon dioxide contents in quenched glassy submarine basalts range from about 0.06-0.2 weight percent with concentrations as high as 0.43 weight percent in island arc basalts, corresponding to $CO_2/H_2O$ ratios mostly around 0.1 (Fig. 3-9) (Moore, 1970; Delaney et al., 1978; Garcia et al., 1979; Muenow et al., 1980; Harris, 1981). However, because of its low solubility, $CO_2$ or other carbon gas species probably dominate the vapor phase at relatively high pressure ($\sim 4$ kb) (Kadik et al., 1972; Anderson, 1975) and thus are the main volatile species in inclusions in phenocrysts crystallized at depth (e.g. Murck et al., 1978). The initiation of $CO_2$-degassing of a basaltic magma at depth is also reflected in much less pronounced changes in $CO_2$ concentration in lava fountains and long-distance lava flows on Hawaii, contrasted with the drastic decrease in $H_2O$ and S concentrations during subaerial transit (Swanson and Fabbi, 1973).

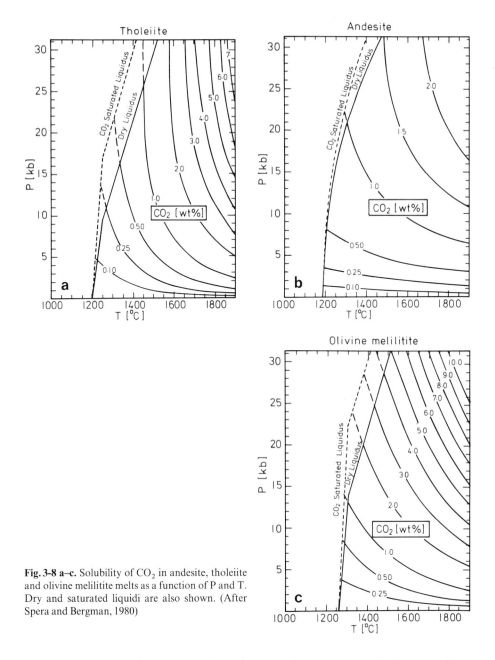

**Fig. 3-8 a–c.** Solubility of $CO_2$ in andesite, tholeiite and olivine melilitite melts as a function of P and T. Dry and saturated liquidi are also shown. (After Spera and Bergman, 1980)

Analysis of gases in unruptured vesicles from glassy selvages of mid-ocean ridge tholeiitic pillow basalts erupted at depths from 700 to 4800 m, indicates that $CO_2$ comprises more than 95 percent of the vesicle gas and ranges from 400 to 900 ppm in the rock (Moore et al., 1977). Volume expansion of the vesicle gas released when the vesicles are pierced at room temperature is about $20\pm5$ times the eruption depth (km), thus allowing an estimate of eruption depth. The ubiquitous presence of vesicles in submarine lavas indicates that basaltic magmas in mid-ocean ridge magma chambers are generally close to volatile saturation (Moore, 1979).

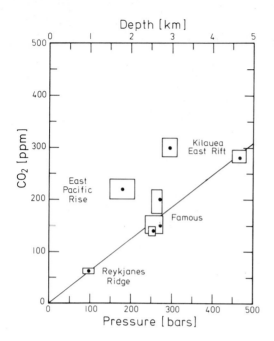

Fig. 3-9. Relationship between depth of eruption and $CO_2$-concentration of glass separates from mid-ocean ridge basalts (East Pacific Rise, Famous, Reykjanes Ridge) and Kilauea tholeiites. (After Harris, 1981)

Vesicularity of deep sea basalts is not simply a function of bulk magma composition and its specific volatile content and eruption depth. Moore (1979) noted significantly higher vesicularity in glassy pillow rinds from the Atlantic (half spreading rate about 10 mm yr$^{-1}$) compared to Pacific (3.8 mm yr$^{-1}$ half spreading rate) samples (Fig. 3-10). He interpreted the more volatile-rich Atlantic basalts to be due to volatile concentration in thin, peaked magma chambers. These vesicle-rich basalts are also enriched in incompatible elements and LREE, and several workers have emphasized the potential role of partitioning LREE and related elements into a $CO_2$-rich fluid (e.g. Moore, 1979; Wendtlandt and Harrison, 1979; Harris, 1981). However, as discussed later, carbonate diffusion is fast in basaltic magmas and $CO_2$ bubbles may form and separate faster from the magma than LREE can diffuse into them (Watson et al., 1982).

The solubility of $CO_2$ in silicic magmas (for which $NaAlSi_3O_8$ melts are a good approximation) is small but finite (Kadik et al., 1972; Holloway, 1976, Fig. 3). The

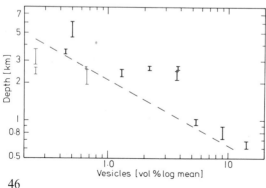

Fig. 3-10. Decrease in vesicularity with depth in 13 suites of samples of glassy rinds of mid-ocean ridge basalts (Atlantic = *thick lines*; Pacific = *thin lines*). Least-squares logarithmic trend line for Hawaiian rift zone samples shown by *dashed line*. (After Moore, 1979)

saturation values of $CO_2$ are much smaller than those for $H_2O$, but the presence of $CO_2$ in silicic magmas appears to be very important: (1) because $CO_2$-solubility is a linear function of $P_{CO_2}$, decreasing total P will decrease $CO_2$-solubility; (2) because $CO_2$-solubility is small even at 10 kb, silicic magmas in the source regions will be saturated with $CO_2$, if carbonate phases become unstable near the granite solidus during crustal anatexis (Holloway, 1976); and (3) because as pressure decreases during magma ascent and/or a magma becomes more differentiated during crystallization, a $CO_2$-rich fluid phase will almost certainly exsolve from the magma. Holloway's (1976) contention that textural, structural, mineralogical, and geochemical features of granitic rocks may be caused by differences in the relative abundance of $CO_2$ and $H_2O$ in their source regions can be extended to the problem of intrusive versus effusive versus pyroclastic mechanisms of magma behavior. Do $CO_2$-rich silicic magmas rise faster during crustal ascent and are therefore more likely to reach the surface than $H_2O$-rich magmas? Is evolution of the increasing amounts of $CO_2$ during ascent of a $CO_2$-saturated silicic magma the cause for explosive eruptions of such magmas, or is the influence of the fluid bubbles on the ascent velocity counteracted by increasing viscosity? Tazieff (1970) and Wilson (1976) even suggest that pyroclastic flows may result from large amounts of $CO_2$ in the magma. Because of its relatively high density, large amounts of $CO_2$ may prevent an eruption column from rising high into the air, but instead cause it to collapse and form a pyroclastic flow. The influence of $CO_2$ on eruptive mechanisms, however, has yet to be demonstrated. Fluid inclusions in silicic rocks, for example, are dominated by $H_2O$.

Sulfur

Sulfur is a major component of magmatic volatiles. Anyone who visits an active volcano is likely to experience this directly by the sharp smell of $SO_2$ or the rotten egg odor of $H_2S$.

The equilibrium proportions of $H_2S$, $S_2$, $SO_2$ and $SO_3$ depend upon temperature, pressure, oxygen fugacity and bulk composition (Anderson, 1975; Gerlach, 1982). Experiments show that the solubility of sulfur in magmas is dependent on temperature and iron content of the melt as well as on oxygen fugacity, sulfur increasing with the degree of fractional crystallization (Katsura and Nagashima, 1974; Haughton et al, 1974; Mathez, 1976).

The principal S-species in magmas may be $HS^-$, whose solution behavior is similar to that of $H_2O$ (Burnham, 1979). At higher $f_{O_2}$, however, abundant $SO_2$ may form. It is less soluble in magmas, and escapes during magma ascent and eruption. Sulfur gas is a minor component in vesicles of pillow basalt rinds erupted at greater water depth (Anderson, 1975). The weight ratio C/S ranges from 6 to 8, and is higher at greater depth (Moore et al, 1977).

The loss of sulfur gases from a magma at shallow depths appears to be accompanied by water loss (Killingley and Muenow, 1975; Moore and Schilling, 1973). Because sulfur can be analyzed more precisely than water, and because of the marked ability of water to diffuse into glass, the sulfur content may be a sensitive guide to monitor loss of volatiles from basaltic magma. Moore and Schilling (1973) show that sulfur is lost only at water depths of less than 200 m. At 43 m on Reykjanes Ridge, half of the sulfur (about 500 ppm) still remains in the K-rich ocean-floor tholeiite. Hawaiian tholeiites at Kilauea contain $700 \pm 100$ ppm S when erupted at depths $>500$ m, but about 80% of the total sulfur is lost by degassing during eruption in shallow water or on land (Sakai et al., 1982). Subaerial Kilauean lavas contain about 50–200 ppm S (Swanson and Fabbi, 1973) and continuous degassing occurs in high lava fountains and in surface flows traveling long distances, as evidenced by the gradual loss of sulfur. Chlorine, in contrast, shows less well-defined trends. During subaerial eruptions, very high fluxes of $SO_2$ in excess of 10,000 t/day have been measured at several volcanoes (Stoiber et al., 1981). The oxidation of this $SO_2$ to $SO_4^{2-}$ is the major process leading to sulfate formation of long-lived stratospheric aerosol clouds that accompany major eruptions.

Other types of volcanic gases, in particular HCl, HF, H, and the noble gases are discussed in detail by Anderson (1974a, b, 1975), Carmichael et al. (1974), White and Waring (1963), Johnston (1980), and Burnham (1979).

**Volatile Distribution in Magma Columns**

Much geologic evidence indicates that in many volcanic eruptions the volatile contents are higher in magma that is erupted first – frequently explosively – and is followed by quiet outpourings of lava in later eruptive stages (Kennedy, 1955). Such observations suggest that volatiles may become concentrated in the upper part of a magma column, although the exact processes by which this might occur is complex and poorly understood. Another way to explain the high energy of initial explosive activity is by sudden vaporization of groundwater as discussed in a later section.

Hildreth (1981) has recently summarized much of the evidence for volatile gradients in silicic magma chambers:

1. The roofward increase in volatiles will lower the liquidus temperature of a magma. Thus, the fact that early erupted tephra commonly contains few or no phenocrysts compared to later tephra of the same eruptive phase may be due to a depressed liquidus rather than to gravitative settling of crystals. The common and rather abrupt transition from phenocryst-poor or aphyric to more phenocryst-rich later products may indicate volatile-rich cupolas, up to hundreds of meters thick in large systems, at the top of highly differentiated magma columns.
2. Volatile elements F and Cl, which are more soluble than $H_2O$, are retained to a greater degree in the erupted tephra and therefore show roofward enrichments.
3. Mineral equilibria (annite-sanidine-magnetite) and T and $f_{O_2}$-estimates from coexisting ilmenite-magnetite pairs led Hildreth (1981) to postulate an $H_2O$-gradient from 2.8 for the lowermost to 4.9 weight percent for the uppermost part of the erupted part of the Bishop Tuff magma chamber.
4. Hydrous silicates (e.g. amphibole, allanite, etc.) are commonly most abundant in, or restricted to, the early erupted tephra.
5. The evolution from high-energy Plinian eruption columns to pyroclastic flows may also be interpreted in terms of decreasing volatile contents.

In summary, the volcanological, chemical, and mineralogical evidence for volatile enrichment within the tops of most magma columns is overwhelming. We now need to know how this concentration develops and how it may lead to explosive eruptions.

The saturation values for $H_2O$ and other volatiles strongly increase with higher water pressure until the mole fraction of $H_2O$ is about 0.5. At higher water pressures, the rate of solubility increase is much reduced. The condition where total confining pressure equals partial pressure of water – while important in understanding pyroclastic processes at low pressure – is the exception, however, because most magmas are undersaturated in volatiles. The general case in which total confining pressure exceeds the partial pressure of water was treated by Kennedy (1955) who pointed out that the condition of water saturation could not be maintained in a magma column because the partial pressure of $H_2O$ would increase rapidly with depth (Fig. 3-11). Thus, in a magma column of uniform $H_2O$ content emplaced at shallow depth, water would tend to diffuse towards the top. If equilibrium is reached, water contents will be much smaller at greater depth because curves for equilibrium partial pressures of water have negative slopes in a $P/H_2O$ diagram. The $P/H_2O$ diagram also shows that the higher the pressure (i.e. extent of magma column) the steeper the gradient. The tendency for water to be concentrated near the top of a deep reservoir is therefore greater than in a shallow reservoir. Water saturation will be reached only near the top of the column. Moreover, the solubility of water will be higher at the top of the magma column because temperatures increase downward owing to cooling from above, and possibly episodic input of heat from mafic magma from below, and because magma compositions generally become more silicic upward.

Such equilibria, and hence such decrease in water content downward, can only be attained if diffusion rates of water in the magma are very high compared to rates

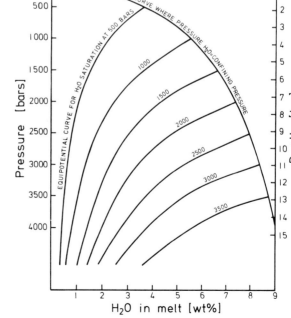

**Fig. 3-11.** Relationship between solubility of $H_2O$ and pressure in extended magma columns. Water content increases with pressure when the $P_{H_2O}$ always equals confining pressure (curve with positive slope). A family of curves with negative slopes for different $P_{H_2O}$ values shows the situation where water would increase in abundance towards the top of a magma column due to diffusion and/or rise of bubbles. (After Kennedy, 1955)

of cooling and crystallization. Matsuo (1961) argued that diffusion rates of water are too low compared with rates of cooling and crystallization to attain equilibrium. McBirney (1973), however, pointed out that saturation at any level is perhaps enhanced by influx of water from wall rocks. The buoyant bubbles rise into increasingly undersaturated parts of the magma and thus might bring about the distribution postulated by Kennedy (1955). Moreover, Shaw (1974) has argued that water need only diffuse into a thin boundary layer. Upward transport by convection would transfer the water to the top and interior of the magma column, thereby constantly replenishing the boundary layer with volatile-undersaturated melt. Watson et al. (1982) have shown that in granitic melts, $H_2O$ diffuses much faster than $CO_2$, while in basaltic magmas the reverse is true.

There are several ways by which volatiles can become concentrated in the upper level of magma column to such a degree that bubbles can nucleate. Perhaps the most general and most accepted mechanism is simply by pressure reduction during nearly isothermal rise of a magma column. This may take place in a wide spectrum of ascending magma systems, from fast-rising basaltic dikes to slowly ascending differentiated larger magma columns. In the latter case, the reverse mechanism could also take place: a decrease in solubility by an increase in temperature due to an influx of hotter mafic into cooler more differentiated magma. Also, volatiles may diffuse upward along concentration gradients in a static magma column.

A second process that potentially can lead to volatile oversaturation is by concentrating volatiles in the residual melt during crystallization of anhydrous mineral phases. This process is called retrograde or second boiling because it is accompanied by a drop in temperature. This process has been treated recently by Burnham (1979), who contends that its importance in explosive volcanism has been recognized for a long time even though it has not received quantitative attention. Two conditions must be met for retrograde boiling to occur in a $H_2O$-bearing magma: the mole ratio $(Al + Fe^{3+})$/alkalies must be unity or greater, and the $H_2O$ initially in the melt must exceed the amount that can be accommodated in hydrous mineral phases. The second condition appears to be met by magmas of approximately granodiorite composition initially containing 0.7 weight percent $H_2O$ and of granitic composition containing about half this amount of $H_2O$. In other words, after crystallization of phases such as hornblende and biotite, there is an excess amount of $H_2O$ partitioned into the residual melt. Burnham argues that the pressure due to the volume increase $(\Delta V_r)$ of the second boiling reaction, melt $\to$ crystals $+ H_2O$ "vapor" can reach very high values. He calculates this volume increase for two pressure conditions:

For total pressure $(P_t) < 2$ kb

$$\Delta V_r \approx (1 - 2.3 \times 10^{-4} P_t) \left(\frac{RT X_w^m}{P_t}\right) - \Delta V_m (1 - X_w^m) \text{ cal b}^{-1} \text{ mol}^{-1}$$

and for $P_t > 2.0$ kb

$$\Delta V_r \approx 0.54 \left(\frac{RT X_w^m}{P_t}\right) - \Delta V_m (1 - X_w^m) \text{ cal b}^{-1} \text{ mol}^{-1}$$

where $P_t$ is in bars and $\Delta V_m$ is the weighted average $\Delta$ of melting of the crystalline phases. For complete reactions, Burnham calculates a volume increase of 60% at

0.5 and 10% at 2.0 kb and suggests that enormous overpressures can be reached if water cannot move out of the magma due to formation of an impermable membrane of early crystallized magma and if the magma body cannot expand.

Well-documented empirical evidence for the importance of retrograde boiling in generating explosive volcanic eruptions is still lacking. Perhaps the greatest difficulty with this model is that magmas of the initial phase of many large pyroclastic eruptions are poor in phenocrysts and have therefore only started to crystallize, perhaps because of the depression of the liquidus due to upward diffusion of volatiles. A second problem is the static view of magma columns from which explosive eruptions are thought to be generated. We visualize magma columns as dynamic systems which have constantly changing highly interdependent variables and which may continue to rise buoyantly during differentiation. There are few examples of large pyroclastic eruptions that contain appreciable quantities of coarsely crystallized rocks that might constitute the crystallized rinds of magma reservoirs. Those that do, such as many trachytic Plinian pumice fallout deposits on São Miguel (Azores) or the rhyolitic pumice deposits of Ascension (Roedder and Coombs, 1967), have not been studied with this problem in mind. The greatest potential for explosive eruptions to have followed from retrograde boiling is in intrusive stocks that are associated with breccias. A third problem is that many of the highly explosive initial phases of volcanic eruptions are probably due to phreatomagmatic processes.

Spera (1984) has recently argued that metasomatizing mantle fluids represent the volatile-saturated residual magmas resulting from alkali basaltic magmas crystallizing at depth.

## Rheology

Viscosity, the single most important and one of the better-studied properties of silicate liquids (Shaw, 1963, 1965, 1969, 1972; Shaw et al., 1968; Murase, 1962; Hess, 1971; Murase and McBirney, 1973; Bottinga and Weill, 1972; Spera, 1980), affects the entire evolution of a magma batch: from the first stage of coalescence of liquid droplets during partial melting through ascent of the magma with concomitant convection and crystal settling or rise within magma chambers; and especially as one control of the mechanism of eruption – whether by degassing and shattering of a viscous rhyolitic magma, or through fountains of low-viscosity basalt whose particles recoalesce to form lava flows. Viscosity is also a major factor in determining the final transport mode of lava flows and pyroclastic flows.

The four basic rheological behaviors of materials are viscous, plastic, elastic, and fracture. In response to stress, a viscous body flows, a plastic body deforms permanently, but when an elastic body is deformed, all of the strain is recoverable. A viscous body is said to exhibit no strength, that is, it will flow at an infinitesimal amount of applied stress, whereas a plastic body exhibits strength (a characteristic of a solid) until too much stress is applied. When the stress exceeds a critical value known as the yield stress, a plastic will flow like a viscous substance. Viscous and plastic behaviors can be illustrated on shear diagrams with shear stress ($\sigma$) plotted against the rate of shear, $du/dt$.

Viscosity, ($\eta$), may be expressed as the ratio of shear stress ($\sigma$) to the corresponding shear strain rate (du/dt):

$$\eta = \frac{\sigma}{du/dt}.$$

When the ratio between shear stress and rate of shear is constant, a fluid is called Newtonian, but if a critical shear stress (the yield value) must be exceeded before the ratio between shear stress to shear rate becomes constant, the fluid is called a Bingham Plastic. The yield strength of magmatic liquids is a major factor in determining the thickness of lava flows (Hulme, 1974) and also the motion of pyroclastic flows (Wilson and Head, 1981b).

**Viscosity**

The viscosity of silicate liquids depends upon several factors including temperature, pressure, chemical composition, dissolved volatiles, and the amount of crystals and bubbles. In recent years our understanding of melt structure has been improved considerably (e.g. Taylor and Brown, 1979; Mysen et al., 1980, 1981) and we can now relate the compositional dependence of viscosity to structural variations in the melts.

Chain type $(Si_nO_{3n})2^{n-}$ anions dominate in the structure of silicate melts at low $SiO_2$ concentrations (or, more precisely, at high ratios of nonbridging oxygens to tetrahedrally coordinated cations) as for instance in picritic or basanitic melts, whereas at higher concentrations of tetrahedrally coordinated cations (Si, but also Al if charge-balanced by Ca, Mg, Na or K, and $Fe^{3+}$ if charge-balanced by Na or K), sheet-like or three-dimensionally connected units predominate. The latter type of unit is of particular importance for rhyolitic and andesitic melts. In the highly polymerized melts, higher activation energies of viscous flow are encountered because the strong T–O bonds (T = network-forming, tetrahedrally coordinated cation) must be broken, whereas in depolymerized melts the flow can be achieved more easily by breaking the weaker M–O bonds (M = network-modifier cation). Figure 3-12 illustrates the range of viscosities for different dry magma compositions at very high temperatures.

The viscosity of a magma also strongly depends on the kind and concentration of volatiles dissolved, but at present appreciable data are only available for $H_2O$, the most abundant volatile component (Fig. 3-13). $H_2O$ acts like a network-modifier, i.e. it breaks the T–O bonds and depolymerizes a melt. Although the nature of the species produced in the breakdown of a highly polymerized melt such as that of albite composition is still a matter of debate (cf. Burnham, 1979; Mysen et al., 1982) it is generally agreed that the addition of 1 mol of $H_2O$ per mol of silicate breaks the three-dimensional network into smaller units having, on the average, only a two-dimensional interconnection. The viscosity is, consequently, drastically reduced.

Burnham argues that this mechanism can be extended to more complex natural silicate magmas. Further addition of water to the melt probably takes place by a hydrolysis-type reaction (Burnham, 1975a, b). This causes the sheets to break up into chains, further reducing the viscosity, although this is much smaller than the above reaction because less energy is required to break the sheets. Bottinga and

**Fig. 3-12.** Calculated viscosity for several anhydrous magma compositions with possible liquidus temperatures. (After Carmichael et al., 1974)

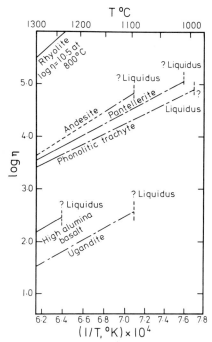

Weill (1972) and Shaw (1972) have developed empirical methods to combine the contributions of individual network-forming and network-modifying oxide constituents to estimate the viscosity of a magma. The method outlined by Shaw (1972) has greater application to geological problems in that it includes an empirical factor for water, whose effect is extremely large even at relatively low concentrations. According to Urbain et al. (1982), the Arrhenius equation used by Shaw is not a generally valid expression for liquid silicate viscosities.

Addition of 4 weight percent $H_2O$ to andesitic magma tends to decrease its viscosity by a factor of about 20, as shown experimentally by Kushiro et al. (1976), in accordance with calculations according to Shaw's (1972) method. At 15 kb and 1350 °C, the viscosity of such hydrous andesitic magmas is about 45 poises. However, significant viscosity changes may occur even in dry melts as a function of pressure. It has been shown by Kushiro et al. (1976) and Mysen et al. (1980) that melts with three-dimensional framework anions, in particular, decrease in viscosity with increasing pressure. The structural reason for this unexpected behavior is still an open question, because the structural change of Al from fourfold to sixfold coordination predicted on theoretical grounds by Waff (1975), and also invoked by Kushiro et al. (1976), has not been proven experimentally (Mysen et al., 1980). One result of such decrease in viscosity with depth, in particular of alkali basalt magmas, may be a rather high velocity of ascent, unless completely counteracted by a simultaneous increase in the density of the melt (Kushiro et al., 1976).

Based on present ideas on the solution mechanism of $CO_2$ as discussed in the preceding section, solution of $CO_2$ tends to polymerize a melt, the opposite effect to the solution of $H_2O$. Since $CO_2$ apparently exsolves out of magmas at much greater depths than $H_2O$, owing to its high vapor pressure, $CO_2$ exsolution will decrease the viscosity of a melt.

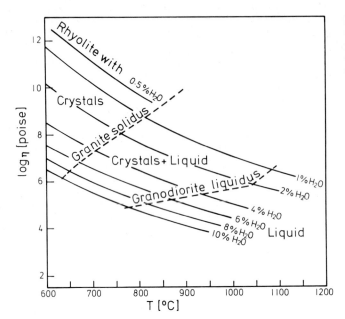

**Fig. 3-13.** Dependence of viscosity of silicic liquids on temperature at various $H_2O$-concentrations. (After Carmichael et al., 1974)

At the time of eruption, natural magmas are rarely free of phenocrysts and bubbles. Both of these factors not only tend to increase the viscosity of a melt, but also change its behavior from that of a Newtonian (at liquidus and super-liquidus temperatures) to a Bingham fluid having a yield value. Unfortunately, there are few data on these realistic conditions, the only measurements under natural conditions (augmented by experiments) being those of Shaw et al. (1968). Einstein's formulation (1906) gives us a method to evaluate $\eta$-change due to the addition of spherical particles

$$\eta_s = (1+2.5)\phi\eta_0$$

$\eta_s$ = viscosity of suspension, $\eta_0$ = viscosity of liquid without crystals, and $\phi$ = fraction of spheres present. The viscosity of Hawaiian tholeiite lava markedly increased with crystal content (Fig. 3-14). The yield values in these lavas at 1130 °C and 1350 °C with about 25% crystals and 2 to 5% bubbles were 1200 and 700 dynes cm$^{-2}$ and the proportionality constants (called plastic viscosities) were 6500 and 7500 poises – compared with about 500 poises at liquidus temperatures (1200 °C) for the same composition. In other words, the calculated viscosities for different magma compositions shown in Fig. 3-12 would be strongly increased with increasing amounts of crystals and bubbles, but this effect would be offset to some degree by the addition of volatiles.

There are, of course, other dynamic changes in the cooling magma that further influence viscosity. For example, some volatiles are lost by degassing, others accumulate in the top part of the magma chamber because of differences in confining pressures and a further increase results during continuing crystallization of mineral phases. Thus, viscosity is a highly dynamic parameter and will continue to change during magma ascent in the magma chamber, and during eruption. Viscosity governs the rate of momentum diffusion just as the thermal diffusivity governs the rate of heat diffusion, and the mass diffusivity governs the rate of mass diffusion.

**Fig. 3-14.** Approximate relationship between apparent viscosity and temperature for Hawaiian tholeiitic lava. The *lines* are for constant total composition, the *solid* one for liquid only, the *broken* for equilibrium liquid-crystal suspensions. The liquidus temperature is at 1,200 °C. (After Shaw et al., 1968)

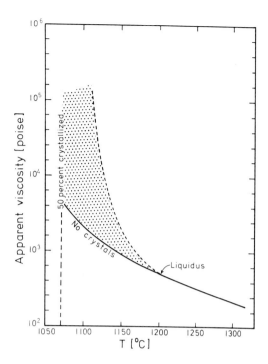

## Vesiculation

The nucleation, growth, and disruption of bubbles is the basic set of processes underlying purely or dominantly magmatic explosive eruptions. The explosivity of volcanic eruptions was traditionally considered to be governed by a few factors, mainly the viscosity of a magma, its gas content and the depth of vesiculation beneath the surface. Today, it is recognized that many of these factors are interrelated and the formation and bursting of bubbles must be analyzed in terms of (1) chemical parameters such as composition of a melt and types, amount and partial pressures of several volatile species, (2) physical parameters such as temperature, viscosity, surface tension, and diffusivity, and properties such as ascent velocity of magma and bubbles, confining pressure and cooling and crystallization rate of a magma.

Modern work on the problem of ash formation started with Verhoogen (1951), who argued that kinetic factors such as the nucleation rate and number of bubbles per unit volume might be more important factors in explosive volcanism. Moreover, he contended that the ratio between rates of bubble growth and vertical rise determines whether or not a magma will lose coherence, principally by coalescence of bubbles.

Present ideas on nucleation, growth, and disruption of bubbles are largely based on the work of Murase and McBirney (1973), McBirney and Murase (1971) and Sparks (1978), and a summary is presented by Williams and McBirney (1979). Nucleation rates are not as important as thought by Verhoogen because, even in highly viscous magma with low water contents and at low levels of volatile supersaturation, rates of nucleation of bubbles are very high. Once formed, gas pressure inside a bubble must balance the surface tension (s) around the circumference of a bubble (2s/r), and then exceed the value in order to grow. Stable bubbles, micrometers in diameter or more, can form at excess pressures of less than 6 b, the value decreasing in volatile-rich melts and when bubbles nucleate on crystals. Decreasing temperature and exsolution of volatiles both tend to increase surface tension.

There are several ways in which bubbles may grow: by coalescence, diffusion of gas into a bubble and decrease in confining pressure. Why bubbles coalesce in some magmas and not in others is not clear. Williams and McBirney (1979) discuss the possibility that small amounts of certain elements such as trivalent transition elements may have the same effect on magma as soap has on water. Such substances could decrease the surface tension and allow many bubbles to form but when diluted during bubble expansion, surface tension would increase. An interpretation of size distribution and types of bubbles of pyroclasts of different types and chemical composition in terms of such processes is a challenging and fruitful avenue of future research.

Growth of bubbles seems to be determined mostly by diffusion in slowly rising magmas and mostly by decompression in those that rise quickly (Sparks, 1978).

Bubble growth is quite rapid. Experiments show, for example, that bubbles about 1 mm in diameter formed at a pressure of about 0.5 kb would grow as they buoyantly rise upward in, or are carried within, a rising basaltic magma to about 1 m in diameter by diffusion and decompression in about $10^4$ to $10^5$ seconds (Blackburn et al., 1976). Similar values are given by Williams and McBirney (1979), who argue that a bubble with a radius of 1 cm will grow radially at a rate of about 0.5 mm s$^{-1}$ in a basalt magma rising about 1 m s$^{-1}$. Growth rates of bubbles in viscous magmas are several orders of magnitude slower.

Recent experimental evidence on diffusion rates of both $H_2O$ and $CO_2$ indicates that diffusional fractionation of $CO_2$ and $H_2O$ in magmas may significantly affect the growth of vapor bubbles (Watson et al., 1982). These authors have applied an approximate quasi-static solution for bubble growth (Epstein and Plesset, 1950):

$$(r/r_0)^2 \approx 1 + \left[\frac{2D(c_i - c_s)}{\varrho r_0^2}\right] t,$$

where r is the bubble radius, t is the time, D the diffusivity of bubble-forming gas in the solution, $c_i$ is the initial uniform and $c_s$ the saturation concentration of the gas and $\varrho$ the density of the gas. Watson et al. (1982) have illustrated $CO_2$ bubble growth rates for a mafic alkalic magma initially containing 2 weight percent $CO_2$. After rapid rise from greater depth, the solubility of $CO_2$ in this magma of 1250 °C will have dropped to 0.7 weight percent at 10 kb, about 30 km, in response to decompression. Thus, for $\varrho = 1.132$ g cm$^{-3}$, $c_s = 0.7\%$ and $c_i = 2\%$, millimeter-sized bubbles will be produced in about 40 h (Fig. 3-15). The above equation is valid only if volumes of melts depleted in dissolved carbonate by growth of bubbles do not overlap. Also, movement of bubbles will slightly accelerate their growth.

Cooling and loss of volatiles cause the viscosity of rising magma to increase, thereby slowing the rate of bubble expansion. Thus, appreciable residual pressures in bubbles over that in the liquid will result. It is this overpressure that leads to bur-

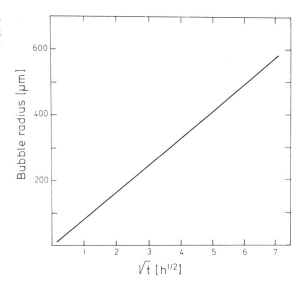

**Fig. 3-15.** Bubble growth in a $CO_2$-oversaturated magma at 10 kb and 1,250 °C. (After Watson et al., 1982)

sting of bubbles and the magnitude of the overpressure determines the energy of some explosive eruptions.

Williams and McBirney (1979) discuss two sets of parameters in governing the bursting of bubbles: (1) the strength of vesiculating magma and (2) several pressure components in a vesiculating magma. At low volume ratios of bubbles to liquid, the strength of the liquid will be the most important factor on short-term stresses, but at high bubble/liquid ratios, the surface tension ($\sigma$) of the gas/liquid interface becomes important. The total pressure from surface tension in a vesiculating liquid is

$$P = \frac{2n^{2/3}\pi r \sigma}{\pi r^2} = \frac{2n^{2/3}\sigma}{r},$$

where n is the number of bubbles per unit volume and r the average radius. With x/n = average volume of bubbles, $r = (3x/4n\pi)^{1/3}$, $\tau = (4E\sigma/\pi c)^{1/2}$, E = Young's modulus and c = circumference of the cross-section across which failure takes place, then the excess pressure of a gas phase $\Delta P$, exerting a force x$\Delta P$ per unit of cross-section area of the vesiculating magma presents the condition of fragmentation, and can be expressed as

$$x\Delta P > \frac{2n^{2/3}\sigma}{r} + \tau.$$

Williams and McBirney (1979) distinguish four components of pressure in a vesiculating magma: (1) the pressure of the overlying magma (liquid-gas mixture) column $\varrho$ gh, $\varrho$, being the density and h the vertical height; (2) the pressure to drive a magma with a viscosity $\eta$ through a conduit with a radius R and a length h at a velocity V; (3) the pressure to balance surface tension as discussed above ($2\sigma/r$); and (4) the pressure of a bubble with the radius r to do work during expansion at a rate dr/dt against the viscous resistance of a liquid:

$$4\eta dr/rdt$$

Thus, the total pressure is

$$P_{total} = \varrho gh + \frac{12V\eta h}{R^3} + \frac{2\sigma}{r} + \frac{4\eta}{r}\frac{dr}{dt}.$$

The first two terms decrease rapidly as the magma ascends and the effect of surface tension is small compared to the other factors. Increasing viscosity during exsolution of gases and decrease in temperature thus become important as the surface of a magma column is approached. Expansion may not proceed beyond bubble/liquid volume ratios of 3:1 to 5:1 as a result of this increase in viscosity. Residual pressures will be especially high in magmas whose viscosity exceeds $10^7$ poise (Sparks, 1978). In these, size and gas pressure of bubbles depend mainly upon the viscosity. Sparks suggests that magma is disrupted by bursting of the larger bubbles within a froth at the free surface of the magma because of a pressure difference across this interface, setting up a wave of bubble disruption by coalescence and bursting (see also Verhoogen, 1951).

# *Chapter 4* Explosive Volcanic Eruptions

Volcanic activity takes many forms, ranging from quiet lava emissions to extremely violent and explosive bursts, many of which can be related to magma composition as discussed in Chapter 3. The kinds of eruptions can be correlated to volcano shapes and sizes, and in this chapter we explore the connection between pyroclastic systems, eruptive mechanisms and their influences upon juvenile particles.

## Types of Volcanic Activity

Eruptive volcanic activity can be viewed on different scales. Styles of activity and type of products may change within minutes or hours, depending upon changes in magma composition, volatiles, or other magma chamber and vent conditions. Volcanic activity that is essentially continuous but of varying intensity and lasting for a few hours to a few days is here called an *eruptive phase* (see Chap. 13). Some types of eruption are characterized by essentially a single eruptive phase (e.g. Plinian activity), while in others (e.g. Strombolian), periods of quiescence may recur over time intervals as small as several days, several weeks to several months or in some basaltic volcanoes up to 2 or more years. We define a group of eruptive phases as an *eruption* (see Chap. 13). The 1902–03 eruption of Mt. Pelée, for example, consisted of many eruptive phases. Layers emplaced during an eruptive phase may differ physically from one another and include tephra originating in several ways as well as lava flows and intrusive rocks, but they are generally related by compositional homogeneity, systematic compositional changes, or the manner by which they were erupted or emplaced.

Volcanic eruptions and eruptive phases are traditionally classified according to a wide range of qualitative criteria; many have been given names from volcanoes where a certain type of behavior was first observed or most commonly occurs (Lacroix, 1904; Mercalli, 1907; Macdonald, 1972). Thus, familiar names, including Peléean, St. Vincent type, or Krakatoan, refer to eruptive phases that have produced hot pyroclastic flows (see Chap. 8). Some authors, for example von Wolff (1913–1914) and Rittmann (1962), have used criteria such as the frequency ratio of effusive or explosive eruptions, type of vent (central pipe or fissure), and rate of activity and duration (paroxysmal or persistent) during an eruption. However, efforts have been made during the last few years to define eruptive phases and eruptions more quantitatively and according to a wider range of criteria. From a geologic viewpoint, this need is critical for answering the question: what aspects of a particular style of eruption are reflected by its deposit?

Three approaches are especially promising with respect to reconstructing eruptive characteristics from pyroclastic deposits. One is quantitative analysis of the mechanics of modern eruptions such as studies by Chouet et al. (1974) and Blackburn et al. (1976) for Strombolian and Self et al. (1979) for Vulcanian eruptions. Another is to analyze the products from a particular kind of eruption, as done for Plinian deposits by Lirer et al. (1973), Walker (1973) and Booth et al. (1978). A third, more theoretical approach, is to define physical parameters of eruption processes and to develop models for predicting the characteristics and distribution of ejected particles, as attempted by Wilson (1976, 1980), Wilson et al. (1978, 1980), Wilson and Head (1981a, 1983), and Kieffer (1982) for explosive volcanic activity. Integration of these kinds of volcanologic, geologic, and theoretical studies should eventually lead to a useful and realistic classification of volcanic eruptions and the ability to reconstruct the type of eruption that produced a specific sequence of pyroclastic rocks in the geological record.

The May 18, 1980 eruption of Mount St. Helens illustrates the value of such studies – and also the complexities of compound eruptive behavior. Within a span of 12 h, the eruptive style of the volcano changed from phreatic, apparently to phreatomagmatic and then to magmatic (Lipman and Mullineaux, 1981), thereby ranging from Vulcanian to Plinian eruptive behaviors, which have distinctive deposits, as discussed in the following paragraphs.

## Types of Eruptions

Traditional names for eruptions and volcano forms such as Hawaiian, Strombolian, Plinian and Vulcanian are in widespread use but for the most part are poorly defined. Walker (1973) has developed a classification scheme for fallout tephra based upon quantitative field and laboratory data which also utilizes observations of eruptions.

Walker's (1973) classification quantifies two arbitrary parameters for the area of dispersal and the degree of fragmentation of fallout ejecta. A measure of the dispersal area is the area enclosed by the isopach contour which represents 1% of the maximum thickness (0.01 $T_{max}$) called D. The value of D ranges from 10 km$^2$ for deposits of cone-building eruptions to more than 1000 km$^2$ for sheet-forming eruptions, such as Plinian eruptions (e.g., Mount Mazama fallout sheet). A measure of the degree of fragmentation, called F, is the percentage value of material finer than 1 mm as determined from samples where the isopach contour representing 10% of the maximum thickness (0.1 $T_{max}$) crosses the main axis of dispersal. The figures 1% and 10% were arbitrarily chosen to represent the respective parameters, and Walker (1973, 1980) has characterized several kinds of pyroclastic fall deposits based upon their D and F values (Fig. 4-1). One of the principal drawbacks to this system is the difficulty in determining the maximum thickness (Chap. 6); another is the necessity of tedious sieve analyses, which limits the use of the classification for eruptions under observation.

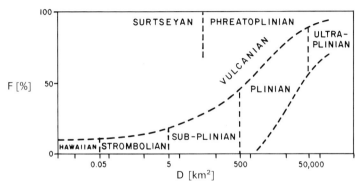

**Fig. 4-1.** Walker's (1973) classification of pyroclastic fall deposits. F% is weight percent of deposit finer than 1 mm along dispersal axis where it is crossed by isopach line which is 10% of the maximum thickness (0.1 $T_{max}$); $D$ is the area enclosed by the (0.01 $T_{max}$) isopach line. Phreatoplinian introduced by Self and Sparks (1978); ultraplinian introduced by Walker (1980). (After Wright et al., 1980b)

# Pyroclastic Eruptive Systems

Pyroclastic eruptive systems consist of (a) a gas/pyroclast mixture extending from the level of the disintegrating magma column to the earth's surface and (b) a visible eruption column that extends from the surface to heights of as much as 50 km. Although major changes occur at the level where a pyroclast-gas dispersion leaves the vent, others, such as the change from the subsonic to supersonic flow in some eruptions, depend on the geometry of the reservoir-conduit-vent system. In attempting to understand the dynamic changes during the course of an eruption and the differences between chemical composition and composition-dependent parameters such as temperature, viscosity, and gas content, both parts of the system have to be considered together. Below, we will discuss three main pyroclastic eruptive systems: Plinian, Hawaiian, and Strombolian. We begin by describing the main features of eruption columns which also apply to the hydroclastic or Vulcanian systems discussed at the end of this chapter.

## Eruption Columns

Pyroclastic as well as explosive hydroclastic eruptions give rise to eruption columns that transport volcaniclastic particles from beneath the ground into the atmosphere or water, or both (Fig. 4-2). The eruption column can be defined as a gas-solid dispersion which is the columnar-shaped part of an eruptive system that extends into the atmosphere from the surface vent prior to lateral dispersal. The physical properties and dynamic processes within eruption columns affect many physical attributes of pyroclastic deposits. Moreover, the different properties of eruption columns define the diverse styles of classically defined pyroclastic eruptions.

An eruption column may be subdivided into two main components: the lower or *gas thrust* part, and the upper or *convective thrust* part (Wilson, 1976; Blackburn et al., 1976; Sparks and Wilson, 1976; Wilson et al., 1978). Expansion of volcanic

**Fig. 4-2.** Eruption column consisting of lower laterally expanding region, a central clast-rich plume and a convective diffuse upper plume. Phreatomagmatic eruption of Taal volcano (1965), Phillippines. (Courtesy Manila Times)

gas and pressure from expanding steam in Vulcanian eruptions is the driving force for the gas thrust part. The initial velocity of the gas thrust ranges from about 100 m s$^{-1}$ to 600 m s$^{-1}$. Its bulk density depends upon the combined densities of the gas and each solid component as well as upon the ratio of solid particles to gas (i.e., particle concentration). Upon leaving the vent, the density of the eruption column is greater than air and its density is proportional to the solids-gas ratio. Fallout of large clasts or admixture of air in the eruption column reduces its density. As shown theoretically, a given volume of the gas thrust can absorb up to four times its own weight of air depending upon the radius of the vent, thereby significantly reducing the density of the column. When the density of the gas thrust becomes less than that of the atmosphere, convective uprise takes over, causing the development of the convective thrust part of the eruption column. The smaller the fragments, the more rapid the heat exchange and the higher the column. The convective thrust commonly makes up 90 percent of the column height.

Temperature inversions, especially at the troposphere-stratosphere boundary (tropopause), will cause convective plumes to mushroom. Lateral displacement is especially pronounced at this level, which is characterized by strong winds (jet

**Fig. 4-3.** Relationship between observed cloud height and calculated volume eruption rates. The *three curves* are for three different values of efficiency of heat use, high values indicating efficient use in the heat released. (After Wilson et al., 1978)

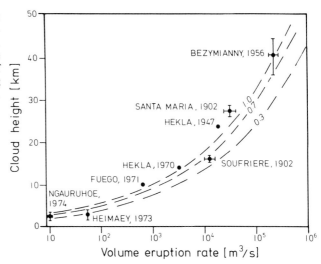

stream). Indeed, the lateral spread and speed of many plumes has been traced during the last few years by satellite photography or remote sensing techniques such as radar (e.g. Harris and Rose et al., 1981). If mass discharge rates are sufficiently high, as is the case in many Plinian eruption, plumes will penetrate this level and ascend to great heights (Fig. 4-3). The dynamics of volcanic plumes, although more shortlived, resemble convection patterns of plumes and thermals in the atmosphere, but spread faster because they are driven by ash-heated entrained air (Sparks and Wilson, 1982).

**Plinian Eruptions and Eruption Columns**

Widely dispersed sheets of pumice and ash are derived from high eruption columns that result from high eruption-rate voluminous gas-rich eruptions, commonly lasting for several hours to about 4 days (Fig. 4-4). These are called Plinian because Pliny the Younger described the famous 3-day eruptions of Vesuvius in 79 AD during which the towns of Pompeii and Herculaneum were buried by several meters of fallout pumice, followed by pyroclastic surges and flows that are an integral part of many Plinian deposits (Figs. 8-21, 8-31). Volumes of Plinian fallout and pyroclastic flow deposits range from less than 1 to more than 3000 km$^3$ and are commonly associated with calderas whose diameters, generally ranging up to about 20 km, may approximate the diameters of the underlying magma chambers. The deposits of such eruptions are discussed in Chapters 6 and 8. Here we summarize some of the research done on the controls of eruptive processes in Plinian and related eruptions (Walker et al., 1971; Walker, 1973; Wilson, 1976, 1980; Sparks and Wilson, 1976; Sparks et al., 1978, Wilson et al., 1978, 1980).

The energy and characteristics of a Plinian eruption depend on many factors, among which gas content of the magma, rheology, vent radius and shape, and volume of magma erupted are especially important. Most Plinian eruptions result from explosions of highly evolved rhyolitic to dacitic, trachytic and phonolitic

Fig. 4-4. Succession of coarse-grained trachytic Plinian fallout deposits alternating with fine-grained fallout and surge tuffs, and soils. Many of the Plinian deposits are chemically zoned. Fogo Volcano (São Miguel, Azores)

magmas with liquidus temperatures from about 750° to 1000 °C. Thus, a mean temperature of 850 °C is assumed in the following discussion (Wilson et al., 1980). The eruption velocity, $U_v$, is nearly proportional to the square root of temperature. This enables adjustments for different temperatures. Magma density is assumed to be $2.3 \times 10^3$ kg m$^{-3}$, and volatile content about 5 weight percent, dominantly water, as discussed in the previous section. The viscosity is about $10^4$ to $10^7$ Pas (rhyolite).

An eruption is initiated by magma rising from the top of a magma column, for example by buoyancy-induced fracturing of the roof, which is commonly accompanied by phreatomagmatic processes. Since the ratio of the surface area to volume of flow is at a minimum in a circular conduit, frictional losses will be a minimum with this geometry and so most conduits beginning as fractures tend to become circular during the course of an eruption.

Two critical pressure levels in a magma column, from which Plinian eruptions are fed, separate (1) a lowermost higher pressure root zone in which all volatiles are still dissolved, separated by an exsolution surface from (2) a middle zone consisting of magma and exsolved gas bubbles separated along a fragmentation surface from (3) an upper zone consisting of a dispersion of liquid to plastic pyroclasts and released gas (Fig. 4-5). The fragmentation surface is characterized by a volume fraction of gas bubbles of about 0.7 to 0.8 (Sparks, 1978). Most particles produced at that fragmentation level are small enough to maintain thermal equilibrium with

**Fig. 4-5.** Schematic diagram of a volcanic system showing different regions and rheological regimes from non-vesiculated magma to eruption plume, corresponding to stage III and IV of Fig. 2-4. (After Wilson et al., 1980)

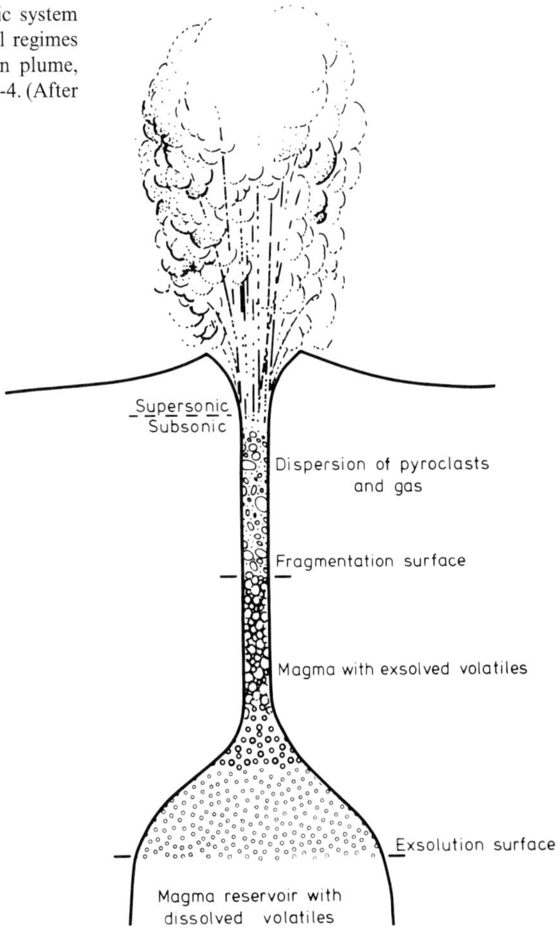

the released magmatic gases while they accelerate. The velocity of the pyroclast-gas mixture is so high that little energy is lost during the few seconds of transit time from the fragmentation level to the surface. Rapid acceleration also prevents much additional gas release. Volatile solubility can thus be related to the pressure, $P_i$, and the amount of released gas, n. Assuming near-lithostatic pressure, eruption velocity can be calculated for total and released volatiles, and depth and pressure of fragmentation (Table 4-1). The eruption velocity is controlled mainly by the gas content at mass eruption rates in $> 10^6$ kg s$^{-1}$ resulting in heights of eruption clouds of about 10 km, most Plinian eruptions ranging up to m of about $1.4 \times 10^8$ kg s$^{-1}$ and column heights up to perhaps 55 km.

The low strength of surface rocks and the high initial exit pressure commonly result in vent erosion. The resulting flared vent shape enhances the velocity of the tephra-gas mixture, leading to a transition from subsonic to supersonic flow at the depth at which the conduit has its minimum diameter. Wilson et al. (1980) have calculated conditions for the following example: a magma with a viscosity $10^5$ Pas and a total gas (H$_2$O) content of 5 weight percent would begin to exsolve gas at

**Table 4-1.** Values for exit velocity ($U_f$), gas pressure at the fragmentation level ($P_f$) and exsolved water content (n) for four different values of total magma water content ($n^1$) (Wilson et al., 1980)

| $n^1$ (wt.%) | n (wt.%) | $P_f$ (b) | $U_f$ (ms$^{-1}$) |
|---|---|---|---|
| 7 | 5.11 | 216 | 534 |
| 5 | 3.45 | 143 | 422 |
| 3 | 1.87 | 76 | 290 |
| 1 | 0.045 | 18 | 116 |

5.4 km depth, reaching a void fraction of 0.77 at 1.6 km. The flow of the mixture above the fragmentation level becomes sonic at 450 m, surface velocity being 160 m s$^{-1}$ and exit pressure 56 b (Fig. 4-6). Depending on the strength of the surface rocks, the vent region will be widened from 8.6 m at a depth of 320 m to a radius of 15 m at the surface rather quickly, the transition to supersonic flow occurring at 320 m, the eruption velocity increasing to 275 m s$^{-1}$ and the exit pressure decreasing to 8 b.

If vent flaring does not take place, the tephra-gas mixture will reach the surface at acoustic velocity with a pressure greater than atmospheric. The mixture rapidly expands to 1 bar and its final velocity, $U_s$, is generally much higher than its initial velocity, $U_i$. Friction can be neglected if conduit diameters exceed 20 m, corre-

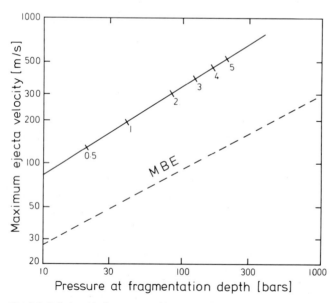

**Fig. 4-6.** Relationship between maximum velocity in the vent in Plinian eruptions as a function of pressure at the fragmentation level showing released gas fractions in weight percent. The *dashed line* shows pressures calculated from modified Bernoulli equation (MBE). (After Wilson, 1980)

sponding to $m > 10^6$ kg s$^{-1}$. Under these conditions, vent pressure are several tens of bars and eruption velocities are $< 200$ m s$^{-1}$. Since the gas density is proportional to the local pressure, and more than compensates for lower velocities where eruption velocities are limited to the sound speed, larger clasts can be transported during the early stages of eruption prior to vent erosion.

Eruption velocity is reduced when additional gas with high molecular weight such as $CO_2$ is present. In rhyolite magmas, only about 10 percent of the gas may be $CO_2$ judging from inclusion studies in the Bandelier Tuff (Sommer, 1977), barring significant diffusional fractionation of $H_2O$ and $CO_2$ (Watson et al., 1982), but it may be higher in highly alkalic magmas such as phonolites.

Wilson et al. (1978) have shown that maximum column height, H, is proportional to the fourth root of the mass eruption rate. If 70 percent of the heat released by the erupted material is used to drive convection, then

$$H = 236.6 \, \dot{m}^{1/4}.$$

Wilson et al. (1980) have discussed three combinations of vent radius and gas content in monitoring exit velocities and column height (Fig. 4-7). They show that while velocity drastically decreases with decreasing gas content, column height is mainly dependent on vent radius. Column collapse, at conditions of constant vent radius equal to 200 m, only occurs when water contents drop below 2.4 percent. Widening vent radius may also lead to reverse grading commonly reported in Plinian deposits. An initial Plinian phase will be followed by pyroclastic flows when either gas content decreases or the vent widens (Fig. 4-8). Salient features of Plinian type eruptions and their products are summarized in Table 4-2.

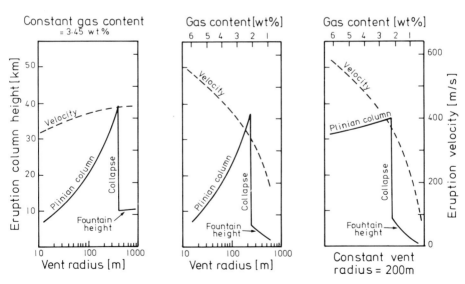

Fig. 4-7. Variations in column height and gas velocity and dependence of collapse of Plinian eruption columns on changes in vent radius and gas content for three different combinations. (After Wilson et al., 1980)

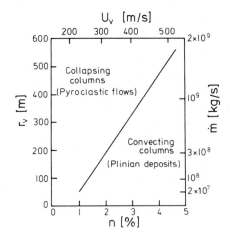

**Fig. 4-8.** Relationship between vent radius ($r_v$) and corresponding mass eruption rates ($\dot{m}$) and exsolved water content ($n$) and corresponding eruption velocity ($U_v$). Column collapse and consequent generation of pyroclastic flows occurs with decreasing contents of exsolved water and/or increasing vent radius. (After Wilson et al., 1980)

**Table 4-2.** Plinian eruptions

| Characteristics of particles and deposits | | Eruptive and transport processes |
|---|---|---|
| *Rock Composition:*<br>Felsic (rhyolitic, trachytic, phonolitic, dacitic) becoming more mafic during course of eruption | indicates → | Rapid emptying of zoned differentiated magma chamber with high gas pressure |
| *Clasts:*<br>Highly vesicular, dominantly essential. Angular | indicates → | Fragmentation by expansion of magmatic gases; cooling by mixing with air prior to deposition |
| *Grain-Size Variation:*<br>Log-normal decrease from eruption center | indicates → | Fallout from high eruption column |
| *Sorting:*<br>Well-sorted | indicates → | Fallout from high eruption columns |
| *Structures:*<br>Massive to poorly bedded thick beds.<br>Inverse to normal grading | indicates → | Single but fluctuating continuous gas blasts for several hours. Fallout from turbulent tephra clouds |
| *Volume:*<br>Moderate to large (up to 1000 km³) | indicates → | High eruption columns |
| *Geometry:*<br>Sheets | indicates → | High eruption columns |
| *Fumarolic Pipes and Alterations:*<br>Absent | indicates → | Extensive cooling and degassing during transit |
| *Welding:*<br>Absent | indicates → | Low temperature during deposition |
| *Association:*<br>Commonly followed by ash or pumice flow deposits. Ballistic deposits near source | indicates → | Collapsing eruption column due to increasing vent diameter and/or decrease in magmatic gas content |

## Hawaiian and Strombolian Eruptions

Observations

Hawaiian eruptions characterize basaltic, highly fluid lavas of low gas content, which give rise to effusive lava flows and less voluminous pyroclastic debris. Thin, fluid lava flows can gradually build up large broad shield volcanoes. Most Hawaiian eruptions start from fissures, commonly beginning as a line of lava fountains that soon concentrate at one or more "central" vents. Most of the vesiculating lava falls back in a still molten condition, coalesces and moves away as lava flows. If fountains are weak, most lava will quietly well out of the ground and move away from a vent as a lava flow. Much lava in shield volcanoes is transmitted through tubes enclosed within lava flows. Small spatter cones and, in some instances, basaltic pumice cones such as at Kilauea Iki may form around vents. Pyroclastic material occurs as bombs, ranging downward in size through lapilli-sized clasts of solidified liquid spatter, to small volumes of glassy Pele's tears and Pele's hair. Pele's hair (Duffield et al., 1977) and broken equivalents are described by Heiken (1972, 1974). The small volume and areal extent, and small amount of fine-grained tephra, places Hawaiian fallout tephra at the lower end of Walker's classification scheme (Fig. 4-1). Wentworth (1938) and Wentworth and Macdonald (1953) describe Hawaiian volcanic forms and pyroclastic debris in considerable detail.

Strombolian eruptions, named after Stromboli Volcano, Italy, are those in which discrete explosions separated by periods of less than a second to several hours occur in magma columns near the surface. Ejecta consist of bombs, scoriaceous lapilli and ash. The deposits of Strombolian eruptions differ from the deposits of Hawaiian tephra in their lack of glassy Pele's tears or hair, their greater dispersal area and the abundance of finely divided tephra. Features of Strombolian and Hawaiian type eruptions and the resulting pyroclastic deposits are summarized in Table 4-3.

Eruptive Mechanisms

Wilson and Head (1981a), extending earlier work by Chouet et al. (1974), Fedotov (1978), Blackburn et al. (1976) and others, have discussed the dynamic factors in Hawaiian and Strombolian eruptions. They treated the dependence of the magma rise-speed and mass eruption rate in open conduits between the magma reservoir and the surface on magma properties (viscosity, density, yield strength, volatile and bulk magma composition) and parameters of the conduit system (fissure width and length, source depth, and lithostatic pressure gradient). The following section is based on their account.

If the magma has sufficient excess pressure possibly as high as 200 b resulting from volume increase during melting, density difference to the surrounding rock or volatile exsolution, then it may ascend in a fissure produced by "magma fracturing" commonly oriented normal to the direction of least principal stress (see also Aki et al., 1977; Shaw, 1980; Spera, 1980). Width to length ratios range from $10^{-2}$ to $10^{-3}$ near the surface to $10^{-3}$ to $10^{-4}$ in the deeper crust. Mass flow rate, i.e. the mass of magma per unit time crossing any horizontal plane cutting the fissure, depends on width and length of the fissure. The largest rise velocity for a given total

**Table 4-3.** Strombolian and Hawaiian eruptions

| Characteristics of particles and deposits | | Eruptive and transport processes |
|---|---|---|
| *Rock Composition* | | |
| Basaltic | indicates → | Low magmatic volatile content, high diffusivity, low viscosity magma |
| *Clasts:* | | |
| Vesicular | indicates → | Ballistic transport from near surface magma pipes; low eruption columns |
| Small glassy particles to large bombs, aerodynamically shaped | | |
| Tachylite common | | |
| Scoria, pumice | | |
| Accidental fragments rare | | |
| Wide range from welded spatter to nonwelded scoria and pumice sheets | | |
| *Grain Size:* | | |
| Moderate; large blocks/bombs, fine ash rare | indicates → | Fragmentation by expansion of large bubbles and gas pockets; short residence time in transporting system prevents attrition |
| *Sorting:* | | |
| Good to excellent | indicates → | Insignificant production of ash-sized particles |
| *Structures:* | | |
| Massive; crude bedding denoted by size changes; alignment of fluid particles by deformation | indicates → | Near-vent accumulations; highly fluid particles; high temperatures |
| *Volume:* | | |
| Relatively small ($\ll 1$ km$^3$) | indicates → | Small magma reservoirs |
| *Geometry:* | | |
| Spatter and scoria cones | indicates → | Ballistic transport; lava fountains; low eruption columns |
| *Fumarolic alteration or pipes:* | | |
| Near-vent alteration common | indicates → | High depositional temperatures of incompletely degassed components |
| *Welding (agglutination):* | | |
| Common in some | indicates → | High temperature ($>$ca. 1000° C) and low residence time in transit; lava fountains |
| *Association:* | | |
| Lava flows | indicates → | High eruption rates leading to complete welding from lava fountains and generaration of lava flows |

mass flux will occur in a circular conduit where cooling and thus freezing of the magma is at a minimum.

If basalt magmas having a minimum viscosity of $10^2$ Pas erupt from a minimum depth of 5 km, the vent radius would be about 0.22 m and the minimum rise velocity would be 0.12 m s$^{-1}$ for a density contrast of 200 kg m$^{-3}$. For a rhyolite with

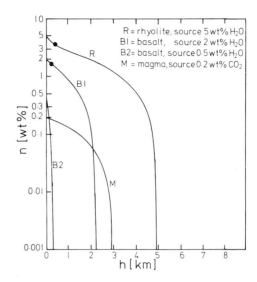

**Fig. 4–9.** Increase in exsolved volatile content (*n*) with decreasing depth, calculated for different magma compositions and volatile compositions in source. *Dots* indicate depth and released gas content at which the gas volume fraction exceeds 0.75 leading to magma disruption. (After Wilson and Head, 1981a)

$\eta = 10^4$ Pas having the same density contrast and the same depth of origin, the minimum vent radius would be 0.7 m and the minimum rise speed 12.5 cm s$^{-1}$

An eruption is initiated when tensile or shear strength of the overlying rocks fails due to magma overpressure. During steady flow of the magma, magma pressure is similar to local lithostatic pressure and wall failure may occur when the pressure is smaller than stress in the surrounding rocks.

During rise, the magma exsolves volatiles and when the gas volume fraction in the magma reaches about 0.75, magma disruption occurs. In Fig. 4-9, several conditions are portrayed and it is evident from the curves presented that magma disruption will generally occur only close to the surface except in very narrow conduits where the magma rise speed is much less than the rise speed of individual gas bubbles in the magma and bubble coalescence can take place. At the stage of magma disruption, the effective viscosity will be greatly reduced, resulting in an acceleration until the gas-pyroclast mixture makes the transition from subsonic to supersonic transport velocity.

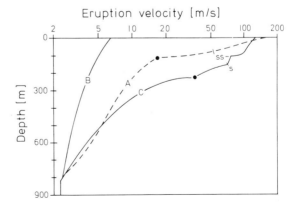

**Fig. 4-10.** Increase in velocity of ascending, vesiculating basaltic magma (see text). *Curve A*: pressure is lithostatic at all depths. *Curves B* and *C* for conduit shapes shown in Fig. 4-12. *Dots* indicate fragmentation level. Flow becomes sonic at *s* and supersonic at *ss*. (After Wilson and Head, 1981a)

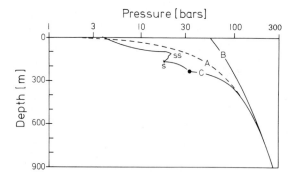

Fig. 4-11. Change of pressure in ascending basaltic magma with depth. For explanation see Figs. 4-10 and 4-12. (After Wilson and Head, 1981 a)

For a magma of density 2800 kg m$^{-3}$, viscosity 10$^3$ Pas, and total water content of 1 weight percent rising in a circular conduit with a radius of 3 m through a crust with a density of 3000 kg m$^{-3}$ at a mass flow rate of $2 \times 10^5$ kg s$^{-1}$, gas will begin to exsolve at 816 m where the rise velocity will be 2.36 m s$^{-1}$. Magma will disrupt at a depth of 108 m when the volume fraction of gas bubbles has reached about 0.75. The subsonic-supersonic transition will occur at a depth of 53 m and the mixture of gas and pyroclasts (<20 mm in size with negligible terminal velocities) will reach the surface with a velocity of 160 m s$^{-1}$. Variations of eruption velocities and pressures for different vent radii are shown in Figs. 4-10 to 4-12. If the pressure of the magma is initially everywhere lithostatic (curve A), the conduit width will become slightly narrower upward to a depth of 100 m, as might be expected in a rising magma. Transition toward a straight-sided shape (curve B) will result in increasing exit pressure but decreasing subsonic exit velocities. These excess pressures might result in blowing out near-surface rocks, generating the funnel-shaped profile (curve C). This will then allow the transition to occur to supersonic exit velocities while resulting in exit pressures close to atmospheric. These changes in vent geometry, exit pressures, and exit velocities will occur rapidly dur-

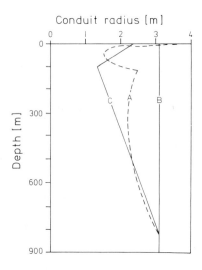

Fig. 4-12. Three different configurations of conduit radius. For explanation see Figs. 4-10 and 4-11. (After Wilson and Head, 1981 a)

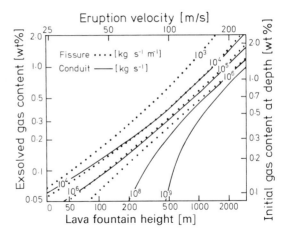

**Fig. 4-13.** Relationship between lava fountain height (eruption velocity) and exsolved gas content (initial gas content) for different effusion rates in circular conduits (*solid lines*) and fissures (*dotted lines*). Calculations are for basalt with density 2,800 kg/m³ and viscosity 300 Pas. (After Wilson and Head, 1981 a)

ing the initial phase of an eruption. In these calculations, $H_2O$ is assumed to behave as an ideal gas, a condition that applies to low pressures ($P \lesssim 500$ b) only.

The amount of gases released during an eruption can be estimated from mass eruption rates and lava fountain heights (Fig. 4-13). The curves closely approximate natural conditions, as discussed in detail by Wilson and Head (1981a). They have also constructed curves that show the dependence of bubble diameters on water content, magma-rise velocity, and mass flow rates for various radii of concentric conduits and fissures for an average basalt with a density of 2800 kg m$^{-3}$ and a viscosity of 300 Pas. As discussed above, magma fragmentation occurs when gas bubbles are closely packed and the size of the largest pyroclast is similar to the largest bubble.

Steady lava fountaining characteristic of Hawaiian style eruptions is expected for the parameters given above when magma-rise velocities at depth are greater than 1 m s$^{-1}$; bubbles will be small under this condition. When the velocity of as-

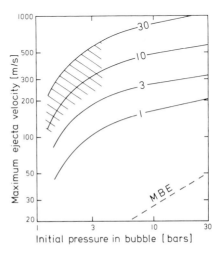

**Fig. 4-14.** Relationship between maximum velocity and total pressure in gas bubbles just prior to bursting in Strombolian eruptions. *Dashed line* shows pressures calculated from modified Bernoulli equation (MBE). *Curves* show different weight percentages of gas in ejecta with *shaded area* indicating most likely range. (After Wilson, 1980)

cending magma is less than about 0.1 m s$^{-1}$, the rise velocity of bubbles becomes an appreciable fraction of the rise velocity of the magma in the crust. Large bubbles will overtake smaller ones and may coalesce. This volume increase will lead to an increase in bubble-rise velocity and a runaway situation may develop. These types of discrete explosions, in which very large bubbles arrive intermittently at the surface of the magma column, are called Strombolian eruptions. In Strombolian eruptions, rise velocities of the magma are less than 0.5 m s$^{-1}$, maximum ejecta velocities range from 100 to 500 m s$^{-1}$ and pressures in bubbles range from about 1 to about 4 b (Fig. 4-14).

## Hydroclastic Eruptive Processes

The formation of pyroclasts involves basically two processes: fragmentation and cooling. In the preceding sections, emphasis has been on fragmentation due to expansion of volatiles. Most of the cooling to low temperatures of pyroclasts that are magmatically produced occurs after disruption of the melt above the vent. In this section, we consider the opposite effect: substantial cooling of a magma followed or accompanied by fragmentation. This occurs where magma and external water meet.

Hydroclastic processes are those causing the breakup of magma by interaction with external water from (1) explosive comminution or (2) nonexplosive granulation, with all gradations between (1) and (2). Vesiculation by magmatic volatiles at shallow confining pressures and/or high partial pressures of magmatic volatiles may accompany, influence or aid these processes at nearly every stage of the explosive to nonexplosive continuum.

When the plutonists became victorious over the neptunists at the turn of the 18th to the 19th century, they had laid a convincing foundation to the theory that volcanoes derive their magmas and their explosive energy from the source area of the magmas deep within the earth. The presence and influence of surface or ground water on the products and eruptive mechanisms of historic volcanic eruptions in the past nearly two centuries has been noted many times but they were generally regarded as minor local modifications or exceptions compared to magmatic mechanisms. The emphatic essay by Jaggar (1949) on the importance of meteoric water, however, is a notable exception.

The recognition by Moore et al. (1966) that contact of magma and lake water had dominated the mechanism of the eruption of Taal volcano in 1965 and that the deposits from this eruption harbored a wealth of information on the nature of eruptive and transport mechanism has triggered a series of studies showing the very widespread occurrence and wide range of eruptive mechanisms due to contact of magma and water. These processes are more complex than those discussed previously because a magmatic system interacts intimately with the environment and generates intricate systems of time-dependent feedback systems. Explosive eruptions occur when magma interacts with water (Fig. 4-2) or when water trapped beneath hot pyroclastic flows is heated (Fig. 4-15). Below, we point out some general aspects of these processes as deduced principally from the deposits. We begin by discussing the effects that these interactions have on the magma, then discuss

**Fig. 4-15.** Craters formed by phreatic explosions of water trapped beneath 1980 hot pyroclastic flow deposits from Mount St. Helens at the margin of Spirit Lake. Large crater on *left* about 50 m in diameter with gray duned rim deposits having formed during phreatic eruptions

processes that occur in the surface environment and conclude with a survey of various eruptive situations.

**Formation of Glassy Pyroclasts**

The cooling of a magma that becomes disintegrated into pyroclasts occurs in several different ways. Exsolution, and especially adiabatic expansion, of a magmatic gas alone can have a considerable cooling effect on a rising magma (Boyd, 1961; McBirney, 1963). Most of the cooling, however, takes place during transit from the point of disintegration to the final site of deposition. Cooling depends on the size of the clast and the height and lateral distance it is transported. Cooling is minimal – a few degrees to at most a few tens of degrees – in lava fountains from which agglutinated spatter cones or lava flows are generated. Cooling is at a maximum in fallout tephra where it can amount to as much as 1200 °C.

The formation of glass is not so much dependent on the *amount* of cooling as it is on the *cooling rate*. Formation and properties of glass formed from natural silicate melts are discussed by Carmichael (1979). Ryan and Sammis (1981) discuss the glass transition in basaltic melts. When silicate melts are cooled rapidly below their crystallization temperature, $T_e$, they will first form a supercooled liquid and then a solid amorphous substance called glass. When a liquid is cooled slowly and crystallizes, its volume decreases by about 5–15 percent, crystallization being accompanied by release of latent heat of crystallization. No heat is released and

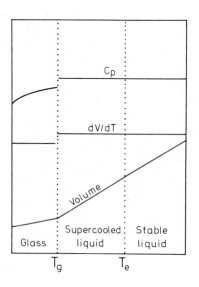

Fig. 4-16. Change in physical properties of a glass-forming liquid, cooled from a stable liquid through a supercooled liquid to a glass. $T_e$ equilibrium crystallization temperature; $T_g$ glass transformation temperature. $C_p$ heat capacity at a constant pressure. (After Carmichael, 1979)

no drastic change in volume nor in heat capacity ($C_p$) occurs at $T_e$ during rapid cooling (Fig. 4-16). Abrupt changes in dV/dT and $C_p$ do occur at a lower temperature, $T_g$, called the glass transformation temperature. The liquid-like structural contributions to mechanical or thermal behavior are thus lost at $T_g$ where a rigid glassy network has formed. In cooling of tholeiitic basalts, the style of stress relaxation changes markedly in the interval of 750°–700 °C. The inferred median glass transition temperature of about 725 °C thus separates enhanced creep at higher from reduced creep regime at lower temperature (Ryan and Sammis, 1981). In the glass industry, $T_g$ is defined as the temperature at which the viscosity of the supercooled liquid exceeds $10^{13}$ poise. The transformation temperature depends on the cooling rate of the supercooled liquid and decreases with increasing amounts of dissolved water. The formation of glass depends not only on cooling rate but also on the viscosity of a stable silicate liquid. Liquids with very low viscosities and those rich in water are difficult to quench to a glass because in these melts rapid nucleation and growth rates will lead to the formation of crystallites. Thus, highly viscous magmas such as rhyolites generally form entirely glassy pyroclasts (except for intratelluric phenocrysts) and may occur as largely glassy lava flows as well. Tephra of basaltic composition, on the other hand, commonly contains quench crystals similar to those found in experimental runs.

There are two varieties of basaltic glass. One, partly or completely charged with microlites of Fe/Ti-oxide, opaque in transmitted light, is called tachylite; the other transparent variety, is called sideromelane. Which variety forms depends on the cooling rate. Because water has a much higher heat capacity and conductivity than air and absorbs large amounts of heat of vaporization, cooling rates at magma/water contacts are much higher than when lava is erupted in air. The abundance of sideromelane is thus a major criterion for recognizing the quenching of basaltic magma by external water.

Such quenching processes affect the magma and the particles in three other ways, each of which contributes to this particular type of clastic process and eruption mechanism.

## Granulation

As shown in Fig. 4-16, the transition of a stable to a supercooled liquid and then to a glass during quenching is accompanied by a volume decrease. The dynamic stresses so produced are relieved by pervasive contraction cracks, causing brittle failure of the glass, decrepitation, and formation of finely comminuted glassy debris.

Granulation was one of the first processes to be considered by geologists in explaining what happens when lava is erupted beneath, or flows into, water. It is still uncertain, however, if sideromelane ash formed under deep sea conditions is due to comminution at the time of eruption or due to later flaking. Small vent dimensions and high eruptive rates may be important prerequisites for primary granulation to occur. Analogous to the experiments by Dullforce et al. (1976), there may be only a fairly restricted "window" of cooling conditions leading to disintegration.

One of the earliest authors to invoke the processes of granulation was Thoulet (1904, 1922). It was subsequently applied by Born (1923), Fuller (1931), Hentschel (1963), Honnorez (1961, 1972) and many others. Born (1923) drew an analogy between the observed granulation of industrial furnace slag (not unlike basaltic magma in composition) when quenched by water, and the formation of submarine clastic debris. The vitrification and granulation by quenching of hot silicate melts in cold water has been used for many years in industry to produce gravel for the cement industry (Laurent, 1956). Drilling into Makaopuhi lava lake on Kilauea Volcano, using a large quantity of water as a coolant, produced granulated sand-sized sideromelane grains when the water flashed into steam (Swanson, 1977, personal communication). However, granulation occurred only when unusually large volumes of water were used. The particles have nonvesicular flake-like to equant shapes (Schmincke et al., 1978), identical to many basaltic hyaloclastic deposits found on the deep sea floor.

## Inhibition of Vesiculation

Another result of rapid quenching of a magma is the inhibition of vesiculation. The very drastic increase in viscosity and the increase in solubility due to the decrease in temperature prevent volatiles from exsolution. As discussed earlier, such quenched magma samples are the most suitable material to determine pristine magmatic volatile contents. There is a natural wide range in the degree of vesiculation, depending upon composition and concentration of the volatiles, the composition of the magma and the confining pressure at which the magma is quenched. At one extreme are submarine lava flows extruded beneath water at great depths where confining pressures are so high that only traces of volatiles have exsolved. At the other extreme are pyroclasts generated in the interior of an eruption column and quenched in a steam cloud. Moore (1970b, 1979), Moore and Schilling (1973) and Moore et al. (1977) have attempted to estimate the confining pressure and thus the water depth of extrusion of submarine basalt magmas by studying the size and modal volume of vesicles and volatile contents within both glass and vesicles as discussed in Chapter 3.

## Transfer of Heat

Still another effect of rapidly quenching a magma is the rapid transfer of heat from the magma to the environment. As a result, pyroclast populations resulting from hydroclastic eruptions are cooled to less than 100 °C during eruptions. This aspect will be considered in more detail below, because it is best discussed in terms of processes that occur within the environment.

The reason for discussing the effects of rapid quenching on magma in detail is that these effects are most unambiguously recorded by the physical properties of the essential clasts derived from hydroclastic eruptions. Indeed, the occurrence and abundance of nonvesicular or poorly vesicular glassy pyroclasts which have fracture-bounded, angular shapes, and which were laid down at low temperatures even very close to the vent, are fundamental criteria upon which interpretations of hydroclastic mechanisms are based. The specific nature of the eruptive process, especially its energy and episodic nature, however, are largely governed by environmental factors. Two main aspects will be considered: the conversion of water to steam and the variables that control the pressure resulting from such eruptions, as well as fragmentation of country rock.

Delaney (1982) has analyzed the behavior of groundwater during rapid intrusion of magma into wet rock. At higher pressures, pore pressure will increase due to thermal expansion of the heated water. Pressures in excess of 100 b are reached for quartz-rich sedimentary rocks with >5 percent porosity and permeability <0.1 mdarcy at T increases of 500 °C above ambient at depths of 2.5 km. For rocks with permeabilities <1 mdarcy, a pressure increase of 100 b is reached at 1 km. Pressure increases are <10 b in highly permeable rocks that are good aquifers. Thus, pressures are greatest when the intrusion is emplaced rapidly and where the intrusive contact is impervious to groundwater contained in impermeable rock. Dilation of pores may thus take place in response to increased pore pressure and failure may occur. Since the overall thermal expansion coefficient, $\alpha$, and the temperature difference, $\Delta T$, between the initial ambient and intrusive T are greater in shallow (<1 km) than in deeper environments, pressure increases due to heating are greatest in shallow environments. For example, a pressure increase of 100 b at a depth of 5 km is only 10 percent of the lithostatic pressure while the same pressure increase at 0.1 km exceeds the lithostatic pressure by 400 percent. The presence of a steam region would serve to maintain high pore pressures if heated rocks fail due to heating of pore water.

## Steam Eruptions

Water is converted to steam by heating to its boiling point. At sea level, this is 100 °C, but the boiling point increases strongly with a slight rise in pressure to about 200 °C at less than 20 b, while rising only slightly with further increase in pressure until the critical point is reached at 374 °C and 218 b (Fig. 4-17). The line delineating the boundary between water and steam, called the boiling curve, is of fundamental importance in understanding hydroclastic eruptions. The conversion of liquid water to steam across the boiling curve below the critical point brings about a very large volume increase. For example, under atmospheric conditions, heating water to 1000 °C increases the volume by a factor of 6000, whereas at

**Fig. 4-17.** Variation of density of H$_2$O with pressure and temperature. *Dashed lines* show three different geothermal gradients. (After Williams and McBirney, 1969)

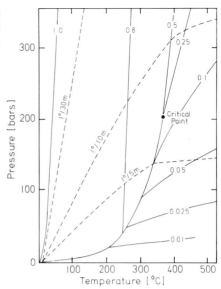

higher pressure, e.g. 1 kb, the volume increases only by a factor of 6. This is because the density of the liquid changes only slightly under isothermal conditions with increasing pressure, while steam, as would be expected, is highly compressible and increases strongly in density with pressure but decreases only slightly with increasing temperature at low pressure.

McBirney (1963) discusses the two most common cases where explosive vaporization of water can arise: (1) When water at a pressure below its critical point is heated through a temperature interval that crosses the boiling point curve for the appropriate subcritical pressure, and (2) when the pressure on water at temperatures below the critical point is reduced isothermally through an interval crossing the vaporization curve (Fig. 4-17). The heating of water by magma through the boiling curve is a necessary but insufficient condition to produce large volumes of steam and large eruption pressures. The generation of large volumes of steam at great water depths during submarine volcanic activity – probably by far the most common type of volcanic activity on earth – will be inhibited by the instantaneous condensation. Explosive steam production will thus be limited to low pressure (shallow water) at which the volume increase of water converted to steam is much higher (Fig. 4-18). But even here it will only be generated in sufficient quantity if the supply rate of heat is high such as during high discharge rates and/or high degrees of fragmentation due to volatile exsolution. The greatest depth from which steam generation has so far been reported was 300 m, measured during the early submarine volcanic activity of Iwo Jima in 1934 (Tanakadate, 1935).

The activity of geysers and steam eruptions associated with geyser fields illustrates some basic principles of hydroclastic mechanisms. Geysers, prior to their eruption, basically are water columns in which the water temperature, while increasing with depth, is below its boiling temperature for each depth. They occur in young volcanic areas with very high geothermal gradients, such as New Zealand, Iceland or Yellowstone (Wyoming). The water is heated by hot gases or by conduc-

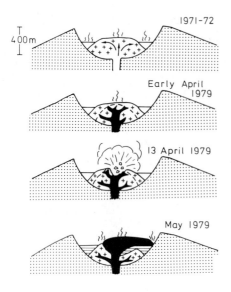

Fig. 4-18. Initiation of phreatomagmatic eruption of Soufrière (May 1979), St. Vincent, by intrusion of new magma into hot 1971 dome surrounded by crater lake. Crater lake filled with debris (*horizontally ruled area*) of phreatomagmatic eruption in May 1979. (After Shepherd and Sigurdsson, 1982)

tion through rocks from a magma source (White, 1967; Williams and McBirney, 1979). When the pressure – but not the temperature – of one part of the column is lowered, as for example by convection, the water can flash to steam when crossing the boiling curve and a rarefaction wave travelling downward may trigger vaporization of the lower part of the column. Geysers will not develop in areas of high permeability because water heated from below will be able to rise and dissipate by convection. On the other hand, if pathways for convecting hot waters become completely clogged ("self-sealed") through precipitation of hydrothermal mineral phases, especially silica, water pressures as much as 30 percent over hydrostatic pressures can develop. Eruptions from such sealed and overpressured hydrothermal systems are called hydrothermal eruptions. They have been described from both Yellowstone and New Zealand (Muffler et al., 1971; Lloyd, 1972; Nairn and Wiradiradja, 1980). Erupted products are steam, mud, and large amounts of angular clasts, and craters as much as 1 km in diameter have formed by such eruptions (Muffler et al., 1971). Unstable sealed hydrothermal systems can be triggered to become explosive by disturbances such as earthquakes.

**Magma-Water Mixing and Fuel-Coolant Interaction**

In the previous sections we described major effects of hydrous quenching on the magma (formation of glass, breakage by thermal contraction and inhibition of vesiculation) and on water (pressurization and water/steam transition). In real situations, however, there are complex time- and pressure-dependent feedback mechanisms at the dynamic interface between magma and water. There are three fundamental ways that the effective surface area of the magma is increased: (a) contraction cracking during quenching; (b) contemporaneous vesiculation at very low confining pressure or high magmatic volatile partial pressure; and (c) mixing of magma and accidental rock fragments, as indicated by the common occurrence of such xenoliths in phreatomagmatic essential pyroclasts. Because of the complexity of these processes, they have not been analyzed in detail, but some light is shed on this problem from experimental work on industrial explosive interactions between hot liquids (of any type = fuel) and cold vaporizable liquids (of any type = coolant) (Buchanan and Dullforce, 1973; Peckover et al., 1973; Buchanan, 1974; Dullforce et al., 1976). These interactions are called fuel-coolant interactions (FCI). The energy of the violent interactions is not derived from chemical reactions but instead comes from the excess heat in the fuel. Transfer of this energy to the coolant causes vaporization of the coolant and explosive fragmentation of both fuel and coolant within fractions of a second.

Many factors influence such reactions: specific heat, thermal conductivity, interfacial surface tension, melting point, density, viscosity, initial temperatures, pressures, etc. Buchanan (1974) develops the following theoretical model for processes at the fuel-coolant interface: (1) an initial perturbation triggers the interaction and causes a vapor bubble to form at the interface; (2) bubbles expand into the coolant, then condense and collapse with jetting; (3) the fuel is penetrated by the liquid jet formed from asymmetric bubble collapse, causing an increase in surface area and disintegration of the fuel; (4) heat is transferred from the fuel to the jet; (5) expansion and formation of a new bubble occurs, which further expands and disperses the disintegrated fuel, and so on. An exponential increase in surface area leads to exponential increase in rate of total heat transfer. The strength of the interaction is reduced as the external pressure is increased and can be inhibited entirely if the pressure is large enough (Buchanan, 1974). Required conditions for fuel-coolant reactions to occur are: (a) A large surface area for effective heat transfer, (b) a pressure threshold, (c) the perturbation and breaking of the initial layer of vaporized coolant that separates the coolant and the fuel, (d) the threshold coolant temperature.

Peckover et al. (1973) noted that heterogeneous bubble nucleation, leading to explosive interactions during submarine volcanism, occurs only below a threshold pressure of about 68 b, corresponding to a water depth of about 700 m. This is much less than the critical pressure of water (216 b), corresponding to a water depth of about 2 km. Moreover, experiments by Dullforce et al. (1976) with water ($T_c$ = varying from 0° to 100 °C) into which molten tin ($T_f$ varying from 300° to 1000 °C) was dropped showed that there is a very restricted zone where disintegration takes place.

These models and experiments cannot be applied directly to the interactions of lava and water principally because of the different physical properties of silicate liquids, but they serve as the best analog available.

Shepherd and Sigurdsson (1982) analyzed the mechanism of the 1979 eruption of Soufrière Volcano which, based on the nature of the deposits and the eruption characteristics, was interpreted to be phreatomagmatic (Fiske and Sigurdsson, 1979). The eruption was thought to be caused by new magma invading a still-hot extrusive lava dome, emplaced in 1971–72, which was surrounded by a lava lake (Fig. 4-18). The mechanism by which lava and water interacted is believed to be due to the high ratio of the coefficient of thermal expansion of water to its bulk compressibility which is about 15 b °C at 100 °C and atmospheric pressures. Thus, at approximately constant volume, pressure rises by 15.1 b for every degree rise in temperature which may generate thermally induced hydraulic fracturing and microearthquakes.

## Phreatic and Phreatomagmatic (Vulcanian) Eruptions

Because the importance of eruptions caused or influenced by magma-water interactions has been recognized only during the last few years, the observational base is insufficient and re-analysis of historic eruptions has not advanced far enough to define important criteria and boundary conditions for hydroclastic eruptions precisely. We know, however, that the deposits of hydroclastic eruptions tend to be well-bedded, poorly sorted and include vesicle-rich layers of tuff. Fragments are commonly nonvesicular to poorly vesicular, vitric, angular and fine-grained. Accretionary lapilli and large bread-crusted to cauliflower-shaped bombs are common. Lithic fragments are very common and dominate many hydroclastic deposits (Chap. 9). A summary of Vulcanian products and processes is given in Table 9-2.

Phreatic eruptions are traditionally defined as steam explosions entirely within the country rock above a magmatic heat source, whereas phreatomagmatic eruptions are those in which new magma is disintegrated as well (Macdonald, 1972; Williams and McBirney, 1979). However, the restriction of these terms to groundwater steam eruptions is impractical because the type of water (ground vs. surface) is difficult to specify especially in older deposits (Schmincke, 1977a).

We consider Vulcanian eruptions to be from hydroclastic processes, although considerable confusion surrounds the name because it has been defined mainly from witnessed eruptions rather than by process-related parameters deduced from the deposits. Many volcanologists use the term Vulcanian for highly explosive, short-lived eruptions that produce black, ash- and steam-laden eruption columns as witnessed during the 1888–90 eruptions of Vulcano, a small volcano in the Eolian Islands, Italy (see e.g. Macdonald, 1972).

The deposits from and features of the 1888–90 eruptions of Vulcano, Italy are similar to those at Surtsey, Iceland in 1963/64. Schmincke (1977a) suggested that the term Vulcanian be redefined to include all explosive eruptions characterized by the fluid dynamic interaction of water and magma and therefore to include as a subset Surtseyan proposed by Walker and Croasdale (1972).

With regard to the mechanism of Vulcanian eruptions, essential features are discrete explosions at intervals varying from minutes to hours, caused by repeated build-up of pressure beneath a plug (McBirney, 1963; Wilson, 1980). The relative importance of (1) exsolution of magmatic volatiles accumulated in noneruptive intervals beneath a plug of solidified magma in the vent vs. (2) vaporization of groundwater surrounding the magma column is, however, still debated. Wilson (1980) suggested that there is a time-scale problem regarding thermal waves from an intrusion. They may not travel fast enough into the surroundings to produce enough steam to generate the high ejection velocities for repeated short-interval explosions. Perhaps, both explosive water/steam transitions and high magmatic volatile pressures are essential features of powerful Vulcanian eruptions.

Based upon the high ejecta velocities observed during some Vulcanian eruptions (up to about 400 m s$^{-1}$; Nairn and Self, 1978), Wilson (1980) calculated the pressures resulting from volatile ($H_2O$) build-up. He suggested that ejecta velocities up to 200 m s$^{-1}$ can be explained by exsolution of magmatic water up to a few weight percent, but speeds over 300 m s$^{-1}$ would imply unreasonably high magmatic volatile contents (>10 weight percent), suggesting the involvement of

**Fig. 4-19.** Relationship between maximum ejecta velocity and initial pressure in plug at the time it fails for Vulcanian explosions. *Curves* are for different wt.% of gas in ejecta, the effect of two different temperatures being shown in 1 wt.% gas. *Dashed line* shows pressures calculated from modified Bernoulli equation (MBE). (After Wilson, 1980)

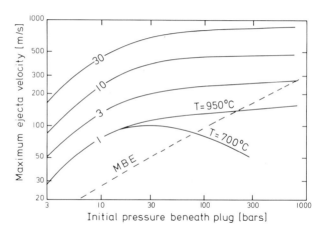

groundwater (Fig. 4-19). Wilson also showed that initial ejecta velocities deduced from the sizes of ejected blocks cannot be calculated from the assumption of high drag forces experienced in still air because initially, drag forces may be at a minimum within the shock wave which immediately precedes ejected debris in sudden volcanic explosions. Thus, deduced initial velocities have been overestimated in the past.

In summary, Vulcanian eruptions – as deduced from the deposits, from the type eruptions and from many other subsequent historic eruptions – generally involve the participation of heated external water. Ejection velocities of up to about 200 m s$^{-1}$, however, are not necessarily caused by explosive vaporization of nonmagmatic water. A re-examination of the products of historic eruptions regarded as Vulcanian is needed to clear up this problem. In Vulcanian eruptions, a large proportion of thermal energy is used to transform water to steam. If the mixture of clasts and steam rises and marked condensation takes place, a substantial increase in density occurs. Thus, other factors being equal, phreatomagmatic eruption clouds should acquire lower altitudes than magmatic eruptions (Wilson et al., 1978), and may collapse to form base surges. This is indeed the case for many observed eruption columns associated with the formation of base surges (Moore, 1967; Waters and Fisher, 1971).

The difficulties in distinguishing the causes of explosive eruptions even if they have been well studied is illustrated by the tragic volcanic blast of the May 18th, 1980, eruption of Mount St. Helens (Washington, USA). The evidence from observation of the eruption, deposits, and theoretical calculations was thought to be most plausibly explained by explosive expansion of a hydrothermal system heated by the rising magma column and unloaded by landsliding of the northern flank of the volcano (Fig. 4-20) (Kieffer, 1981, 1982; Christiansen and Peterson, 1981). Eichelberger and Hayes (1982) on the other hand reasoned that the observed ejecta velocities of 100–250 m s$^{-1}$ of the blast could be explained by rapid expansion of a water-saturated magmatic system, requiring a landslide thickness of a few hundred meters and only about 1 weight percent total magmatic water within the system (Fig. 4-21).

The question of what happens to eruption columns formed beneath water is less well-known. Depending upon depth (pressure due to the weight of the water col-

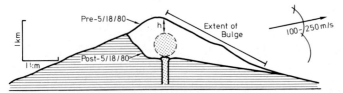

**Fig. 4-20.** Schematic N-S cross-section of Mount St. Helens, showing pre-eruption (May 18th, 1980) and post-eruption cross-section (horizontally ruled). Inferred position of new magma shown by stippled pattern. (After Eichelberger and Hayes, 1982)

umn) and partial pressure of magmatic volatiles, eruption columns may not form at all, even from the most volatile-rich magmas. However, if pressure conditions allow the formation of a subaqueous eruption column, the greater (than air) drag effects of water will enhance turbulent mixing and consequent rapid energy loss at the column margin. It is likely that subaqueous eruption columns quickly increase in density and thereby undergo immediate collapse and initiate pyroclastic flows as postulated by Fiske (1963) and Fiske and Matsuda (1964) (Chap. 10). Under such conditions, thermal energy must be rapidly lost, and thus subaqueous pyroclastic flows may be emplaced at elevated temperatures beneath the water but welding will not occur within deposits from subaqueous eruption columns.

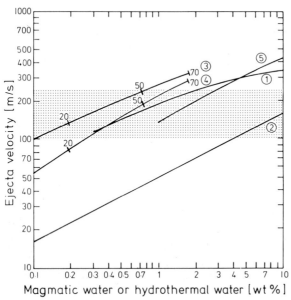

**Fig. 4-21.** Relationship between water vapor (magmatic) or water (hydrothermal) content and peak ejecta velocity. Predicted velocities are: (1) Hydrothermal reservoir unloaded by landslide 500 m thick; (2) same according to Kieffer (1981); (3) Magmatic reservoir unloaded by landslide 500 m thick with varying bubble content (in vol.%); (4) Plinian eruption with fragmentation at 500 m depth (model I of Wilson, 1980); (5) Volcanic blast due to failure of plug 500 m thick (model III of Wilson, 1980). Shaded zone indicates observed or inferred velocities for Mount St. Helens blast. (After Eichelberger and Hayes, 1982)

## Volcanic Energy

Volcanoes are abnormally hot spots where thermal energy, centralized within the crust and mantle, is dramatically released; elsewhere, thermal energy is slowly but pervasively moving up from the interior to be dissipated into the atmosphere. The total heat flow ($Q_t$) from the earth's surface can be expressed as follows:

$$Q_t = Q_{co} + Q_{cr} + Q_{cl} + Q_m \quad (1)$$

$Q_{co}$ is the average heat flow that may correspond to a radiogenic source, $Q_{cr}$ and $Q_{cl}$ are the regional and local excess conducted heat flow from magmatic intrusions and circulation of hydrothermal systems, and $Q_m$ is the heat flow caused by mass transfer from extrusions onto the surface (Fujii, 1975). The global heat flow from extrusive volcanic material has been estimated to be $6.6 \times 10^9$ W (Fujii, 1975) compared with a global flow of about $1.8 \times 10^{13}$ W. Heat transfer by extrusive volcanic material ($Q_m$) is comparable to the conducted heat flow near lithospheric plate boundaries, which is higher than in non-volcanic areas elsewhere within plates. Global heat flow from hot springs is estimated at about $6.6 \times 10^8$ W. Considering the less well-known but intense hydrothermal activity at mid-ocean ridges, heat flow from hot spring activity might attain a magnitude comparable to that of eruptive volcanic activity.

Volcanic energy is released in the form of kinetic energy represented by the mass ejection velocity of pyroclastic ejecta, potential energy represented by the change in level of the magma (energy acquired by the magma), volcanic tremors and sympathetic earthquakes, airwaves and seismic sea waves, energy required to comminute solid basement rocks, and thermal energy (Yokoyama, 1956). However, the thermal energy carried by volcanic solid and gaseous materials is by far the most important source of volcanic energy (Verhoogen, 1946; Yokoyama, 1956, 1957a,b, 1972; Hedervari, 1963; Nakamura, 1965, 1974; Polak, 1967; Shimozuru, 1968; Fujii, 1975).

Because most of the energy of volcanic eruptions is heat energy, estimates of total energy (E) may be made in terms of mass transfer (i.e., total volume) of lava and pyroclastic material to the surface (Yokoyama, 1956, 1957a,b, 1972):

$$E_{th} = E_{ths} + E_{thw} + E_{thf}, \quad (2)$$

$$E_{ths} = M(\Delta T \times C + H)J, \quad (3)$$

$$E_{thw} = M_w(\Delta T \times C_w \times H_w)J, \text{ and} \quad (4)$$

$$E_{thf} = M(\Delta T \times C)J, \quad (5)$$

where $E_{th}$ = total thermal energy; $E_{ths}$ = thermal energy of essential lava; $\Delta T$ = temperature difference; C = specific heat; H = latent heat of fusion; $M_w$ = mass of water; $C_w$ = specific heat of water; $H_w$ = latent heat of fusion of water; J = equivalent work; and $E_{thf}$ = thermal energy of pyroclastic ejecta. $E_{thw}$ = amount of thermal energy of volatile material in the lava (assumed to be water vapor). M (=mass) is the only measurable parameter for eruptions recorded only by the deposits. Other types of energy spent in expansion of volatiles, mainly water, is through heat lost by underground conduction (Nakamura, 1965). By assuming various values for T, C, H, J and $M_w$, then $E_{th}$ (~ E) can be estimated by determining the mass of the eruptive products, a measurable field quantity which directly relates geologic parameters to theoretical considerations.

Energy calculations ($E_{th}$) based upon ejected volumes using Yokoyama's formulas were made by Hedervari (1963) for 94 volcanic eruptions arranged into an intensity scale from 0 to IX. Some attempts have also been made to analyze the different types of energies in individual volcanoes (e.g. Nakamura, 1965) and during a volcano's history (Yokoyama, 1972) (Fig. 4-22). Such energy calculations are at best only approximations because of the several assumptions that must be made, yet they serve to provide energy limits for various eruptions based upon the measured estimates of erupted solids and they give a method for comparison of energy output by different volcanoes. The studies by Nakamura (1965) and Yokoyama (1972) illustrate how volcanologic information can be obtained from stratigraphic analysis of pyroclastic rocks, thereby paving the way for volcanologic studies of ancient volcanic fields. Volcano-energy studies, in combination with heat-flow studies and the recognition that plate boundaries such as mid-oceanic ridges are important areas of magma production, enable more realistic assessments of worldwide production rates of magma (Nakamura, 1974; Fujii, 1975) (Table 1-4).

Based mainly upon field and laboratory study of erupted products, Walker (1980) distinguishes other less theoretical parameters of practical use for geologists

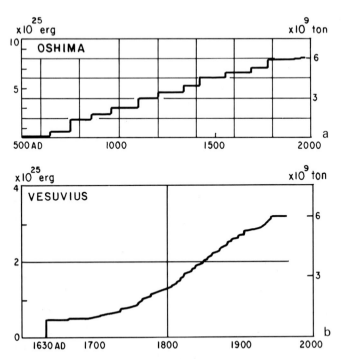

**Fig. 4-22 a, b.** Estimated cumulative energy release: **a** Oshima Volcano (Japan) since 500 A.D. *Left ordinate:* energy, *right ordinate:* cumulative mass. **b** Vesuvius (Italy) since 1630 A.D. Ordinates as in *a*. (After Yokoyama, 1972, from previous authors)

to compare the relative size of eruptions, namely magnitude, intensity, dispersive power, violence, and destructive potential. *Magnitude* refers to the total estimated volume of erupted products (in dense rock equivalents) and is therefore related to total energy as discussed above. Hedervari's (1963) intensity scale is thus a magnitude scale in Walker's classification. *Intensity* is the rate of release of material or energy which can be directly observed by the filming of eruptions (e.g., Self et al., 1979), or roughly estimated from deposits by using known fall velocities of fine- and coarse-grained material and estimating the times of fall necessary to produce a bed having a characteristic graded sequence (or lack thereof) at a particular distance from the source. *Dispersive power* refers to the extent of the dispersal area which increases as column height – and also wind speed – increases. *Violence* is applied to eruptions in which the distribution of ejecta is determined by their momentum, for example, a pyroclastic shower which has a high concentration of falling fragments. *Destructive potential* refers to the actual or potential areal extent of destruction. This classification opens the way for further quantification of volcanic energy as deduced from field studies of pyroclastic products.

Newhall and Self (1982) have integrated sparse quantitative data using Walker's (1973; 1980) classification with the subjective classical method of observation to develop a "volcanic explosivity index (VEI)"; this is adopted by Simkin et al. (1981) for their catalog of eruptions of the world's volcanoes (Table 4-5).

**Table 4-4.** Estimated energies of different forms as released during a single volcanic eruption (Nakamuru, 1964)

| Energy released by | Oshima Volcano, Izu | | Usu Volcano (1943–1945) (Showa-sin-zan) ($\times 10^{24}$ ergs) |
|---|---|---|---|
| | 1953–1954 ($\times 10^{21}$ ergs) | 1950–1951 ($\times 10^{23}$ ergs) | |
| Heat transferred by solid product (Eths) | $7.1$–$6.7 \times 10^0$ | $9.6$–$7.5 \times 10^0$ | $3.2 \times 10^0$ |
| Ground vibration (Ev) | Tremors $1 \times 10^{-2}$ Local shocks $5 \times 10^{-4}$ | $2 \times 10^{-4}$ $2 \times 10^{-5}$ | $>1.8 \times 10^{-5}$ |
| Explosive ejection (Ek) | $<7 \times 10^{-6}$ | $<7.3 \times 10^{-4}$ | $5 \times 10^{-4}$ ($>3.6 \times 10^{-4}$) |
| Work done against gravity (Ep) | $1.3$–$2.5 \times 10^{-1}$ | $>4.2$–$3.5 \times 10^{-1}$ | $2.9 \times 10^{-1}$ |
| Energy supplied to the realm of volcanic activity | $1.9$–$3.4 \times 10^3$ | $>1.1 \times 10^3$ | |

**Table 4-5.** Criteria for volcanic explosivity index (VEI). (After Newhall and Self, 1982)

| VEI | 0 | 1 | 2 | 3 | 4 | 5 | 6 | 7 | 8 |
|---|---|---|---|---|---|---|---|---|---|
| Description | Non-explosive | Small | Moderate | Mod-large | Large | Very large | — | — | — |
| Volume of ejecta (m$^3$) | $<10^4$ | $10^4$–$10^6$ | $10^6$–$10^7$ | $10^7$–$10^8$ | $10^8$–$10^9$ | $10^9$–$10^{10}$ | $10^{10}$–$10^{11}$ | $10^{11}$–$10^{12}$ | $>10^{12}$ |
| Column height (km)[a] | $<0.1$ | 0.1–1 | 1–5 | 3–15 | 10–25 | $>25$ | | | |
| Classification | Hawaiian | | Strombolian | Vulcanian | Plinian | | Ultra-Plinian | | |
| Duration (hours of continuous blast) | | $<1$ | | 1–6 | | 6–12 | $>12$ | | |
| Tropospheric injection | Negligible | Minor | Moderate | Substantial | | | | | |
| Stratospheric injection | None | None | None | Possible | Definite | Significant | | | |
| Eruptions (total in file)[b] | 443 | 361 | 3108 | 720 | 131 | 35 | 16 | 1 | 0 |

[a] For VEI's 0–2, given as km above crater; for VEI's 3–8, given as km above sea level
[b] Catalogue of active volcanoes (Simkin et al., 1981)

## Chapter 5   Pyroclastic Fragments and Deposits

### General Components

*Pyroclastic* fragments, also known as *pyroclasts* (Schmid, 1981), are produced by many processes connected with volcanic eruptions. They are particles expelled through volcanic vents without reference to the causes of eruption or origin of the particles. *Hydroclastic* fragments are a variety of pyroclasts formed from steam explosions at magma-water interfaces, and also by rapid chilling and mechanical granulation of lava that comes in contact with water or water-saturated sediments (Chaps. 4 and 9). We focus upon pyroclastic material in this chapter, but much of the discussion also applies to hydroclastic deposits.

Volcanic fragments are also found in deposits other than pyroclastic and hydroclastic. Among these are *epiclastic* (weathering and erosion of older volcanic rocks), *autoclastic* (fragmentation by mechanical friction or gaseous explosion during movement of lava, or gravity crumbling of spines and domes), and *alloclastic* (disruption of pre-existing volcanic rocks by igneous processes beneath the earth's surface, with or without intrusion of fresh magma). Granulation by tectonic processes is another way that volcanic fragments are formed. The general term *volcaniclastic* introduced by Fisher (1961a) includes all clastic volcanic materials formed by any process of fragmentation, dispersed by any kind of transporting agent, deposited in any environment or mixed in any significant portion with nonvolcanic fragments. *Tephra* (Thorarinsson, 1954) is applied to pyroclastic accumulations irrespective of size and is synonymous with "pyroclastic material".

A common problem in naming fragmental volcanic rocks concerns mixtures of pyroclasts and other kinds of clasts, especially epiclastic fragments. Resolving this problem is far from an exercise in semantics, its significance rests in determining penecontemporaneous volcanism. Thus, attention must be given to the physical aspect, composition, and relative percentage of clast types. A *pyroclastic* fragment is an "instant" fragment, produced directly from volcanic processes. Irrespective of later processes that may recycle such fragments, such as by water or wind, they remain pyroclastic. An *epiclastic* volcanic fragment is produced by weathering and erosion of volcanic rocks and thus may, but commonly does not, record penecontemporaneous volcanism. It is possible for epiclastic fragments, however, to be derived from the erosion of older lithified tuffs; individual epiclastic fragments may thus be composed of still smaller pyroclastic fragments within them, but the origin of the larger fragments is epiclastic. Reworking or recycling of unconsolidated pyroclastic debris by water or wind does not transform pyroclasts into epiclastic fragments. Reworked pyroclastic fragments are derived from the remobilization of loose materials. Especially important in resolving such questions are glass shards and pumice. As a general rule, epiclastic accumulations contain smaller amounts of glass shards or pumice, if any, because weathering readily alters metastable glass to clays and zeolites, and the rigors of transportation destroy original textures of the glassy clasts.

In addition to the above-mentioned difficulties, there are two problems of recognition that may not be resolvable in the ancient geologic record. One is that ash beds, which remain poorly consolidated over millions of years with little alteration of many glass shards (e.g. upper Miocene Mascall Formation, central Oregon), may become reworked into streams, lakes or other environments. Such "delayed reworking" could produce abundant debris interpreted to indicate contemporaneous volcanism. Another situation difficult to interpret is pyroclastic deposits composed mainly of lithic pyroclasts (e.g., 1902 Mt. Peléc deposits, Fisher and Heiken, 1982) that become reworked in streams, beaches, and offshore environments. Slight rounding of fragments produces lithic sands identical to sands composed of epiclastic fragments. In both of the above situations, facies analyses or rock associations might be the only way to resolve problems of identification, if this is at all possible.

Three main varieties of pyroclastic ejecta according to origin are *juvenile* (or essential), *cognate* (or accessory), and *accidental*. Juvenile pyroclasts are derived directly from the erupting magma and consist of dense or inflated particles of chilled melt, or crystals that were in the magma prior to eruption (pyrogenic crystals). Cognate particles are fragmented co-magmatic volcanic rocks from previous eruptions of the same volcano. Accidental fragments are derived from the subvolcanic basement and therefore may be of any composition. Pyroclasts are named according to a large variety of criteria but the fundamental basis is grain size (Table 5-1) giving three main types – ash (<2 mm) lapilli (2–64 mm) and bombs or blocks (>64 mm) (Fisher, 1961a; Schmid, 1981). Other criteria include composition, ori-

Table 5-1. Granulometric classification of pyroclasts and of unimodal, well-sorted pyroclastic deposits. (After Schmid, 1981)

| Clast size | Pyroclast | Pyroclastic deposit | |
|---|---|---|---|
| | | Mainly unconsolidated: tephra | Mainly consolidated: pyroclastic rock |
| 64 mm | Block, bomb | Agglomerate, bed of blocks or bombs or block or bomb tephra | Agglomerate, pyroclastic breccia |
| 2 mm | Lapillus | Layer, bed of lapilli or lapilli tephra | Lapillistone |
| 1/16 mm | Coarse ash grain | Coarse ash | Coarse (ash) tuff |
| | Fine ash grain (dust grain) | Fine ash (dust) | Fine (ash) tuff (dust tuff) |

**Table 5-2.** Terms for mixed pyroclastic-epiclastic rocks. (After Schmid, 1981)

| Pyroclastic[a] | | Tuffites (mixed pyroclastic-epiclastic) | Epiclastic (volcanic and/or nonvolcanic) | Average clast size (mm) |
|---|---|---|---|---|
| Agglomerate, agglutinate pyroclastic breccia | | Tuffaceous conglomerate, tuffaceous breccia | Conglomerate, breccia | 64 |
| Lapillistone | | | | |
| (Ash) tuff | coarse | Tuffaceous sandstone | Sandstone | 2 |
|  | fine | Tuffaceous siltstone | Siltstone | 1/16 |
|  |  | Tuffaceous mudstone, shale | Mudstone, shale | 1/256 |
| 100% | 75% | 25% (increase) | 0% by volume | |

← ———————— Pyroclasts ————————

(increase)

———— Volcanic + nonvolcanic epiclasts (+ minor amounts ————→
of biogenic, chemical sedimentary and authigenic constituents)

[a] Terms according to Table 5-1

gin, and vesicularity. Rock names for mixtures of pyroclastic and epiclastic fragments are given in Table 5-2.

Volcanic ash is composed of various proportions of vitric, crystal or lithic particles of juvenile, cognate or accidental origin forming 75 volume percent or more of an aggregate. *Tuff* is the consolidated equivalent of ash and is subdivided into fine- and coarse-grained varieties according to the size of component particles (Table 5-1). Further classification is made according to environment of deposition (lacustrine tuff, submarine tuff, subaerial tuff) or manner of transport (fallout tuff, ash-flow tuff). Reworked ash is commonly named according to the transport agent (fluvial tuff, aeolian tuff).

As with ash-sized particles, lapilli may be juvenile, cognate or accidental; lithified accumulations containing greater than 75% lapilli may be termed *lapillistone* although many workers prefer *lapilli-tuff* (Schmid, 1981). We reserve the name lapilli-tuff for lithified mixtures of ash and lapilli, where ash forms 25–75% of the pyroclastic mixture (refer to Fig. 5-1 for mixture terms and percentages). Lapilli are commonly angular to subround. Subrounded forms are commonly of juvenile origin, but explosively broken cognate and accidental fragments may become rounded by repeated extrusion and fall back into a vent before final ejection. On the other hand, juvenile fragments may break upon impact as they land, especially pumice (Fig. 5-2). *Accretionary lapilli* are special kinds of lapilli-sized particles that form as moist aggregates of ash in eruption clouds, by rain that falls through dry eruption clouds or other processes (Figs. 5-3, 5-4). *Armored lapilli* (Waters and Fisher, 1970, 1971) form when wet ash becomes plastered around a solid nucleus such as crystal, pumice or lithic fragments during hydroclastic eruptions (Fig. 5-5).

Blocks are commonly angular to subangular fragments of cognate and accidental origin derived from the edifice of the volcano or from its basement. Blocks of juvenile origin may also develop from the disruption of contemporaneously developed domes or by the breakage of congealed bombs upon impact (Fig. 5-6). *Pyro-*

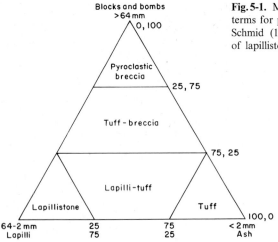

**Fig. 5-1.** Mixture terms and end-member rock terms for pyroclastic fragments (Fisher, 1966c). Schmid (1981) recommends lapilli tuff instead of lapillistone

*clastic breccia* is a consolidated aggregate of blocks containing less than 25% lapilli and ash. The term *volcanic breccia* has a broader meaning, applying to all volcaniclastic rocks composed predominantly of angular volcanic particles greater than 2 mm in size (Fisher, 1958, 1960); thus it parallels the term "breccia" widely used for non-volcanic sedimentary rocks.

Bombs are thrown from vents in a partly molten condition and solidify during flight or shortly after they land. Thus, they are composed almost exclusively of juvenile material. The molten clots are shaped during flight and modified upon impact if still plastic. Bombs are named according to shape, such as ribbon bombs, spindle bombs (with twisted ends), cow-dung bombs, spheroidal bombs etc. (Wentworth and Macdonald, 1953; Macdonald, 1972). The chilled crusts of bread crust bombs are fractured by continued expansion of gas within the still-plastic core. They are most commonly produced from magma of intermediate and silicic compositions. Basaltic bombs usually show little surface cracking, although some may have fine cracks caused by stretching of a thin, glassy surface over a still-plastic interior upon impact. Bombs with surfaces that have cauliflower shapes with dense to vesicular interiors are characteristic of hydroclastic processes. *Agglomerate* is a welded aggregate consisting predominantly of bombs. It contains less than 25% by volume of lapilli and ash.

*Cored bombs* have rinds of lava enclosing cores of cognate or accidental fragments; their composition is quite variable (Fig. 5-7). Most attention has been given to ultramafic nodule cores derived from the mantle that occur in some scoria cones and maar volcanoes and many kimberlite breccias. In some cases, solid fragments become plastered with wet ash in steam-rich eruption clouds and may be called *armored bombs* to distinguish them from cored bombs. Their origin is similar to that of armored lapilli. In some instances, block-size clots of wet ash lack cores and become flattened upon impact to form layers composed of aligned lensoid tuff bodies.

Three common names which depend in part on the degree of vesicularity are *scoria, cinders,* and *pumice.* They are named without reference to size, but most

Fig. 5-2. Dacitic fallout pumice lapillus (10 cm), broken on impact. 1976 Augustine eruption (Alaska)

Fig. 5-3. Layers of rhyolitic accretionary lapilli (fine-grained white rim, dark core) in pure vitric tuff. Pliocene Ellensburg Formation (Sentinel Gap, Washington, USA). See also Fig. 5-4

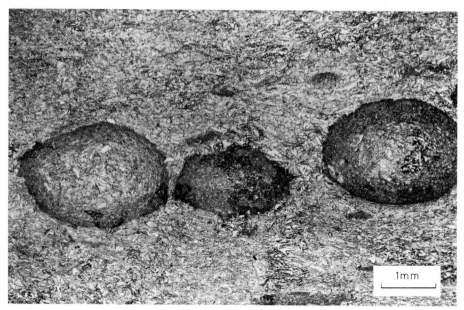

**Fig. 5-4.** Photomicrograph of vitric tuff containing accretionary lapilli with coarse glass shards in core and fine-grained tangentially aligned platy shards forming rim. Pliocene Ellensburg Formation (Sentinel Gap, Washington, USA)

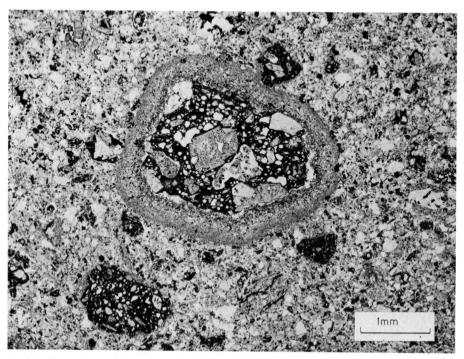

**Fig. 5-5.** Photomicrograph of armored lapillus consisting of a basalt lava fragment with clinopyroxene and olivine phenocrysts mantled by rim of fine-grained ash. Quaternary Boos volcano (Eifel, Germany)

**Fig. 5-6.** Bread-crust block of dacitic composition expanded after transport in pyroclastic dome collapse (?) flow. 1976 eruption of Augustine volcano (Alaska)

**Fig. 5-7.** Cored bomb of angular lava fragment rimmed by lava. Quaternary Herchenberg volcano (Eifel, Germany)

commonly are in the lapilli or larger size range. Pumice is a white or pale gray to brown, highly vesicular, silicic to mafic glass foam which will commonly float on water; bubble walls are composed of translucent glass. Scoria and cinders (essentially synonymous terms), usually of mafic composition, are highly inflated juvenile fragments but less so than pumice. They readily sink in water. They are generally composed of tachylite, that is, glass rendered nearly opaque by microcrystalline iron oxides. Another term, *spatter,* applies to bombs, usually basaltic, formed from almost completely molten material that readily welds (agglutinates) upon impact and contrasts with cinders which do not stick together. Cinder cones, for example, are composed largely of loose cinders; spatter cones are composed mainly of agglutinated bombs.

Large fragments are useful for interpreting some aspects of pyroclastic layers (such as graded bedding) and may be used much like lava flows in problems involving chemical composition, but the dominant components are commonly ash-sized vitric, lithic, and crystal fragments. The following discussion emphasizes the juvenile ash-sized material.

## Vitric Particles and Pyrogenic Crystals

### Glass Shards

Many glass shards consist of the walls of tiny broken bubbles or the junctions of bubbles developed by the vesiculation of silicic magma. A variety of nonvesiculation processes such as (1) thermal shock or spalling from rapid quenching during hydroclastic eruptions, (2) attrition of the glassy margins of moving domes, plugs or lava flows, and (3) attrition by wave action on floating pumice rafts (Chap. 7) may form similar small glass particles but lack bubble wall texture (Fig. 5-8). The morphology of shards and pumice attracted the attention of early petrographers (e.g. Pirsson, 1915); Swineford and Frye (1946) published a rather detailed account of shard shapes. Modern studies include those by Heiken (1972, 1974) using scanning electron microscopic (SEM) techniques, Perlaki (1966) on the origin and structure of pumice, and Walker and Croasdale (1972) who compared basaltic pyroclasts of magmatic and hydroclastic origin as did Heiken. Ross and Smith (1961) petrographically described the glassy textures of welded tuffs in great detail, and Schmincke (1974b) called attention to the dependence of textures in peralkalic silicic welded and nonwelded tuffs on the viscosity of the melt. Characteristic features of volcanic ash particles and associated features are given in Table 5-3.

### Shards Formed by Vesiculation

Perhaps the most familiar kinds of glass shards are silicic varieties formed from shattered bubbles (Pirsson, 1915; Ross et al. 1928; Swineford and Frye, 1946; Ewart, 1963; Heiken, 1972, 1974; and others). There are essentially three end-member types: (1) cuspate or lunate-shaped fragments of broken bubble walls that are commonly Y-shaped (in cross-section) representing remnants of three bubble junctions, or double-concave plates that formed the wall between adjoining bubbles; (2) flat plates from the glass walls separating large flattened vesicles; and (3) small

Fig. 5-8. Photomicrograph of sideromelane tuff consisting dominantly of angular vesicle-free sideromelane shards. Miocene Obispo Formation (California, USA)

pumice fragments with a fibrous or cellular structure composed of minute elongate or circular cavities enclosed by glass walls (Figs. 5-9, 5-10). Types (1) and (2) may be called bubblewall shards and type (3) pumice shards (Fig. 5-11). Spheroidal bubbles completely enclosed in glass were found following the Katmai (Novarupta) eruption of 1912 (Fig. 5-12) (Perret, 1950), but such bubble shards are not common. Spheroidal bubbles also have been described from low-viscosity peralkalic magmas (Schmincke, 1974b; Hay et al., 1979).

Although there are many variables affecting the shape of glass shards, Izett (1981) presents evidence to show that pumice shards tend to develop from relatively high viscosity rhyolitic magmas with temperatures $<850\,°C$, whereas bubble wall and bubble junction shards tend to develop from lower viscosity rhyolitic magmas at temperatures $>850\,°C$.

There are few descriptions of glass shards of intermediate composition. Those observed have irregular lumpy forms with some botryoidal surfaces and contain ovoid or spherical vesicles. Some are twisted, possibly from twirling around a central axis during flight (Heiken, 1972, 1974). Vitric particles from hydroclastic processes commonly have blocky shapes and few vesicles (Fig. 5-8).

Shards from mafic Strombolian and Hawaiian-type eruptions form by rapidly expanding gases which disrupt low-viscosity magma to produce an incandescent spray or lava fountain of coarse-ash sized droplets, and lapilli and block-sized clots (spatter). Gases expanding within individual fluid particles result in varying degrees of vesicularity but ordinarily do not cause the particles to explode into

**Table 5-3.** Some features of common volcanic ash. (After Heiken, 1974)

| Composition | Basaltic | Andesitic | Dacitic and rhyolitic |
|---|---|---|---|
| *A. Magmatic eruptions*[a] | | | |
| Volcanic features associated with the ash | Scoria cones, lava lakes, basalt flows, and ash interbeds in stratovolcanoes | Stratovolcanoes, domes, thick short lava flows, lahars, and pyroclastic flow units | Domes and lava flows, some stratovolcanoes, calderas, and inferred fissure vents which are sources for ash flows; pyroclastic flow sheets |
| Petrography | Complete and broken droplets of sideromelane (basaltic glass) and tachylite (submicrocrystalline basalt); phenocryst volume is variable | The main components are variable volumes of vitric, crystal and lithic grains; vitric grains are colorless glass shards and pumice generally with oriented microlites; lithic fragments include andesite of varying textures and alteration stages and accidental igneous and sedimentary xenoliths; broken plagioclase and pyroxene crystals and opaque minerals common | Common are large volumes of colorless glass with variable amounts of microlites and phenocrysts of quartz, sanidine, biotite, and small amounts of other ferromagnesian minerals; small amount of lithic (rhyolitic or accidental) fragments |
| Morphology of grains | 1. Irregular droplets with fluidal shapes (spheres, ovoids, dumbbells, and teardrops)<br>2. Broken droplets; some with original droplet surface present<br>3. Long, thin strands of glass (Pele's hair)<br>4. Polygonal, lattice-like network of glass rods; highly vesiculated froth<br><br>Shapes vary from spheres and droplets in very low viscosity lavas to irregular elongate, often broken droplets from lavas of slightly higher viscosities; all are vesicular | 1. Vitric grains: equant to elongate pumice fragments, dependent on the vesicle shape in the fragment; elongate pumice contains elongate, ovoid to tubular vesicles; fragment surfaces are irregular; individual vesicle walls are smooth. Flat, pointed shards with smooth or conchoidal fracture surfaces are probably broken vesicle walls<br>2. Lithic fragments: generally equant; surface features dependent on texture and fracture of the rock type; some fragments are rounded<br>3. Crystal fragments: shape governed by fracture of the mineral; most appear to have been broken during eruption | 1. Vitric grains; elongate to equant pumice (grain shape dependent on vesicle shape) with thin vesicle walls; curved, Y-shaped, or flat thin shards which are smooth vesicle walls<br>2. Lithic fragments are generally equant<br><br>Surfaces of pumice are rough, broken vesicles; no smooth surfaces other than vesicle walls; smooth fluidal surfaces are absent |

| Composition | Basaltic | Rhyolitic | Basaltic (littoral)[c] |
|---|---|---|---|

B. *Hydroclastic eruptions*[b]

| | | | |
|---|---|---|---|
| Volcanic features associated with the ash | Maar volcanoes: tuff rings, tuff cones, explosion pits | Tuff ring with central dome | Littoral cone |
| Petrography | Vitric ash; angular sideromelane fragments, generally free of crystals except phenocrysts; lithic component of some ash depends on composition of basement rocks | Most ash particles are equant or elongate colorless glass; traces of rhyolitic lithic fragments; glass generally free of, or contains, very few microlites | Vitric to vitric-lithic ash: sideromelane droplets; tachylite, and fragments of aphanitic basalt |
| Morphology | Equant blocky glass shards with few vesicles. Smooth, flat fracture surfaces or scalloped where fractures intersect vesicles | Sharply pointed elongate shards and flat elongate pumice fragments. Conchoidal to irregular fracture surfaces; smooth vesicle walls | 1. Crystalline basalt, equant lithic fragments<br>2. Sideromelane grains with few vesicles; blocky or crescent-shaped; grain shape may be controlled by vesicle shape<br>3. Nonvesicular pyramidal glass fragments |

[a] Eruptions are caused by rapid gas release resulting in (1) large-scale ash eruption (Plinian) with resulting fallout tephra and pyroclastic flows; high-viscosity lavas; associated lahars or volcanic mudflows if the eruption is into a crater lake or is accompanied by rainstorms; and (2) lava fountaining and ash eruptions (Hawaiian and Strombolian types) of low-viscosity lavas

[b] Eruptions caused from steam explosions where rising magma comes in contact with surface water, ice or ground water

[c] Formed when lava flows into the sea

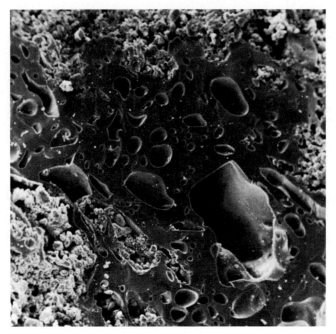

**Fig. 5-9.** Scanning electron photomicrograph of slightly collapsed pumice lapillus. Subrecent phonolitic ash flow (Mt. Suswa, Kenya). Diameter of lapillus 1 mm (Schmincke, 1974b)

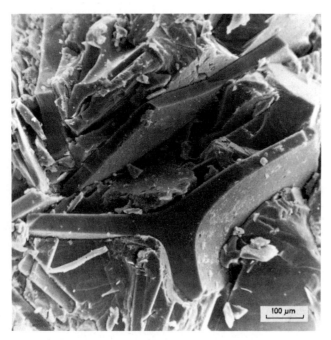

**Fig. 5-10.** Scanning electron photomicrograph of rhyolitic bubble junction shard in matrix of bubble wall shards from center of accretionary lapillus (compare Fig. 5-4). Pliocene Ellensburg Formation (Washington, USA)

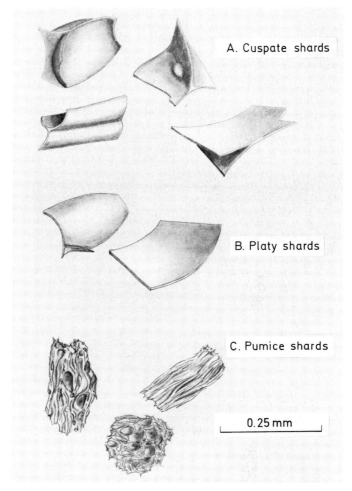

Fig. 5-11 A-C. Diagrammatic representation and terms for common glass shards

smaller pieces. Shards include smooth-skinned, slightly vesicular spheres, ovoids, teardrops, dumbbells and other attenuated shapes, commonly known as Pele's tears, up to several centimeters in diameter, and the broken shards of these droplets (Fig. 5-13). Similar shards occur on the moon (Heiken et. al, 1974). Highly drawn-out threads of glass known as Pele's hair attain lengths of up to a meter and commonly have circular cross sections between 15 µm and 500 µm (Heiken, 1974). Mafic shards are described by Wentworth (1938), Wentworth and Macdonald (1953), Macdonald (1967), Heiken (1972, 1974), Walker and Croasdale (1972), Duffield et al. (1977) and others.

**Fig. 5-12.** Bubble shards found floating at sea following the 1912 eruption of Novarupta (Katmai). Maximum length of shards is 1 mm (Perret, 1950, Fig. 89)

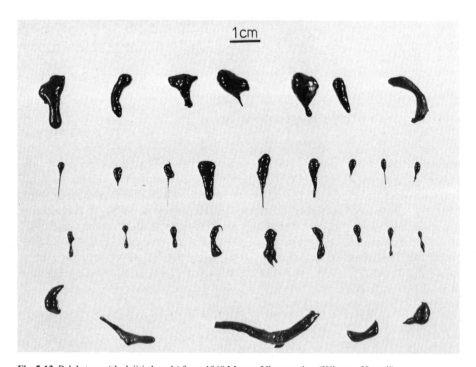

**Fig. 5-13.** Pele's tears (tholeiitic basalt) from 1969 Mauna Ulu eruption (Kilauea, Hawaii)

## Pumice

Pumice is composed of highly vesiculated volcanic glass (Fig. 5-14). The vitric ash-sized component of many tuffs consists dominantly of bubble-wall shards mixed with pumice shreds and lapilli (Fig. 5-15). Bubble-wall shards are mostly the broken bubble or vesicle walls of pumice.

Silicic pumice usually has a high porosity (up to 90%) and density of $<1.0$ g cm$^{-3}$, and low permeability, and therefore floats on water, a useful property for interpretations involving subaqueous deposits. Silicic pumice occurs as two end-member types: (1) fibrous fragments with tubular, subparallel vesicles and (2) fragments with spherical to subspherical vesicles. Fibrous varieties have vesicle length/diameter ratios $>20$. Distortion and stretching of vesicles occur during vesiculation, extrusion, and flowage, although the smallest vesicles tend to be spherical. Silicic pumice with spherical vesicles suggests conditions of high vapor pressure during eruption, whereas fibrous pumice typifies eruptions of lower pressures (Ewart, 1963).

Mafic (basaltic) pumice (or scoria) (Fig. 5-16) commonly contains small, spheroidal vesicles. Unless they are abundant, most vesicles are isolated without impinging upon one another. Unlike vesicles in silicic pumice, vesicles in basalt that touch do not indent neighboring walls, instead opening into one another with little distortion. The junctions of such vesicle intersections are cuspate, with points varying from sharp to highly rounded; relatively large vesicles have smoothly scalloped edges where several bubbles have joined. Rounding of cuspate bubble intersections is caused by liquid withdrawal due to surface tension. The relative isolation of vesicles in basalt glass as compared with vesicle abundance and spacing in silicic pumice is a function of the rapid freezing of the silicic melts upon vesiculation.

An unusual type of mafic pumice is reticulite ("threadlace scoria") (Dana, 1890, p. 163–166) with maximum porosity values between 98 and 99 (Wentworth, 1938). Reticulite consists of an open network of three-dimensional polygonal rings (diameters 0.25 to 2.0 mm) outlined by nonvesicular triangular glass rods and therefore readily sinks in water.

## Pyrogenic Minerals

Pyrogenic minerals consist of intratelluric phenocrysts and microlites that are deposited as whole or broken crystals enclosed by, or separated from, quenched magma (glass). The fact that pyroclastic material is cooled quickly – or is even quenched during eruption – allows the study of arrested states of crystallization better than is possible with more slowly cooled lavas. Moreover, liquidus phase compositions can be better assessed, and growth forms of minerals and resorption of phases that were out of equilibrium at surface pressures can be better studied in pyroclastic systems than in lavas. However, the size distribution and abundance of various mineral species attained within a magma easily become modified by sorting during eruption and dispersal.

Relatively free growth of minerals within a magma allows the development of euhedral crystals, a well-known characteristic useful for identifying pyroclastic materials in otherwise nonpyroclastic sediments (Ross, et al. 1928; Pettijohn et al., 1972). Characteristic minerals include bipyramidal quartz, euhedral pseudo-hexag-

**Fig. 5-14.** Photomicrograph of poorly sorted ash flow tuff showing highly vesicular phonolitic pumice lapilli/shards set in matrix of fine-grained vitric-lithic-crystal ash. Quaternary Laacher See "Trass" (Eifel, Germany)

**Fig. 5-15.** Photomicrograph of rhyolitic ash flow tuff showing fibrous pumice lapillus with dark rim of very fine-grained ash set in matrix of medium to fine-grained glass shards and feldspar and quartz phenocrysts. Top of Quaternary rhyolitic Bishop Tuff (California, USA)

Fig. 5-16. Highly vesicular pumice lapilli (ca. 1 cm in diameter) of tholeiitic basalt formed during lava fountaining of 1969 Mauna Ulu eruption. Kilauea Volcano (Hawaii, USA)

onal crystals of biotite, high temperature (disordered) forms of plagioclase and K-feldspar (sanidine), and euhedral olivine, pyroxene, and amphibole. The rapid decline of pressures during explosive eruption commonly causes breakage upon ejection or impact; thus, completely unbroken euhedral crystals of phenocryst size are relatively uncommon in fine-grained ash. Broken crystals are especially characteristic of ash-flow tuffs.

**Surface Textures of Minerals and Shards**

Minute pits and other imperfections, or glass crusts on otherwise smooth surfaces of pyrogenic minerals and glass shards, hold promise for interpreting some aspects of volcanism. For example, submicrometer pits resembling lunar impact on glass shards from deep sea sediments led Huang and Watkins (1976) to conclude that microfeature analysis of glass shards by scanning electron microscopy (SEM) is a promising means for determining relative volcanic explosivity of ancient volcanic eruptions. Glass rinds coating crystal surfaces may be smooth or scalloped with bubble impressions (bubble-wall texture, Fisher, 1963). These are highly characteristic of pyroclastic materials. The bond between the glass and crystal is strong enough so that rinds can survive the abrasive rigors of beach, sand dune, and river environments, as well as transport within pyroclastic flows. Glass rinds can only be effectively removed by chemical corrosion (Fisher, 1966d, p. 713). Meyer (1972) found that the amount of glass attached to non-reworked pyrogenic minerals is inversely proportional to eruptive violence; the abrasion index (% mineral/% glass) is greatest in the most voluminous tephra layers, which are presumed to be from eruptions of greatest violence (Fig. 5-17). Freundt (1982) found the abrasion index to be significantly higher in pyroclastic flow deposits than in interbedded fallout tephra.

Micropits in the micrometer to submicrometer size range are common on the surface of glass shards (Huang et al., 1980). Those due to chemical etching are generally spherical to ovoid. They are more nearly symmetrical and have triangular and other geometric shapes depending upon internal inhomogeneities.

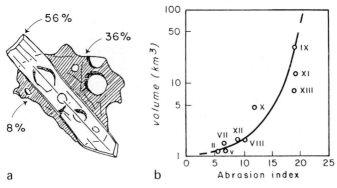

**Fig. 5-17 a, b.** Abrasion index. **a** Crystal with glass crust showing estimated areal percentages. **b** Abrasion index versus log of volume for 9 fallout layers from Coatepeque volcano (El Salvador). Distribution of the layers shown in Fig. 6-16. (After Meyer, 1972)

The amount of abrasion on glass fragments and crystals from both the pumice flow and pumice fall deposits of pre-historic Crater Lake (Mount Mazama), Oregon is similar. This suggests that a significant amount of attrition of the pyroclasts took place within the vent prior to ejection into the atmosphere (Fisher, 1963). Modern experimental and theoretical work on micropits on glass shards (Huang et al., 1980) provides quantitative evidence that pits (hence abrasion) form by collisions in the conduit system; the calculated velocity range to produce particles from a magma clot (from 10's up to 120 m s$^{-1}$) is similar to the range determined from experiments (25–140 m s$^{-1}$) to produce observed morphology and frequency (density) of pits.

Mechanical abrasion is indicated by pitting of the surface edges of glass shards which have unpitted re-entrants or interior vesicle walls. If, on the other hand, vesicle walls are as pitted as the grain surface and corners are not excessively rounded, chemical corrosion is most likely. Impact pits commonly occur on the broken surfaces and edges of rounded pyrogenic minerals; those grains rounded by resorption may have percussion pits, but the pits are unrelated to roundness. In some cases, highly rounded (resorbed) mineral grains with thin, chemically frosted glass coatings closely resemble grains that are rounded and frosted by wind action (Fisher, 1966d).

The surface textures on glassy particles also depend in part upon the viscosity of the magma. The surfaces of glass droplets from Hawaiian lava fountains are smooth or have dome-shaped blisters, circular depressions, and micro-craters caused by bursting bubbles, or they are spattered by smaller blebs of glass (Heiken et al., 1974). The surface markings of glassy particles from magmas of higher viscosities tend to be those that develop on broken vesicle walls. Such markings include concavities, troughs, and other irregularities.

SEM has been successfully used to relate surface textures of sand grains to mode of deposition and environment of deposition (Krinsley and Margolis, 1969), and can discriminate between glass shards emplaced by pyroclastic flow, pyroclastic surge and fallout processes (Sheridan and Marshall, 1983), and, in many instances, between pyroclasts of Hawaiian, Strombolian and Plinian origin (Heiken and Wohletz, 1984).

# Structure

Structural features useful for interpretation and field identification of pyroclastic rocks include their layered geometry, the relationship of layers to one another and to underlying deposits, internal features such as laminations and cross bedding, surface markings and others (Table 5-4). Features characteristic of particular kinds of deposits are discussed in separate chapters. Some of the descriptive and interpretive aspects of texture, especially grain-size parameters, are dealt with in the following section.

## Pyroclastic Bed or Stratum

A bed or stratum is a single layer visibly set apart from contiguous layers by distinctive structural, textural or compositional properties. A bed may be defined as

Table 5-4. Some structural features of pyroclastic deposits

| | Feature | Deposits in which they are characteristic |
|---|---|---|
| Overall geometry of deposit | *Areal distribution*<br>1. Fan-shape or lobate<br>2. Valley fill (shoestring shape) | Fallout and pyroclastic flow[a]<br>Pyroclastic flow |
| | *Vertical (cross sectional) distribution*<br>1. Wedge (in direction of transport)<br>2. Lensoid (perpendicular to transport)<br>3. Valley profile shape | Fallout and pyroclastic flow<br>Fallout and pyroclastic flow<br>Pyroclastic flow |
| Primary stratification (individual beds) | *Relations of upper and lower surfaces*<br>1. Flat top, base follows underlying surface<br>2. Top parallel to base | Pyroclastic flow<br><br>Fallout and plateau pyroclastic flow |
| | *Basal relationships*<br>1. Draping over or against obstacles<br>2. Structures in the lee of obstacles | Fallout<br>Pyroclastic flow |
| | *Internal structures*<br>1. Graded bedding<br>2. Cross bedding<br>3. Massive beds<br>4. Alignment and orientation bedding | Fallout and pyroclastic flow<br>Pyroclastic surge<br>Pyroclastic flow<br>Pyroclastic flow |
| | *Bed forms*<br>1. Plane bed<br>2. Antidune<br>3. Chute-and-pool | Pyroclastic surge<br>Pyroclastic surge<br>Pyroclastic surge |
| Post-depositional structures | *Upper surface structures*<br>1. Bedding sags<br>2. Convolute bedding<br>3. Load casts and bedding sags<br>4. Mudcracks<br>5. Rills | Base surge<br>Base surge<br>Base surge<br>Base surge<br>Base surge |

[a] The term pyroclastic flow is used here to include hot or cold pyroclastic flows and lahars

a layer greater than 1 cm thick; laminae are layers less than 1 cm thick (Table 5-5). A single bed may be internally structureless ("massive"), cross bedded, show progressive vertical grading, have internal continuous laminations, or show discontinuous planar or wavy elements such as trains of lapilli, blocks or oriented fragments. A single bed is distinct enough to be measured and described. The field definition of a bed may be straightforward if it is internally structureless or bounded by sharp bedding planes, but becomes subjective and dependent upon the scale of observation and judgment of the observer if boundaries are gradational or if the bed is composed of many texturally distinct smaller units. The causes of bedding and internal layering differ according to the mode of transport (fallout versus flow) and are discussed in the chapters of different kinds of pyroclastic deposits.

Table 5-5. Nomenclature and stratification thickness (Ingram, 1954)

| Name | Thickness |
| --- | --- |
| Very thickly bedded | > 1 m |
| Thickly bedded | 30–100 cm |
| Medium bedded | 10–30 cm |
| Thinly bedded | 3–10 cm |
| Very thinly bedded | 1–3 cm |
| Thickly laminated | 0.3–1 cm |
| Thinly laminated | <0.3 cm |

Descriptively, a sequence of beds with distinct internal structures, textures or compositions that set them apart from other beds above and below are *bedding sets* (Fig. 5-18). Two or more subsets may be defined within a single set if there is internal layering (beds or laminations). Different bedding sets develop in response to changing conditions at the site of eruption or during transportation, or both, and therefore have genetic significance. Co-sets are defined on the basis of larger-scale similarities and may include more than one set.

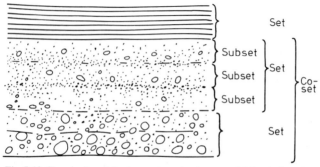

**Fig. 5-18.** Terminology of sets, cosets and subsets useful for designating essentially continuous vertical pyroclastic sequences

## Graded Bedding

Graded bedding is defined as the progressive vertical change in particle size (size grading) or fragment density (density grading) within a single bed. Density grading may occur without significant changes in particle size. The matrix and large clasts are commonly graded in the same direction, but in some beds, large clasts are graded but the matrix is not. Figure 5-19 illustrates several kinds of graded beds and also shows how subjective the recognition of bedding can be. In some cases (Fig. 5-19A, B), grading clearly develops from emplacement of a single flow or "pulse" of material, but in others (Fig. 5-19D, G, H) it can be argued that grading developed from more than one distinct depositional event or closely spaced "pulses" within a single event. In the case of G and H (Fig. 5-19), if grading (hence bedding) were less distinct than shown, which is often the case, argument for dividing the units into three distinct beds becomes less convincing. In many instances, graded beds within a single unit may disappear laterally into a single bed where separate depositional events are indistinguishable.

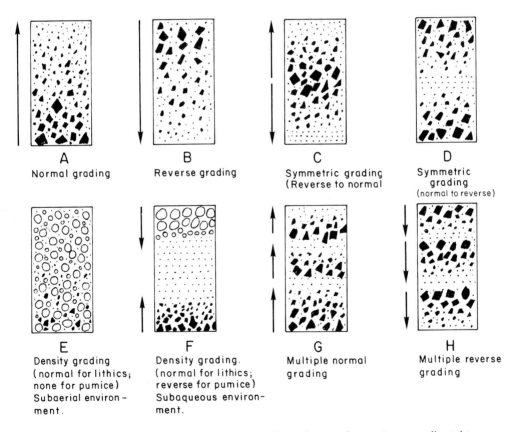

**Fig. 5-19 A–H.** Diagrammatic examples of graded bedding and nomenclature. *Arrows* are directed toward finer-grained sizes in sequence. *Open circles* pumice; *solid irregular forms* lithic fragments; *dots* ash-sized material. See text for explanation

*Normal grading* is the upward decline in grain size or density within a single bed (Fig. 5-19A). *Reverse grading* (also called inverse grading) is the upward increase in grain size or density (Fig. 5-19B), but it is commonly reverse only within a narrow zone at the base of a bed, reverting upward to normal grading. Such beds rarely show a totally systematic increase in fragment sizes from bottom to top. *Symmetric grading* may be reverse to normal or normal to reverse (Fig. 5-19C, D). If there is only one gradation within a bed, it shows *single grading* (Fig. 5-19A, B); *multiple grading* refers to more than one graded layer within a bed (Fig. 5-19G, H). Grading in fallout tephra is used to deduce settling rates and also magnitude, explosivity and duration of explosive eruptions (Chaps. 6, 7).

Density grading in fallout deposits containing pumice and lithic fragments are illustrated in Fig. 5-19E, F (also see Chap. 6). The presence of pumice in fallout deposits greatly aids in determining subaerial vs. subaqueous environments of deposition (Fiske, 1969), because density separation of pumice and lithic fragments in water is greatly accentuated (Chap. 7). Processes operating within pyroclastic flows results in a great variety of graded types (Chap. 8). Unlike aerially dispersed material, particle concentration in the flows is more important in determining structures and textures than the settling velocity of particles.

**Cross Bedding**

Cross bedding (and cross lamination) may be defined as internal stratification that lies at an angle to the contacts of the bed in which it occurs, or, on a larger scale, stratification of beds that are at an angle to the overall layering of several contiguous beds. Cross bedding develops by the movement of grains across a surface, swept along by currents of wind or water, to develop ripples and dunes of various sizes. In pyroclastic rocks, reworking by wind or water is common and we refer the reader to definitive works on cross bedding in sedimentary rocks such as McKee and Weir (1953) and Allen (1963, 1982) or to standard texts on sedimentary petrology (e.g. Pettijohn, 1975). Our concern here is with cross bedding that develops directly by volcanic action. Cross beds and ripple- or dune-like forms in pyroclastic surge deposits resemble those in other kinds of sedimentary deposits, but can be identified geologically by gradational contacts, lateral continuity and associations with pyroclastic layers that have not been reworked, by characteristic shape differences and by grain size parameters discussed in Chapters 8 and 9.

Cross bed geometry in pyroclastic beds is complex and commonly records the internal structures of asymmetrical mound-shaped dunes and ripples composed of ash- to lapilli-size tephra, in places accompanied by block-size debris. Internal unconformities which modify or destroy dune forms are common. Complexities may be simplified descriptively by organizing layers into sets, etc. A *set* is a single group of cross beds bounded by bedding planes; two or more sets form a *coset* (Fig. 5-20). Cross beds may approach the lower and/or upper boundaries of a set tangentially or at high angles. Cross beds that dip away from the source on the lee-side of dunes and ripples are *foreset* beds or laminations; those that dip toward the source on the stoss side are *backset* beds or laminations. In pyroclastic deposits, strata inclined to lower bedding surfaces also develop by draping of fallout tephra over surface irregularities. These may closely resemble cross beds, but genetically cross bedding is a current-induced structure.

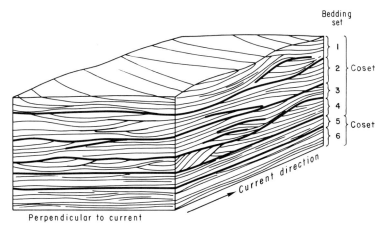

**Fig. 5-20.** Some terms for cross beds produced by pyroclastic surges (see Chaps. 8, 9). Bedding sets are manifested by some kind of physical boundary (color, texture, erosional). Sets may be divided into subsets; cosets are 2 or more sets (see Fig. 5-18). Internal laminae may be tangential, parallel or at angles to set boundaries. Individual sets may be tabular, wedge-shape, dune-shape, etc., as are individual laminations

The various kinds of cross beds in pyroclastic surge deposits may be related to regimes of flow, a concept originally developed for alluvial sands in flumes and rivers. Flow regimes are defined by sedimentary structures such as ripples, dunes and antidunes which are related to hydraulic conditions of flow – velocity (U), shear stress imparted to the bed by the flowing medium ($\tau_0$) and grain size (Simons and Richardson, 1961) (Fig. 5-21). Another variable that changes the configuration of flow regime fields is depth of the flowing medium (Southard, 1971).

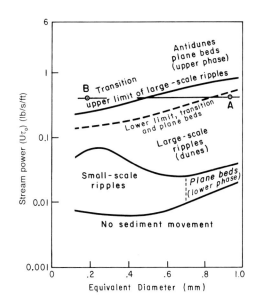

**Fig. 5-21.** Hydraulic criteria used to define fields for bedforms described in text. Line $AB$ indicates influence of grain size on bedforms developed at equivalent flow power: decreasing grain size increases the flow regime in the case indicated. $U$ velocity, $\tau_0$ shear stress. (After Blatt et al., 1972)

Ripples and dunes are classified as lower flow regime bed forms; transition plane (flat) beds and antidunes are upper flow regime bed forms. The term "bed form" was originally applied only to the surface configuration of the bed but is now extensively used to describe profiles or parts of profiles shown by laminations or beds that have geometric shapes similar to the profiles described in experimental studies. Under equilibrium conditions with net deposition or erosion equal, changing variables cause bed forms to change, and, as a current wanes, the higher flow regime bed forms revert to lower-order bed forms and are destroyed. Thus, high rates of sedimentation are required for high flow regime bed forms to be preserved in the sedimentary record. Additionally, the presence of cohesive sediment inhibits the interaction between fragments of the bed and the passing current which increases the likelihood of preservation. Research on flow regime bed forms has been concerned mainly with open channel flow, but high flow regime forms characteristic of flow in open channels can be duplicated with turbidity currents (Hand, 1974). Inasmuch as pyroclastic flows are one kind of density current, Hand's experiments may be applicable to them.

Ripples and dunes that form in the low flow regime are destroyed and the bed becomes flat in the transition to the high flow regime. If deposition occurs at this stage, internal laminations become parallel, and are called *plane-parallel* bed forms, even though subtle cross bedding features may be internally preserved. Plane-parallel beds closely associated with contiguous high flow regime cross bedded structures in pyroclastic surge deposits may closely resemble fallout tephra, but in some places they grade into cross bedded sequences and others may have very subtle cross laminations.

At higher regimes of flow, the flat bed becomes a wave or antidune which is roughly symmetric in cross section, and it may migrate upstream, downstream or

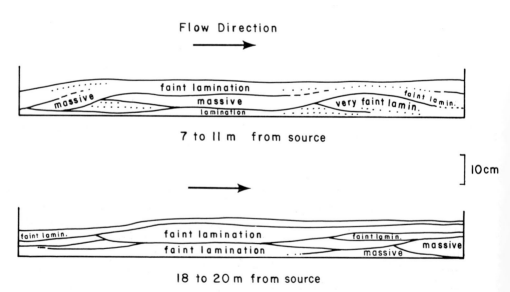

**Fig. 5-22.** Bedforms produced experimentally from water flowing over cohesionless sand within the antidune flow regime field (see Fig. 5-21). (After Middleton, 1965)

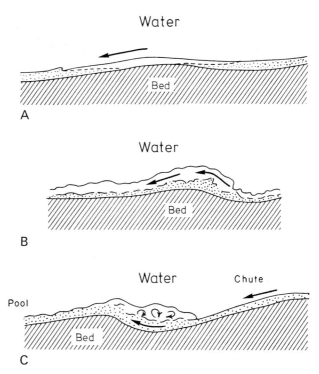

**Fig. 5-23 A–C.** Flow of density current over erodible bed. *Stipple* indicates suspended sediment. **A** Smooth flow over antidune. **B** Breaking antidune. **C** Chute-and-pool flow. (After Hand, 1974)

remain stationary (Gilbert, 1914; Simons and Richardson, 1961; Kennedy, 1961). The word antidune comes from the fact that *some* migrate upstream opposite to migration directions of ordinary dunes. Deposits with bed forms described as antidunes from ancient turbidites have profiles or internal laminations shaped like antidune bed forms that develop in flumes (Hand, 1969; Skipper, 1971). However, laminations that were deposited under high flow regime, open channel conditions in flume experiments by Middleton (1965), were discontinuous cross-cutting lenses that formed during the migration of bed materials, rather than showing preserved sand wave forms (Fig. 5-22). Most instructive are Hand's (1974) experiments with turbidity currents; the high flow regime bed forms were similar to many of the structures that occur in pyroclastic flow deposits at Laacher See, Germany (Schmincke et al., 1973).

Chute-and-pool flow also occurs in the upper flow regime (Simons and Richardson, 1961). Such flow consists of a long chute of water where the flow accelerates, becomes unstable and terminates in a turbulent zone. As the flow decelerates with a consequent increase in water depth, a pool forms within the downstream end of the turbulent zone (Fig. 5-23). The transition from chute to pool is known as a hydraulic jump. The transport capacity of the flow in the chute is high, but in the turbulent zone where depth increases and velocity decreases, the carrying capacity is reduced and deposition occurs. Steep backset laminations on dune-like structures are characteristic of chute-and-pool structures.

113

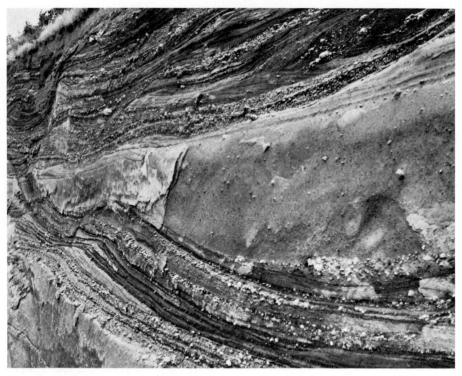

**Fig. 5-24.** Massive pyroclastic flow deposit over- and underlying cross bedded surge deposits. Flow levels out sand wave morphology. Basal pumice clasts are concentrated on downstream side of troughs between dunes. Transport direction from *left* to *right*. Flow deposit 1.5 m thick in valleys. Laacher See (Eifel, Germany)

## Massive Beds

The term "massive" is generally applied to thick beds without visible internal lamination or obvious manifestations of fragment organization such as grading or aligned fragments. Massive beds appear homogeneous and are commonly but not always poorly sorted (Fig. 5-24). Massive pyroclastic beds may consist entirely of ash-sized material. More commonly, they contain abundant lapilli and block-sized fragments scattered irregularly within a finer-grained matrix. Such coarse-grained deposits are said to be "chaotically sorted", and some authors use the term "unsorted", meaning that the emplacement process was unable to sort the material according to size, density or other variables. Strictly speaking, however, all aggregates have a sorting coefficient; thus, the term "unsorted" means "extremely poorly sorted". This kind of massive pyroclastic bed typically occurs as thick pyroclastic flows, lahars, and vent breccias (Fig. 5-25).

## Alignment and Orientation Bedding

The lining-up of isolated fragments or bands of fragments within an otherwise structureless bed may be termed *alignment bedding* (Fig. 5-26). Other terms used

**Fig. 5-25.** Coarse-grained poorly sorted laharic breccia flow showing concentration of large rocks about 1 m above base. Pliocene tephritic Roque Nublo Formation (hammer at *right* for scale). Gran Canaria (Canary Islands)

to describe alignment bedding in pyroclastic rocks are lithic or pumice "trains" or "swarms". Platy or tabular fragment swarms that are oriented approximately parallel to contacts (called an anisotropic fabric in the following section) impart an *orientation bedding* to otherwise massive layers. Both alignment bedding and pumice swarms commonly occur in otherwise massive pyroclastic flow deposits but are rarely continuous over long distances.

**Penecontemporaneous Deformation Structures**

Internal and external structures, such as contorted layers and brecciated beds, which formed prior to consolidation of the clastic material, are called penecontemporaneous deformation structures. They are distinct from structures formed by processes of deposition (e.g. cross bedding), by resedimentation from density currents, or from structures produced after lithification. They may be abundant in base surge and associated deposits and are discussed in Chapter 9.

**Fig. 5-26.** Alignment bedding in pyroclastic flow deposit developed from base surges, Ubehebe Craters, California. (From Crowe and Fisher, 1973, Fig. 9)

## Texture

Texture refers to the size, shape and fabric (pattern of arrangement) of the particles that form a rock. The size characteristics and aggregate properties of clastic rocks result from the processes that form the grains, and that bring them together in a deposit. The *absolute size* of individual pyroclasts is determined by explosive and vesiculation processes and the type and length of transportation, whereas the *size distribution* of an aggregate is determined mainly by how the particles were transported and deposited. *Shape and roundness* of particles are determined by abrasion and fragmentation during both extrusion and transportation. Initial shapes and sizes of pyrogenic minerals, however, also depend upon crystallization processes within the magma. *Fabric* in pyroclastic rocks is determined almost exclusively by depositional mechanics; but a major exception is the flattening of shards and pumice during sintering or welding of pyroclastic flow or fallout deposits. New fabrics are frequently imprinted over primary ones by the diagenetic crystallization of shards and pumice into clays, zeolites and opal.

### Grain Size and Size Distribution

The size of a particle may be defined by its length, breadth, thickness, and volume. Other ways to measure size, however, are needed for special purposes. The hydrodynamic characteristics, for example, are measured by fall velocity rather than by

**Fig. 5-27.** Plinian fallout pumice lapilli showing internal fractures in 15 cm pumice lapillus. Scale in cm. Quaternary phonolitic Laacher See tephra (Eifel, Germany)

sieve diameters, or length and breadth measured with a microscope or tape. Methods of measurement also must be chosen to bring out the degree of consolidation, and the kinds of fragments present in the tephra. Methods that depend on the settling of fragments in water, for example, obviously are not practical if pumice fragments are present because some will float and others sink (Fisher, 1965; Walker et al., 1971). Moreover, if pumice size is determined by sieving, care must be taken not to break delicate outer edges. Some workers hand-sieve samples that contain abundant pumice fragments mixed with abrasive lithic and mineral clasts to avoid excessive breakage. In Plinian fallout tephra, pumice fragments of greater than about lapilli size are commonly broken (Fig. 5-27), and size parameters of such deposits are generally minimum values. Techniques of size measurements, statistical parameters, and ways that data may be presented are thoroughly explored in many texts and reports on sedimentary rocks and so are not detailed here (see Folk, 1966; Pettijohn, 1975; Müller, 1967; Griffiths, 1967; Blatt et al., 1972; Füchtbauer, 1974).

Measurements of the areal distribution of maximum pyroclast sizes are used to (1) estimate eruption energy, (2) determine direction to source, or (3) gain insight about carrying capacity of the transport system. Size distribution measurements to determine sorting and other aggregate characteristics are used to interpret (1) the hydrodynamics of flow processes, (2) how particles are carried and deposited, and therefore (3) different kinds of transport processes, transport (and source) directions and environments of deposition. Vertical and lateral changes in composition are influenced by sorting processes, so that studies where composition is important (correlation, petrologic or chemical changes in magma and others) must take size parameters into account.

**Distribution Curves**

The size distribution of samples is commonly illustrated by cumulative frequency curves, or numerically by parameters of central tendency (median, mean or mode), or by other parameters that describe the shapes of the curve (sorting, skewness or kurtosis; see Pettijohn, 1975) (Table 5-6). Numerical parameters are graphically read from frequency curves or calculated directly as moment measures (Krumbein and Graybill, 1965). Cumulative curves are constructed by plotting cumulative weight percentages against grain diameter recorded in millimeters (Fig. 5-28) or in phi units (Table 5-7). Many sediments sorted during transportation develop a normal (Gaussian) distribution which typically plots as an S-shape on a cumulative curve and a straight line on normal probability paper. Subpopulations have been inferred with limited success from angular discontinuities on probability paper (Sheridan, 1971; Kittleman, 1964). Cumulative curves, tables or scatterplots of pyroclastic deposits may be compared and analyzed to show differences (Figs. 5-29 to 5-32).

Krumbein and Tisdel (1940) suggested that Rosin's size frequency distribution, developed to describe artifically crushed coal, is more suitable than the normal distribution to describe clastic material produced by mechanical disintegration or by volcanic explosion; Rosin size distributions plot as straight lines on special probability graph paper (Rosin law paper). Fallout tephra and pyroclastic debris worked by water or wind commonly follow a normal distribution; pyroclastic flows should more closely follow a Rosin distribution (Kittleman, 1964). Pyroclastic flow deposits (159 samples) examined by Murai (1961), however, plot closer to straight lines on normal probability paper than on Rosin law paper (Fig. 5-33), indicating closer affinities to a Gaussian distribution and suggesting that sorting took place during eruption and transportation. Which size distribution law better

Table 5-6. Descriptive measures of sediment size distribution. (From Blatt et al., 1972)

| Measure | Trask (1932) | Inman (1952)[a] | Folk and Ward (1957) |
|---|---|---|---|
| Median | $Md = P_{50}$[b] | $Md\phi = \phi_{50}$[c] | $Md\phi = \phi_{50}$ |
| Mean | $M = P_{25} + P_{75}/2$ | $M_\phi = \phi_{16} + \phi_{84}/2$ | $M_z = \phi_{16} + \phi_{50} + \phi_{84}/3$ |
| Sorting (dispersion measure) | $S_0 = P_{75}/P_{25}$ | $\sigma_\phi = \phi_{84} - \phi_{16}/2$ | $\sigma_I = (\phi_{84} - \phi_{16}/4) + (\phi_{95} - \phi_5/6)$ |
| Skewness | $Sk = P_{25} \cdot P_{75}/Md^2$ | $\alpha_\phi = M_\phi - Md_\phi/\sigma_\phi$<br>$\alpha_{2\phi} = \frac{1}{2}(\phi_5 + \phi_{95}) - Md/\sigma_\phi$ | $Sk_I = \phi_{16} + \phi_{84} - 2\phi_{50}/2(\phi_{84} -$<br>$+ \phi_5 + \phi_{95} - 2\phi_{50}/2(\phi_{95} -$ |
| Kurtosis | $K = P_{75} - P_{25}/2(P_{90} - P_{10})$ | $\beta_\phi = \frac{1}{2}(\phi_{95} - \phi_5) - \sigma_\phi/\sigma$ | $K_G = 0/_{95} - \phi_5/2.44(\phi_{75} - \phi_{25})$ |

[a] See Fig. 5-28
[b] P is a percentile measure in millimeters
[c] $\phi$ is a phi ($\phi$) percentile, $\phi = -\log_2 d$

**Table 5-7.** Terminology and class intervals for grade scales. (After Wentworth, 1922)

| U.S. Standard Sieve Mesh | Phi | mm | Wentworth (1922) | National Research Council[a] |
|---|---|---|---|---|
| | −12 | 4096 | Boulder gravel | VL boulders |
| | −11 | 2048 | | L boulders |
| | −10 | 1024 | | M boulders |
| | −9 | 512 | | S boulders |
| | −8 | 256 | Cobble gravel | L cobbles |
| | −7 | 128 | | S cobbles |
| | −6 | 64 | Pebble gravel | VC gravel |
| | −5 | 32 | | C gravel |
| | −4 | 16 | | M gravel |
| 5/16 | −3 | 8 | | F gravel |
| 5 | −2 | 4 | Granule gravel | VF gravel |
| 10 | −1 | 2 | VC sand | VC sand |
| 18 | 0 | 1 | C sand | C sand |
| 35 | 1 | 1/2 | M sand | M sand |
| 60 | 2 | 1/4 | F sand | F sand |
| 120 | 3 | 1/8 | VF sand | VF sand |
| 230 | 4 | 1/16 | Silt | C silt |
| | 5 | 1/32 | | M silt |
| | 6 | 1/64 | | F silt |
| | 7 | 1/128 | | VF silt |
| | 8 | 1/256 | Clay | C clay-size |
| | 9 | 1/512 | | M clay-size |
| | 10 | 1/1024 | | F clay-size |
| | 11 | 1/2048 | | VF clay-size |
| | 12 | 1/4096 | | |

[a] VL = very large, L = large, M = medium, S = small, VC = very coarse, C = coarse, F = fine, VF = very fine

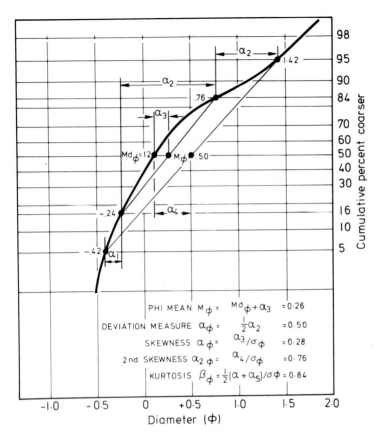

**Fig. 5-28.** Graphic illustration of descriptive measures of size-distribution obtained from a cumulative frequency curve. (After Inman, 1952)

describes most kinds of flow deposits, however, remains unanswered. Although the Rosin fit appears to be slightly better in some cases, the statistical curve-fits of both size distributions are good (Sheridan, 1971). Many workers prefer the normal probability plots over the Rosin distribution plots because they are easier to work with.

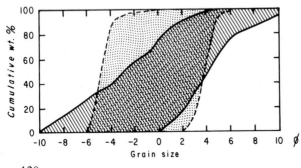

**Fig. 5-29.** Comparison of cumulative frequency curve fields for 38 fallout samples (*stippled*) and 202 pyroclastic flow samples (*crosshatched*) from Japan. (Data from Murai, 1961)

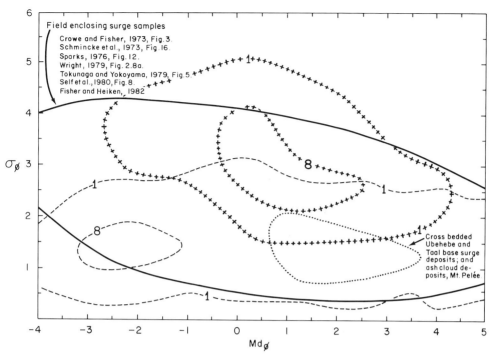

**Fig. 5-30.** $Md_\phi$–$\sigma_\phi$ diagram showing fields for fallout, flow and surge deposits. Fallout (*dashed lines*) and flow field (*crosses*) of Walker (1971) shown by 1% and 8% contours. *Solid line* encompasses surge samples from sources indicated. *Dotted line* encloses cross bedded surge deposits as indicated

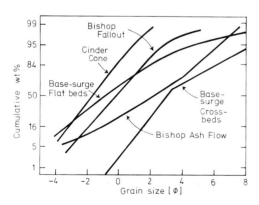

**Fig. 5-31.** Composite grain-size frequency distributions (log-normal plots on probability paper) of several kinds of pyroclastic deposits. (After Sheridan, 1971)

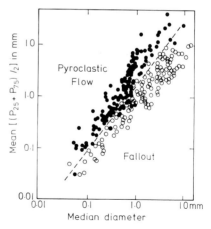

**Fig. 5-32.** Mean-median diameter plot showing boundary between pyroclastic flow and fallout deposits. (After Buller and McManus, 1973)

121

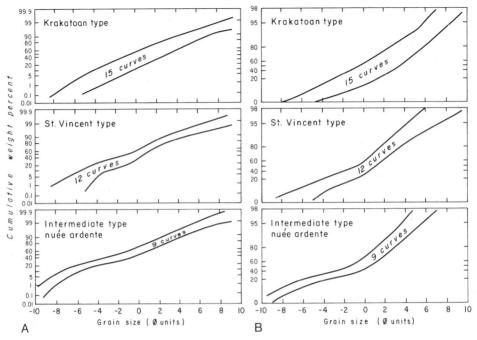

**Fig. 5-33 A, B.** Fields of cumulative grain size curves, pyroclastic flow deposits. **A** Normal probability plots compared with **B** Rosin Law plots of the same samples. (Classification and curves modified from Murai, 1961, Figs. 8 and 9)

Walker's $Md_\phi$-$\sigma_\phi$ diagram (Fig. 5-30) is in widespread use for discriminating flow and fall deposits. The original diagram, however, was constructed using data from many authors who use different sizing techniques; fallout samples of Walker's diagram include those deposited in different environments (subaerial and subaqueous); flow deposits include those deposited by different processes (surge, pyroclastic flow, debris flow). Moreover, $Md_\phi$ and $\sigma_\phi$ of samples depend upon distance from source, type of eruption, and mechanics of transportation and deposition. It is not surprising therefore that there is considerable overlap between the flow and fallout fields, and data collected since Walker (1971) presented his diagram show even more overlap. The field of pyroclastic surge deposits of different kinds (Fisher, 1979) broadly overlaps the fallout and pyroclastic flow fields (Fig. 5-30). The diagram generally shows that pyroclastic flow deposits are more poorly sorted than fallout deposits with pyroclastic surge deposits possibly being intermediate. It is clear that the $Md_\phi$-$\sigma_\phi$ diagram should be used with caution.

**Shape and Roundness**

The shape of particles is defined by the relative lengths of particle intercepts and by their angularity or roundness (Krumbein, 1941). Pyroclastic particles are shaped initially by the forces of explosion, vesiculation and surface tension and by drag effects on semi-molten juvenile liquid. For minerals, crystal habit is important. Many particles – accidental or cognate, and juvenile pumice, glass shards,

**Fig. 5-34.** Zingg's (1935) classification of particle shapes

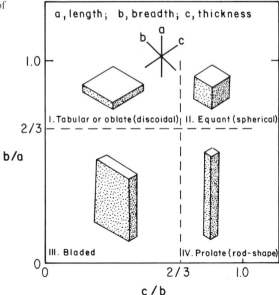

and pyrogenic minerals – are initially angular, and some may break upon impact, especially pumice. Original shapes of previously solid rocks comminuted by explosions are controlled by fractures and other planes of weakness. Pyroclastic fragments tend to be relatively equant or somewhat elongate (prolate), but platy or bladed pyroclastic fragments may be common, using Zingg's (1935) classification (Fig. 5-34). In terms of settling velocity of pyroclasts, however, it has been found that Zingg's shape classification is not a good expression of particle shape because irregular volcanic particles tumble as they fall in air (Wilson and Huang, 1979).

Fragment rounding is generally more important than shape for interpreting the history of transport, although shape partly controls the hydrodynamic behavior of particles during flow transportation. Some fragments become rounded during repeated ejection and fallback in vents, and others (liquid droplets) become rounded by surface tension. Pumice fragments in pyroclastic flow deposits are invariably better rounded than in fallout deposits; this is especially obvious where the two kinds of deposits occur together. Rounding of pumice lumps in the Mount Mazama pumice flow (Crater Lake, Oregon) increases farther from the source (Williams, 1942). Attrition of pumice fragments during flowage may partly account for both the rounding of fragments and the high proportion of fine-grained ash in pyroclastic flow deposits. Murai (1961) shows that pumice fragments within pyroclastic flow deposits are commonly better rounded than lithic fragments in the same ash flow deposit, with differences in roundness not more than about 0.1 (roundness values of Wadell, 1932).

**Fabric**

Fabric is defined as the orientation, or lack of orientation, of the fragments that comprise a clastic rock (Pettijohn, 1975). An *apposition* fabric is one that formed at the time of deposition; a *deformation* fabric, exemplified in metamorphic rocks, is produced by external stress on the rock. If the orientation of the fabric elements is random, the fabric is *isotropic;* if fragments have a preferred orientation, the fabric is *anisotropic*. A different aspect of fabric is *packing* – the spacing or density pattern of the constituent fragments. The manner by which grains are packed controls the density, porosity, and permeability of a sediment. Discussions of packing may be found in Pettijohn (1975), Johansson (1965), and Blatt et al. (1972).

The orientation of fragments in pyroclastic rocks also includes *dimensional* and *magnetic* orientation. Dimensional orientation is determined by particle shape. Spherical particles, for example, cannot be dimensionally oriented although they may be strung out parallel to bedding. Dimensional orientation is also achieved where lava or "mud" bombs and accretionary lapilli become flattened upon impact, or in welded or sintered tuffs where shards and pumice deform by compaction and welding at high temperatures. Individual fragments that come to rest at temperatures greater than the Curie point of the contained ferromagnetic minerals develop a preferred orientation of remnant magnetism (magnetic orientation). Thus, magnetic orientation or lack of it is a useful tool for distinguishing hot pyroclastic flow deposits or near-vent agglutinates of welded spatter from ash falls or pumice-rich lahars deposited under lower temperatures (Aramaki and Akimoto, 1957; Chadwick, 1971; Crandell, 1971; Crandell and Mullineaux, 1973).

## *Chapter 6* Subaerial Fallout Tephra

The transport modes of subaerial fallout tephra are (1) by ballistic trajectory and (2) by turbulent suspension. Energy is supplied initially to fragments by the eruption and later by wind. Tephra that falls from the atmosphere onto land is called *subaerial fallout* or airfall tephra. Tephra deposited in standing water is called *subaqueous fallout* tephra and includes (1) submarine or marine tephra, and (2) sublacustrine tephra. Subaqueous fallout tephra originating from eruptions on land is discussed separately (Chap. 7) from pyroclastic materials originating wholly beneath water (Chap. 10). Island arc and oceanic island flank environments generally include both kinds.

Parameters useful for identifying the airfall process are rock associations, geometry of an entire deposit, and the sorting and grading in individual layers (Table 6-1). Field mapping of lateral distribution, isopach maps and size parameters, in particular maximum sizes of clasts, are especially useful. Such data are fundamental for further calculations regarding eruptive centers, ancient wind patterns, volume and shape of magma chambers, vent radius, dynamics and height of eruption columns, terminal velocity of ballistic fragments, and eruptive energy.

Wide-spread ash layers are important stratigraphic marker horizons because they are deposited within a short period of time. The fields of tephrochronology and tephrostratigraphy are concerned chiefly with dating and correlating geologically young ash deposits (Chap. 13). Airfall tephra also is increasingly studied to analyze eruptive processes. We concentrate here on some of the descriptive parameters of tephra sheets and their geologic and volcanologic interpretations.

Vertically directed subaerial eruptions are characterized by the formation of an eruption column consisting of pyroclasts, expanding gases, and entrained air. Size and density sorting of particles occurs within the column, because particles of least settling velocity are carried to greater heights and spread over greater distances than those with large settling velocities. Many eruption columns consist of a lower gas-thrust region, and an upper convective-thrust region (Chap. 4) in which thermal energy converts to mechanical energy, thus developing an expanding turbulent column (eruption cloud) which may rise to great heights. Eruption clouds are commonly pushed laterally *en masse* by the wind (Fig. 6-1), but individual fragments within the cloud are suspended by turbulence. Some eruptions, however, show multiple behavior, for example the May 18, 1980 Mount St. Helens eruption with a directed blast as well as vertical columns which produced complex patterns of ash deposition (Waitt and Dzurisin, 1981).

Three dominant threshold settling velocity values for a given set of conditions affect the distribution and sorting of airborne tephra: (1) fragments with large settling velocities follow ballistic trajectories that are little affected by wind and only

**Table 6-1.** Characteristics of subaerial fallout deposits

*Comment:* The characteristics of subaerial fallout deposits are highly variable depending upon the kind of eruption (e.g. Strombolian, Plinian, etc.), changes in eruptive style during an eruption, composition, and distance to source. Some of the features listed below are generally applicable, but others can only be used for certain eruptive types.

*Distribution (fallout pattern) and thickness*

Circular or fan-shaped (regular to irregular) distribution with respect to source. Secondary thickness maxima may occur far downwind.

Flat wedges that decrease systematically in thickness along fan axes. Some have displaced or multiple thickness maxima

Thickness may be skewed to one side, perpendicular to fan axis.

Azimuth of fan axis may change with distance from source.

Apex of fan axis may not be on volcano (e.g., Mount St. Helens).

*Structures*

Plane-parallel beds drape over gentle topography and minor surface irregularities. Ash layers wedge out against cliffs or other steep surface irregularities.

Laminations and thicker beds reflect compositional changes or textural changes; either of these may cause overall color changes.

Minor lenticularity may occur close to source.

Grading may be both normal and reverse in various combinations depending upon variations in wind and/or eruption energy, vent radius or eruption column density.

Reverse grading in beds on cinder cones and other steep slopes commonly develops by downslope rolling or sliding of dry granular material.

Fabric in beds is commonly isotropic because elongate fragments are uncommon. Exceptions: phenocrysts such as biotite, amphibole, etc. and platy shards.

Bedding planes may be sharp if there are abrupt changes in eruptive conditions, wind energy or directions, or in composition.

Bedding planes are distinct where deposition is on weathered or erosional surfaces, or different rock types. May be gradational if deposition is slow by small increments so that bioturbation and other soil-forming processes dominate, and also if deposition is fast whereby small and large particles rain down continuously from a fluctuating eruption column.

*Textures*

Sorting: moderate to good: Inman parameters, $\sigma_\phi$, 1.0 to 2.0 most common. This applies to relatively coarse-grained as well as to fine-grained tephra.

Median diameter, $Md_\phi$: Highly variable depending upon type of volcanism and distance to source. $Md_\phi$ commonly $-1.0$ to $-3.0$ (2 mm to 8 mm) or smaller (phi values) close to source, but farther from source, $Md_\phi$ may vary from 0.0 (1 mm) to 3.0 (1/8 mm) or more.

Size and sorting parameters vary geometrically with distance in single layers.

*Composition*

Any composition, but silicic or intermediate fallout more widespread than mafic fallout, due to usually greater explosivity and discharge rate of the eruptions.

Intermediate composition is commonly associated with large composite volcanoes.

Mafic composition commonly associated with scoria cones and lava flows.

Bulk composition generally becomes slightly more silicic away from source due to eolian fractionation.

*Rock Associations and Facies*

Close to source (within vent or on steep volcano slopes): lava flows, pyroclastic flows, domes, pyroclastic tuff breccias, avalanche deposits and debris flows.

Intermediate to source: coarse-grained tephra, some lava flows, pyroclastic flows, ash falls and reworked fluvial deposits. The coarser-grained pyroclastic deposits gradually decrease, and reworked pyroclastic deposits gradually increase away from source.

Far from source: Fallout tephra, most easily recognized in marshy, lacustrine, wind blown environments. Rock associations depend on environment of deposition; no coeval coarse-grained pyroclastic rocks or lava flows.

**Fig. 6-1.** Eruption of Mt. Asama, Japan (1:09 p.m., December 5, 1958) illustrating fallout of particles from a turbulent eruption column. The turbulent column is being pushed *en masse* by the wind. (Courtesy of I. Murai)

slightly by the expanding eruption cloud, (2) particles suspended by turbulence in the eruption cloud but which are too heavy to be suspended by atmospheric winds and (3) those light enough to be suspended by wind independently of the eruption cloud. Wind may modify the trajectories of all but the largest fragments; fragments held in turbulent suspension within an eruption cloud begin to fall according to their settling velocities as the energy within the cloud dissipates; fragments with fall velocities that are small compared to wind strength may circle the earth many times before settling (Lamb, 1970). It is becoming increasingly clear, however, that the very small particles commonly become agglutinated by moisture in eruption clouds to form accretionary lapilli which are larger and heavier than the individual particles that make up the lapilli. Thus, large volumes of very small particles may fall out prematurely (Brazier et al., 1982; Carey and Sigurdsson, 1982). Wide dispersal of airborne tephra is strongly dependent upon wind vectors at different altitudes (Baak, 1949; Waitt and Dzurisin, 1981).

The sorting of tephra deposits either by size or density is rarely perfect. The finest of particles, for example, may (1) fall as accretionary lapilli along with large particles, (2) come down in rain drops ("rain flushing"), (3) become trapped within a densely crowded fall of coarse-grained ash and lapilli or (4) at long distances, form porous clusters by mechanical interlocking and electrostatic attraction which fall like snow flakes (Sorem, 1982). Thus, large and small particles may be deposited together within any single layer close to the source. With the exception of

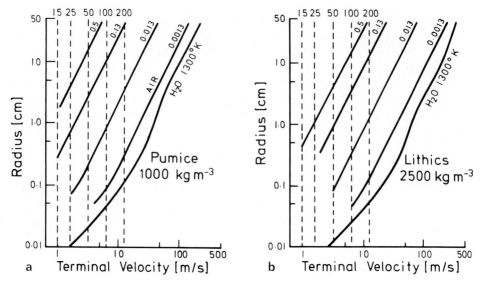

**Fig. 6-2 a, b.** Terminal velocities of **a** pumice and **b** lithic particles of different sizes (to 50 cm) in fluids of different densities. *Lower two curves* are for air at room temperature and steam at 1,300 K. *Vertical dashed lines* represent shear stress velocities of 15 to 200 m s$^{-1}$ and a drag coefficient of 0.01 to define particle sizes that can be suspended in pyroclastic flows at specific velocities. (After Sparks et al., 1978)

aberrations caused by premature fallout from particle agglutination, however, maximum particle sizes and sorting coefficients decrease in a general way as distance from the source increases.

The fall velocities of large ballistic fragments in air have been investigated by Minakami (1942), Fudali and Melson (1972), Chouet et al. (1974) and others; those of smaller fragments by Walker et al. (1971), Wilson (1972) and Wilson and Huang (1979) in air, and in water by Fisher (1965) (Chap. 7). Graphs useful for plotting terminal settling velocities of pumice and lithic fragments in air are presented by Sparks et al. (1978) (Fig. 6-2). Settling velocity data derived from grain size data has been used to construct isopleth maps to estimate eruption column heights and mean wind velocity, and to estimate intensity (mass emission rate of material) as determined from grading characteristics of Plinian deposits (Walker, 1980, Fig. 2 and p. 77, 81).

## Components of Subaerial Fallout

Three main types of tephra components must be distinguished when considering size-distance parameters and therefore compositional variations in tephra (Walker, 1971): crystals, lithic or dense vitric fragments, and pumice – including glass shards (Chap. 5). These different components generally occur in different but overlapping grain size ranges leading to pronounced lateral changes in the overall composition of an ash layer (Walker and Croasdale, 1971; Lirer et al., 1973; Booth et

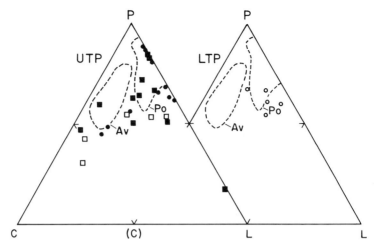

**Fig. 6-3.** Ternary diagrams of pumice (*P*), crystals (*C*) and lithics (*L*) for Upper Toluca Pumice (*UTP*) and Lower Toluca Pumice (*LTP*), Mexico. *Solid circles* on UTP triangle represent the Lower Member; *open squares* and *solid squares* represent the different stratigraphic parts of the Upper Member. *Dashed lines* enclose fields of Avellino (*Av*) and Pompeii (*Po*) pumice deposits of Somma-Vesuvius (Lirer et al. 1973). (After Bloomfield et al., 1977)

**Fig. 6-4.** Basaltic Plinian fallout lapilli forming a relatively homogeneous bed 1 m thick, becoming bedded in its upper part. Late Quarternary scoria cone (Eifel, Germany). (From Schmincke 1977a)

al. 1978). Tephra from Plinian eruptions may be characterized from ternary diagrams (Fig. 6-3).

We distinguish here between basaltic and felsic composition of essential tephra particles to contrast end member varieties in a rather simplified scheme. Close to the source, basaltic fallout tephra deposits consist of vesicular to dense lapilli with minor ash and highly variable proportions of bombs and blocks (Figs. 6-4, 6-5) Toward the vent, they become increasingly welded and may grade into agglutinates (Fig. 6-6). An origin by phreatomagmatic processes is likely if the tephra includes large amounts of lithic components associated with abundant fines, an increase in sorting and pronounced bedding.

Felsic Plinian fallout deposits are much more widely dispersed than those of basaltic composition. Angular pumice lapilli (Fig. 6-7) interspersed with lithic fragments occur in discrete layers or with gradational contacts (Fig. 6-8). Bedding is defined by changes in grain size and composition. A well-sorted matrix of ash-sized crystals, pumice and shards generally forms a minor part of such deposits. Felsitic tephra deposits with overall much finer grain size than pumice-rich Plinian tephra may be attributed to the interaction of water and silicic magma (phreatoplinian) (Self and Sparks, 1978). Most unusual are carbonatite tuffs (Dawson, 1962; Hay, 1978; Keller, 1981). Some resemble limestone beds and calcrete of nonvolcanic origin.

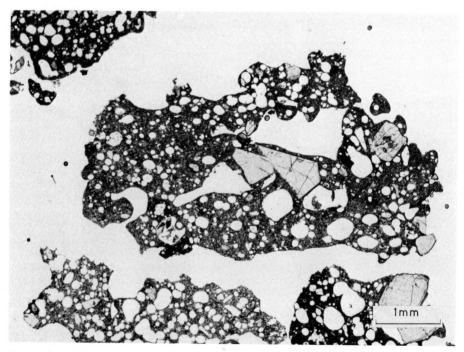

**Fig. 6-5.** Photomicrograph of lapilli of Fig. 6-4. Note irregular shape of lapilli with round smooth edges and clinopyroxene phenocrysts

**Fig. 6-6.** Basaltic fallout spatter (agglutinates) at crater rim of Vesuvius (Italy)

**Fig. 6-7.** Trachytic fallout pumice deposit showing very angular lapilli (3 to 5 cm in diameter) and absence of fine matrix. Fogo volcano (São Miguel, Azores)

**Fig. 6-8.** Phonolitic Plinian fallout tephra showing enrichment in Devonian slate fragments below surge deposits. Quaternary Laacher See tephra (Eifel, Germany). Scale in 10-cm intervals

## Areal Distribution

### Distribution and Thickness

Fallout tephra sheets are distributed in two end-member patterns with many variations; circular patterns form from rather low eruption columns during calm winds, and elliptical or fan-shaped patterns from high eruption columns that encounter strong unidirectional winds. The most characteristic pattern of single tephra sheets is fan-shaped (Eaton, 1964) with the apex at or near the source. Multicomponent sheets radiate from a common source or overlap with slightly different distribution trends, depending on shifting wind directions (Fig. 6-9). Contemporaneous sheets can be distributed in opposite directions by contrasting winds at different altitudes (Fig. 6-10). Unusually powerful eruptions give rise to sheets easily recognizable 1000 km or more from their source (Fig. 6-11). The ash from such eruptions may be dispersed world-wide, extending far beyond the recognized limits of the dispersal fan, as was illustrated by the May 18, 1980 eruption of Mount St. Helens, Washington (Lipman and Mullineaux, 1981).

Fallout tephra sheets tend to be gentle wedges that systematically thin along an axis away from the source, and are best defined by an isopach map constructed from numerous thickness measurements. In detail, differential compaction, contemporaneous erosion and local topographic irregularities may introduce errors (Aramaki, 1963; Nairn, 1972; Walker, 1980). The source is inferred to lie within

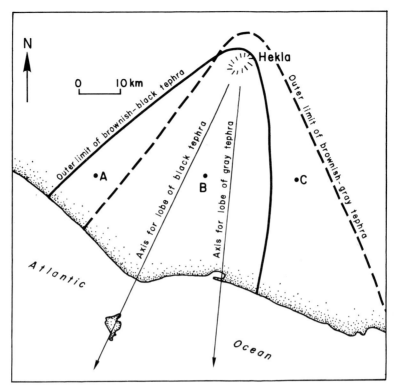

**Fig. 6-9.** Black and gray tephra fans from 1947 Hekla eruption deposited one-half hour apart. The layer at point *A* is dark andesitic tephra and at *C* it is light-colored and dacitic. The andesitic tephra overlies dacitic tephra at *B*. (After Thorarinsson, 1954)

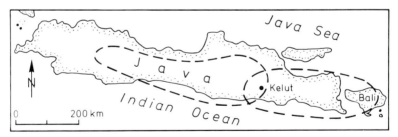

**Fig. 6-10.** Map of Java showing distribution of fallout tephra from the 1919 eruption of Kelut. (After Wilcox, 1959)

the maximum thickness contour, but caution must be exercized because the maximum might be displaced downwind (Walker, 1980; Sarna-Wojcicki and Shipley et al., 1981; Waitt and Dzurisin, 1981), or more than one maximum may occur within a single sheet (Fig. 6-12) (Larsson, 1937; Bogaard, 1983). Recent data suggest that displaced maxima may be more common than previously supposed, and that de-

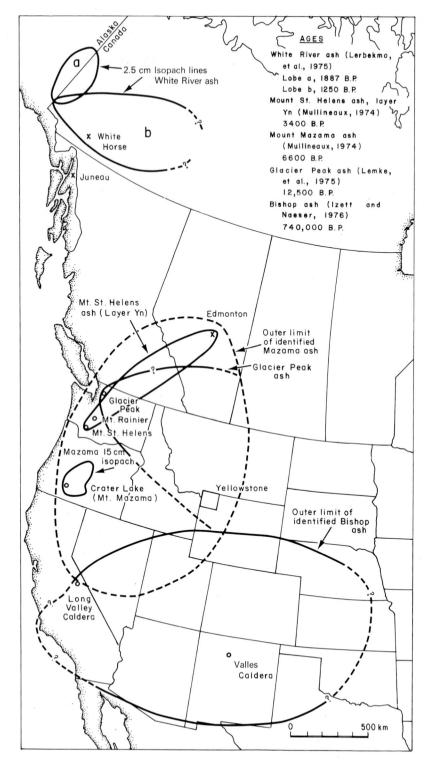

Fig. 6-11

termining tephra volumes based upon the near-source exponential thinning of deposits to large distances must be questioned (Brazier et al., 1983).

With fan-shaped sheets, source may be inferred by projection of the axis but must be done with care because fan axes may curve, as illustrated at Laacher See, Germany (Bogaard, 1983), where fan axes bend at about 3 km from source (Fig. 6-13). Sections at right angles to the axes of elliptical fans may be steeper on one side than on the other (Suzuki et al., 1973). Asymmetry of single tephra sheets is caused by different wind velocities and directions at different altitudes imposed upon an eruption column of fluctuating heights (Fisher, 1964b).

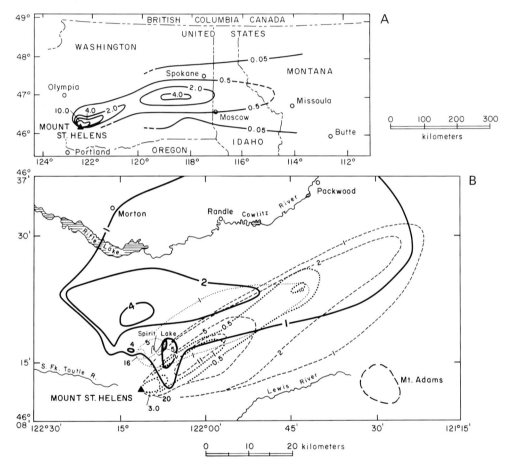

**Fig. 6-12 A, B.** Thickness distribution (in cm) of May 18, 1980 fallout ash from Mount St. Helens. **A** Distal ash, **B** proximal ash. *Solid line:* unit A; *dashed line:* unit B; *small dots:* unit C; *large dots:* unit D. See text for explanation. Contours in cm. (Sarna-Wojcicki and Shipley et al., 1981; Waitt and Dzurisin, 1981)

◀ **Fig. 6-11.** Five widespread western North American tephra layers <0.7 m.y. old. Outer limit of Mazama ash compared to area of main tephra fan illustrates difficulties in estimating volume; volume estimates of Mazama ash (Table 6-3) exclude fine-grained tephra distributed worldwide. Map is a Chamberlin Trimetric projection; major state or province boundaries, and Alaska-Canada boundary extend north-south

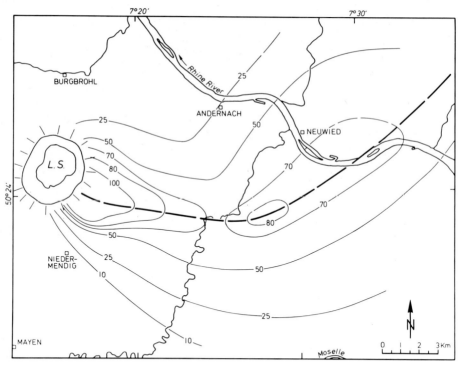

**Fig. 6-13.** Isopach map of Quaternary Laacher See fallout tephra (layer MLST C1) showing change in direction of pattern 5 km from source. Thickness in cm. (After Bogaard, 1983)

Further complexities of fallout deposition are illustrated by the May 18, 1980 eruption of Mount St. Helens, Washington. In a 12 h period, four main fallout units of different origins were deposited within a few tens of kilometers from the volcano (Fig. 6-12) (Waitt and Dzurisin, 1981). Unit A, the lowermost unit, is a basal gray lithic ash containing abundant organic material which is derived from the directed initial pyroclastic surge (terminology of Moore and Sisson, 1981), thus its thickness distribution is an irregular fan that does not have an apex at the volcano's crater. Unit A is composed of three layers, the upper one of which (A3) extends 50 km to the north and hundreds of kilometers to the northeast. Moreover, according to Waitt and Dzurisin (1981), the downwind axis of thickness of layer A3 has its apex 15 km north of the volcano within the 4-cm contour area (Fig. 6-12). Unit B, next in succession, is composed of tan, pumice, and lithic lapilli derived from vertical eruption plumes which came after the directed pyroclastic surge and shows a regular pyroclastic fan shape with its apex at the crater. Unit C is pale-brown vitric ash which has a fan-shape distribution but its apex is about 9 km north of the crater. This unit was derived from hydroclastic explosions through a hot pyroclastic flow deposit near Spirit Lake but also may in part be derived from fallout elutriated from the pyroclastic flow during its emplacement. Unit D, the uppermost and thinnest unit, is fine gray ash from the settling of fines left in the atmosphere after the events of the day, but this unit also shows a fan-shaped distribution with its apex at the crater.

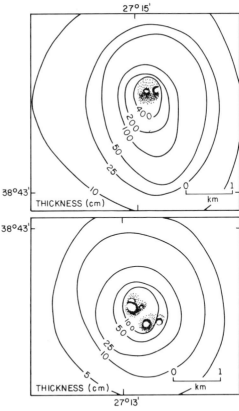

Fig. 6-14. Isopach maps of two small basaltic scoria cones (Terceira, Azores). (After Self, 1976)

Nearly circular isopach patterns are a common feature of low intensity basaltic eruptions (Fig. 6-14) that have low eruption columns, but individual basaltic fallout layers may be asymmetrical. Even Plinian fallout deposits may have circular patterns as at Fogo A (Azores) (Walker and Croasdale, 1971). Isopach maps may mask the existence of several eruptive events if composite sheets that form cones are not distinguished from single sheets. Cones are composed of several radiating overlapping fans within which may be a single mappable fallout unit of particular interest (Fig. 6-15). An isopach map of the cone may be a slightly elongate composite (average directions of dominant prevailing winds) of many tephra blankets. An example is the 9-year eruption of Paricutin Volcano, Mexico (Segerstrom, 1950), in which dry-season winds were generally from the east and wet-season winds from the west. Overlapping but mappable tephra fans (Fig. 6-16) from Coatepeque Volcano (El Salvador) formed the basic data used by Meyer (1972) to calculate eruptive energy (Table 6-2).

The classification scheme proposed by Walker (1973) for different eruption types (Chap. 4) relies in part upon determining the maximum thickness ($T_{max}$) of tephra sheets at their source. This is rarely measurable and therefore is commonly extrapolated graphically by assuming an exponential increase, but whether or not thickness increases exponentially is not clear. Thorarinsson (1954), for example, showed that thickness (T) is an exponential function of distance (x) from the source ($T = ae^{-kx}$; straight line on semi-log plot), but Porter (1973) shows that it is a

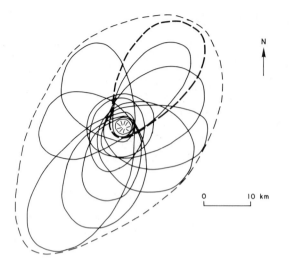

**Fig. 6-15.** Hypothetical distribution pattern of individual tephra fans distributed around a central vent. *Heavy dashed line* is a mappable fall unit. The ovate composite form of the tephra pile shown by *outer dashed line* reflects dominant annual wind direction

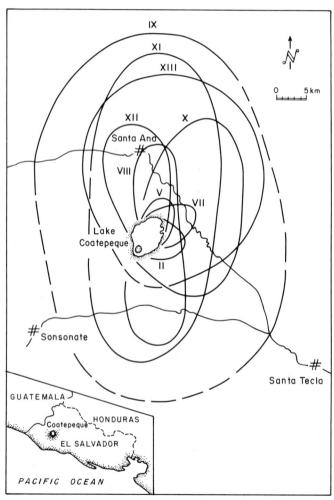

**Fig. 6-16.** Three-meter isopach lines of nine tephra layers, Coatepeque volcano (II, V, VII, VIII, IX, X, XI, XII, XIII) as designated by Meyer (1972)

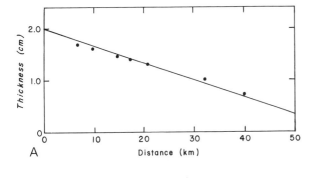
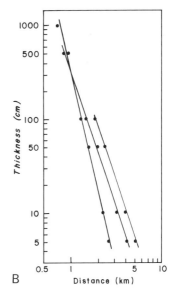

Fig. 6-17. **A** Distance/thickness plot, 1947 Hekla ash (Thorarinsson, 1954); **B** log-log distance/thickness plot, three Holocene Hawaiian ash layers (Porter, 1973)

power function (T = $ax^{-k}$; straight line on log-log plot). The fact that Thorarinsson's (1954) data were from 2 to 40 km and Porter's (1973) from three tephra sheets 1 to 4 km from source (Fig. 6-17), suggests that the change, if it indeed exists, could be caused by a change from a dominantly ballistic transport mode to a dominantly turbulent transport mode. The thickness of Mazama ash also decreases exponentially with distance (Williams and Goles, 1968), but at about 90–100 km from source there is a marked change in slope (Fig. 6-18). Williams and Goles (1968) suggest that the tephra was carried mainly by turbulence in the eruption cloud out to about 100 km, but at greater distances it was carried by normal wind patterns. Sparks and Walker (1977), however, argue that this "fine tail" is a co-ignimbrite ashfall (Chap. 8) and not related to the Plinian fallout layer. The Laacher See, Germany, tephra layers show similar breaks in slope, the main northeast lobe at about 100 km from source and a minor southern lobe at about 10 km from source (Bogaard, 1983). At Laacher See, the volume of pyroclastic flows compared to fall-out tephra is much too small to have generated such large amounts of co-ignimbrite ash, supporting a change in dispersal mechanism proposed by Williams and Goles (1968).

## Volume

Volumes of recent and prehistoric fallout deposits (Tables 6-2 and 6-3) are easily estimated with isopach maps, but many estimates of witnessed eruptions have been made without benefit of thickness measurements and thus are suspect. Early volume estimates of the 1935 Coseguina (Nicaragua) eruption, for example, varied from 1 km³ to 150 km³ and were later revised to 10 km³ or less (Williams, 1952). Volume data have been used to estimate volumes of magma chambers (Smith, 1979) and are used to calculate eruption energy and to construct intensity and magnitude scales of eruption (Tsuya, 1955; Yokoyama, 1957a; Hedervari, 1963; Izett,

**Table 6-2.** Area and volume of nine tephra fans within the 3-meter isopach, Coatepeque Volcano, El Salvador (Meyer, 1972)

| Area (km$^2$) | Volume (km$^3$) |
|---|---|
| 42 | 1.2 |
| 43 | 1.2 |
| 59 | 1.5 |
| 166 | 1.7 |
| 201 | 1.7 |
| 288 | 4.7 |
| 532 | 7.9 |
| 718 | 13.1 |
| 1,388 | 32.0 |

**Table 6-3.** Areal and volume estimates of tephra sheets determined from isopach maps

| Location and age | Area (km$^2$) | Volume of fallout deposits (km$^3$) | Dense-rock equivalent (km$^3$) |
|---|---|---|---|
| Nevado de Toluca volcano, Mexico | | | |
| 1. Upper Toluca Pumice, 11,600 yrs B.P. | 2,000 (within the 40 cm isopach) | 2.3 | 1.5 |
| 2. Lower Toluca Pumice, 24,500 yrs B.P. (Bloomfield et al., 1977) | 400 (within the 10 cm isopach) | 0.33 | 0.16 |
| Laacher See volcano 11,000 B.P. | | | |
| NE-Lobe | 170,000 | 12 | 3.3 |
| S-Lobe | 50,000 | 3.5 | 1.1 |
| SW-Lobe | 5,000 (within the 0.5 cm isopach) | 0.9 (?) | 0.3 (?) |
| (Bogaard, 1983) | | | |
| Mount Mazama, Oregon, 6,600 yrs B.P. (Williams and Goles, 1968, Fryxell, 1965) | 900,000 | 29–37 | |
| White River Ash, Yukon Territory North Lobe, 1887 yrs B.P. East Lobe, 1250 yrs B.P. (Lerbekmo et al., 1975) | ~320,000 | 25 | |
| Pompeii, Italy, 79 A.D. (Lirer et al., 1973) | | 2.6 (est.) (only 0.80 deposited on land) | 0.53 |
| Santa Maria, Guatemala, 1902 (Sapper, 1904, Rose, 1972) | 150 (within the 1 m isopach) | 5.5 | |
| Hekla, Iceland, 1947 (Thorarinsson, 1954) | (3,130 on land) 70,000 total (within 0.06 mm isopach) | 0.18 | 0.045 |

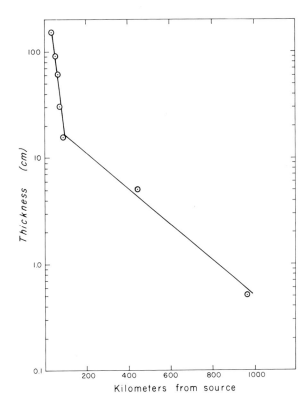

**Fig. 6-18.** Variations in thickness of Mazama tephra with distance. (After Williams and Goles, 1968)

1981). The values have been generally underestimated, however, because the amount of finer-grained material transported great distances can exceed the volume of the mappable fan deposit (Williams and Goles, 1968) (Fig. 6-11). To determine the volume of the Taupo pumice, New Zealand, and several other Plinian deposits, Walker (1980) plots the area enclosed by isopachs on a log-log area-thickness graph (Fig. 6-19); volume is then determined from the area beneath the resulting curves. This and other methods for determining volume are discussed by Froggatt (1982).

Several workers have drawn attention to the similarity between fallout patterns of tephra and other types of material, notably radioactive aerosol particles, and have developed theoretical models to account for the distribution patterns of both types of components (Knox and Short, 1963; Slaughter and Hamil, 1970; Scheidegger and Potter, 1968; Eaton, 1964; Shaw et al., 1974).

## Structures

### Bedding

In general, bedding in near-source fallout deposits consists of alternating gradational coarse-grained to finer-grained layers without sharp bedding planes (Fig. 6-20), but whether bedding is sharp or gradational depends upon several factors. These are: (1) the duration and energy fluctuations of an eruption; (2) vol-

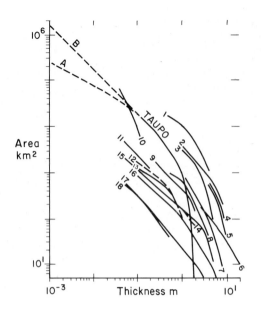

Fig. 6-19. Plot of area enclosed by isopachs vs. thickness for Taupo pumice, New Zealand and other Plinian pumice deposits: *1* Ontake (Kobayashi et al., 1967), *2* Shikotsu (Katsui, 1959), *3* Askja 1875 (G.P.L. Walker, unpublished data), *4* Tenerife J (Booth and Walker, unpublished), *5* Granadilla (Booth, 1973), *6* Pompeii (Lirer et al., 1973), *7* La Primavera B (Clough, Wright, and Walker, unpublished), *8* Toluca upper member (Bloomfield et al. 1977), *9* Fogo 1563 (Walker and Croasdale, 1971), *10* Fogo A (Walker and Croasdale, 1971), *11* Crater Lake (Fisher, 1964b), *12* Hekla 1947 (Thorarinsson, 1954), *13* Tenerife L (Booth and Walker, unpublished), *14* Mangaone (Howorth, 1975), *15* La Primavera J (Clough et al., in Walker 1980), *16* Hekla H3 (Thorarinsson, 1967b), *17* Furnas C (Booth et al., 1978), *18* Avellino (Lirer et al., 1973). (After Walker, 1980)

ume and discharge rate of tephra; (3) direction and strength of winds during an eruption; and (4) the length of time between eruptions during which erosion, weathering or winnowing of the depositional surface may take place (Fig. 6-21, 6-22). Moreover, a change in composition of material being erupted may result in the deposition of two or more layers that have narrow gradational boundaries (Thorarinsson, 1954; Waitt and Dzurisin, 1981; Bogaard, 1983).

Individual beds of fallout tephra commonly have few or no internal structures or laminations. They are set apart from one another by color differences caused by different compositions (or varying percentages of different juvenile fragments) and by post-depositional diagenetic changes, by sorting, or by grain size or grading differences (Mullineaux, 1974; Waitt and Dzurisin, 1981). A continuous, voluminous, and irregularly pulsating eruption of high intensity (Plinian) lasting several hours or days may deposit sequences composed of numerous alternating and gradational coarse- to fine-grained beds with a combined thickness of many meters (Fig. 6-23). Such an eruption, however, may also produce a continuous rain of fragments of different sizes deposited together and at random, resulting in a single, thick, poorly sorted layer without distinct internal laminae. Walker (1980) relates such beds to fluctuating discharge rates, column height and therefore particle fall velocities as shown in Fig. 6-24).

Material suspended by wind currents is deposited as thin tephra blankets over long distances, as witnessed during the May 18, 1980 eruption of Mount St. Helens, Washington (Sarna-Wojcicki and Shipley et al., 1981). Such thin deposits, if preserved, commonly lack internal bedding. Farther from source, later and earlier eruption fragments in the eruption cloud may even become mixed (Salmi, 1948). Areas receiving periodic light ash falls from distant volcanoes may gradually build up thick homogeneous ash deposits lacking bedding (caused by bioturbation and other soil-forming processes) that resemble wind-blown dust (loess) in nearly every aspect (Taylor, 1933; Fisher, 1966d) (Fig. 6-25).

**Fig. 6-20.** Near-vent fallout deposits consisting of pumice lapilli and abundant basanite and sandstone blocks. Note angularity of pumice lapilli and scarcity of fine-grained matrix. About 1 km from vent, Laacher See tephra (Eifel, Germany)

**Fig. 6-21.** Erosional unconformity between two cycles of late Quaternary trachytic eruptions dominated by coarse Plinian fallout in the lower and fine-grained, in part massive, accretionary lapilli-bearing surge and fallout in the upper part. Rim of Sete Cidades Caldera (São Miguel, Azores)

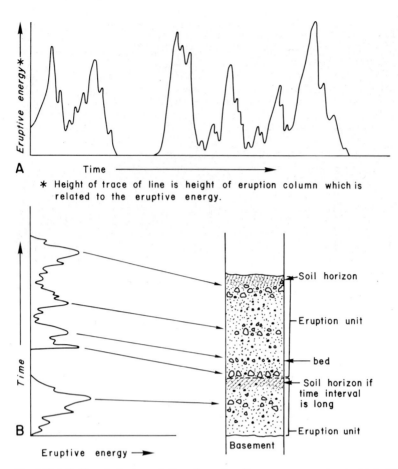

**Fig. 6-22 A, B.** Hypothetical correlation of eruption diagram energy peaks, **A** with bedded sequence, **B** of pyroclastic rocks. Other factors that affect bedding are wind direction and velocity, characteristics of eruption column and distance to source (presence or absence of ballistic fragments). Development of bedding also depends upon pulse time of eruption column versus fall time of fragments

## Mantle Bedding

Fallout tephra commonly mantles all but slopes greater than about 25–30 degrees, a feature that has long been recognized as characteristic of fallout tephra (Wentworth, 1938). Tephra drapes over hills and valleys, into small depressions and channels, and over any other roughness elements of the landscape (Figs. 6-23 to 6-27). Wind, water, and gravity immediately begin the downhill transfer of material into low areas, but rapidly buried ash may retain high initial dips. The rate of downhill transfer depends upon (1) the transporting agent, (2) the size and shape of fragments, (3) angle of slope, (4) rate of deposition, (5) amount and type of vegetation and (6) climate (humid versus arid, etc.). Sheet flooding or rill development is inhibited in freshly deposited, coarse-grained, highly permeable tephra, but in fine-grained deposits light rain causes relatively hard crusts to develop on fresh ash;

**Fig. 6-23.** Lateral termination of pyroclastic flow deposit interbedded with Plinian pumice lapilli layers. Note mantle bedding where fallout tephra layers overlie basal light-colored pyroclastic flow deposit. Bedding in fallout sequence is due to differences in grain size, thin ash cloud layers and change in composition, the deposit becoming darker and more mafic upwards. Laacher See tephra (Eifel, Germany), about 5 km from source. Scale in 10-cm intervals

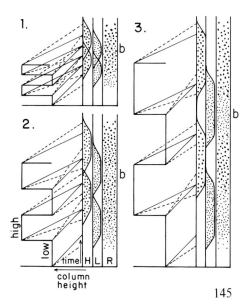

**Fig. 6-24.** Diagrammatic representation of the development of gradational bedding from a pulsating eruption column. $H$ deposition from high column; $L$ deposition from low column; $R$ resultant. *Dashed line* is a standard deviation in fall time equal to one-third to the fall time. *1* column pulsations are shorter than fall time; deposition from second low column does not produce a discrete bed at $b$. *2* column pulse time is comparable to fall time; discrete bed beginning to appear at $b$. *3* column pulse time is longer than fall time; well-defined bed at $b$. (After Walker, 1980)

**Fig. 6-25.** Lateral termination of pyroclastic flow deposit showing flat base and curved top interbedded with pumice lapilli and lower pyroclastic flow deposit. Quaternary Laacher See tephra (Eifel, Germany). Fine-grained bedded surge and massive fallout tephra at top represent terminal phase of eruption. Scale in 10-cm intervals

**Fig. 6-26.** Mantle bedding of late Quaternary phonolitic fallout tephra overlying basaltic lava flow (end of road cut). Tenerife (Canary Islands)

this lessens permeability and accelerates erosional processes (Fig. 6-22) (Segerstrom, 1950; Waldron, 1967). Thick ash deposits at Paricutin Volcano, Mexico, disrupted or destroyed the original drainage pattern and killed vegetation. With changes in slope caused by growth of the volcano, erosion was accelerated and new streams quickly cut to levels below the old stream-bed levels.

Fragments falling directly from above roll, slide or avalanche down steep-sided channels to deposit wedge-shaped layers in cross-section (Fig. 6-27). Depending upon channel width, these deposits may be thicker inside than outside the channel. With equal volumes, individual layers progressively decrease in thickness upward as the cross sectional area of the channel increases. Over undulating topography, layers thicken and thin depending upon slope angles and trajectory of the path of the falling particles.

Broad-scale topographic smoothing caused by long-term ($\sim 5$ million years) deposition and reworking of pyroclastic debris is well illustrated by the Oligocene-Miocene John Day Formation of central Oregon (Fisher, 1964a). The lowermost member covers ancient hills with slopes of up to 25 or 30 degrees; in low areas, the ash accumulated in lakes and marshes. Later members filled broad, gentle valleys several kilometers wide; they thin toward broad structural highs. The changes in thickness occur over such broad areas that they are only revealed by detailed mapping of facies and measured sections (Fisher, 1967; Fisher and Rensberger, 1973).

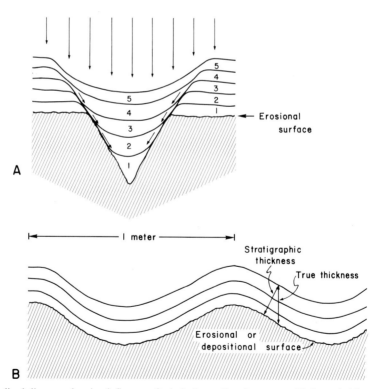

**Fig. 6-27 A, B.** Idealized diagram showing influence of relatively small-scale topographic irregularities on shape and thickness of fallout layers

In some places high initial dips on pre-existing hilly topography might be mistaken for tectonic deformation.

**Graded Bedding**

It is expected that particles falling out of the atmosphere would form density- or size-graded layers with the largest and heaviest particles at the base, but several factors work against perfect grading. These include fluctuations in eruption energy, wind direction and strength, turbulence in the eruption column, and rain flushing. As distance from the volcano increases and particle transport is by wind rather than by fluctuations in eruption energy, settling velocity becomes a more important factor and therefore normal grading becomes increasingly common.

Normal and reverse grading in Plinian and sub-Plinian pumice deposits has been described by Booth (1973), Lirer et al.(1973), Self (1976), and Bloomfield et al.(1977) (Fig. 6-28). Possible causes listed by Self (1976) for reverse grading are: (1) A progressive increase in initial gas velocity during the eruption ejects larger fragments to greater heights in the later phases, promoting a wider dispersal by the wind (Lirer et al., 1973; Booth, 1973); (2) A change in vent morphology from cylindrical to conical as the eruption proceeds enables particles to be ejected at lower angles and therefore to travel farther (Murata et al., 1966). This effect is observed only very near the vent; (3) For a constant gas velocity, an increase in eruption column density by an increase in the percentage of very fine dust in the eruption column can increase the release height of individual large clasts from the column and therefore the range of dispersal of large clasts (Wilson, 1976); (4) Widening of the conduit radius as the eruption proceeds may reduce the frictional drag acting on the high velocity flow of gas and particles, thereby giving an increased exit velocity if the same mean gas velocity is maintained. An additional cause is erratic changes in wind velocity and direction during eruption.

Slope instability at the time of deposition also effects grading. Reverse-graded pumice beds of the Coso Range, California are attributed to rapid accumulation of pumice during eruption on steep surfaces causing downslope grain flows (Duffield et al., 1979). Reverse grading develops from grain-dispersal pressures where larger grains move (upward) toward the zone of least shear strain within a grain flow (Bagnold, 1954b) such as occurs in avalanche deposits on the slip face of sand dunes (Bagnold, 1954a). Reverse-graded beds also develop by dry grain flows on cinder cones where slopes attain steep angles (Wentworth and Macdonald, 1953) (Figs. 6-29, 6-30) and by frost-heaving in cold climates.

**Fabric**

Particle orientation in fallout tephra is commonly isotropic, but anisotropic fabric may occur if particles are elongate or platy. Close to the eruptive source, however, rapidly deposited large and small fragments may remain supported at various angles in a crowd of neighboring particles (Fig. 6-20). At long distances, elongate or platy particles falling from suspension tend to land parallel to depositional surfaces.

**Fig. 6-28.** Reverse grading in one meter thick phonolitic Plinian fallout tephra overlain by darker more mafic fallout and pyroclastic flow deposit. Scale in 10-cm intervals. Laacher See (Eifel, Germany)

**Fig. 6-29.** Inversely graded deposits of basaltic composite lapilli. Quaternary Herchenberg volcano (Eifel, Germany)

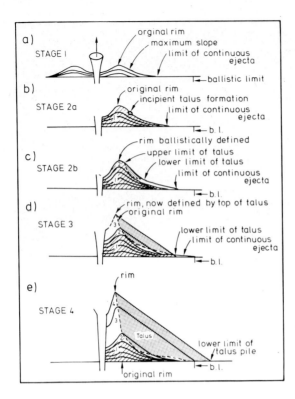

Fig. 6-30 a–e. Development of cinder cones in four stages. **a** Stage 1, low rimmed scoria and tuff ring. **b,c** Stage 2, build-up of rim and development of exterior talus. **d** Stage 3, slumping and blast erosion causes destruction of original rim. **e** Stage 4, talus extends beyond ballistic limit of ejecta as volume of cone grows. (After McGetchin et al., 1974)

## Size Parameters

Although structural aspects are most useful for identifying fallout layers in the field, textural parameters determined by laboratory granulometric analysis provide information about eruption mechanics, vent location, height of eruption column, energy of eruption, wind conditions, etc. when used in conjunction with isopach maps. Parameters most commonly used are maximum size of components, median diameter and sorting values.

### Maximum Size of Components

The maximum diameters of lithic and pumice fragments in fallout deposits have been measured to infer vent location, relative volcanic energy, inclination of eruption column, and wind directions (Minakami, 1942; Kuno et al., 1964; Fisher, 1964b; Walker, 1971; Walker and Croasdale, 1971; Suzuki et al., 1963; Lirer et al., 1973; Self, 1976; Schmincke, 1977b; Bloomfield et al., 1977; Booth et al., 1978; Walker, 1980; Bogaard, 1983). The maximum diameter of a selected number of the largest pumice and lithic fragments is measured in the field; the arithmetic mean of each set of measurements may be then plotted graphically or as isograde maps. The number of fragments measured per outcrop varies from 10 (e.g. Kuno et al., 1964) to 5 (e.g. Minakami, 1942) to as few as 3 (e.g. Walker and Croasdale, 1971). All studies show an exponential decrease in maximum diameter of both types of

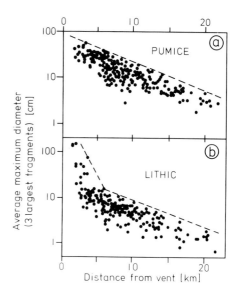

Fig. 6-31 a, b. Average maximum diameter of pumice fragments (**a**) and lithic fragments (**b**) vs. distance from vent for Fogo A (Azores) pumice sheet. (After Walker and Croasdale, 1971)

components away from the source (Fig. 6-31). Some authors note that maximum lithic fragment sizes decrease sharply in the first 2 km or so and then more gradually away from the source. This is attributed to ballistic transport of the largest lithic fragments near the source (Walker and Croasdale, 1971; Self, 1976; Schmincke, 1977b; Booth et al., 1978) (Fig. 6-32). The maximum size of lithic fragments is commonly about half that of pumice in the same outcrop depending on their relative densities. Equidimensional lithic fragments give more consistent results than platy components because factors influencing settling velocity of platy fragments are more complex. Brittle pumice can break upon impact (Fig. 5-2), and, if not recognized, this can lead to significant errors. Distribution patterns of isograde maps that differ from isopach maps of the same deposit near the source may be attributed to directional wind patterns at different altitudes (Fisher, 1964b; Waitt and Dzurisin, 1981) or to inclined eruption columns (Fig. 6-33).

**Median Diameter**

Median diameters of material within pyroclastic layers exponentially decrease overall away from the source but in detail the decrease is rarely systematic. Median diameters may provide a more sensitive indicator of inclined eruptions or wind variations during an eruption than do isopach maps alone (Fisher, 1964b; Walker, 1980) but require more laboratory time to determine than field measurement of maximum diameters. Also, the sampling from thick tephra layers with vertical variations in size parameters (Fig. 6-34) may introduce large errors, and because tephra layers are polycomponent, further complications may occur. The relationship between median diameter, distance to source and relative amounts of crystals-lithics-pumice is given by Walker and Croasdale (1971) and Booth et al., 1978 (Fig. 6-35). A general median diameter-distance plot for polycomponent tephra samples (Fig. 6-36) is useful to infer broad limits for travel distance where the source is un-

**Fig. 6-32.** Basalt block emplaced ballistically into pumice lapilli layers. Zone of compacted and faulted lapilli extends for about 1 m beneath block. Late Quaternary Laacher See tephra (Eifel, Germany)

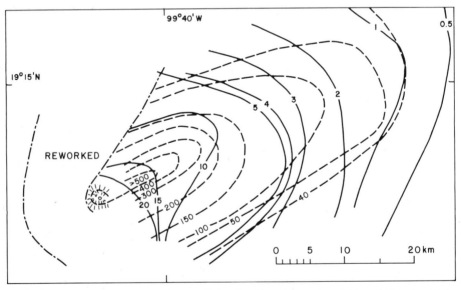

**Fig. 6-33.** Isopach map (*dashed line*) of Upper Toluca Pumice (thickness in cm), Nevado de Toluca Volcano, Mexico, upon which are superimposed contours (in cm) of average maximum diameter of the five largest pumice fragments from Lower Member (*solid lines*). Illustrates different axes of distribution. Vent is *hachured*. (After Bloomfield et al., 1977)

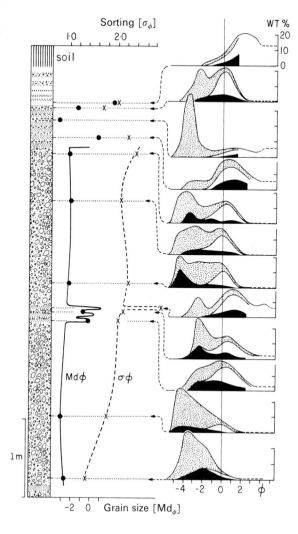

Fig. 6-34. Median size (Md$_\phi$), Inman sorting coefficient ($\sigma_\phi$) and frequency distribution of pumice (*stippled*), crystals (*unornamented*) and lithics (*solid*) for Fogo A tephra, 3.7 km southeast of Lake Fogo (São Miguel, Azores). (After Walker and Croasdale, 1971)

known, but cannot be used to determine exact limits because of wide scatter. For example, fallout tephra with Md$_\phi$ = $-3.0$ (8 mm) is probably within 60 km or less of its source.

### Grain-Size Distribution (Sorting)

In general, good sorting is an important criterion for distinguishing fallout ash from pyroclastic flow deposits (Walker, 1971). For example, fine-grained dust is less abundant in coarse-grained pumiceous fallout layers than in nonwelded pyroclastic flow deposits. Sorting coefficients ($\sigma_\phi$, Inman, 1952) determined granulometrically for many samples have been published by Murai (1961), Fisher (1964b), Walker and Croasdale (1971), Walker (1971; 1980), Sheridan (1971), Kittleman (1973), Lirer et al., (1973), Schmincke (1977b), Self (1976) and Booth et al., (1978).

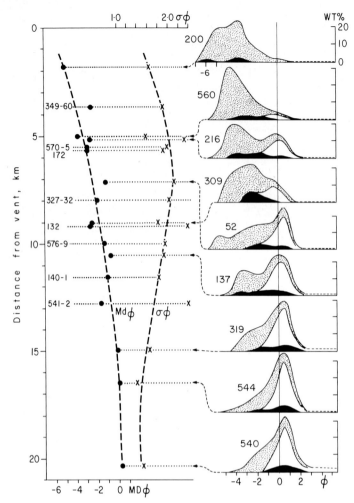

**Fig. 6-35.** Lateral changes from Lake Fogo (Azores) in median diameter (Md$_\phi$), Inman sorting coefficient ($\sigma_\phi$) and frequency curves for pumice (*stippled*), crystals (unornamented) and lithics (*solid*) for nine samples, Fogo A pumice deposit. (After Walker and Croasdale, 1971)

From the summaries of Fisher (1964b) and Walker (1971), it appears that most fallout tephra deposits have sorting values between 1 and 2, with most around 1.4. Overall, sorting becomes better with increasing distance from the vent (Fisher, 1964b) but the pattern is complex in detail. For example, small-particle agglutination to form accretionary lapilli can cause sedimentation of fine-grained material alongside coarse material (Carey and Sigurdsson, 1982). And, at small distances from the vent (about 30 km, depending on grain size), sorting values may increase prior to a decrease (Walker, 1971; Walker and Croasdale, 1971; Suzuki et al., 1973; Schmincke, 1977b). Much of this variation at a few kilometers from the vent is due to pronounced differences in density between pumice, crystals, and rock fragments. Thus, in tephra deposits on Fogo, Azores (Walker and Croasdale, 1971) the grain size distribution is unimodal (mode $-3_\phi$ to $-4_\phi$) and positively skewed close to the

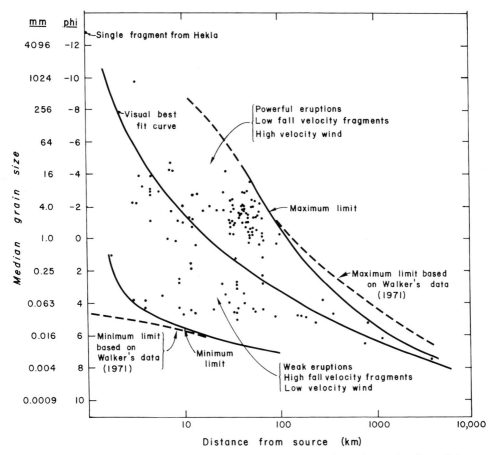

**Fig. 6-36.** Median diameter (Md$_\phi$) plotted against distance from source for various tephra sheets. (After Fisher, 1964b; Walker, 1971)

source, but becomes bimodal (mode between 0.0$_\phi$ and 1.0$_\phi$) at about 16.5 km from the source. The modes closer to the source are dominated by pumice, while those at large distances are dominated by sanidine crystals (Fig. 6-37).

Another factor influencing sorting is the relative increase in density of pumice with decreasing grain size, owing to the decrease of the ratio of vesicle volume to the volume of actual glass. Thus, at very small grain sizes (depending on vesicle size and degree of vesiculation), densities of glass, lithic and felsic crystals become similar.

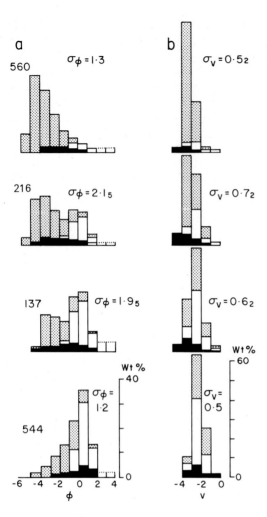

**Fig. 6-37 a, b.** Variation in pumice (*stippled*), crystals (*blank*) and lithics (*block*) for four samples from Fogo A tephra, Azores. Sample 560 is from nearest the vent (5.0 km) and 544 farthest (16.5 km). In histograms *b*, weight percentages are plotted against a function of fall velocity instead of grain size. (After Walker, 1971)

## Eolian Fractionation

Long ago, Murray and Renard (1884, p. 478) (also see Judd, 1888, p. 40) pointed out that the andesitic tephra nearest to the 1883 Krakatau eruption was more mafic than farther away. They suggested that differences in settling velocity between crystals and glass shards were the main cause – the denser crystals settling first, and the more silicic, less dense glass shards and pumice being carried farther away, possibly because they went higher and therefore were carried farther downwind. This same explanation is given for downwind changes in composition of the fallout ash from May 18, 1980 Mount St. Helens ash (Sarna-Wojcicki and Meyer et al., 1981), which shows downwind increases in $SiO_2$, $K_2O$ and decreases in $Al_2O_3$, CaO, and $TiO_2$ from bulk samples.

Larsson (1937), who compared the bulk chemical composition of ash from three widely spaced samples from the 1932 eruption of Quizapu Volcano, Chile, with lapilli deposited near the source, noted that near-source lapilli were andesitic to dacitic (about 64% $SiO_2$), whereas ash became more silicic with increasing distance from the source (Table 6-4), the most "differentiated" sample having about 70% $SiO_2$. He attributed the increase in $SiO_2$ content to a progressive loss of crystals (plagioclase, hornblende, pyroxene, apatite, magnetite) which contain less silicon than the glass. Larsson called this process of chemical change with distance in volcanic ash *eolian differentiation*. We prefer the term *eolian fractionation* of Lirer et al. (1973) to emphasize that different types of components in a particulate system can be fractionated relative to each other.

Table 6-4. $SiO_2$ Content, distance from source and original composition (lapilli), Quizapu ash (Larsson, 1937)

| Analyzed material | Distance (km) | $SiO_2$ (wt. %) |
|---|---|---|
| Lapilli | 85 | 63.8 |
| Lapilli | 85 | 64.6 |
| Ash | 230 | 67.5 |
| Ash | 1,120 | 69.8 |
| Ash | 780 | 70.2 |

An assessment of eolian fractionation must distinguish three sets of variables: (a) magma chamber variables; (b) eruption column variables; and (c) transfer system variables. Most large silicic-alkalic magma chambers, for example, are compositionally zoned both chemically and mineralogically (composition and types, ratios and volumes of mineral phases). Moreover, the products are commonly erupted in order of their position in the magma column, proceeding from silicic to more mafic in both Plinian fallout and pyroclastic flow eruptions, as is shown in the deposits. In some cases the more mafic parts of such a zoned eruptive phase are distributed farther (Lirer et al., 1973), whereas in others they are confined near the volcanic source (Bogaard, 1983). Thus, studies of eolian fractionation must assess progressive lateral as well as temporal vertical changes, keeping in mind the possibility of compositionally zoned magma chambers.

Although dynamic processes within eruption columns are poorly known, several workers have noted that crystals are commonly enriched in pyroclastic flow deposits because the vitric lighter components are winnowed out, that is, fractionated during upward rise, collapse, and dispersal of the eruption column, and also during transport (Hay, 1959a; Lipman, 1967; Walker, 1972a; Sparks and Walker, 1977). Thus, the bulk composition of fallout tephra associated with pyroclastic flows at any single locality (co-ignimbrite airfall tephra) does not truly represent the composition of the original magma.

Fractionation within a traveling ash cloud results from settling velocity differences of crystals, small pumice particles, glass shards and rock fragments. Their modal size distribution in the deposits is also determined by the initial size distribution in the magma chamber or eruption column. In any case, bulk chemical analyses of samples from fallout layers is not very useful for studies of original magma

**Table 6-5.** Fragment percentages, calculated ratios and size parameters with respect to distance for 1919 Kelut ash samples (Java). Percentages of fragmental components are determined for total of sample fractions in the following grade sizes: 2–1 mm, 1.0–0.5 mm, 0.5–0.2 mm, 0.2–0.1 mm, and 0.1–0.05 mm (Atterberg scale) (Baak, 1949, Table 9). $Md_\phi$ and $\sigma_\phi$ were determined from curves drawn from Baak's (1949) data on his Table 8 for total samples

| Sample No. | Distance (km) | Relative percentage of fragments | | | | | Ratios | | Size parameters (whole samples) | |
|---|---|---|---|---|---|---|---|---|---|---|
| | | Op.[b] | Fm.[c] | Plag. | Lithic | Shards | Op./Op. + Fsp. | Lithic/glass | $Md_\phi$ | $\sigma_\phi$[d] |
| 7038[a] | 36 | 6 | 13 | 51 | 13 | 18 | 0.11 | 0.72 | 3.29 | 1.80 |
| 7038 | 43 | 5 | 7 | 45 | 10 | 33 | 0.10 | 0.30 | 3.90 | 1.38 |
| 8561 | 56 | 2 | 8 | 48 | 8 | 34 | 0.04 | 0.24 | 4.40 | 1.60 |
| 7050 | 66 | 8 | 9 | 56 | 12 | 18 | 0.13 | 0.66 | 4.30 | 1.61 |
| 8272 | 154 | 4 | 8 | 43 | 9 | 36 | 0.09 | 0.25 | 4.65 | 1.30 |
| 7039 | 166 | 15 | 5 | 39 | 7 | 34 | 0.28 | 0.21 | 4.68 | 1.28 |
| 7042 | 4 | 6 | 9 | 38 | 13 | 34 | 0.14 | 0.38 | 4.25 | 2.10 |
| 7046 | 9 | 6 | 13 | 54 | 15 | 12 | 0.10 | 1.25 | 2.70 | 1.82 |
| 7043 | 36 | 3 | 11 | 27 | 12 | 43 | 0.10 | 0.28 | NA[e] | NA |
| 8501 | 92 | 3 | 7 | 27 | 9 | 54 | 0.10 | 0.17 | 4.70 | 1.30 |
| 8271 | 360 | 19 | 6 | 44 | 4 | 27 | 0.30 | 0.15 | 4.55 | 0.53 |

[a] Average of four samples: Nos. 7049, 8503, 8506, 8505
[b] Op. = opaque oxides, Fsp. = Feldspar
[c] Fm. = Ferromagnesian minerals (augite, hypersthene, green and brown hornblende)
[d] Phi $(\phi) = -\log_2 \phi$ in mm; $\sigma_\phi = 84_\phi - 16_\phi/2$ (Inman, 1952)
[e] NA = not analyzed

compositions; it is necessary to know the size and distribution of different components, prior to detailed chemical analyses by microprobe and other methods. The composition of glass within ash, however, represents the composition of the liquid fraction at the time of eruption.

Minerals are generally most abundant between about 2 mm ($-1_\phi$), and 1/16 mm ($4_\phi$) and are practically absent below about 10 μm, whereas glass shards can be much smaller. An ash-fall deposit from a homogeneous magma batch should therefore become increasingly silicic with distance because the interstitial liquid of a magma is generally more silicic than the crystals, except, of course, for quartz.

Minerals and glass shards in ash samples from the 1919 eruption of Kelut volcano (Java) showed the following trends (Table 6-5): lithic fragments, amphibole, pyroxene, and feldspar crystals decrease in size and amount away from the source relative to glass shards and opaque minerals (dominantly magnetite). The relative increase in glass shards is apparently the reason for an increase in silica content of the bulk ash, despite a relative increase in opaque oxide minerals. Although feldspar tends to decrease in absolute amount with distance from Kelut, the grade size between 0.1 mm and 0.02 mm usually contains the largest volume of feldspar in any given sample (Table 6-6) reflecting primary size fractionation during dispersal. At the same time, the size of feldspar in any given sample (Table 6-6) reflects only the primary size of phenocrysts in the magma prior to eruption. The total volume of feldspar crystals decreases with distance in the coarser fractions, but remains about the same in the finer fractions.

Although the settling velocities of heavy and light minerals differ significantly, Kittleman (1973) shows that their relative proportions in Mazama ash does not change significantly with distance up to 400 km from the source, even though grain size of the samples decreases from a median diameter of $-4_\phi$ to $-5_\phi$ near the source to about $2_\phi$ at about 200 km (Fig. 6-38). This rather puzzling feature has not been ex-

Table 6-6. Volume percent feldspar crystals, 1919 Kelut ash. Figures in italics are maxima. (Data from Baak, 1949, Table 11)

|  | Distance (km) | Grade size (mm) | | | |
| --- | --- | --- | --- | --- | --- |
|  |  | 0.5–0.2 | 0.2–0.1 | 0.1–0.02 | 0.02–0.05 |
| East of Kelut | 36 | *12.0* | 11.0 | 10.0 | 3.5 |
|  | 36 | *13.0* | 10.0 | *14.0* | 4.3 |
|  | 36 | 7.5 | 12.2 | *14.0* | 2.2 |
|  | 36 | –0– | 8.3 | *16.0* | 1.7 |
|  | 43 | 3.5 | 6.0 | *13.0* | 3.3 |
|  | 56 | 7.8 | 7.5 | *12.0* | 3.5 |
|  | 66 | 8.8 | *10.0* | 9.3 | 3.8 |
|  | 154 | –0– | 5.7 | *16.3* | 2.0 |
|  | 166 | –0– | 2.3 | *15.0* | 4.3 |
| West of Kelut | 4 | *10.0* | 8.5 | 9.0 | 3.3 |
|  | 9 | 11.5 | 11.5 | *12.2* | 2.5 |
|  | 36 | 4.0 | 9.0 | *12.0* | 3.0 |
|  | 92 | –0– | 3.7 | *11.7* | 4.0 |
|  | 360 | –0– | –0– | *19.0* | 4.5 |

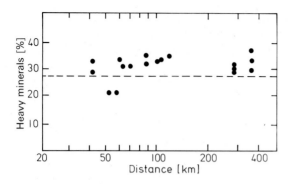

**Fig. 6-38.** Percent heavy minerals among free crystals in Mazama fallout tephra plotted against distance from source. *Dashed line* is percentage of heavy minerals among crystals enclosed in pumice. (After Kittleman, 1973)

plained and requires further study. Other data indicating that eolian fractionation is not a simple issue come from the 1947 eruption of Hekla, where silica percentage of glass shards did not vary significantly with distance despite a difference in $SiO_2$ content between the earliest erupted, brownish-gray tephra and the closely following and more mafic, brownish-black tephra during the continuous eruption (Table 6-7).

One of the most complete studies of eolian fractionation is on the Recent White River ash in Alaska and Canada (see Fig. 6-11) (Lerbekmo and Campbell, 1969; Lerbekmo et al., 1975) that originally extended over an estimated area of >300,000 $km^2$ and had a minimum volume of about 25 $km^3$ (Fig. 6-39). The chemical com-

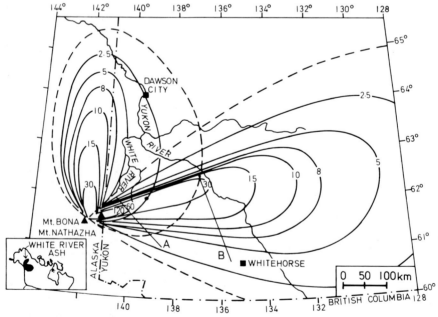

**Fig. 6-39.** Isopach map of the two lobes of White River ash. Lines A (west) and B (east) are approximate locations of traverses for data in Fig. 6-40 and Table 6-8. (After Lerbekmo and Campbell, 1969; Lerbekmo et al., 1975)

**Table 6-7.** $SiO_2$-variation with distance, 1947 Hekla (Iceland) brown-gray and brown-black tephra compared with original composition (bomb). (Data from Thorarinsson, 1954, Plate 2, Table 8)

| Analyzed material | Distance (km) | $SiO_2$ |
|---|---|---|
| Brown-gray bomb | – | 60.9 |
| Brown-gray tephra | 32 | 61.9 |
| Brown-gray tephra | 50 | 61.7 |
| Brown-gray tephra | 70 | 61.5 |
| Brown-black tephra | 4 | 56.3 |
| Brown-black tephra | 6 | 57.5 |
| Brown-black tephra | 32 | 57.7 |
| Brown-black tephra | 283 | 57.1 |
| Red-brown dust | 3,800 | 56.4 |

**Table 6-8.** Averages of chemical analyses of ash samples collected from two traverses 190 km apart across east lobe of White River ash, Yukon Territory, Canada (Lerbekmo and Campbell, 1969)

| Traverse A Western (proximal) traverse (mean of samples 30–46, 66) | | Traverse B Eastern (distal) traverse (mean of samples 1–27) | |
|---|---|---|---|
| | wt. % | | wt. % |
| $SiO_2$ | 62.2 | $SiO_2$ | 70.2 |
| $Al_2O_3$ | 17.8 | $Al_2O_3$ | 13.9 |
| MgO | 2.5 | MgO | 1.6 |
| $Na_2O$ | 5.2 | $Na_2O$ | 3.9 |
| $K_2O$ | 1.9 | $K_2O$ | 2.8 |
| CaO | 6.3 | CaO | 3.3 |
| FeO | 2.1 | FeO | 2.2 |
| $Fe_2O_3$ | 1.5 | $Fe_2O_3$ | 1.5 |
| $TiO_2$ | 0.5 | $TiO_2$ | 0.5 |
| (Rb | 21 ppm) | (Rb | 39 ppm) |
| (Sr | 1,269 ppm) | (Sr | 624 ppm) |
| K/Rb | 762 | K/Rb | 589 |
| Ca/Sr | 35 | Ca/Sr | 38 |

position of samples from two traverses 190 km apart across the east lobe shows large differences in many oxides and in Rb and Sr (Fig. 6-40, Table 6-8). The differences are interpreted as caused by an increase in glass relative to the crystal components as a function of size and distance from the source. The spread of values around the best-fit curves at equal distances (Fig. 6-40) is probably because many samples were collected off the main dispersal axis.

In a later, more detailed mineralogical study, Lerbekmo et al. (1975) noted that the age of the northern lobe is about 1887 years B.P. while that of the eastern lobe is 1250 years B.P. Although the chemical composition of glass and most pheno-

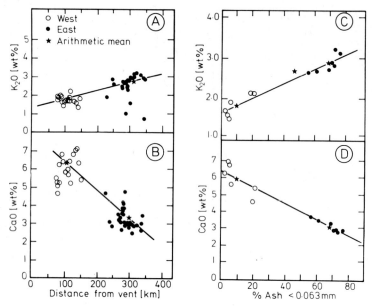

**Fig. 6-40 A–D.** $K_2O$ and CaO contents versus distance from vent for traverses **A** and **B** (in Fig. 6-39) across east-trending lobe of White River ash. **C** $K_2O$ versus grain size for east and west traverses across east-trending lobe. **D** CaO content for same samples as in **C**. (After Lerbekmo and Campbell 1969)

crysts is identical in the two lobes, there are significant differences in ilmenite composition (particularly Fe, Ti, and Mg). Calculated temperatures of crystallization and oxygen fugacities from the composition of co-existing magnetites and ilmenites suggest that magmas supplying ash for the northern, older lobe crystallized at about 830° to 1005 °C while those from the younger eastern lobe had higher temperatures (about 1035° to 1125 °C). Curiously, temperatures calculated from mineral composition in vertical sections of the northern lobe ash show a decrease with time and distance indicating that higher temperature magmas (from a zoned magma chamber?) were erupted first, whereas lower temperature magmas were erupted last and the products transported farther from the source (also Lirer et al., 1973).

# *Chapter 7* Submarine Fallout Tephra from Subaerial Eruptions

Following classic studies of deep sea exploration in the 1940's (Bramlette and Bradley, 1942; Neeb, 1943; Norin, 1948), a great deal of modern research has been done on marine ash layers, much of which is summarized by Kennett (1981). Marine ash layers were originally studied for their value as widespread stratigraphic markers, but deep penetration of the sea floor by drilling from the Glomar Challenger, compared to piston cores or dredging, has allowed assessment of paleovulcanicity extending as far back as Jurassic time, with major implications for understanding rates of sea floor spreading and the evolution of island arcs. Additionally, marine ash layers have supplied information about the cyclicity of volcanism, volcanic production rates and volumes, and the influence of large explosive eruptions on climate.

Most well-studied *marine fallout* tephra deposits and lacustrine tuffs were derived from eruptions on land and were initially dispersed by wind. Some of their features are outlined in Table 7-1. *Lacustrine* tuffs resemble their marine counterparts but we emphasize marine tephra. Volcanic products formed by submarine (underwater) eruptions are treated in Chapter 10.

## Chemical Composition

Determining the chemical composition of submarine glass shards aids in correlation of marine ash beds, and in tracing ash layers from the sea floor to their source volcanoes on land. Since volcanic glass is quenched magma, the chemical composition of glass shards provides important information on magmatic evolution in the source areas. Although tephra is readily altered to clay minerals and zeolites, some is preserved in the marine environment with only minor chemical modifications in layers as old as Cretaceous.

Refractive index (R.I.) measurements are traditionally used to characterize shards and to estimate their chemical composition based on the inverse relationship between R.I. and $SiO_2$-content (e.g. Mathews, 1951; Huber and Rinehart, 1966). This method provides useful approximations but glasses become hydrated with increasing age prior to their breakdown to diagenetic mineral phases and the R.I. generally decreases with increasing $H_2O$-content (Ross and Smith, 1955) (Fig. 12-10). Moreover, the R.I. is also influenced by other chemical parameters such as the oxidation state of iron. The suggestion has been made to calculate the R.I. of glass from a rock analysis (Church and Johnson, 1980) but the reverse step cannot be taken.

**Table 7-1.** Characteristics of submarine fallout tephra

*Distribution and Thickness*

Distribution of fallout sheets may be modified by water currents. Mostly regular to irregular fan-shaped close to source. Tend to become thicker toward source, but may be highly irregular.
Thickness of single layers commonly < 50 cm unless thickened by currents in low places. Thick layers with many thin laminae may be multiple fall units.

*Structures*

Plane parallel beds extend for hundreds of km$^2$.
Normal grading from crystal and lithic-rich bases to shard-rich tops.
Basal contacts sharp; upper contacts diffuse due to reworking by burrowing animals.
May be inversely graded if pumice is present. Presence of abundant pumice suggests restricted circulation, more common in lacustrine than marine environments.
Structures on land-based outcrops may include post-depositional thickening, thinning and flow structures, especially if diagenetically altered; may include water escape structures and load or slump structures.

*Textures*

Sorting: good to poor depending upon amount of bioturbation.
Inman sorting, $\sigma_\phi$, generally > 1.0 and < 2.5.
Median diameter, Md$_\phi$: commonly > 3.0 – fine-grained sand-size and smaller.
Size and sorting parameters vary irregularly with distance from source but overall size tends to decrease.

*Composition*

Subaqueous tephra ranges from mafic to silicic with silicic ash most widespread.
Composition generally related to composition of nearest volcanic sources.
$SiO_2$ content of glass shards may range 10% within a single layer.
Bulk samples are more $SiO_2$-rich near top than bottom of single layers because of grading.
Ancient layers in terrestrial geologic settings are usually altered to clays (dominantly montmorillonite) and zeolites, and are commonly known as bentonite.

*Rock Association and Facies*

Tephra is commonly interbedded with pelagic calcareous or siliceous oozes, or with terrigenous muds and silts depending upon proximity to land. Terrigenous materials are commonly turbidites.
Ancient tephra layers on land are commonly interbedded with nonvolcanic or tuffaceous shale or siltstone.

---

Broad relationships between color and chemical composition have been noted for some time with colorless shards being of silicic and deep brown shards of mafic composition (e.g. Horn et al., 1969). Moreover, the degree of vesicularity is generally higher in the more silicic shards and a combination of all four parameters is shown in Fig. 7-1.

Within the last decade, microprobe analysis of individual shards has emerged as the most powerful tool for characterizing shards because of the wide spectrum of elements which can be analyzed, provided certain precautions are taken in order to minimize loss of elements during analysis (Fig. 7-2). Chemical analyses of bulk ash samples are also used in favorable cases, and in unaltered ash, results of R.I., microprobe and whole rock analytical methods generally yield similar results (Scheidegger et al. 1978).

Using 28 major and trace elements of glass shards from 90 submarine ash layers, Scheidegger et al. (1980) evaluated the explosive volcanic history of the

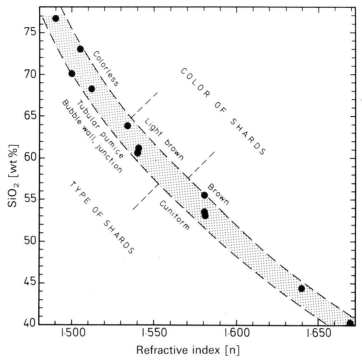

**Fig. 7-1.** Relationship of color and shape of shards to $SiO_2$-content and refractive index of glass. (After Schmincke, 1981) Solid dots represent refractive indices calculated from bulk rock compositions. (Church and Johnson, 1980)

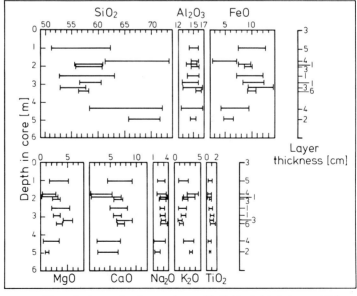

**Fig. 7-2.** Depth *versus* composition of microprobe glass analyses of individual shards in core 70 from the Fiji Plateau showing very large spread in some oxides within shards from the same ash layers. Note especially the large variation in $SiO_2$. (After Jezek, 1976)

North Pacific volcanic arcs from the late Miocene to Recent time. One of their principal conclusions is that during the last 10 m.y., there has been no overall transition from arc tholeiite to calc-alkaline magma in parentage of ash derived from the North Pacific volcanic arcs. Instead, they note short-term fluctuations of 0.1 to 0.5 m.y, intervals during which particular types of volcanism are prevalent, e.g., a transition from arc tholeiite to calc-alkaline composition in ash layers from Kamchatka during the last 0.8 m.y.

The problem of heterogeneity was also addressed by Huang, Watkins and Wilson (1979), but many problems remain. There is great need for regional and systematic studies of single shard chemical variability from deposits of known eruptions.

## Structures of Submarine and Lacustrine Ash Layers

The lower contacts of submarine fallout ash layers are generally quite sharp in contrast to the more diffuse upper boundaries. The abundance of shards relative to nonvolcanic components decreases either gradually or irregularly upward, as was first pointed out by Bramlette and Bradley (1942), and detailed by Norin (1958) (Fig. 7-3). This diffuse upper boundary is explained by most workers as due to bioturbation. Ruddiman and Glover (1972), for example, show that significant amounts of ash can be biologically mixed through a thickness of sediment ranging up to 40 cm. Hein et al. (1978) caution that dispersed ash particles are equal in importance to discrete ash layers and to pods of ash in evaluating volcanic activity. If dispersed pyroclastic material forms more than 15 percent of the deposit, they consider it to represent significant explosive volcanic activity. A 3400-year-old ash off the coast of New Zealand studied by Lewis and Kohn (1973) appears to be preserved intact where its thickness exceeds about 2 cm, but where it is thinner it was

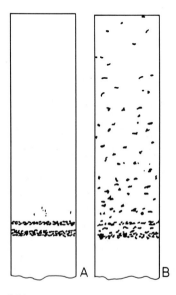

Fig. 7-3 A, B. Distribution of shards in marine sediments. A Ash accumulated in quiet water from two explosive eruptions; B ash from two eruptions where mud-feeding animals continually reworked the sediment during accumulation of ash-free sediment. (After Bramlette and Bradley, 1942)

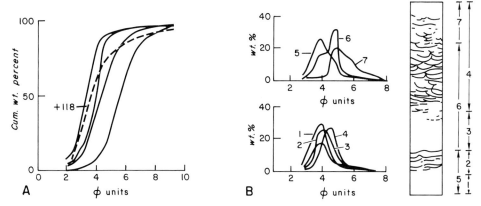

Fig. 7-4 A, B. Grain-size distribution and structures in Eocene ash layers, Denmark. **A** Cumulative curves of submarine tephra layers including the layer with dish structures (+118); **B** histogram curves from samples taken at varying intervals through layer +118. Despite disturbance by water-escape to produce dish-structures, curves show excellent normal grading typical of marine fallout tephra. Sorting is best at base and decreases as grain size decreases. (After Pederson and Surlyk, 1977)

mixed by animals and preserved only in burrows. Rate of mixing relative to thickness of deposits, however, depends upon the biologic environment.

Important criteria used by Norin (1958) for interpreting the mode of transportation of ash and its deposition as layers are: (1) the nature of basal and upper contacts, either sharp or diffuse; (2) vertical grading relative to composition and density of pyrogenic crystals; (3) the relative amounts and kinds of pumice with respect to position in the beds; (4) the kind and amount of admixed nonvolcanic debris; (5) the type of internal lamination; (6) the morphology of the grains. Norin described two dispersal processes: emplacement by air transport from land and fallout into water; then, dispersal and redeposition by turbidity currents. Some fragments and minerals may be derived from attrition by floating pumice. Norin (1958) used the following criteria to distinguish marine tephra mainly transported by wind: fallout (his beds 1 and 2, Division D) is characterized by a sharp base and diffuse upper contacts and by grading (concentration of minerals – feldspar, quartz, biotite, and pyroxene – at the base, and concentration of fibrous pumice at the top). The diffuse upper contact is due to bioturbation after deposition and/or mixing with non-pyroclastic debris during settling. Norin attributed wavy laminations on the top of bed 2 to reworking or to turbidity currents. Another bed (5, Division D) was interpreted to be due to mass flowage as evidenced by: no density or size grading; mixing with foraminifera and terrigenous debris throughout; sharp upper and lower contacts.

In Denmark, an extensively studied assemblage of 179 numbered ash layers, 1 to 18 cm thick, occurring within a 60-m-thick section of Lower Eocene marine diatomite (Bøggild, 1918; Norin, 1940; Pederson et al., 1975) have vertical grading typical of marine fallout tephra (Fig. 7-4), and lack current-induced sedimentary structures. In addition, Pederson and Surlyk (1977) report well developed water-escape structures (i.e. dish structures, see Wentworth, 1967; Stauffer, 1967; Lowe and LoPiccolo, 1974; and Lowe, 1975) which are interpreted to form from the stabilization and extrusion of water from layers with an unusually high porosity and a metastable grain framework that developed under conditions of very rapid sedimentation.

Studies of pre-Tertiary submarine tephra show similar sharp basal and gradational upper contacts and grading. Suites of tuff beds (bentonite) in marine lower Cretaceous dark shales, Wyoming (Slaughter and Early, 1965), range from <2.5 cm to as much as 7.5 m thick. Internally, the thick beds are multiple units containing several visually discernible planar strata, each of which are coarsest at the base (sand-size pyrogenic crystals and altered glass shards) and grade upward to dominantly clay-size material that is fissile in many beds. A remarkable feature in some multiple units is the presence of laterally persistent tuff beds a few centimeters thick; one such bed, 3-cm-thick, was traced up to a distance of about 160 km over an area of 2540 km$^2$ (Slaughter and Early, 1965). They did not describe bedding structures other than grading, but because of abundant diagenetic clays, there are post-depositional thickening features, thinning, folding and various flow structures. Schiener (1970) and Brenchley (1972) report loadcasts and slump structures in Ordovician ash-shower deposits from Wales interpreted to have been deposited wholly or partly in shallow marine water environments.

Lacustrine tephra, described in detail by Hansen et al. (1963), is similar to submarine tephra in that the beds are uniform over a wide lateral extent and cross bedding or scour-and-fill structures are absent. Graded bedding is common, and where pumice is present, beds are inversely graded (see also Chesterman, 1956). Where pumice and lithic lapilli-sized fragments occur in the same bed, the pumice is inversely graded and the lithics are normally graded. This same relation holds for ancient marine tephra (Fiske, 1969), but we suspect that marine tephra containing abundant large lapilli-sized pumice fragments was deposited where marine currents were restricted, because large pumice fragments are relatively rare in present-day marine tephra layers (Lisitzin, 1972).

In many cases, accretionary lapilli (Chap. 5) may break down upon deposition even under subaerial conditions (Sisson, 1982). Contrary to expectations, however, some accretionary lapilli may survive within subaqueous depositional environments (Grange, 1937; Bateson, 1965; Self and Sparks, 1978) although they are generally interpreted as reworked (Bateson, 1965) or else the deposits in which they occur are, *ipso facto,* subaerial (Brenchley, 1972; Francis et al., 1968). Because the outer skins of accretionary lapilli are commonly composed of very fine-grained (hence cohesive) ash, they might not readily fall apart in sinking through water.

## Areal Thickness Distribution and Volume

Most well-defined deep sea ash layers are less than 10 cm thick, and only rarely do they reach a thickness of more than 0.5 m. For example, young ash layers (less than 0.3 m.y. old) from cores within the upper 16 m of ocean-bottom sediments in the eastern equatorial and southeastern Pacific Ocean are <1 cm to 46 cm thick (Bowles et al., 1973). Ninkovich et al. (1964) report ash layers close to the South Sandwich Islands that range from 1 to 4 cm; those farthest from the islands (about 650 km distant) are 0.2–0.5 cm thick. Thirty-one ash layers in cores from the Fiji

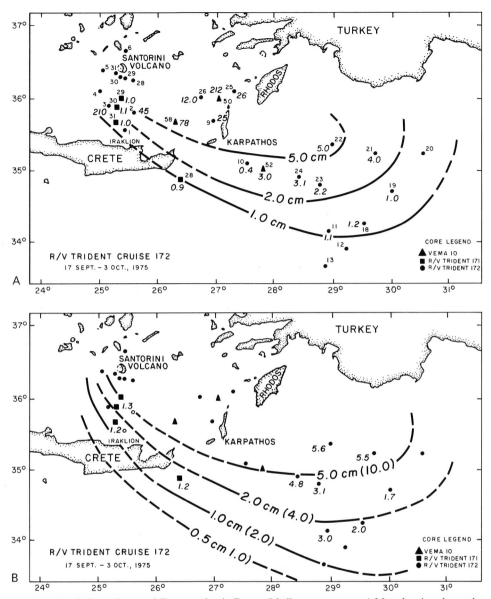

**Fig. 7-5 A, B.** Isopach maps, Minoan tephra in Eastern Mediterranean cores. **A** Map showing observed thickness (in cm); core numbers given by small letters. **B** Adjusted thickness (in cm) to include modification of dispersal by bioturbation; numbers in parentheses on isopach contours is thickness of fresh tephra fallen on land before compaction. (After Watkins et al., 1978)

Plateau range in thickness from <1 cm to 15 cm (Jezek, 1976). The thickness of 163 layers of white ash in the north Pacific ranges from 1 to 29 cm with an average of 6.5 cm, while the thickness of 82 layers of brown ash varies from 1 to 13 cm, with an average of 3.9 cm (Horn et al., 1969). Post-depositional compaction, however, can substantially alter thickness, and hence affect the isopach map (Watkins et al., 1978).

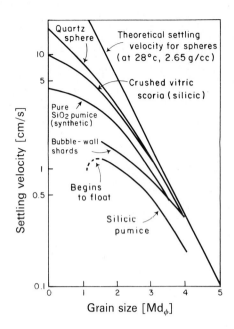

Fig. 7-6. Settling velocity, in water, of sand-size pumice, bubble-wall shards, and shard-like material (bulk samples) compared with settling velocity of quartz spheres and theoretical values from Stokes Law. Bubble wall shards are from Pearlette Ash (Kansas); pumice is from Mount Mazama (Oregon) pumice flow deposit. (After Fisher, 1965)

Watkins et al. (1978) show that compacted tephra in a deep-sea core can be as little as one-half the thickness of equivalent dry tephra on land (Fig. 7-5), and further, that up to 65 percent of the thickness of some visible tephra layers have been mixed upward by bioturbation. The overall distribution of submarine ash layers derived by subaerial eruptions is governed by eruption column height and wind vectors, modified by marine currents and postdepositional redistribution as grain flows and ash turbidites, and by bioturbation.

The dominance of atmospheric conditions over marine currents in the initial fallout distribution of tephra is largely due to the relatively rapid settling velocities of fragments in water. Some marine tephra layers, for example, systematically decrease in thickness away from the source, indicating that fallout patterns from the atmosphere were not significantly distorted by marine currents (e.g., Ninkovich et al., 1964; Ninkovich and Shackleton, 1975; Carey and Sigurdsson, 1980). Contrary to what might be expected, the settling velocity of volcanic particles is sufficiently high (Fisher, 1965) (Fig. 7-6) to carry particles through the relatively thin surficial water currents in only hours or days. Also, sinking rates are greatly enhanced in the upper hundred meters of the ocean where fine particulate matter organically aggregates into coarser fecal pellets which settle out rapidly before final disintegration back to the original fine fraction (Sonayada, 1971). In some areas, however, thickness changes are not systematic (Nayudu, 1964b; Hampton et al., 1979). Ninkovich and Shackleton (1975), for example, document that strong current action affected the distribution of ash L, a recognized marker horizon in the Eastern Pacific (Fig. 7-7).

Few isopach maps have been constructed for deep sea tephra because adequate spacing and numbers of samples are prohibited by the high cost of obtaining data from the sea floor. In lieu of isopach maps, Ninkovich et al. (1978) plot thickness

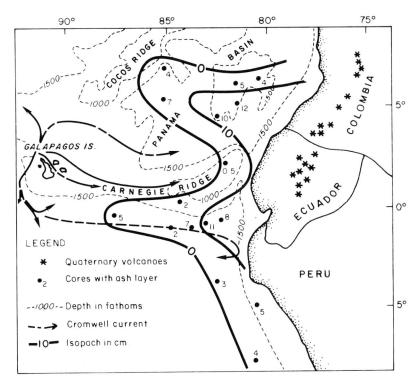

**Fig. 7-7.** Ash layer L (of Bowles et al., 1973), 0.23 m.y. old, 0.5 to 12 cm thick. Estimated area = ~300,000 km$^2$; volume = ~19 km$^3$. Source from volcanoes in Colombia and/or Ecuador. W-shaped distribution corresponding to branches of Cromwell current suggest that currents redistributed much of the ash. (After Ninkovich and Shackleton, 1975)

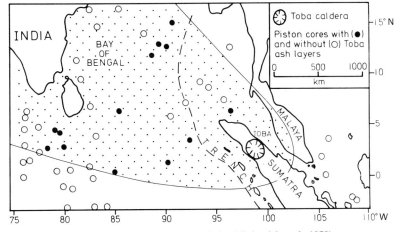

**Fig. 7-8.** Distribution of Toba (Sumatra) tephra in deep sea cores. (After Ninkovich et al., 1978)

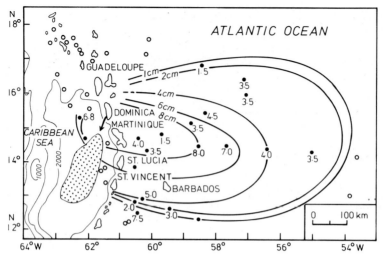

**Fig. 7-9.** Thickness of Roseau fallout tephra determined from sea bottom cores. Stippled area shows distribution of subaqueous pyroclastic debris flow in Grenada Basin with arrow showing path from Dominica. Submarine depth contours in meters. Open circles are cores that lack tephra. (After Carey and Sigurdsson, 1980)

of the Toba tephra layer (75,000 yr. B.P.) (Sumatra) (Fig. 7-8) graphically against distance. Using this as a guide, they subdivide the known dispersal area into two parts: within 1000 km of source they estimate an average thickness of 30 cm, and within 2500 km of source they estimate an average thickness of 10 cm. They conservatively estimate a total volume of 1000 km³ for the Toba deep sea tephra layer which was derived from the eruption that produced Toba caldera, Sumatra, one of the world's largest (Williams, 1941; Van Bemmelen, 1949). Together with comparable estimates of tephra volume from Toba on land, Ninkovich et al. (1978) estimated a total dense rock equivalent of rhyolitic magma to be 1000 km³.

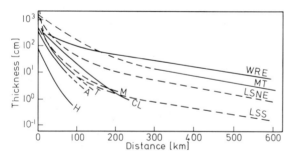

**Fig. 7-10.** Thickness and distance from source along dispersal axis for several fallout tephra layers. Volumes are: Hekla (*H*), Iceland, 0.17 km³; Avellino (*A*), Vesuvius, Italy, 2.1 km³; Toluca (Mexico) (*T*), 3.5 km³; Crater Lake, Oregon (*CL*), 15 km³; White River Eastern lobe (*WRE*), Canada and Alaska, 25 km³; Minoan Santorini, Eastern Mediterranean (*M*), 28 km³, Minoan Plinian deposit curve is based upon compacted thickness of total Minoan tephra layer (*MT*) using adjusted isopach contours of Fig. 7-5. (After Watkins et al., 1978) Laacher See northeastern (*LSNE*) and southern (*LSS*) lobe (16 km³). (After Bogaard, 1983)

One of the few isopach maps of submarine tephra layers (Fig. 7-9) is of Pleistocene Roseau ash which lies east of the Lesser Antilles island arc and was derived from the island of Dominica (Carey and Sigurdsson, 1980). Cores containing the visible tephra layer define a tephra fan covering $3 \times 10^5$ km$^2$; beyond 650 km east of the arc, the tephra occurs only as a dispersed layer (Huang and Carey et al., 1979). Carey and Sigurdsson (1980) calculate the volume within the 1 cm contour to be 13.2 km$^3$ based upon observed thickness, and estimate an additional volume of 12.2 km$^3$ was transported beyond the observed fan based upon mass balance calculations using crystal/glass ratios, giving a total volume of 25.4 km$^3$. Thickness and volume estimates for other known large eruptions are given in Fig. 7-10.

## Grain Size and Sorting

The size characteristics and consequently sedimentary structures of marine fallout layers are governed initially by both the atmospheric and subaqueous media through which they fall. Extensive size fractionation tends to develop better size grading in marine tephra than their exact equivalents on land (Ledbetter and Sparks, 1979), but several factors operate against perfect sorting. These include aggregation and mixing with pelagic and hemipelagic materials as they fall through the water column together with bioturbation processes after deposition. Moreover, diagenetic alteration creates additional difficulties in determining initial size parameters, particularly in ancient submarine tephra layers.

Evaluation of size data must also take into account contamination from floating pumice with distribution patterns unrelated to atmospheric fallout (Lisitzin, 1972). Large pumice fragments commonly float because many of the vesicles are intact, whereas tiny pumice particles smaller than 1/4 mm ($>2\phi$) readily sink because most bubble walls are broken or cracked (Fig. 7-6); large blocks can float for several years (Binns, 1967; 1972). Floating pumice rafts, aggregated from large eruptions, may travel over 12,000 km (Fig. 7-11), and during transport the grinding action between pumice lumps results in a constant rain of sand- and silt-size pumice, shard and crystal particles along their pathway (Murray and Renard, 1884). Their trace far exceeds that of similar-sized tephra transported aerially. Pumice rafts are sufficiently common to have delivered enormous volumes of tephra to the sea floor through geologic time (Bryan, 1968; Coombs and Landis, 1966; Gass et al., 1963; Richards, 1958; Sutherland, 1965).

The downwind grain size patterns of submarine tephra have been used by Shaw et al. (1974), Huang et al. (1973, 1975), Huang and Carey et al. (1979), Watkins and Huang (1977) and Ninkovich et al. (1978) to estimate eruption column heights, magnitude, and energy of eruptions. Ninkovich et al. (1978) compared the relationship between median diameter and distance to source for samples from the basal parts of three deep sea tephra layers and subaerial samples of the 1947 Hekla ash (Fig. 7-12), but point out that the coarsest one percentile ($\phi_1$) is a more useful grain-size parameter than Md$_\phi$. This is because $\phi_1$ is a measure of maximum grain size within a given deposit as discovered by Suzuki et al. (1973). The $\phi_1$ parameter is easier and faster to determine, is less sensitive to sorting in the water column and to contamination by fine-grained sediment that can substantially modify Md$_\phi$,

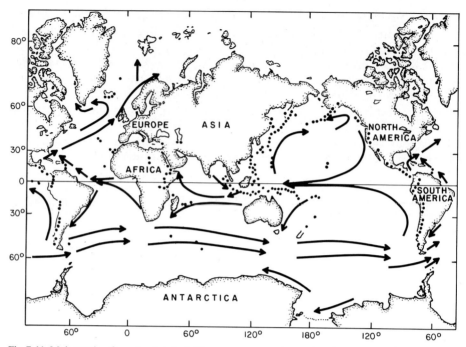

**Fig. 7-11.** Major paths of pumice distributed by ocean currents (*arrows*) and relation to major subaerial volcanoes (*dots*) in world oceans. (After Lisitzin, 1972, Fig. 169)

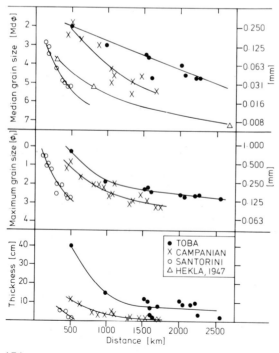

**Fig. 7-12.** *Upper diagram:* Median grain size and distance for submarine fallout layers, Toba (Sumatra), Campanian and Santorini layers (Eastern Mediterranean) compared with 1947 Hekla (Iceland) tephra. *Middle diagram:* Maximum particle size from same layers as in upper diagram. *Lower diagram:* Thickness plotted against distance from source. (After Ninkovich et al., 1978)

Fig. 7-13. Model of development of grading in a deep sea tephra layer to estimate duration of eruption. See text for explanation. (After Ledbetter and Sparks, 1979)

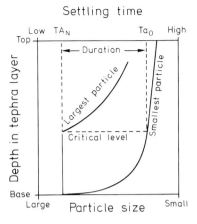

thereby giving a convenient method by which to compare relative intensity of eruptions. Comparing size data from the Toba deep-sea ash layer with known eruption column height, wind velocity and size data from the 1947 Hekla eruption, Ninkovich et al. (1978) estimate an eruption column height during the Toba eruption of greater than 50 km. The same order of magnitude for column height was obtained using the model of Huang and Carey et al. (1979) for variation of downwind particle size with eruption column height and wind velocity in conjunction with maximum size ($\phi_1$) data from the Toba tephra.

An innovative use of grain size data was introduced by Ledbetter and Sparks (1979) and used by Ninkovich et al. (1978) to predict duration of large magnitude explosive eruptions from the vertical size grading of nonvesicular ash particles (e.g. feldspars) in deep sea tephra layers. The size grading is a function of the duration of eruption, rate of release, residence time in the atmosphere, settling velocity and water depth at the site of deposition. The model predicts a zone at the base of a deep sea tephra layer where the coarsest particles remain constant in size with height above the base (Fig. 7-13). The coarsest particle size decreases with height above this critical level, with the break in slope corresponding to the level in the layer where the last large particle, ejected at the end of the eruption ($TA_N$), was deposited. They demonstrate that the finest particle deposited together with the last largest particle was erupted at the beginning of the eruption ($Ta_0$). Thus, the duration of the eruption is predicted to be the difference in settling time between the largest and smallest particles at this level. Ledbetter and Sparks (1979) calculate a duration of 20 to 27 days for the eruption that produced the Worzel D Layer (Bowles et al., 1973); Ninkovich et al. (1978) calculate a duration of 9 to 14 days for the eruption which produced the Toba deep sea ash.

The significance of determining duration times of eruptions is that, used in conjunction with computed ash volumes, estimates of average volume eruption rates can be made. Thus, for the Toba tephra, if 1000 km$^3$ of magma were erupted in about 10 days, the average magma discharge would exceed $10^6$ m$^3$ s$^{-1}$. Wilson et al. (1978) and Settle (1978) independently show that eruption column heights are directly controlled by the volume discharge rate of magma. The theoretical limit of 55 km for the height of a convective eruption column in a standard atmosphere suggested by Wilson et al. (1978) corresponds to a volume discharge rate of $10^6$ m$^3$ s$^{-1}$. This in turn corresponds to a column height during the Toba eruption of over 50 km deduced by them on other grounds.

## Regional Distribution and Tephrochronology

### Source

Three main approaches have been followed in attempts to determine sources of deep-sea tephra. On the most general level, one or more ash layers are recorded in two or more cores and an attempt is made to correlate distinctive ash layers between adjacent cores and a plausible source area (e.g. Bramlette and Bradley, 1942). Also, vitric volcanic ash dispersed within ocean bottom sediments (dispersed layers) has been used to designate stratigraphic horizons even though discrete ash layers may not be present (e.g. Hein et al., 1978).

A second approach is to scan systematically a large number of cores for the presence and distribution of ash layers. Horn et al. (1969, 1970), for example, surveyed 300 piston cores from the North Pacific available at that time and identified two major provinces occurring in a zone about 1000 to 1300 km wide parallel to Japan, Kuriles, Aleutians and Alaska. The western province contains layers with dominantly white (silicic) shards, whereas the northern and northeastern province contains layers of dominantly less silicic brown shards (Fig. 7-14). Horn et al. (1969) did not attempt to correlate individual ash beds, but they demonstrated the

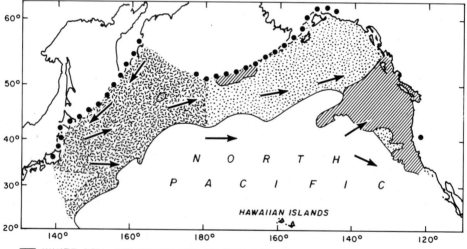

Fig. 7-14. Distribution of submarine tephra in the North Pacific relative to water and wind currents shown by *arrows*. (After Horn et al., 1969)

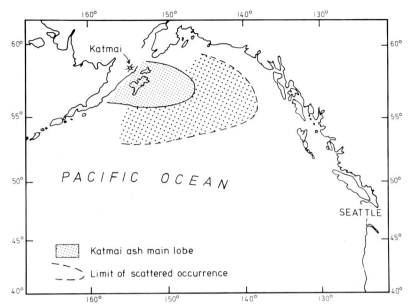

Fig. 7-15. Distribution of 1912 Katmai ash (from Novarupta) on sea floor, Gulf of Alaska. (After Nayudu, 1964b)

very wide distribution of ash layers downwind from the most likely sources of active island arcs and continental margin volcanism.

A third approach deals with regional correlation of individual ash layers. It is usually difficult to determine sources of layers where there are no historical records of eruptions, but isopach maps, chemical composition, and parameters of directional significance may be used to infer relatively restricted source areas (e.g. Huang et al., 1973, 1974, 1975; Huang and Carey et al., 1979; Ninkovich and Shackleton, 1975; Thunell et al., 1979). Recent work, however, has defined specific sources for (1) the Toba ash (75,000 yr B.P.), Sumatra (Ninkovich et al., 1978), (2) the Y-8 layer and Worzel D layer from the Lake Atitlan caldera (84,000 ± 5000 yr B.P.), Guatemala (Hahn et al., 1979), (3) the Y-5 ash layer in the Mediterranean from the Neapolitan volcanic province of Italy (Thunell et al., 1979); and (4) the Roseau Ash (Pleistocene) from Micotrin Volcano, Dominica, in the Lesser Antilles (Carey and Sigurdsson, 1980). Exact sources are definitively located where volcanic eruptions are historically documented. Probably none are more famous than the tephra layers from Santorini volcano (Mellis, 1954; Ninkovich and Heezen, 1965; Keller et al., 1978; Watkins et al., 1978). Another example is ash from the 1912 eruption of Novarupta (Katmai), Alaska, which has been identified and described from within the upper 10 cm of the bottom sediments on the sea floor as far south as 1120 km from Novarupta (Nayudu, 1964b) (Fig. 7-15).

## Correlation and Age

Deep sea ash layers from different cores are correlated by a number of properties including refractive index, color and shape of shards, mineral content, chemical composition of shards and mineral phases or bulk composition, stratigraphic position relative to other distinctive layers, and so on. An additional tool is seismic reflection (Fig. 7-16), although coring is needed to verify a reflector as an ash bed (or several beds) before it can be used for correlation. Correlation is especially difficult in areas of many active volcanoes, especially near the circum-Pacific island arcs and continental margins because of the likelihood of many overlapping ash lobes from adjacent volcanoes. Jezek (1976), for example, who studied some 30 ash layers in three cores only 130 to 400 km apart, had great difficulty in making any clear-cut correlations.

Electron microprobe techniques to analyze individual glass shards and mineral phases have been used for some time for subaerial tephra deposits (Chap. 13), and are increasingly used for deep sea tephra (e.g. Carey and Sigurdsson, 1978; 1980; Watkins et al., 1978). The chemical equivalence of the Los Chocoyos Ash (Guatemala) and the Worzel D Layer in the eastern Pacific was established on the basis of major and trace element composition (Hahn et al., 1979) (Fig. 7-17).

Ash layers intercalated in marine sediments can be dated by several methods. Once an ash layer is found to be diagnostic and is also well dated by independent methods, its proven presence in other cores can then be used as a precise time horizon. In the early work, distinctive fossils or faunal assemblages were used to give the relative age of an ash layer. From the mid-sixties onward, paleomagnetic methods were employed to date both fauna and ash layers (e.g. Ninkovich et al., 1966). Several attempts have been made to date glass shards by K/Ar techniques, but these often gave spurious results. Commonly the ages given were too old, due to inherited argon, and to potassium loss during alteration of the glass, or for other reasons (Ninkovich and Shackleton, 1975). Hogan et al. (1978) have used the $^{40}$Ar-$^{39}$Ar dating technique with success. U/Th and fission track dating techniques have been applied to glass shards (e.g. McCoy, 1974), and $^{14}$C age determinations have been made on young ash (<45,000 yr B.P.) which are associated with organic calcareous debris. Steen-McIntyre (1975) has suggested that it may be possible to date tephra layers by relative amounts of superhydration (amount of water in closed vesicles) and presented data to indicate that superhydration curves from deep sea ash are similar to those of subaerial shards of approximately similar ages (Fig. 7-18).

Oxygen-isotope stratigraphy using $^{18}$O/$^{16}$O ratios from calcareous organisms is another stratigraphic tool useful for correlating marine sequences (Shackleton

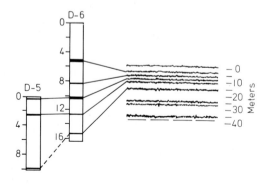

Fig. 7-16. Ash layer reflections in the eastern Pacific correlated by 3.5-KHz echogram. (After Bowles et al., 1973)

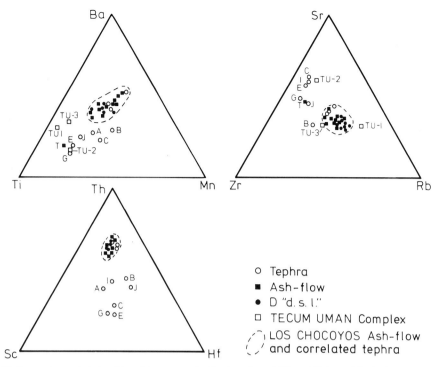

Fig. 7-17. Trace element plots of Los Chocoyos ash, Guatemala and Worzel D layer (*d.s.l.*, deep sea layer) in the eastern Pacific compared with other upper Quaternary ashes (tephra and Tecum Uman Complex) in Guatemala. *Letters* and *numbers* designate samples and stratigraphic locations in Hahn et al. (1979). (After Hahn et al., 1979)

and Opdyke, 1973, 1976) and is used by Keller et al. (1978) and Thunell et al. (1979) to establish a tephrochronologic framework and date tephra layers within the Mediterranean Sea.

Results of some studies in the Pacific and Atlantic Oceans are briefly discussed in the following section. Students interested in marine ash layers should also investigate numerous excellent studies of deep-sea tephra in the Mediterranean Sea, some of which are Mellis (1954), Ninkovich and Heezen (1965), McCoy (1974), Richardson and Ninkovich (1976), Watkins et al. (1978), Keller et al. (1978), Thunell et al. (1979) and Sparks and Huang (1980).

## Pacific Region

As shown in earlier pages, considerable research on marine tephra has been accomplished seaward of the most active volcanic zone in the world – the Pacific island arcs and continental margins (Fig. 7-19). One of the earliest studies on widespread ash layers in the Pacific was that by Worzel (1959), who discovered a submarine ash from sub-bottom seismic reflections within a 500 km-wide zone extending 2500 km between 11° N and 12° S latitude off the coast of Central and South America. Originally the ash field was thought to be a single layer and called the

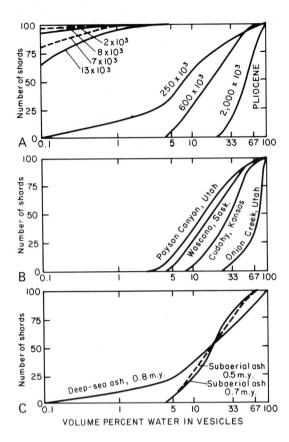

Fig. 7-18 A–C. Superhydration curves for naturally occurring shards of pumiceous fine ash. Each *curve* plots average volume of water in closed, spindle-shaped cavities that occur in 100 shards. Cavities are 10–50 μm long. A Curves for ashes of progressively older ages as designated. B All Pearlette type 0 ash samples, 600,000 years. C Inverted triangle deep sea ash, northwest Pacific, 800,000 years. Compared with subaerial ash of comparable age. (After Steen-McIntyre, 1975)

Worzel ash. Recovered ash from the very few cores that penetrated the reflection zone contained colorless bubble-wall shards with R.I. = 1.50 and sizes ranging between 0.07 mm and 0.2 mm. Ewing et al. (1959) inferred a correlation of the Worzel ash with the lower tephra layer described by Bramlette and Bradley (1942) from the north Atlantic, as well as to an ash layer found in the Gulf of Mexico. Nayudu (1964b) pointed out that correlation based on similarity of refractive index is not sufficient; he suggested that the "single ash" was composed of several layers derived from different volcanic eruptions.

From trace and minor element studies, Bowles et al. (1973) suggested that one of the Worzel ash layers (Layer D) might be equivalent to an ash in the Gulf of Mexico reported by Ewing et al. (1959). Indeed, Hahn et al. (1979) discovered that the composition of samples from Worzel Layer D coincide for all 15 trace elements analyzed by them with the Plinian fallout and pyroclastic flow sequence of the Upper Pleistocene Los Chocoyos Ash erupted from Lake Atitlan cauldron in the Guatemalan Highlands (Hahn et al., 1979) (Fig. 7-17). Moreover, Drexler et al. (1978), using neutron activation analysis of bulk ash samples and major element composition, index of refraction and morphology of glass shards, show that the Los Chocoyos Ash extends into the Gulf of Mexico and Caribbean Sea where it is preserved as the Y-8 tephra, and therefore is correlated with the Worzel D Layer.

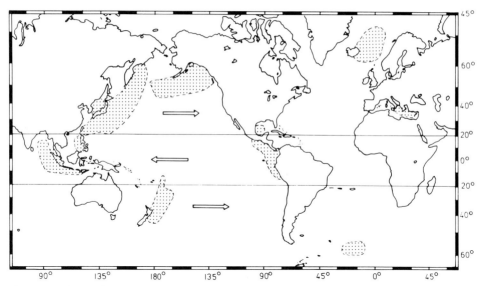

Fig. 7-19. Main provinces of coarse ash in the world's oceans (*stippled*) derived from land volcanoes. Dominant wind directions shown by *arrows*. (After Ninkovich, unpublished; in Kennett, 1981)

Thus, the Los Chocoyos Ash may well have a distribution exceeding $6 \times 10^6$ km$^2$ (Fig. 7-20). A significant aspect of the correlations is that the submarine ash allows dating of the land-deposited Los Chocoyos Ash. Based upon sedimentation rates, the Worzel Ash is estimated to be between 50,000 and 60,000 years old (Bowles et al., 1973); the Y-8 tephra in the Gulf of Mexico and eastern Pacific is $84,000 \pm 5000$ yr B.P. on the basis of oxygen-isotope stratigraphy. This places the age of the Los Chocoyos Ash within an age bracket which is otherwise difficult to date radiometrically on land.

A discussion has arisen concerning the importance of deep sea ash layers in the Pacific to infer changes in rates of explosive volcanic activity that may lead to further implications concerning global control of volcanic episodicity as well as the influence of volcanic dust in the atmosphere on climatic changes. Kennett and Thunell (1975) suggested that maxima in explosive volcanism resulted in greatly increased continental glaciation. Ninkovich and Donn (1976), on the other hand, argued that the increase in ash layers in late Pliocene to Pleistocene sediments in the Indian Ocean west of the Indonesian arc was due primarily to eastward migration of the plate, with the uppermost part of the sedimentary pile reaching the zone of ash accumulation much later than the older part. An additional argument was that there is no significant change in the frequency of occurrence of ash layers on fixed plates east of the arc. Ninkovich and Donn (1976) did not deny potential influence of volcanic dust veils in the atmosphere on climate, but argued that volcanic dust erupted during historic times settled out quickly. Thus, only explosions of unusual magnitude during critical times of climatic evolution might modulate the climate. Kennett and Thunell (1977) rejected the criticism by Ninkovich and Donn (1976) and their postulate of uniform volcanic activity during the Neogene. They insisted

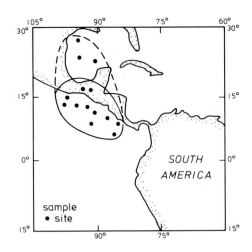

Fig. 7-20. Inferred aerial distribution of Los Chocoyos Ash and submarine equivalents. (After Hahn et al., 1979)

that worldwide peaks in volcanic activity, especially in the Pleistocene, are not only recognizable in the ocean basins, but also in many continental areas.

Hein et al. (1978) argue that earlier workers have overlooked altered (bentonite) ash beds. Their study, which includes consideration of the bentonite beds together with unaltered ash beds, indicates that volcanism on the Aleutian Ridge and Kamchatka Peninsula appears to have increased every $2.5 \times 10^6$ years for the past $10 \times 10^6$ years and every $5.0 \times 10^6$ years from 10 to $20 \times 10^6$ years ago. They conclude that the middle and late Miocene and Quaternary were times of increased volcanic activity in the North Pacific and elsewhere around the Pacific Basin. Furthermore, they hold that the apparent absence of a volcanic record before late Miocene time at Deep Sea Drilling Project Site 192 is due to masking of the record by diagenetic alteration and not the result of plate motions as was suggested by Stewart (1975) and Ninkovich and Donn (1976). Hein et al. (1978) point out that although major global volcanic episodes appear to have occurred at least twice in the last 20 m.y., only one occurred during the Pleistocene. Thus, the increased volcanism itself may not have been the primary cause of global cooling. Indeed, loading of ice in polar regions during cool periods may cause changes in the earth's spin axis, or stress changes on the earth's crust that might lead to increased volcanism (Fairbridge, 1973). The problem of volcanological control of climate is further addressed by Porter (1981) and in a special issue of the Journal of Volcanology and Geothermal Research (Newell and Walker, 1981).

**Atlantic Region**

Bramlette and Bradley (1942) first discovered two volcanic ash horizons in sediment cores in the North Atlantic between latitudes 48° and 50° N with silicic shards (R.I. = 1.51), which they thought were derived by subaerial fallout, most likely from Iceland or Jan Mayen. Ruddiman and Glover (1972) found the same ash-rich zones in five additional cores in the North Atlantic taken east of the Mid-Atlantic Ridge at latitudes between about 50° and 54° N, dated them as 9,000 yr B.P. and 65,000 yr B.P. respectively and found a third horizon dated at about 340,000 yr B.P. They found no shards in a core at 42° N latitude, strengthening the conclusion of Bramlette and Bradley that Iceland or Jan Mayen were the most likely source area and not the Azores. Ruddiman and Glover reject the notion that the shards were transported through the air, however, arguing that (1) the shards were ice-rafted to their present position because the ash is mixed with terrigenous debris, (2) the ash includes mafic shards and vesicular basaltic lithics and (3) the coarse size of many shards (between 0.07 and 1 mm) precludes aerial transport as much as 1,800 km, the distance of the southernmost core to Iceland, or 800 km to the Azores. They estimated that the total ash volume carried out to sea for the three ash-rich zones to be 0.94 km$^3$, 7.59 km$^3$, and 6.27 km$^3$.

Examination of a large number of piston and DSDP cores in the North Atlantic Ocean near Iceland (Fig. 7-21) (Donn and Ninkovich, 1980) indicates increased explosive volcanism during middle Eocene and Pliocene times, with the Eocene activity about twice the volume of the Pliocene. These are followed in decreasing abundance by the Pleistocene, Miocene, and Oligocene (Fig. 7-22). Distribution of cores containing ash layers, and prevailing wind patterns, suggests that the Cenozoic ash layers originated subaerially during the growth of Iceland.

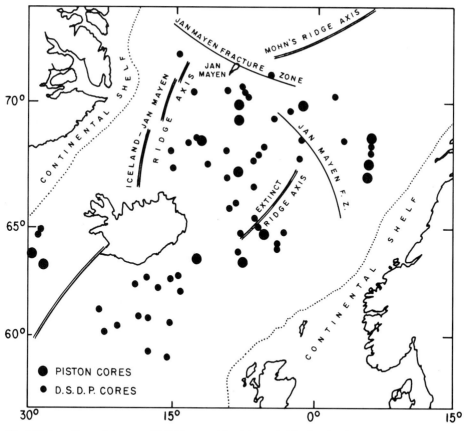

Fig. 7-21. Locations of piston and DSDP cores with ash layers in North Atlantic Ocean around Iceland. (After Donn and Ninkovich, 1980)

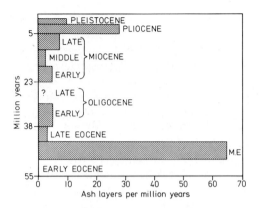

**Fig. 7-22.** Frequency of extrapolated ash layers per million years in the North Atlantic Ocean around Iceland determined from DSDP cores. (After Donn and Ninkovich, 1980)

From cores in the Norwegian and Greenland seas (east and northeast of Iceland), Sigurdsson and Loebner (1981) can distinguish two series of Cenozoic tephra layers: a high potash series ranging from quartz trachytes to alkali rhyolites and a low potash series ranging from icelandites to rhyolites (Fig. 7-23). Ashes from both series were erupted from sources in Greenland, Iceland, and Britain throughout Cenozoic time. Major elements alone do not allow precise source locations, but Iceland can be ruled out as a source for the Tertiary high potash series because alkali rhyolites older than Quaternary are unknown from Iceland. Moreover, five major episodes of Cenozoic explosive volcanism are reflected in ash abundances: late Paleocene, middle Eocene, middle Oligocene, early to middle Miocene, and Plio-Pleistocene (Fig. 7-24), agreement with the data of Donn and Ninkovich (1980) being best for the middle Eocene and Pleistocene. These episodes, and periods of high magmatic production rates on the Canary Islands

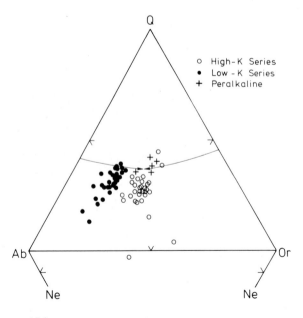

**Fig. 7-23.** Normative projection of North Atlantic tephra in the granite system showing the contrast between a low potash and a high potash to peralkaline group. The quartz-feldspar boundary curve is for $P_{H_2O} = 1,000$ kg cm$^{-2}$. (After Sigurdsson and Loebner, 1981)

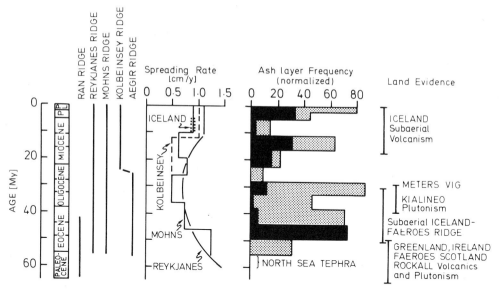

**Fig. 7-24.** Comparison of active rift axes, spreading rate, ash layer frequency and subaerial volcanism in the North Atlantic during the past 65 m.y. Ash layers studied in detail are shown in *black*, those reported in DSDP-legs are *shaded*. (After Sigurdsson and Loebner, 1981)

during the Neogene, especially mid-Miocene and Pliocene/Pleistocene (Schmincke, 1982a), appear to be synchronous with periods of increased volcanic and tectonic activity as indicated by frequency of tephra layers around the Pacific and K/Ar data of volcanic provinces (McBirney et al., 1974; Kennett et al., 1977; Kennett, 1981) thereby suggesting some kind of global control of volcanic episodicity during Cenozoic time (Vogt, 1979).

Petrographic, granulometric, chemical and morphological data of ash in 37 cores distributed east of the Azores Islands were studied in detail to infer frequency of volcanism, accumulation rates and total volumes, estimates of eruption cloud heights and eruptive energies (Huang et al., 1979).

In the South Atlantic, several tephra layers that extend about 640 km from the South Sandwich Islands toward the Mid-Atlantic Ridge have been described by Ninkovich et al. (1964). Rothe and Koch (1978) and Schmincke and von Rad (1979) were able to assign most of the 15 or so ash layers of Miocene to Pleistocene age, found in cores of Glomar Challenger drill holes (369, 397, 397a) (Legs 41, 47A) about 100 km southeast of the Canary Islands, to specific magmatic stages on different islands of the Canarian archipelago based on age and chemical and mineralogical composition of the ashes. The excellent agreement of the biostratigraphic age of unique peralkalic rhyolitic ashes in these cores with the K-Ar age of the unusual peralkalic rhyolitic ash flow tuffs on Gran Canaria is a powerful argument in support of K-Ar dates, indicating a middle Miocene age for the main phase of subaerial volcanic activity on Gran Canaria.

# *Chapter 8* Pyroclastic Flow Deposits

Pyroclastic flows are volcanically produced hot, gaseous, particulate density currents. Their deposits offer unparalleled opportunities to estimate minimum volumes of near-surface magma chambers as well as vertical chemical, mineralogical, and thus temperature and pressure distributions within the magma columns immediately prior to eruption. The modes of origin and transport of pyroclastic flows have been the subject of intense debate ever since the 1902 eruption of Mt. Pelée produced nuées ardentes which destroyed the town of St. Pierre. Because pyroclastic flows are emplaced rapidly and may flow for long distances, they are particularly useful for intrabasinal correlations.

**Fig. 8-1.** Trachytic welded tuff "Eutaxite" from type locality at Arico (Tenerife). Dark clasts are collapsed pumice blocks (obsidian). Note decrease in grain size in lower 1 m. Cooling unit 4 m thick

## Historic Development of Concepts

Puzzling rocks that have the combined features of pyroclastic rocks and lava flows have been known for more than 100 years. Von Fritsch and Reiss (1868) published perceptive descriptions of tuffs from Tenerife (Canary Islands), believed to have been emplaced by flowage, calling them *eutaxites* (Fig. 8-1). Independently, Abich (1882) named widespread sheets of rocks from Armenia with structural attributes typical of lava flows as well as pyroclastic rocks *tufolavas* (Fig. 8-2). Wolf (1878) described pyroclastic flow eruptions of Cotopaxi volcano (Ecuador) from eye witness accounts as "foam from a boiling-over rice pot" and provided the first map of pyroclastic flow deposits and their reworked aprons (Fig. 8-3).

Pyroclastic flows were described in some detail from the famous 1902 eruptions of Mt. Pelée (Martinique) and La Soufrière (St. Vincent, British West Indies) by Anderson and Flett (1903) and Lacroix (1904) in accounts that became milestones in the volcanological literature. The term *nuée ardente* introduced by Lacroix (1904) to describe these pyroclastic flows continues to be used for small *observed* flows from central vents (Fig. 8-4). Until a few decades ago, it was not realized that the bulk of material in pyroclastic flows was transported in glowing avalanches, hidden from view by the billowing glowing clouds (nuée ardente) (Smith, 1960a). The term nuée ardente may be retained as a term for witnessed flows as a general approximation of eruptive events, although it does not specify the origin of the flow or the complexities of flowage and deposition (Fig. 8-5). At Mt. Pelée, for ex-

**Fig. 8-2.** Rhyolitic "Tufolava" (welded tuff) from Armenia. Dark fiamme are collapsed pumice lapilli

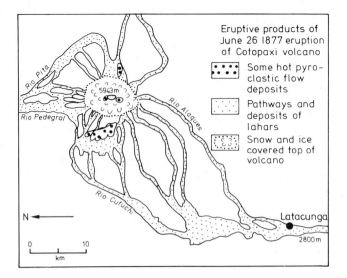

**Fig. 8-3.** Distribution of pyroclastic flow deposits and their reworked laharic apron of 1877 eruption of Cotopaxi (Ecuador). (After Wolf, 1878)

**Fig. 8-4.** Nuée ardente from the 1968 eruption of Mayon Volcano (Philippines). Origin by column collapse. (Courtesy of W. G. Melson)

ample, different types of nuées ardentes originated by (1) column collapse, (2) lateral projection, (3) dome collapse, and (4) "boiling-over" without a vertical column to produce deposits more aptly called block-and-ash flow (Fig. 8-6) and ash-cloud surge deposits (Fisher et al., 1980; Fisher and Heiken, 1982).

Application of the nuée ardente concept was not long in following its discovery at Martinique. Dakyn and Greenly (1905) suggested a "probable Peléean origin" for some of the Ordovician volcanic rocks in North Wales, Great Britain. Völzing (1907) used it to explain the emplacement of the Late Quaternary *Trass* deposits of the Laacher See area (Germany) (Fig. 8-7). *Trass* became the common name in the older literature for unwelded massive ash and pumice flow deposits similar to the way ignimbrite is used today by many workers. In 1911, Yamasaki (see Aramaki, 1956, p. 217) applied the nuée ardente concept to explain a pyroclastic flow from the 1783 eruption of Asama Volcano, Japan, and in 1923, Fenner interpreted emplacement of the 1912 "sand flow" at the Valley of Ten Thousand Smokes, Alaska, in terms of nuées ardentes.

Marshall (1935) was the first to realize that widespread lava-like pyroclastic rocks (welded tuff) in New Zealand were actually composed of glass shards, pum-

**Fig. 8-5.** Pumiceous nuée ardente from the July 22, 1980 eruption of Mount St. Helens (USA). Main flow is northward through the breached crater wall. (Courtesy of H.X. Glicken)

**Fig. 8-6.** Block and ash flow of dacitic composition from February 1976 eruption of Augustine volcano (Alaska). Flow formed by dome collapse. Note new steaming dome in center of volcano and steaming deposit in notch eroded during passage of pyroclastic flows

**Fig. 8-7.** Unwelded poorly sorted phonolitic ash flow deposit, "Trass", from type locality at Laacher See (Eifel, Germany). Thin surge deposits overlying pyroclastic flow. Scale 1 m long

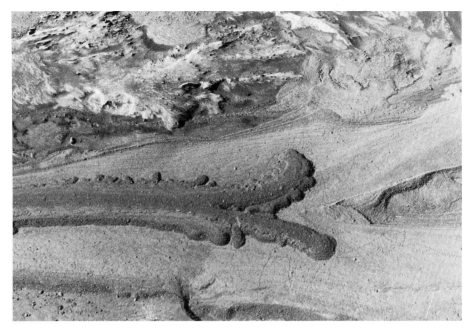

**Fig. 8-8.** Pumice flow deposits from June 1980 Mount St. Helens (Washington, USA) eruption. Note levees and steep flow fronts and margins. Main flows 10–20 m wide

**Fig. 8-9.** Inversely graded dacitic pumice ash flow deposit 50 cm thick from May 1980 eruption of Mount St. Helens (Washington, USA)

ice, and crystals deposited at high temperature. He thought the rocks – which he called *ignimbrite* – were laid down by "fiery showers" rather than by gravity flows. Shortly following Marshall's work, Gilbert (1938) provided the first detailed description of such rocks and established the term *welded tuff* for welded parts of the Bishop Tuff, California. In their classic papers, Smith (1960a, b) and Ross and Smith (1961) provided thorough summaries, and some of the subsequent work has been reviewed by Peterson (1970), Fisher and Schmincke (1978), Williams and McBirney (1979), Heiken (1979) and Sheridan (1979). The 1980 eruption of Mount St. Helens (Lipman and Mullineaux, 1981) has provided volcanologists with abundant new data on different types of pyroclastic flows and their deposits which as yet are only partially evaluated (Figs. 8-8, 8-9).

## The Deposits

Deposits of pyroclastic flows are diverse and reflect different types of eruptions and depositional regimes. Viewed from the perspective of flow mechanics to be discussed later, there are essentially two kinds of deposit – *pyroclastic flow deposits* which are commonly poorly sorted and massive, and *pyroclastic surge deposits* which are better sorted, finer-grained, thinner and better bedded than pyroclastic flow deposits. These two kinds of deposit occur alone or in close association and are described in later appropriate sections.

### Volume

Pyroclastic flows originate in different tectonic and volcanic settings and have vastly different volumes. Eruptions producing pyroclastic flow deposits on the order of 0.001 to 1.0 km$^3$ are from small central vent volcanoes typical of, but not confined to, magmatic arc systems such as 1902 Mt. Pelée, Martinique, 1968 Mount Mayon, Philippines, 1976 Augustine Volcano, Alaska, 1980 Mount St. Helens, Washington, and 1982 El Chichon, Mexico. More voluminous flows, 1–100 km$^3$, originate from larger stratovolcanoes such as 1883 Krakatau, Java, and Mount Mazama at the site of Crater Lake, Oregon. Volumes of 100 to 1000 km$^3$ are associated with the formation of large calderas such as the Long Valley (California) and Yellowstone (Wyoming) calderas, which develop as a consequence of eruption of such large volumes and not necessarily at the site of a previous volcano. Larger volumes have been derived from eruptions of Toba, Sumatra, and in one case (La Garita Caldera, San Juan Mountains, Colorado) the volume of a single pyroclastic flow sheet exceeds 3000 km$^3$ (Steven and Lipman, 1976). In general, small- to intermediate-volume flows range from rhyolitic to basaltic in composition, whereas large-volume flows are most commonly rhyolitic to dacitic.

The different types of pyroclastic fragments comprising pyroclastic flow deposits are related to how the flows originate. Small-volume flows produced by dome collapse or explosions associated with dome formation commonly contain abundant poorly vesiculated products of the domes, although small-volume flows consisting dominantly of pumice also occur. Intermediate- to large-volume flows are usually composed entirely of highly vesiculated materials derived from the rapid vesiculation of magma.

Some features of pyroclastic flow deposits are related to the total volume of the deposit and thus to the magnitude of the eruption. As volume increases, for example, so does distance that flows travel (Smith, 1960a), and, in some cases, vesicularity of juvenile fragments (Aramaki and Yamasaki, 1963).

**Relationship to Topography**

Pyroclastic flows may completely drain from upper slopes and only be preserved in the lower parts of valleys, thereby becoming initially thicker away from the source (Gorshkov, 1959; Taylor, 1958). In areas of rugged topography, small-volume pyroclastic flows may be confined to valleys. On the upper slopes of volcanoes, pyroclastic flows drain down their centers, leaving levees or "high water marks" and larger rock fragments on both sides of a valley, or along the outer edge of a sinuous channel because of the momentum of flow (Fig. 8-10). Pyroclastic flows spread out in fan-like lobes beyond the mountain slopes much like lahars (Cap. 11). Surface features such as levees and lobate forms and their implications are discussed by Wilson and Head (1981b) at Mount St. Helens. Pyroclastic surges can spread over topography of moderate relief and override the sides of a valley, and their deposits may mantle topography similar to fallout tephra, but unlike fallout tephra they can become ponded and thin toward valley margins (Crandell and Mullineaux, 1973; Wilson and Walker, 1982).

**Fig. 8-10.** Aerial view of lobate dacitic block-and-ash flow deposits from February 1976 eruption of Augustine volcano (Alaska). Note central channel in main lobe and younger lobe at head of channel. Compare with Fig. 8-6. Channel about 30 m wide

Confinement of flows in valleys or channels affects the mechanics of emplacement by flow as illustrated by deposits at Laacher See, Germany (Schmincke, 1970, 1977b; Schmincke et al., 1973). There, between pre-existing channels close to the vent, where flows were unconfined, the deposits are medium- to thin-bedded, laterally continuous and occur as cross-bedded pyroclastic surge deposits. Within channels, however, equivalent beds thicken 10-fold, are massive, poorly sorted and lack cross bedding (Fig. 8-11). This relationship suggests that the flows became confined and concentrated within the channels, thereby increasing the particle concentration which caused mass flowage. A similar interpretation is given by Crowe and Fisher (1973) for pyroclastic surge and equivalent massive beds at Ubehebe maar volcano, California. Walker et al. (1980) interpret thin beds that mantle topography which are traceable into ponded thick beds as a "tail deposit" left in the wake of a pyroclastic flow based upon arguments of volume, lateral extent, and grain-size. They call them "ignimbrite veneer deposits". During the May 18, 1980 eruption of Mount St. Helens, Washington, the initial blast surmounted multiple barriers, some up to 600 m, and left debris-mantled slopes. Deposits of "rootless" hot pyroclastic flows, that may have been derived from the base of the blast as it swept across topographic barriers, or else from slumping and flow of blast deposits from steep slopes, occur as pond and valley fills (Hoblitt et al., 1981).

Widespread sheet-like ignimbrite layers associated with large calderas and other volcanotectonic depressions commonly have sufficient volume and thickness to smooth out underlying terrain. They thicken and thin according to topographic irregularities underneath but maintain a nearly flat, horizontal, or gently sloping surface and gradually thin toward their distal edges. Successive flows of great vol-

**Fig. 8-11.** Late Quaternary pyroclastic flow deposits showing transition from paleovalley (*right*) to overbank facies (*left*). Paleovalley cut into fallout deposits (*lower part of scale*) by lower pyroclastic flows (*central part of scale*). Plinian fallout deposits (*top of scale*) interlayered with flow deposits. Scale 5 m long. Laacher See (Eifel, Germany)

ume may completely mask previous topography, constructing thick widespread sheets later dissected into plateaus such as on North Island, New Zealand, an ignimbrite plateau that covers 25,000 km².

**Flow Units and Cooling Units**

A basic stratigraphic and field distinction that must be made for intermediate- to large-volume pyroclastic flow deposits is the difference between flow units and cooling units (Smith, 1960b). A *flow unit* is a depositional unit that represents a single pyroclastic flow deposited in one lobe. The thickness of individual flow units can vary from a few centimeters to many tens of meters, and the lobes may follow one another within minutes or hours. The boundaries between flow units are marked by changes in grain size, composition, fabric, concentration of pumice lapilli or block accumulations, cross-bedded zones, etc. (Fig. 8-12). When several very hot flow units pile rapidly one on top of the other, they may cool as a single *cooling unit*. A simple cooling unit forms when a single flow or successive flows cool as a unit with no sharp changes in the temperature gradient. A *compound cooling unit* forms when there is an interruption in temperature that disturbs the continuous coolingunit zonation of successive hot flows. Cooling from emplacement- to ambient temperature may take tens of years, depending on the thickness of the deposit and the emplacement temperature. Thus, many ash flow deposits are mapped as cooling units (Smith, 1960b) even though they are composed of several flow units.

**Fig. 8-12.** Fine-grained top flow unit (about 1 m thick) of pantelleritic ignimbrite, underlying highly welded base of ignimbrite C. Gran Canaria (Canary Islands). Arrow points to hammer

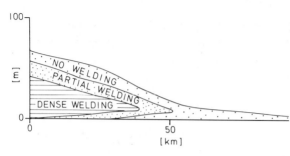

Fig. 8-13. Idealized lateral and vertical configuration of welding zones in a simple cooling unit. (After Smith, 1960b)

A cooling unit is marked by a more or less systematic pattern of zones of rock differing in degree of welding and thus density resulting from different cooling regimes (Fig. 8-13). In young deposits not lithified by diagenesis, compaction, and metamorphism, the top and bottom parts of cooling units are commonly composed of friable unwelded pyroclastic material: the basal layer is unwelded because it cools quickly against the cold rock basement, and the top because of relatively rapid heat convection and radiation into the atmosphere. The area of densest welding, in the lower half of a cooling unit, is the zone that remains longest at the maximum emplacement temperature (Jaeger, 1968; Riehle, 1973). At high emplacement temperatures and slow cooling rate, partial or complete crystallization (primary devitrification) of the hot and compacting glassy pyroclasts occurs in the interior

Fig. 8-14. Bandelier Tuff. Three main flow units comprise a compound cooling unit in this photograph. The tuff forms a widespread plateau around Valles Caldera (New Mexico, USA) (background)

of thicker cooling units. Such zones grade into poorly welded zones lithified by crystallization of high temperature vapor phase crystals, typically silica polymorphs and alkali feldspar. Simple and compound cooling units thus have characteristic cooling zones (Smith, 1960b) (Figs. 8-13, 8-14).

**Components**

Pyroclastic flow and surge deposits are composed of crystals, glass shards and pumice, and lithic fragments in highly variable proportion, depending upon the composition of the magma and the origin of the flows. In some deposits, a significant portion of the crystal and lithic components may be xenoliths. Flow deposits derived from the explosive disruption or collapse of domes or lava flows contain mixtures of nonvesicular to partially or wholly vesicular juvenile lithic particles.

Ash-flow tuff, by definition, is composed of more than 50 percent of components in the ash size range ($<2$ mm). These form a matrix into which varying amounts of pumice lapilli or lithic lapilli may be set. The most common ash-sized clasts are glass shards, usually accompanied by smaller amounts of pumice particles. The ash- and lapilli-sized pumice fragments are characterized by either spheroidal or long subparallel tubular vesicles a few millimeters to micrometers in diameter. Tubular pumice is thought to develop as the vesiculating magma rapidly rises in the conduit, thereby stretching the vesicles as they form.

Crystals are the next most common ash-size component. In contrast to phenocrysts in lavas, those in ignimbrite are commonly broken. Phenocrysts within accompanying comagmatic pumice lapilli or blocks, however, are largely nonfragmented, which indicates that breakage takes place during eruption and transportation. Breakage may continue even during compaction, as shown by slightly separated crystal fragments with glass-filled fractures, "boudinaged" feldspar tablets, and crumpled mica flakes. Crystal abundance ranges from near 0 to about 50 percent in ignimbrites and may be higher than in associated lava flows of the same composition. Crystals are generally more abundant in the matrix than in pumice lapilli and bombs, strong evidence of preferential concentration in the matrix relative to glass shards during transport (Hay, 1959a; Walker, 1972a; Sparks and Walker, 1977).

Because most large-volume pyroclastic flow deposits are calc-alkalic dacite to rhyolite, most phenocrysts are quartz, sanidine, and plagioclase with minor amphibole, pyroxene, biotite, Fe/Ti-oxide, and accessory phases such as zircon and sphene. In trachyte, phonolite, and peralkalic rhyolite, anorthoclase takes the place of the two feldspars. The common upward increase of phenocryst abundance and also the change to more mafic composition within single depositional units is discussed later, and in Chapter 2.

Lithic fragments rarely exceed 5 volume percent of intermediate- to large-volume and some small-volume pumiceous pyroclastic flows. There are three major sources for these lithic fragments: (1) slowly cooled and crystallized magma "rinds" from chamber margins (Schmincke, 1973), (2) rocks from the conduit walls (Eichelberger and Koch, 1979), and (3) rock fragments picked up along the path of the pyroclastic flow. From the first two sources, information may be obtained on the depth of the magma chamber if the regional stratigraphy is sufficiently well known.

**Fig. 8-15.** Giant stranded block transported by pyroclastic flow of 1976 eruption of Augustine volcano (Alaska). Pyroclastic flow in which block was transported was deposited on lower flanks of volcano (background)

Many small-volume pyroclastic flow deposits are composed almost entirely of juvenile lithic fragments (Fig. 8-15) and broken crystals with adhering matrix derived from the explosive disruption of the neck and dome of a volcano. Examples are the 1974 eruption of Ngauruhoe Volcano, New Zealand (Nairn and Self, 1978) and the 1902 eruption of Mt. Pelée, Martinique (Fisher and Heiken, 1982).

**Primary Structures in Unwelded Deposits**

Most unwelded pyroclastic flow deposits are poorly sorted and massive, but may show subtle grading, alignment bedding or imbrication of oriented particles. In contrast, most pyroclastic surge deposits are thinner, finer-grained and better sorted than flow deposits, and wavy- or cross-bedded structures may be common.

Internal Layering

*Pyroclastic Flow Deposits.* The layering within pyroclastic flow deposits is manifested by graded basal zones, discontinuous trains of large fragments, alternating coarse- to fine-grained layers, crude orientation of elongate or platy particles, and by color or composition changes. Many of the features, including graded bedding, give evidence of emplacement as high-concentration laminar flows.

Grading within a single flow unit can be normal, inverse, symmetrical, or multiple (Chap. 5). Pumice fragments may be inversely size graded and lithic fragments

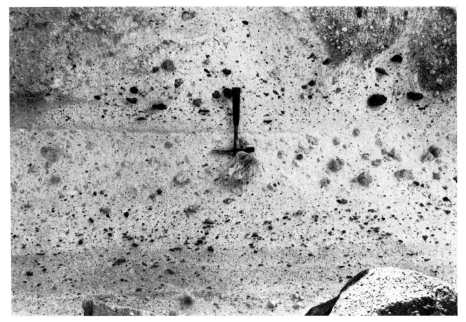

**Fig. 8-16.** Grading in flow unit of Pleistocene trachytic pyroclastic flow deposits. Note concentrations of dark, smaller dense trachyphonolitic rock fragments at base and large pumice lapilli and blocks at top of flow unit (hammer). Medano (Tenerife, Canary Islands) (Schmincke, 1974a)

normally graded because of their wide differences in density (Fig. 8-16). Crystal- and lithic-fragment concentrations in the basal zones of pyroclastic flow deposits, caused by sorting processes during flow, are discussed in the section on texture.

An example of crude stratification is given by Kuno (1941) from the 1929 Komagatake pumice flow, Japan (Fig. 8-17). Slight differences in size of the fragments in different layers give an irregular and indistinct stratification to the deposit. Some layers are a haphazard mixture of fragments and are without internal

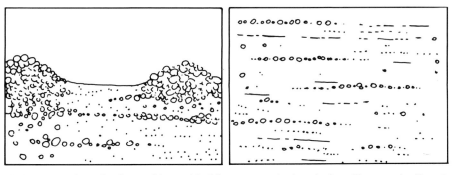

**Fig. 8-17.** Leveed pumice flow and internal bedding structure in deposits from Komagatake (Japan). (After Kuno, 1941)

layering and some are composed of more or less uniformly sized fragments that either thin out laterally or pass into an unstratified mass. Still other layers composed chiefly of fine-grained particles contain somewhat coarser fragments arranged in ill-defined zones parallel to the general bedding (Figs. 8-17, 8-18). Kuno (1941) attributes the origin of these structures to differential movements between a number of intertonguing streams within each pumice flow, which overlap and merge into one another.

Flat fragments within pyroclastic flow deposits may be strongly oriented parallel to depositional surfaces or may become imbricated (Mimura and MacLeod, 1978), particularly near the basal parts of the deposits (Schmincke et al., 1973). Orientation results from shearing forces within a flow during laminar movement, similar to that in debris-flow deposits (Johnson, 1970; Fisher, 1971).

Multiple grading may result from (1) a continuous recurrent surging within a single flow (Smith, 1960a), (2) mechanical differentiation due to shearing during laminar motion within a moving flow, or (3) separate flows repeated at relatively short time intervals (Sparks, 1976). Thus, multiple grading is difficult to interpret in terms of flow dynamics. The most common interpretation is that separate layers within the unit are actually separate flow units, but unless there are marked compositional changes, soil horizons, or erosional irregularities between the units, such an interpretation should be suspect.

*Pyroclastic Surge Deposits.* Pyroclastic surge deposits are thinly to thickly laminated and many have planar but slightly wavy bedded structures. Perhaps their most characteristic feature is wavy-, lenticular- or low angle cross bedding (Walker, 1971; Walker and Croasdale, 1971; Sparks et al., 1973; Crowe et al., 1978; Rowley et al., 1981; Fisher, 1979; Fisher et al., 1980) (Fig. 8-19).

Surge deposits may be interbedded or grade into the overlying ignimbrite (Crowe et al., 1978), suggesting an intimate association with the pyroclastic flow, but in many instances there is a sharp break above the ground surge deposits. This led Sparks et al. (1973) and Sparks (1976) to propose a separate origin for the surge which they compared to the ash hurricane of Taylor (1958). Fisher (1979) suggests that surges can develop from the margin of a collapsing eruption column; the surge precedes the pyroclastic flow which is derived from the collapse of the central main part of the column. On the other hand, C.J.N. Wilson (1980) and Wilson and Walker (1982) argue that surge deposits may develop from extreme turbulent action at the base of a flow as air is infolded beneath the advancing front (see also McTaggert, 1960).

Gas-Escape Structures

Gases escaping from pyroclastic flow deposits after deposition caused the famous fumaroles in the Valley of Ten Thousand Smokes (Griggs, 1922). The fumaroles were once believed to be magmatic gases released from an underlying magma chamber, but were later found to represent degassing of the pyroclastic flow deposit itself, which was erupted from Novarupta near Katmai volcano (Curtis, 1968). The geochemistry of these fumaroles is described in detail by Zies (1924, 1929). Degassing pipes may be recognized as oxidized zones or as fines-depleted pipes in lower-temperature deposits. Concentration of certain elements may lead

**Fig. 8-18.** Stratification and accumulation of pumice lapilli in lower part of distal Bishop Tuff ignimbrite overlying preceding Plinian pumice fall deposits. Bishop (California, USA)

**Fig. 8-19.** Cross bedded ash cloud surge tuff on top of lower ash flow cooling unit, Tshirege Member, Bandelier Tuff. Valles Caldera (New Mexico, USA)

to the formation of ore deposits along such fossil fumaroles. Fumaroles were evidently responsible for the fumarolic mounds and ridges of the Bishop Tuff, California (Sheridan, 1970). These mounds are relatively large features that stand 0.5 to 15 m above their surrounding terrain and are up to 60 m in diameter; the ridges are as long as 600 m. They occur at the top of the vapor-phase zone in the Bishop Tuff and are most numerous where crystallization of the sheet is intense. They are absent from areas where the sheet is thick, densely welded, and vitric. Smith (1960b) pointed out that the zone of fumarolic development occurs in the upper part of cooling units where intense vapor-phase activity is localized by deep joints (Fig. 8-20).

Some small fumarole pipes, however, lack evidence of vapor-phase crystallization but show that gases moved with sufficient upward velocity to transport particles. Such pipes commonly are more numerous near the tops of unwelded deposits and are not related to fracture systems (Fig. 8-21). The pipes are enriched in crystals and lithic fragments and depleted in dust-size tephra (Walker, 1972a). They may occur in single or branching patterns ranging up to 10 m long and 30 cm wide (Yokoyama, 1974) or as small as 10 cm long and 1 cm wide as we have observed at the base of the 1902 nuée ardente deposits of Mt. Pelée, in the top of 1976 Augustine, Alaska, ash flow deposits and some of the 1980 pyroclastic flow deposits at Mount St. Helens (Hoblitt et al., 1981). Some are lenticular, curvilinear and crescentic masses or segregation layers of coarse pumice or lithic fragments. Experiments by Wilson (1980) suggest that such segregation structures can form

**Fig. 8-20.** Cylindrical degassing pipes of Bishop ignimbrite (California), ca. 20 cm in diameter

**Fig. 8-21.** Degassing pipe in pyroclastic flow deposit at Herculaneum 79 A.D. eruption of Vesuvius (Italy). Note round pumice lapilli

*during* flowage under high degrees of fluidization (gas flow). Once formed, they seem to resist further mechanical mixing. In unconsolidated lithic-rich block-and-ash flow deposits that are difficult to distinguish from cold lahars (Chap. 11), the presence of numerous gas pipes suggests high-temperature emplacement. However, such gas pipes are described from some lahars emplaced at temperatures of < 400 °C and in other lahars which seem to have been cold (Crandell, 1982, personal communication).

## Emplacement Facies

A useful approach to understanding Plinian ignimbrite sheets is a facies model still in its early stages of development (Smith, 1960b; Sheridan, 1979; Wright, 1979; Wright et al., 1980a; Wilson and Walker, 1982). Volcanologically, a facies can be considered as an eruptive unit with distinct spatial lithologic relationships and dis-

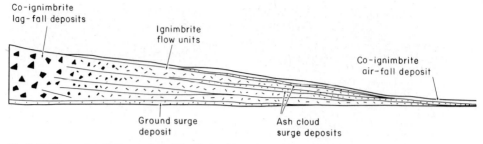

**Fig. 8-22.** Diagrammatic illustration of the kinds of deposits and lateral facies changes in a Plinian eruption unit. (After Wright et al., 1980a)

tinct internal structures and textures within vertical stratigraphic sequences; a facies model is a generalized summary of the organization of the deposits in time and space (Wright et al., 1980a) (Fig. 8-22; also see Chap. 13).

Wilson and Walker (1982) use the term facies for ignimbrite units defined by their morphology, relative superposition, composition and grain size characteristics. Based upon a fluidization model they infer the above features to be caused by different depositional regimes within a single pyroclastic flow which consists of a head, body and tail. They apply this concept to the "standard ignimbrite flow unit" (Plinian) introduced by Sparks et al. (1973). Fluidization caused by air ingestion occurs at the head of a flow and generates "layer 1" deposits (Fig. 8-23). The body and tail represent the bulk of the flow and generate "layer 2" deposits. Different

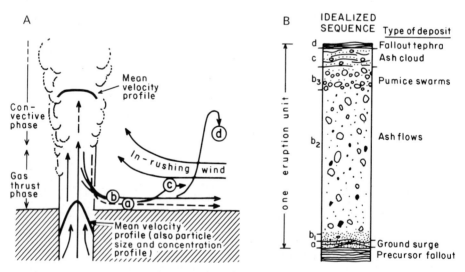

**Fig. 8-23. A** Collapse of Plinian eruption column. **B** Related to idealized eruption unit. Outer part of eruption column collapses to form a pyroclastic surge (*a*) followed by continued collapse and feeding of the interior parts of the column to form a pyroclastic flow (*b*). Ash cloud (*c*) is formed by elutriation of material from top of pyroclastic flow and finest material (*d*) settles from atmosphere onto surface after flow passes. Flow processes cause segregation of pyroclastic flow into various parts ($b$, $b_2$, $b_3$) as discussed by Sparks (1976) and Fisher (1979). Layer *a* equals layer *1*, layer *b* equals layer *2* and layer *d* equals layer *3* of Sparks et al. (1973) (After Fisher, 1979)

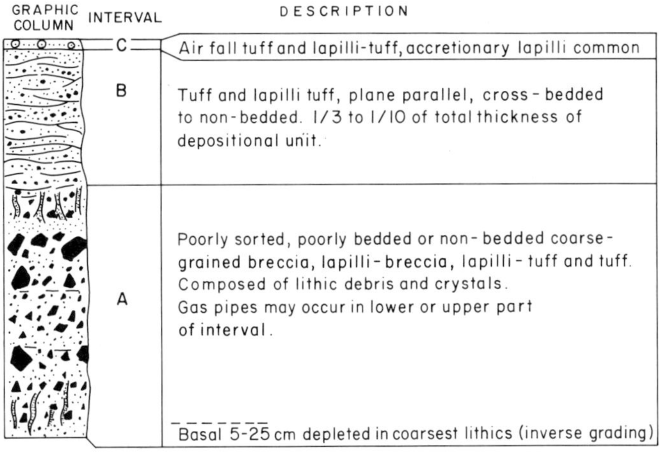

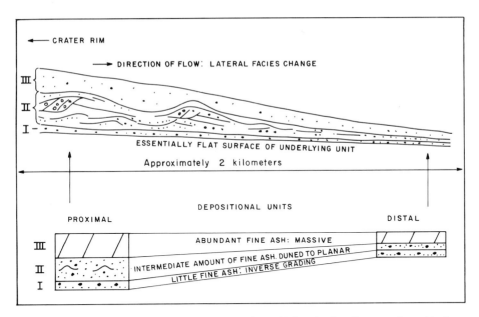

Fig. 8-24. Stratigraphic sequence of secondary block-and-ash flow and ash cloud surge deposits from 1902 Mt. Pelée eruption (Martinique). (After Fisher et al., 1980)

Fig. 8-25. Products of a single eruptive phase (an eruption unit), Laacher See, Germany, formed by flow. *I* Thin (5–25 cm), low-aspect ratio, silt-poor (fine ash) breccia interpreted to represent an initial high eruption column, followed by *II*, a more voluminous but denser, pulsating and lower eruption column giving rise to wavy bedding and antidune structures, and *III*, a very dense, short (or "boiling-over") eruption column with abundant fine ash giving rise to massive pyroclastic (or hydroclastic) flows. Such units, in various stages of completeness, are repeated more than 15 times in the upper pyroclastic sequence at Laacher See (11,000 yrs. B.P.). They are interpreted to form by repeated contact of ground water with magma, each eruption beginning with a small amount of water which increases in amount through a single eruption

kinds of pyroclastic flows, however, produce different vertical stratigraphic sequences. For example, sequences developed during the 1902 eruption of Mt. Pelée (Fig. 8-24) differ from the "standard ignimbrite flow unit" (Plinian) introduced by Sparks et al. (1973) and its modifications (Fisher, 1979; Sheridan, 1979). A different sequence occurs at Laacher See, Germany (Fig. 8-25), which appears to be related to fluctuations in eruption column height and column density resulting from the influx of water (Fisher et al., 1983). The May 18, 1980 blast deposits of Mount St. Helens exhibit still a different sequence (Hoblitt et al., 1981). Thus, there is no single standard flow unit to cover all types of eruptions.

## Texture

Most pyroclastic flow deposits have sorting values ($\sigma_\phi$) greater than 2.0, and sorting values tend to decrease, as do median diameter values ($Md_\phi$), with length of transport (Murai, 1961; Sheridan, 1971; Walker, 1971; Sparks, 1976). As previously noted (Chap. 5), pyroclastic flow and surge deposits are more poorly sorted than fallout deposits although there is considerable overlap (Fig. 8-26).

In the textural analysis of pyroclastic flow deposits, it is important to know the relative proportions of pumice, lithics, and crystals (Walker, 1971; Sparks, 1976)

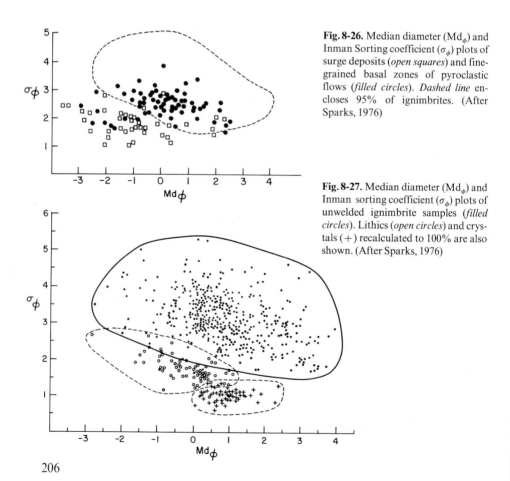

**Fig. 8-26.** Median diameter ($Md_\phi$) and Inman Sorting coefficient ($\sigma_\phi$) plots of surge deposits (*open squares*) and fine-grained basal zones of pyroclastic flows (*filled circles*). Dashed line encloses 95% of ignimbrites. (After Sparks, 1976)

**Fig. 8-27.** Median diameter ($Md_\phi$) and Inman sorting coefficient ($\sigma_\phi$) plots of unwelded ignimbrite samples (*filled circles*). Lithics (*open circles*) and crystals (+) recalculated to 100% are also shown. (After Sparks, 1976)

because the size distributions, sorting, and other parameters of these three subpopulations may differ for other reasons than sorting in the eruption column and during flow. Lithic fragments, for example, can be derived from magmatic stoping, by fragmentation of the walls of a magma chamber and vent, or fragmentation of a plug or dome within the vent, and they also may be picked up from the ground during flowage. The size distribution of crystal fragments is a function of original phenocryst sizes in the magma and of breakage during explosive eruptions. Moreover, different mineral species have different size ranges (e.g., feldspar vs. magnetite). Pumice has low mechanical strength and therefore may be reduced in size during eruption and flow, causing a preponderance of pumice dust in the fine-grained fraction of a deposit (Walker, 1972a). Figure 8-27 compares whole-rock grain-size parameters with those of lithics and crystals.

Pyroclastic Flow Deposits

In attempting to discriminate between different types of pyroclastic flow deposits (Krakatoan, St. Vincent, etc.), Murai (1961) showed that the size distribution curves tend to follow a Gaussian distribution (Chap. 5), which indicates that sorting takes place during eruption and transport. Poor sorting is characteristic throughout the length of a single pyroclastic flow sheet, but sorting varies vertically at any single locality and tends to improve slightly with distance. Use of grain size characteristics alone to verify origin of pyroclastic flow deposits is questionable (Sparks, 1976), but may have genetic implications if used in conjunction with the geometry and depositional structures determined during field studies (e.g. Hoblitt et al., 1981).

Vertical size grading of clasts is highly variable in different pyroclastic flow units, but is not always present. Apart from the fine-grained basal layer of pyroclastic flow deposits, however, vertical size-grading of large clasts is common. Pumice clasts, for example, are commonly reversely graded (Kuno, 1941; Self, 1972, 1976; Sparks et al., 1973; Sparks, 1976; Wilson and Head, 1981 b) but also may be normally graded (Smith, 1960a; Fisher, 1966a); lithic fragments tend to be normally graded (Noble, 1967; Sparks et al., 1973; Yokoyama, 1974), but they can

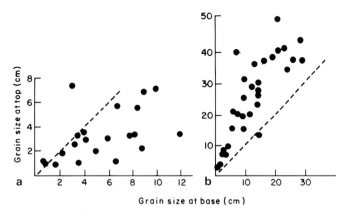

**Fig. 8-28 a, b.** Average maximum diameters of the five largest lithic (**a**) and pumice (**b**) clasts at top and base of several pyroclastic flow deposits. (After Sparks, 1976)

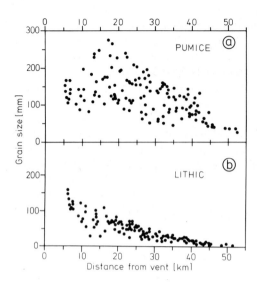

**Fig. 8-29 a, b.** Average maximum grain size of pumice (**a**) and lithic (**b**) clasts in the Towada pumice flow deposits plotted against distance from crater. (After Kuno et al., 1964)

be inversely graded (Sparks, 1976), especially in the basal layer. Maximum pumice size may be up to three times larger at the top of a flow than at the base (Fig. 8-28). In some cases, extreme concentration of pumice clasts occurs at the top of a flow unit (Sparks, 1976), but they have also been observed near the base (Crowe et al., 1978). Maximum size of both lithics and pumice also decreases with distance in subaerial deposits (Fig. 8-29) (Katsui, 1963; Kuno et al., 1964; Fisher, 1966a; Moore and Sisson, 1981).

Pyroclastic Surge Deposits

Pyroclastic surge deposits are better sorted than their associated pyroclastic flow deposit, although median diameters may be larger or smaller. Cross-bedded ground surge deposits are generally reported to be more poorly sorted than those composed of a single bed, but this may be caused by sampling difficulties involving several laminae (Sparks, 1976). Surge deposits commonly are enriched in crystal and lithic fragments compared to their associated pyroclastic flow deposits (Walker, 1971; Self, 1972, 1976; Sparks, 1976).

Segregation of Crystals and Lithics

Segregation and elutriation of particles occur within the conduit and the eruption column, and during flowage, causing enrichment of crystals and lithics, and depletion of fine-grained vitric particles within pyroclastic flow deposits. Fragments with relatively low settling velocities are carried high into the atmosphere and do not enter the flow, and others escape from the top of the flow while it is moving. These processes give rise to a distinctive type of fine-grained fallout tephra which has a crystal/vitric ratio systematically lower than that from manually crushed pumice taken from the associated ignimbrite (Sparks and Walker, 1977).

Hay (1959a) showed that the pyroclastic flow at St. Vincent in the Lesser Antilles is richer in crystals than juvenile pumice blocks in the same deposit. He accounted for this by the loss of light-weight pumice in the eruption column above the vent. A similar relationship was illustrated by Lipman (1967) in ash-flow deposits at Aso caldera, Japan. Walker (1972a) and Sparks (1975, 1976) later demonstrated crystal and lithic enrichment in the fine-grained base of ash flows and within ground surge deposits. The degree of crystal enrichment is expressed by an enrichment factor,

$$E.F. = (C_2/P_2)/(C_1/P_1),$$

where $C_1/P_1$ = weight ratio of crystals to pumice fragments from artificially crushed pumice, and $C_2/P_2$, is the same ratio from the matrix of the ignimbrite. Crystal enrichment is greater in basal layers (and also gas pipes) than in the middle of the same ignimbrite. This difference may be attributed to a selective loss of glass shards from the flow and to a reduction in pumice size by abrasion during flow. Rounding of pumice, which is a common feature seen in ash flow deposits, suggests that abrasion is the most significant cause.

Because of sorting processes that take place during eruption and emplacement of an ignimbrite layer and associated fallout, the chemical composition of bulk ignimbrite samples may depart notably from that of the magma. More reliance can be placed upon chemical analyses of large pumice fragments, or of fiamme within ignimbrite layers.

**Chemical Composition**

Most intermediate- to large-volume ignimbrites are erupted from felsic calc-alkalic and alkalic magmas: latite-dacite-rhyodacite-rhyolite and phonolite-trachyte and alkalic- to peralkalic rhyolite. Small-volume ignimbrites may range from rhyolitic to basaltic in composition (Johnson et al., 1972; Taylor, 1956; Williams and Curtis, 1964). Many intermediate- to large-volume ignimbrite sheets show a vertical compositional zonation due to eruption from an initially zoned magma chamber, independent of the sorting that occurs within the eruption column and during flow.

Many pyroclastic flow sheets composed of several flow units also show pronounced upward increase in crystals, together with changes in the kind and composition of the minerals and other chemical changes attributed to zoned magma chambers (Williams, 1942; Lipman and Christiansen, 1964; Lipman et al., 1966; Fisher, 1966a; Smith and Bailey, 1966; Schmincke, 1969b, 1976; Gibson, 1970). For example, the Topopah Spring Member in southern Nevada (Lipman et al., 1966) shows a zonation from basal crystal-poor rhyolite (77% $SiO_2$, 1% phenocrysts) to a capping crystal-rich quartz-latite (69% $SiO_2$; 21% phenocrysts). The chemical variation trend closely follows fractionation curves for the liquid line of descent in the experimentally determined system $NaAlSi_3O_8$-$KAlSi_3O_8$-$SiO_2$-$H_2O$ at about 600 b water pressure, suggesting that the compositional variation resulted from fractional separation of crystals from the liquid under conditions of near equilibrium. Phenocrysts in the quartz latite appear not to have accumulated from the rhyolite magma but crystallized in situ in a previously zoned magma body. Commonly, the highly differentiated magma is largely or completely erupted, whereas the more mafic underlying magma is only partially erupted. Vertical mineralogical and chemical zonation have provided an important argument in the discussion on the origin of peralkalic silicic rocks (i.e., crystal fractionation versus volatile transfer or direct mantle melting): lower flow units of peralkalic rhyolites (pantellerites and comendites) for example, are commonly overlain by flow units of peralkalic trachytes (Schmincke, 1969a, b, 1976; Gibson, 1970, 1972; Noble and Parker, 1974), indicating that the rhyolitic magmas were underlain by trachytic magmas from which they were probably derived by differentiation. Chemically heterogeneous pumice lapilli occur side by side in some ash-flow sheets, giving evidence of magma zonation and mixing prior to eruption, such as at Katmai, Alaska (Curtis, 1968), Aso, Japan (Lipman, 1967), Gran Canaria, Canary Islands, and Augustine Volcano, Alaska (Johnston and Schmincke, 1977). The topic of magma chamber zonation is discussed in more detail in Chapter 2. Young (Quaternary to Tertiary) ignimbrites commonly have a densely welded basal zone (vitrophyre) that may at first sight appear particularly useful for chemical analyses. However, such vitrophyres are commonly hydrated during diagenesis with concomitant loss (especially Na) and gain of some elements (Ross and Smith, 1955) and are thus not reliable representatives of the original magmatic composition (Chap. 12).

**Temperature Effects**

Measured Temperatures

Pyroclastic flows, although they are particulate systems, are an amazingly good heat-conserving mechanism. Boyd (1961) calculated that mixing of hot pyroclastic material with cold air during flowage is minimal and is restricted to a thin surface layer of the flow. Thus, hot pyroclastic flows may be nearly at magmatic temperatures during movement and shortly after deposition. Liquidus temperature, height of the material lifted in the eruption column (and therefore the amount of admixed cold air during upward movement of the eruptive column, and that trapped during collapse of the column, and total volume of a flow determine the emplacement temperature of pyroclastic flows.

Temperatures measured within hours to days, to depths of several centimeters to meters within observed pyroclastic flow deposits, generally range from about 500° to 650 °C (Kienle and Swanson, 1980) (Fig. 8-30). At Mount St. Helens em-

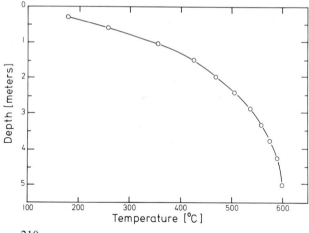

**Fig. 8-30.** Temperature profiles of pyroclastic flow deposits measured a few weeks after emplacement. St. Augustine Volcano (Alaska). (After Kienle and Swanson, 1980)

placement temperatures of May 18 and 25, June 12, July 22, August 7 and October 17, 1980 (pumiceous) pyroclastic flow deposits near the vent ranged from 750° to 850 °C and 300° to 730 °C farther away, with the later eruptions in general being hotter than the earlier ones (Banks and Hoblitt, 1981). Temperatures within individual pyroclastic flow deposits did not decrease substantially along their flow path. Banks and Hoblitt (1981) conclude that after initial and rapid cooling of several hundred degrees by adiabatic expansion and incorporation of air during the eruption and the development of flowage, little air entered the main body of the pyroclastic flows along the flow path. Casadevall and Greenland (1981) show that gases emitted from pyroclastic flow deposits were almost entirely hot atmospheric air.

Cooling rate is initially rapid on the surface, followed by a slow decline toward the center. As in other igneous bodies, if deposits are several meters thick, the temperatures can remain high in the interior for several years. For example, temperatures of up to 645 °C were measured in fumaroles of the ash flow deposits in the Valley of Ten Thousand Smokes, Alaska, 7 years after emplacement (Allen and Zies, 1923). These deposits are poorly welded.

Temperatures in excess of 400 °C occur around the margins of most recent pyroclastic flow deposits, as shown by zones of charred vegetation and the softening or melting of glass and plastic objects.

Inferred Temperatures

In older deposits, temperatures can be estimated by a variety of techniques such as softening and welding of glass shards and pumice, charring of wood, thermoremanent magnetism, certain mineral geothermometers, presence of degassing pipes, and others. Charred wood within and at the base of the 79 A.D. pyroclastic flows from Vesuvius indicates that they were about 300 °C (Maury, 1973) (Fig. 8-31).

Because of different temperatures, there is a complete spectrum of deposits that range from (1) completely unwelded and uncemented rocks (sometimes referred to as *trass* or *pozzolan*) that may have formed by thorough quenching in air, (2) through rocks that are cemented by spot welding of shards at points of contact (called "sillar" by Fenner, 1948), to (3) the wide-spread sheets of calc-alkalic partially welded tuffs; and, at the other end of the spectrum, to (4) peralkalic ash-flow tuffs that show extreme degrees of welding and crystallization and may resemble lava flows (Schmincke, 1974b) (Fig. 8-32).

Many small-volume pumice flows and block-and-ash flows closely resemble deposits of lahars, cold-rock avalanches or glaciers, but there are a few criteria which may distinguish them (also see Chap. 11). Aside from thermoremanent magnetism, criteria for high-temperature emplacement include a pinkish-gray to hematite-red zone in the upper part of a pumice-rich deposit (Williams, 1960; Crandell and Mullineaux, 1973), although such reddish zones are absent from many pyroclastic flow deposits. High temperature emplacement is indicated by prismatically jointed blocks resembling a three-dimensional jigsaw puzzle (Perret, 1937; Francis et al., 1974); many such blocks deposited on the surface of a flow rapidly disintegrate into a pile of rubble after emplacement. Charcoal within deposits also suggests high temperature emplacement. In some cases, heating of wood in place causes gas-escape pipes depleted in fines above the charcoal (Mullineaux and Crandell, 1962).

**Fig. 8-31.** Massive to vaguely bedded (*top*) pyroclastic flow deposits overlain by surge deposits of 79 A.D. Vesuvius (Italy) eruption cover house ruins (*in front of person*) at beach near Pompeii

**Fig. 8-32.** Strongly welded pantelleritic ignimbrite from Gran Canaria (Canary Islands)

## Welding and Compaction

Some of the most characteristic features of ignimbrites deposited at high temperatures are the plastic deformation and welding together of glass shards (Figs. 8-33, 8-34). In a detailed analysis of compaction of hot pyroclastic flow deposits, Sheridan and Ragan (1977) point out two kinds, mechanical compaction and welding compaction. Mechanical compaction takes place as the result of simple loading without significant change in particle shape. Particles maintain their relative positions after coming to rest except that elongate particles tend to be rotated toward the horizontal. The effect of mechanical compaction on depositional texture is relatively minor but at the same time can significantly reduce porosity and produce a compact rock. Pumice fragments generally maintain a random orientation; significant foliation without welding is rarely, if ever, produced.

Welding compaction results from viscous deformation of vitric fragments (Riehle, 1973). There are all transitions from completely undeformed shards typical of fallout ash deposited at low temperatures to nearly homogeneous solid glass typical of obsidian in which there are only ghost-like outlines of former shards (vitroclastic texture) in a continuous glass matrix (see illustrations in Ross and Smith, 1961). The main control for producing vitroclastic texture is the amount of time that temperatures remain above the threshold for welding. A number of authors have attempted to reproduce welding of glass shards experimentally (Boyd and Kennedy, 1951; Friedman et al., 1963; Yagi, 1966). Below about 550 °C there is

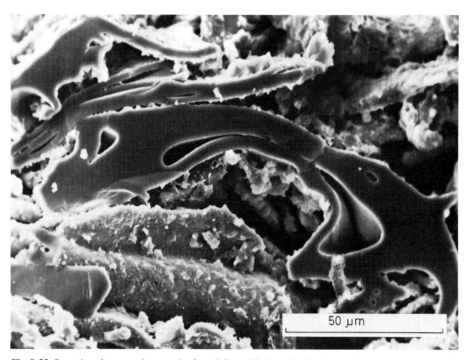

**Fig. 8-33.** Scanning electron micrograph of partially welded pumice shard. Note incompletely collapsed large vesicle in center of shard. Green ignimbrite (Pantelleria, Italy) (Schmincke, 1974b)

**Fig. 8-34.** Photomicrograph of welded tuff showing stronger compaction and homogenization near edges of phenocrysts. Sardinia

**Fig. 8-35.** Columnar jointing in highly welded Bishop Tuff ignimbrite (California, USA). Columns 1 m in diameter

negligible deformation of glass shards, but the threshold temperature depends on volatile content and chemical composition of the glass. Complete welding of shards may cause a "homogeneous" black glass to form, as does the collapse of porous pumice to form dense, black, glassy "fiamme". The degree of welding is also dependent on load pressure (Smith, 1960a) but this is probably less important than temperature, viscosity and gas content, as suggested by the existence of fused tuff (Lipman and Christiansen, 1964; Schmincke, 1967a) welded fallout tuffs (Sparks and Wright, 1979; Wright, 1980), and highly welded peralkalic tuff (Schmincke, 1974b). Columnar jointing is common in moderately to highly welded tuff (Fig. 8-35).

Structures Related to Temperature and Viscosity

Several structural features of welded to partly welded tuffs are related to the effects of viscosity (hence temperature and chemical composition) that develop during movement and deflation of a pyroclastic flow and are especially useful for determining flow directions. There are three main groups of features: (1) those induced during inflated movement, (2) structures caused by dense, lava-like flow or creep during deflation after emplacement and shortly before coming to complete rest, and (3) those caused by compaction after the flow has ceased forward movement. Aside from cooling-unit zones easily visible in the field, however, the most characteristic structural feature of welded tuffs are flattened pumice fragments (fiamme) generally of darker color than the surrounding matrix.

**Fig. 8-36.** Highly stretched and pulled-apart welded pumice lapillus, gash fractures dipping in flow direction (arrow) in comenditic ignimbrite. Gran Canaria (Canary Islands) (Schmincke and Swanson, 1967)

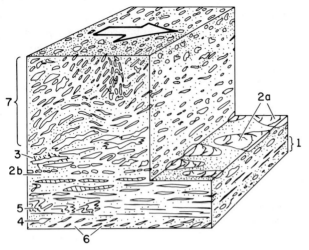

**Fig. 8-37.** Flowage structures in peralkalic ignimbrites, Gran Canaria, Canary Islands. Arrow is in direction of flow. *1* Zone of densest welding and maximum stretching of pumice. *2* Stretched and pulled apart pumice; *2a* shows cracks convex toward flow front; *2b* broken and rotated segments of pumice lapilli with rotation toward flow direction. *3* Tension fractures in welded matrix adjacent to unbroken pumice clast; cracks dip in direction of movement. *4* Spindle shape structure around rotated inclusion and strongly developed imbrication. *5* Folds with axial planes that dip sourceward; dips of axial planes may dip in opposite directions. *6* Imbricated stretched pumice dipping up-flow. *7* Ramp structure showing asymmetry. (After Schmincke and Swanson, 1967)

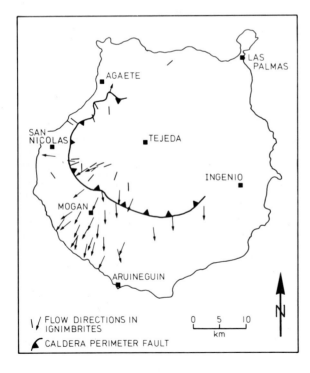

**Fig. 8-38.** Flow directions in comenditic to pantelleritic welded ignimbrites from Gran Canaria, each arrow representing the average of several flows at single localities (After Schmincke, 1974b)

Schmincke and Swanson (1967) pointed out several kinds of directional structures: (1) stretched and lineated pumice fragments, (2) broken, pulled-apart pumice fragments (Fig. 8-36), (3) tension cracks in the matrix, (4) hollows around inclusions, (5) folds, (6) imbricated pumice fragments and (7) ramp structures (Fig. 8-37). Measurement of these structures yielded consistent movement directions of the ignimbrite away from the source caldera (Fig. 8-38).

Similar directional features are described by Chapin and Lowell (1979), showing a close correspondence between a paleovalley and directional features in a 36 m.y. old deposit. Foliation planes, analogous to flow banding formed by laminar flow in lava flows, are parallel to the bottom and sides of the paleovalley. Lineation of gas pockets, stretched pumice, and elongate solid particles (including crystals) lie approximately in the plane of foliation; the long axes of solid particles are imbricated and dip sourceward relative to foliation planes. Flow folds are at right angles to lineation directions. Deflation or runout of the mass within the center of the paleovalley resulted in U-shaped cross-channel profiles with foliation planes dipping toward the valley center. Flow folds and growth faults with axes parallel to lineation and to valley sides formed locally by creep downslope after the flow had been emplaced.

The flow foliation is manifested by flattened and elongate gas cavities. Evidence that the foliation was caused by flow rather than by static post-depositional compaction includes (1) pervasive lineation on foliation planes, (2) folding of the lineation planes, and (3) planes adjacent to the paleovalley walls dip more steeply than the slope of the walls; in many places foliations dip nearly vertical even where valley walls slope less than 34 °C.

Post-depositional compaction of high temperature pyroclastic flows causes flattening and welding of vitric shards and pumice. Measurements of the flattening ratio (F = length/height) of pumice fragments in one ignimbrite sheet (Peterson, 1979) reveals a steady increase of the flattening ratio from the top downward into the body of the sheet despite changes in density of the deposit (Fig. 8-39). Disruptions in the systematic increase of the ratio provide a guide to determine whether

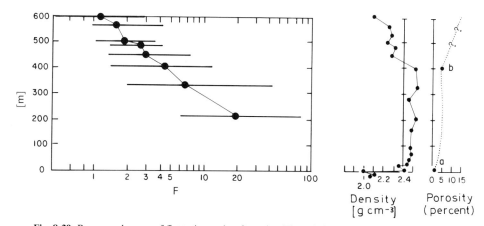

Fig. 8-39. Range and mean of flattening ratio of pumice (F), and density and porosity of whole-rock samples within a single cooling unit. Zero datum is base of densely welded zone of 600 m thick cooling unit (After Peterson, 1979)

or not a sheet is a complex cooling unit, and uniform change without disruptions can be used to determine original thickness where a part has been lost by erosion. Abrupt changes in flattening ratio have been used to detect faults and estimate their throws. Because of post-emplacement crystallization and diagenesis, porosity values may indicate higher degrees of welding than can be inferred from flattening ratios. Pumice-flattening ratios also serve as an approximate guide to relative viscosity of pumice during emplacement which, in turn, is a function of temperature, chemical composition, volatile content, and crystallinity. Relative viscosity is shown by slopes of flattening ratio curves downward within a deposit; they are steepest where viscosity is highest.

**Thermoremanent Magnetism**

The presence or absence of magnetic orientation of fragments in unwelded pyroclastic flow deposits in some instances has proved to be a useful tool to distinguish them from lahars (Aramaki and Akimoto, 1957; Wright, 1978; Hoblitt and Kellogg, 1979). Individual fragments within flows that come to rest at temperatures greater than the Curie point (about 400 °C) or higher (the blocking temperature of constituent ferromagnetic minerals) may have similar, well-grouped directions of thermoremanent magnetism. Random directions suggest that clasts were rotated by transport after the clasts were cooled below the Curie point. Crandell and Mullineaux (1973), however, have reported deposits with randomly oriented directions of remanent magnetism believed by them to have been deposited by hot pyroclastic flows because of reddish tops that signify high temperature emplacement. Also, hot pyroclastic flow deposits (block-and-ash flows) derived from shattered domes or the explosively disrupted side of a volcano composed of cognate volcanic rocks (e.g., Mount St. Helens, May 18, 1980) may be partly emplaced at temperatures below the Curie point. It is possible for elongate fragments to become dimensionally oriented by flowage processes, but magnetic orientations would be random if they cooled to below the Curie point prior to deposition.

# Classification and Nomenclature

As expected in an area of vigorous on-going research, the classification and nomenclature of pyroclastic flows and their deposits have been the subject of much confusion and debate. The wide range in physical properties of the erupting magmas, the different ways that flows originate, differences induced by transport processes, and differences in compaction and cooling structures and textures have given rise to a large array of names for the flows and their deposits. Moreover, a distinction is not always made between eruptive processes, flow processes or a deposit and a rock type. One common denominator is that pyroclastic flows are emplaced by a range of gravity-driven mechanisms.

Traditionally, different kinds of pyroclastic flow have been named according to the volcano where they were first observed, and a number of classifications have been proposed (Lacroix, 1930; Escher, 1933; Fenner, 1937; van Bemmelen, 1949; Macgregor, 1952, 1955; Aramaki, 1957; Murai, 1961). Many of these classifications are reviewed by Smith and Roobol (1982). The classification shown in Table 8-1 follows Williams (1957), who arranged the different types of flows in order of increasing gas content, increasing volume, and diminishing viscosity of the initiating magmas. However, few authors today follow such classifications largely because characteristic field criteria are poorly defined.

Wright et al. (1980b) and Smith and Roobol (1982) name the flows and their deposits according to field criteria such as relative amounts of pumice, scoria, poorly vesiculated blocks, and ash (Table 8-2). They emphasize (1) vesication of essential fragments (related to gas content, viscosity, and rate of gas release) which corresponds to types I-VI in Table 8-1, and (2) eruptive mechanisms, thereby bridging the gap between the classic names and descriptive parameters that we prefer.

Another set of criteria deals with flow and emplacement mechanisms which depend upon particle concentration. Thus, there are (1) pyroclastic flows, defined as high-concentration semifluidized bodies, which move with essentially laminar motion (Sparks, 1976) and (2) pyroclastic surges which are low-concentration flows that are turbulent (Sparks and Walker, 1973; Sparks, 1976; Fisher, 1979) and give rise to deposits with characteristic sedimentary structures. In many instances, pyroclastic flows and pyroclastic surges are derived from the same initial flow of material but become separated by various processes of gravity segregation (Fisher, 1983) and a range of fluidized gas-flow velocities through the moving mass, (C.J.N. Wilson, 1980; Fisher and Heiken, 1982; Moore and Sisson, 1981; Wilson and Walker, 1982).

Pyroclastic surge deposit is the general name for surge deposits of any type, named according to origin or place in sequence. They include (1) ground surge deposits which underlie the deposits of many small- to intermediate-volume pyroclastic flow deposits, (2) ash cloud deposits that overlie and extend beyond the margins of pyroclastic flow deposits, and (3) base surge deposits that form from hydroclastic eruptions. Base surge deposits appear to develop largely from eruption column collapse. Ground surge deposits may originate (1) by eruptions that predate a pyroclastic flow, (2) from eruption column collapse and (3) from infolding of air at the front of a pyroclastic flow. Ash-cloud deposits apparently originate from elutriation from the top of a moving pyroclastic flow or by gravity segregation of coarse- from fine-grained material during flowage.

Descriptively, pyroclastic flow deposits may be termed ash-flow deposits if more than 50 volume percent of the deposit is of ash size or less. The term block-and-ash flow deposit has been used for small-scale, coarse-grained deposits with abundant juvenile lithics and with less than 50 percent ash (Perret, 1937). The name pumice flow deposit is commonly used for deposits containing abundant pumice blocks and lapilli ( < 50% ash). The term welded tuff is used when features of welding are observed, which is distinct from welded fallout tuff, and from sintered tuff that has become reheated and fused at the contact with another extrusive or intrusive igneous body.

The general term ignimbrite, coined by Marshall (1935), has been used by several workers as the deposit of a pyroclastic flow whether welded or unwelded (Fisher, 1966c; Williams and McBirney, 1979; Walker, 1971; Sparks et al., 1973). The name is often criticized because it has been used for the eruptive and transport process as well as for the rock, but is useful because it does not refer to grain size

Table 8-1. Classification of hot pyroclastic flows. (After Macdonald, 1972)

| Type | Gas | Observed velocities | Origin | Grain size | Fragments | Welding | Comments |
|---|---|---|---|---|---|---|---|
| I. Merapi (Stehn, 1936) | Gas-poor, non-explosive hot avalanches | Up to 110 km h$^{-1}$ | Collapse or spalling of rising spines and domes | Coarse-grained | Angular rocks, vesicular blocks from interior of dome or spine | Unwelded | May travel farther than cold avalanches. Some authors do not consider these to be true pyroclastic flows |
| II. Peléan (Lacroix, 1904) | Range from gas-poor to gas-rich. Explosive to non-explosive | Up to 160 km h$^{-1}$ or more | Laterally directed during rise of spine or dome | Coarse-grained | Almost wholly lithic (dense to partly vesicular) | Unwelded | Associated with dome formation |
| III. St. Vincent (Anderson and Flett, 1903) | More gas-rich than Peléan type | "Hurricane velocity" | Subsidence of vertical eruption column | Medium-grained | Rich in crystals and ash. Lithic blocks and lapilli are less common than vesicular juvenile material | Unwelded to poorly welded | Flow down all sides of volcano slopes |
| IV. Krakatoan (Verbeek, 1886) | More gas-rich than above types | High velocities | Probably subsidence of vertical eruption column | ? | Pumice-rich, almost exclusively juvenile material | ? | Issue from summit craters in late stage of long-established composite volcanoes |
| V. Valley of Ten Thousand Smokes (Fenner, 1923) | Gas-rich | Not observed | Possible subsidence of vertical eruption column | Fine- to medium-grained | Juvenile ash | Unwelded to highly welded | Flowed on average gradient of 1°08' for a distance of 20 km. Came out of Novarupta Volcano |
| VI. Volcano-tectonic subsidence (Williams, 1957) | Gas-rich | Not observed | From arcuate fissures that accompany large subsidence structures | Fine- to medium-grained | Mostly juvenile | Unwelded to highly welded | Sheet-like deposits of vast dimensions. Several orders of magnitude larger than above types. Emplaced at high temperatures |

**Table 8-2.** Genetic classification of pyroclastic flows. (After Wright et al., 1980b)

| Essential fragment | Eruptive mechanism | Pyroclastic flow | Deposit | Comments |
|---|---|---|---|---|
| Vesiculated ↑ | Eruption column collapse or "boiling-over" | Pumice or ash flow | Ignimbrite; pumice and ash flow deposit | Intermediate-volume deposits formed by continuous collapse of a Plinian eruption column as envisaged by Sparks et al. (1978). Silicic in composition |
|  |  |  |  | Small-volume deposits probably formed by interrupted column collapse. Intermediate to silicic in composition |
|  |  | Scoria flow | Scoria and ash deposit | Small-volume deposits probably formed by interrupted eruption column collapse produced by short explosions (see Nairn and Self, 1978). Basalt to andesite in composition. Some deposits contain large unvesiculated blocks, e.g. those of Ngauruhoe (1975) |
| Decreasing average density of juvenile clasts | Lava or dome collapse, or lateral projection — Explosive | Lava debris flow — nuée ardente | Block and ash deposits | Small-volume deposits, usually andesitic or dacitic in composition |
|  | Gravitational | Lava debris flow — nuée ardente | Block and ash deposits | Small-volume deposits usually andesitic or dacitic in composition. These are the hot avalanche deposits of Francis et al. (1974) |
| Nonvesiculated |  |  |  |  |

as does ash-flow tuff or block-and-ash flow deposit. In general, the context will clarify whether a process, deposit or rock is meant so long as sufficiently detailed descriptions are provided. Sparks et al. (1973) confine the term ignimbrite to deposits of pumice-rich pyroclastic flow deposits (welded or unwelded). Because of the many transitional varieties, we recommend that the term ignimbrite be used for all deposits formed by the emplacement of pyroclastic flows.

Variants of pyroclastic flow deposits recently introduced include "co-ignimbrite lag-fall deposits" consisting of coarse lithic-rich breccias close to vent areas (Wright and Walker, 1977) which laterally grade to a crystal- and lithic-rich "ground layer". The ground layer lies beneath ignimbrite layers, commonly in sharp contact, and appears to have formed by gravity segregation from the overlying flow or by ingestion of air at the front of a flow (Walker, Self and Froggatt, 1981). Another variant is the "low-aspect ratio" ignimbrite which consists of a thin "ignimbrite veneer deposit" that mantles topography and a valley-ponded part (Walker, Heming and Wilson, 1980; Walker, Wilson and Froggatt, 1980; Walker and Heming et al., 1981). Low-aspect ratio ignimbrite deposits give evidence of rapid emplacement from eruptions of unusually high discharge rates of magma. "Fines-depleted ignimbrites" are interpreted to be developed from turbulent, highly fluidized pyroclastic flows with a high throughput of gases (Walker, Wilson and Froggatt, 1980). These relatively new concepts, not yet fully tested or evaluated, have suggested new avenues of study of ignimbrites.

## The Flows

Concepts of how pyroclastic flows originate and the processes of flowage are based upon limited observations of small-volume eruptions including Cotopaxi, Ecuador (Wolf, 1878), Mt. Pelée, Martinique (Anderson and Flett, 1903; Lacroix, 1904; Perret, 1937), Mount Lamington, Papua (Taylor, 1958), Mayon Volcano, Philippines (Moore and Melson, 1969), Fuego Volcano, Guatemala (Davies et al., 1978), St. Augustine Volcano, Alaska (Stith et al., 1977; Kienle and Swanson, 1980) and Mount St. Helens, Washington (Lipman and Mullineaux, 1981), and upon the study of deposits and theoretical considerations (Sparks and Wilson, 1976; Sparks et al., 1978). Ideas of origin and emplacement of large-volume ignimbrite sheets have been extended from these limited observations by analogy and from deductions made from sedimentological and stratigraphic studies by Hay (1959a), Smith (1960a, b), Murai (1961), Fisher (1966a, b, 1979), Schmincke and Swanson (1967), Walker (1971, 1972a), Walker, Heming and Wilson (1980), Walker, Self and Froggatt (1981), Walker, Wilson and Froggatt (1980) and Sparks (1975, 1976).

### Origin

Pyroclastic flows form in several ways (Fig. 8-40): (1) by an inclined blast from the base of an emerging spine or dome (Perret, 1937); (2) by collapse of a growing dome (Escher, 1933; Schmincke and Johnston, 1977); (3) by the "boiling-over" of a highly gas-charged magma from an open vent (Wolf, 1878; Anderson and Flett, 1903; Taylor, 1958); (4) by gravitational collapse of an overloaded vertical erup-

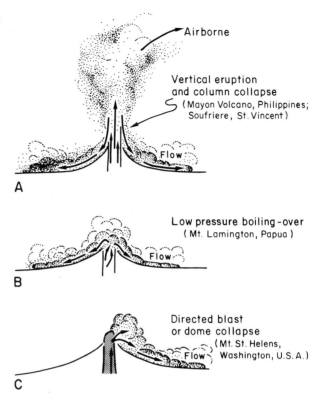

Fig. 8-40 A–C. Some ways that pyroclastic flows can originate. (After Macdonald, 1972)

tion column (Hay, 1959a; Smith, 1960a; Sparks and Wilson, 1976; Sparks et al., 1978); and (5) the explosive disruption (eruption) from the front of a lava flow as observed at Santiaguito Volcano, Guatemala in 1973 (Rose et al., 1977). More than one process may occur during a single eruptive episode.

A current controversial aspect of the origin of ash-rich pumice flows (as distinguished from block-and-ash flows formed by disruption of domes and from disintegrating lava flows) is whether they form by flowage of vesiculating magma from the crater lip ("boiling-over") or by collapse of vertical eruption columns, or both. Observations of small-volume central-vent eruptions at Mount St. Helens (Rowley et al., 1981) confirm that both processes can occur, but whether observations of small eruptions can be applied to large-volume eruptions associated with large calderas and volcano-tectonic depressions is still an unresolved question.

The idea of collapse of vertical eruption columns as a major – if not universal – process in the evolution of pyroclastic flows appeared firmly established by the sedimentological data of Hay (1959a), and by the recognition that pyroclastic flows, such as those produced by the 1929 eruption of Komagatake, Japan, appear to develop from vertical eruption columns (Smith, 1960a). Pyroclastic flows have frequently been observed to be associated with vertical eruption columns, but in some cases their formation appears to precede columns such as in 1980 during the eruption of Mount St. Helens, Washington (Rowley et al., 1981).

During the early stages of eruption at Komagatake, a vertical eruption column extended upward 12 km above sea level and deposited pumice as fallout in well-stratified beds downwind from the volcano (Kozu, 1934). The early stage was followed by vertical eruption columns to lesser altitudes which contained great quantities of pumice that fell back and flowed down the slope and valleys as far as 6 km from the vent. Williams (1942) suggests that the pumice flows of ancient Mount Mazama (Crater Lake, Oregon) formed in the same way. Another often-quoted example is that of Mayon Volcano, Philippines in 1968 (Moore and Melson, 1969). During daylight hours, several nuées were observed to appear rapidly at the base of vertical eruption columns but the process of development, however, could only be observed at night. On the night of May 1–2, explosions occurred every few minutes, hurling incandescent blocks and some fine ejecta to heights of only 600 m, falling back to produce a glowing collar around the summit crater, which then moved downslope as nuées ardentes. The nuées apparently formed by fallback of abundant material along ballistic trajectories similar to fountaining processes.

During the 1974 eruption of Ngauruhoe Volcano, New Zealand, several small nuées were formed. Nairn et al. (1976) state that the most voluminous eruptive intervals did not commence explosively, but rather, emission steadily increased until the dense eruption column widened to spill over the crater rim and form pyroclastic avalanches. During the eruption of February, 1975, collapse of vertical eruption columns appeared to occur, although observation of the actual process was obscured by billowing clouds of ash (Nairn and Self, 1978). They state that material comprising the avalanche possessed momentum from free fall onto the upper slopes, which, combined with a very rapid accumulation rate, caused an immediate dense flow of hot ejecta (Nairn and Self, 1978, Fig. 4). The eruption which produced the largest avalanches, however, was not observed by geologists. Eyewitness accounts and their photographs suggest that avalanches were initiated by rapid accumulation of "... pyroclasts continually falling from the eruption column onto the summit."

These observations, coupled with the theoretical work of Wilson and co-workers (Wilson, 1976; Sparks and Wilson, 1976), indicate that column collapse should occur – and produce pyroclastic flows – under conditions such as decline in gas pressure or widening of the vent, and have led many workers to the widespread assumption that most, if not all, pyroclastic flows are initiated by the collapse of eruption columns. However, during the 1951 eruption at Mount Lamington, Papua (Taylor, 1958), some of the pyroclastic flows originated without vertical eruption columns as "massive disgorgements" of fragmental material that flowed down the mountain. Several times, successive explosions filled the crater with convoluted clouds having little tendency to rise; instead, the heavier parts flowed through low gaps in the crater wall, whereas the lighter parts poured over the crater rim. These events occurred without distinctive seismic activity, interpreted by Taylor (1958) to mean that eruptions occurred from a shallow level. Taylor suggested that pervasive vesiculation of the magma gave it a mass buoyancy and hydrostatic properties, enabling it to flow as a fragmental mass. Wolf (1878) reported an eruption at Cotopaxi (Ecuador) which "boiled over" from the crater with "fluid", "glowing lava" that flowed with furious velocity in all directions down the slopes, a description similar to that given by Taylor (1958) of some flows at Mount Lamington.

Activity similar to the "massive disgorgements" at Mount Lamington also occurred during at least one eruptive episode on July 9, 1902 at Mt. Pelée. Anderson and Flett (1903, p. 492–493) state that a very dark, globular cloud with a nodular surface unlike an ordinary steam jet appeared in the crater. "It did not rise into the air, but rested there, poised on the lip of the fissure, for quite a while it seemed, and retained its shape so long that we could not suppose it to be a mere steam cloud. Evidently it had been emitted with sufficient violence to raise it over the lip of the crater, but it was too heavy to soar up into the air like a mass of vapor, and it lay rolling and spouting on the slopes of the hill. The wind had no power over it, fresh protuberances spurted out from its surface, but it did not drift leeward any more than if it had been a gigantic boulder. It then began moving downhill, and it seemed that the farther the cloud travelled the faster it came ...". From their further description (p. 494) it took about 30 min after it appeared in the crater for the nuée to clear (deposit its load).

At Mount St. Helens most of the pumiceous pyroclastic flows formed when bulbous masses of inflated ash, lapilli, and blocks erupted to a few hundred meters above the inner crater and then spilled out through the open crater to the north (Rowley et al., 1981). These upwellings took place before or during the development of the gas thrust of the Plinian column. The flow first moved slowly over the gently sloping floor of the crater but accelerated to as much as 100 km/h on the steeper northern flank of the volcano. If the crater had not been open, it seems likely that only some of the upper parts of the low eruption columns would have spilled over low parts of the rim as at Mount Lamington. Moreover, the description by Anderson and Flett (1903) of a globular cloud which appeared in the crater and accelerated as it moved downhill is similar to the description given in Rowley et al. (1981).

**Transport and Mobility**

The manner of transport and deposition from pyroclastic flows are inferred mainly from sedimentological and stratigraphic studies of their deposits, but direct observations aid our understanding of emplacement processes. For example, in addition to observed velocities, ranging from 14 km/h (Tsuya, 1930) to about 230 km/h (Moore and Melson, 1969), pyroclastic flows have been observed to separate gravitationally (gravity transformation of Fisher, 1983) into a lower part containing most of the solid mass (the basal avalanche or the underflow) and an overriding expanding cloud of ash derived from the flow (Anderson and Flett, 1903; Lacroix, 1904; Perret, 1937; Van Bemmelen, 1949; Taylor, 1958; Smith, 1960a; Davies et al., 1978) (Fig. 8-41). A dramatic example of their dual nature is provided by the 1902 Mt. Pelée eruption where the main part of the flow ("glowing avalanche") was confined to the Rivière Blanche and the ash cloud generated above it (the "glowing cloud") billowed outward north and south of Rivière Blanche along the lower coastal region of the mountain (Anderson and Flett, 1903); the southern margin of the expanding ash cloud overwhelmed the town of St. Pierre. The ash cloud continued across the water and upset, sank or burned boats in the harbor. Several meters of pyroclastic debris was deposited in Rivière Blanche, whereas ash deposits in the south part of town were only about 30 cm thick.

The high velocities and ability of witnessed flows to move over and around obstacles testifies to the great mobility of pyroclastic flows, but most impressive are the long distances (>100 km) over which ignimbrite sheet deposits have been traced on low slopes or the high barriers that they can surmount (>600 m). Their

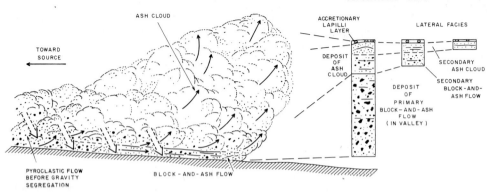

**Fig. 8-41.** Diagram showing development of block-and-ash flow and ash cloud surge by gravity segregation from turbulent flow from source. Wide downward arrows show trends of bulk movement of greatest volume of fragments. Lateral facies illustrate continued gravity segregation of ash cloud surge into laminar-flow underflow. Illustrates process described at Mt. Pelée (Martinique) by Fisher and Heiken (1982)

great mobility has been explained by: (1) exsolution of gas from juvenile (glassy) particles which buoys up particles and reduces friction between them (Fenner, 1923; Perret, 1937; Sparks, 1979), (2) breakage of congealed, hot magma fragments (lithic) with resultant release of heated gases contained within the lithic material (Fisher and Heiken, 1982), and (3) the heating and expansion of air engulfed at the front of the flow (McTaggart, 1960, 1962; C.J.N. Wilson, 1980; Wilson and Walker, 1982). Magmatic gases are mixed with solid and liquid particles prior to and during eruption, but there is evidence that gas is also released during transport (Moore and Melson, 1969). Gas that provides mobility to flows consisting dominantly of pumice and bubble-wall shards may be exsolved from liquid and plastic particles during flow. That which provides mobility to flows consisting dominantly of lithic debris may be partly derived from breakage of the fragments which releases hot gas from their interiors. In both instances, engulfment, heating and expansion of air may aid mobility of turbulent parts of the flows (C.J.N. Wilson, 1980). Data from the Mount St. Helens eruption strongly indicate that air within pyroclastic flows is ingested mainly within the eruption column at the source (Banks and Hoblitt, 1981). Exsolution of magmatic gases from liquid particles, release of gases from particle breakage and expansion of engulfed air probably all aid mobility of flows to varying extents, as suggested by the height/distance ratios of cold lahars and avalanches compared with pyroclastic flows (Fig. 8-42). Regardless of how gases originate, fluidization experiments (Eden et al., 1967) suggest that sedimentation is retarded if there are abundant silt-size particles (in the 40 micrometer range) with spacing between particles of less than 10 microns. This results in a low permeability and a slow outward diffusion of interstitial fluid thereby prolonging flow. Sparks (1976) argues that pyroclastic flows are semi-fluidized high-concentration dispersions comparable in some ways to the flow behavior of non-volcanic debris flows and high-concentration turbidity currents (Fisher, 1971; Middleton, 1967) but they have an overall lower fluid density. The turbulence mod-

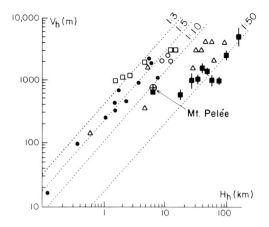

Fig. 8-42. Height of origin (Vh in m) and travel distance (Hh in km) for various kinds of debris flows, avalanches and hot pyroclastic flows. *Closed circles:* cold avalanches; *open circles:* flows from dome collapse; *open squares:* dense clast pyroclastic flows; *closed squares:* ignimbrites with bars to show uncertainty of height or origin; *open triangles:* mudflows. (After Wright, 1979, from several authors)

el for emplacement of pyroclastic flows (Fisher, 1966b) is more applicable to pyroclastic surges and is used in modified forms by Hoblitt et al. (1981) and Waitt (1981) to explain the emplacement of the laterally directed blast which occurred on May 18, 1980 at Mount St. Helens, Washington.

Intermediate-volume, pumice-rich pyroclastic flows are known to have surmounted topographic barriers of considerable height (Yokoyama, 1974; Miller and Smith, 1977; Koch and McLean, 1975; Rose et al., 1979). A spectacular case is the 22,000 yr B.P. Ito pyroclastic flow, Japan which travelled 70 km over topographic barriers as high as 600 m (Aramaki and Ui, 1966; Yokoyama, 1974). Recently, Froggatt et al. (1981) and Wilson and Walker (1981) report a low aspect ratio ignimbrite (Taupo, New Zealand) that mantles mountains up to 1500 m higher than the inferred vent and as far as 45 km from the source; trees were felled over a 15,000 km$^2$ area. Another example is provided by pyroclastic flows from Aniakchak and Fisher calderas, Aleutian Islands, that travelled as far as 50 km over mountainous barriers between 250 and 500 m high (Miller and Smith, 1977). Miller and Smith present compelling evidence that momentum rather than gaseous expansion of the flows caused topographic overtopping (Fig. 8-43). Sparks et al. (1978) argue that the momentum of flows with velocities of 100 m/s can overcome barriers several hundred meters high (Fig. 8-44). According to Sheridan (1979), the slope of the "energy-line" (Hsü, 1975) traces the potential flow head from the top of the gas-thrust region of an eruption column to the distal toe of a flow along the line of transport; a pyroclastic flow could surpass all topographic barriers that do not extend above the line. This agrees with estimated heights of eruption columns (Sparks et al., 1978), calculated velocities of flows (Sparks, 1976), barrier heights and lateral extent of deposits (Yokoyama, 1974; Miller and Smith, 1977).

**Tufolavas, Froth Flows, Foam Lavas and Globule Flows**

The apparent obliteration of vitroclastic structures and textures by thorough welding and devitrification of many ash flow tuffs resulting in a lava-like rock has led a number of workers to believe that there are all transitions in the mode of eruption of lava and hot pyroclastic flows. These speculations have resulted in a voluminous

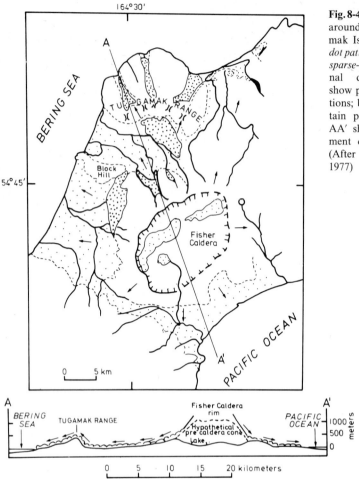

Fig. 8-43. Pyroclastic flows around Fisher Caldera, Unimak Island, Alaska. *Close-dot pattern* indicates outcrop; *sparse-dot* is inferred original distribution; *arrows* show postulated flow directions; brackets show mountain passes. Cross section AA' shows inferred movement of pyroclastic flows. (After Miller and Smith, 1977)

literature and a profusion of special names such as froth flows (Boyd and Kennedy, 1951; Kennedy, 1955; Boyd, 1961; McCall, 1964); tufolavas (tufflavas) and clastolavas of Soviet authors (many authors in Cook, 1966); foam lavas (Locardi and Mittempergher, 1967); ignispumites (Pantó, 1962); and globule lavas (Johnson, 1968).

Perhaps the most common alternate theory to the ash flow mechanism is that an initially homogeneous lava flow can disintegrate during flowage and become clastic. Deposits from such systems are believed to resemble, but be genetically distinct from, welded or unwelded tuffs deposited by pyroclastic flows. According to some authors (e.g. Sheimovich, 1979) "ignimbrite magmas" especially rich in volatiles are responsible for the high degree of vesiculation and partial fragmentation of such "lava flows".

Although very complicated rocks can result when both lava flows and very hot pyroclastic flows erupt from a zoned magma chamber, the evidence for a special origin of the rock types listed above is not compelling. Martin and Malahoff (1965) hold that the rocks called tufolavas or clastolavas, especially those from Armenia

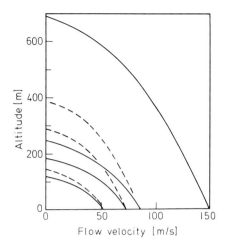

Fig. 8-44. Altitude-velocity curves for movement of pyroclastic flows with density = 1.0 and thickness = 10 m up a 15° slope. Dashed curves are friction free cases; solid curves with frictional drag accounted for. (After Sparks et al., 1978)

(Cook, 1966), could be explained as pyroclastic flow deposits. Field and microscopic studies of major areas of such "transitional" rocks in Armenia, Italy and East Africa indicate that all can be interpreted either as pyroclastic flow deposits, some showing pervasive high temperature devitrification and thorough welding, or as welded fallout tuffs (agglutinates) of especially hot and low viscosity magmas (Schmincke, 1972, 1974b). More detailed work on individual occurrences has led to similar conclusions (Sparks, 1975; Hay et al., 1979). The famous piperno from the Phlegrean fields, Italy, once believed to be a pyroclastic flow deposit, appears to have formed as an agglutinate (Fig. 8-45).

Fig. 8-45. Piperno (agglutinate) from type locality in the Phlegrean Field (Italy)

Although eruptive and flowage processes of very hot and very low viscosity fallout and pyroclastic flows, such as those of peralkalic trachytic, phonolitic and rhyolitic composition can generate deposits difficult to distinguish from lava flows, processes that generate "pyroclastic flow-like" deposits from lava flows have not been unequivocally documented. The reverse situation ... i.e., the gradation from a lava flow into near-vent spatter deposits ... is, on the other hand, a common phenomenon in deposits with a wide range of chemical compositions.

## Ignimbrite Vents: Speculation

An intriguing and as yet unresolved problem concerning the origin of large-volume ignimbrites revolves around their source and bears directly upon the column collapse idea for their origin. Most of the described voluminous ignimbrite sheets can be related directly or indirectly to calderas (Smith, 1979) and most workers believe that they were erupted from ring fractures concomitant with caldera collapse. But which comes first, the development of ring fractures and collapse with magma rising through the fractures, or the eruption followed by collapse and the development of ring fractures? On the other hand, the idea of collapse of a Plinian eruption column to produce ignimbrite sheets (Sparks and Wilson, 1976; Sparks et al., 1978) requires a central vent eruption.

There are some accounts of exposed vents filled with welded tuff (Cook, 1968; Korringa and Noble, 1970; Noble et al., 1970; Almond, 1971; Quinlivan and Rogers, 1974; Ekren and Byers, 1976), but such exposures are extremely rare worldwide. Ekren and Byers (1976) list possible reasons to account for their general scarcity: (1) most vents are covered by their own ejecta, (2) vents close off at shallow depths and subsequent erosion destroys them, (3) vents that fed ash flows were later filled with liquid lava and (4) vents were destroyed by cauldron collapse. We add a fifth possibility, that is, most large ignimbrite-producing eruptions are from central-vent type of eruption (Plinian) which initially create a void that initiates caldera subsidence. Following subsidence, some vesiculating magma may leak upward through the ring fractures produced by collapse, but downward movement of the central block would tend to close such fractures. It is possible therefore that the general loss of pressure and the most voluminous loss of material is through the central part of the caldera and would account for the scarcity of ignimbrite-filled fissures and the presence of comagmatic Plinian pumice and fallout deposits beneath many ignimbrite sheets.

# Chapter 9   Deposits of Hydroclastic Eruptions

Many volcanic eruptions result from the interaction of magma and external water (Table 9-1), but few volcanologists (e.g., Jaggar, 1949) have emphasized the importance of nonmagmatic water in volcanic eruptions. In our view, the importance of external water in explosive eruptions is still underestimated. Wood (personal communication) even holds that maars, which most commonly develop from hydroclastic eruptions, are the second most common volcanic landform on earth next to scoria cones.

The recognition of the influence of external water on volcanic eruptions has opened a Pandora's box of new problems that offer a broad field for future studies. To name only a few: How do (1) the geometry and size of vent, (2) magmatic parameters such as chemical composition, temperature, viscosity, and difference between solidus and liquidus temperatures, and (3) the amount of external water, as well as reservoir properties such as porosity and permeability, interact to produce steam explosions? What is the mixing mechanism that causes what appears to be virtually instantaneous incorporation of country rock, thorough fragmentation and vent-coring? How can the relative roles of magmatic and external volatiles be estimated in deposits that contain vesiculated tephra indicating at least some exsolution of magmatic volatiles? Can the claim of some authors be verified that many explosive eruptions are actually favored by the interaction of magma and external water? We have outlined present ideas on the main processes believed to govern hydroclastic eruptions in Chapter 4. Here we will concentrate on the deposits.

## Definition of Terms

We use the name *hydroclastic* (introduced by Fitch in Walker and Blake, 1966), almost in parallel with pyroclastic, as an inclusive term for products of *hydroexplosions* which are defined as "explosions due to steam from any kind of water" (Wentworth, 1938, p. 22). Many types of explosively produced hydroclastic particles are expelled through vents, however, and therefore are *sensu stricto* pyroclastic, but if it can be determined that they are derived from steam explosions, the term hydroclastic is more appropriate than pyroclastic. A major reason that we generalize hydroclastic is that some workers have equated the terms phreatic and phreatomagmatic (Gary et al., 1974) and have used them for nearly any kind of explosion caused by the interaction of magma or lava with external water (e.g. Daly, 1933, p. 311).

**Table 9-1.** Different ways that hydroclastic eruptions can occur. (After Schmincke, 1977a)

1. Magma erupted into shallow ocean or lake waters
2. Water flowing into an open vent containing magma
3. Ascending magma intersecting a ground water horizon
4. Water flowing into a partially emptied magma chamber
5. Magma erupted beneath a glacier
6. Lava or hot pyroclastic flows travelling over wet ground or entering a body of water
7. Groundwater heated by, but not mixed with, magma

**Table 9-2.** Hydroclastic surface eruptions and corresponding attributes of deposits. (After Schmincke, 1977a)

| Characteristics of particles and deposits | | Eruptive and transport processes |
|---|---|---|
| *Chemical composition:* predominantly basaltic | indicates → | Low magmatic volatile content, high temperature, low viscosity |
| *Clasts:* Slightly vesicular; Sideromelane; Bread crust/cauliflower bombs | indicates → | Quenching and granulation at magma-water contact; minor degassing; steam explosions |
| *Grain size:* Small. May contain large lithic clasts and broken pillows | indicates → | Thorough breakage by thermal stress and absence of separation of fine clasts in eruptive column |
| *Sorting:* Poor | indicates → | Abundant water (vapor) in system |
| *Structure:* Penecontemporaneous deformation; Vesicular tuff; Good bedding; Mud cracks; Accretionary lapilli (vesicular); Deposition on vertical planes; Poor sorting | indicates → | Abundant water (vapor) in transporting system |
| *Abundance of lithic clasts in deposits:* | indicates → | Sudden steam generation within country rock and spalling |
| *Large size of lithic and essential clasts:* | indicates → | High energy due to abundant water (vapor) |
| *Abundant evidence for horizontal transport:* | indicates → | Base surge transport |
| *Geometry:* Low to very low angle cone | indicates → | Ballistic and base surge transport |
| *Fumarolic alteration or pipes:* Absent | indicates → | Deposition at low temperatures |
| *Sintering or welding:* Absent | indicates → | Low temperature ($<100\,°C$) |
| *Association:* Strombolian deposits | indicates → | Fluctuating supply of external water or sealing of conduit walls |

Stearns and Macdonald (1946, p. 16–17) divide hydroclastic (hydro-explosion) eruptions into four categories: (1) *Phreatic eruptions* (explosions) are driven by the conversion of ground water to steam. Such steam explosions have a low temperature and do not expel juvenile ejecta because ground water is vaporized only by heat or hot gases and not by direct contact with fresh magma. More recently, Muffler et al. (1971) introduced the term "hydrothermal explosion" for steam explosions that do not produce fresh magmatic tephra, but occur when ground water is heated by an igneous source and flashes to steam which violently disrupts the near-surface confining rocks. We consider the terms hydrothermal eruptions and phreatic eruptions to be synonymous. (2) *Phreatomagmatic eruptions* (explosions) (Stearns and Vaksvik, 1935) occur when ascending magma contacts ground water; resulting eruption products include juvenile as well as cognate ejecta. Limiting such eruptions to ground water environments, however, is too restrictive. Here, we generalize phreatomagmatic to include magma and water interactions within any environment – submarine and sublacustrine as well as ground water, ice and wet-sediment environments. They differ from magmatic explosions, which are caused by internal magmatic gases.

As originally defined (Stearns and Vaksvik, 1935), phreatomagmatic refers to magma interacting only with ground water. Therefore, Walker and Croasdale (1972) proposed the name "Surtseyan" for eruptions of magma (commonly basaltic) through sea water. Schmincke (1977a) interpreted the deposits of the 1888–1890 eruption of Vulcano (Italy) as phreatomagmatic and therefore suggested that the term Vulcanian be used synonymously with phreatomagmatic (Chap. 4). (3) *Submarine explosions* occur when magma rises into the shallow sea, producing abundant juvenile glassy fragments. Sublacustrine and subglacial explosive activity give rise to similar kinds of ejecta; therefore we prefer the more inclusive term, *subaqueous*, rather than submarine, and regard explosive submarine eruptions as one variety of phreatomagmatic eruptions. (4) *Littoral explosions* (Stearns and Clark, 1930) occur where subaerial lava flows or hot pyroclastic flows meet water. As with underwater explosions, littoral explosions are regarded here as a variety of phreatomagmatic eruptions.

Self and Sparks (1978) defined "Phreatoplinian deposits", within Walker's (1973) classification, as formed by the interaction of water and silicic magma. Such deposits differ from Plinian deposits principally by a higher degree of fragmentation.

Rittmann (1958; 1962) introduced the term *hyaloclastite* for rocks composed of sideromelane clasts produced by essentially nonexplosive spalling and granulation of rinds of pillow lavas by increase in diameter of pillow lava tubes during growth. Since then, the term has been expanded to include vitric tuff from shallow-water explosive volcanism (Tazieff, 1972) as well as sideromelane-bearing tuff produced by lava flowing into water (Fisher, 1968a) and occurring in maar volcanoes (Fisher and Waters, 1970; Heiken, 1971, 1972, 1974; Schmincke, 1974a). Thus, we use the term to include all vitroclastic tephra produced by the interaction of water and hot magma or lava whether or not the interaction is associated with venting. The general characteristics and origin of hydroclastic deposits are given in Table 9-2.

Fine-grained material believed to have formed by nonexplosive granulation that is commonly associated with pillow lavas has been called *aquagene tuff* by Carlisle (1963). Honnorez and Kirst (1975), emphasizing the need to distinguish between the clastic products of nonexplosive and explosive origin, introduce the term *hyalotuff,* a glassy *pyroclastic* rock resulting from phreatic or phreatomagmatic explosion.

## Components of Hydroclastic Deposits

### Grain Size Distribution

Deposits from hydroclastic eruptions are characteristically fine-grained, although coarse-grained lapilli- and tuff-breccias are common in some deposits. Grain size analyses of phreatomagmatic deposits have been reported by Waters and Fisher (1971), Sheridan (1971), Sheridan and Updike (1975), Crowe and Fisher (1973), Schmincke et al. (1973), Yokoyama and Tokunaga (1978), Nairn (1979) and Self et al. (1980). Data on tuff ring deposits are summarized by Walker and Croasdale (1972) and Walker (1973). Walker (1973), in a study of 88 samples from tuff rings in the Azores and Iceland, shows that the median diameter is less than 1 mm in

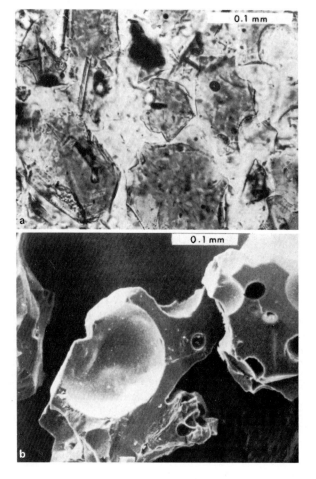

**Fig. 9-1a, b.** Photomicrograph of hydroclastic shards. **a** Rectangular and polygonal mafic shards with notable scarcity of vesicles (Heiken, 1974; pl. 24 B), **b** Blocky mafic shards showing breakage across vesicles (SEM photo from Heiken, 1974, pl. 24A). Shards are from Surtsey (Iceland)

about 75 percent of the samples. Qualitative inspection of many hydroclastic volcanoes composed wholly or partly of phreatomagmatic deposits leads us to suspect that there are essentially two main groups with many gradations in between. One group, represented by the ash and tuff cones, probably results from shallow explosions. Deposits of this group appear to be finer-grained and much better sorted than those of a second group represented by many maars which result from more powerful eruptions. The median diameters and sorting coefficients of the second group occupy a field between most flow and fallout deposits (Schmincke et al., 1973).

**Characteristics of Essential Components**

Vitric shards from hydroclastic eruptions are mostly mafic but silicic varieties also occur (Heiken, 1972, 1974). Most characteristic are blocky, nearly equant shapes with fracture-bounded surfaces transsecting few vesicles (Fig. 9-1). Some vitric hydroclastic shards have abundant mosaic cracks (Fig. 9-2) indicating rapid chilling.

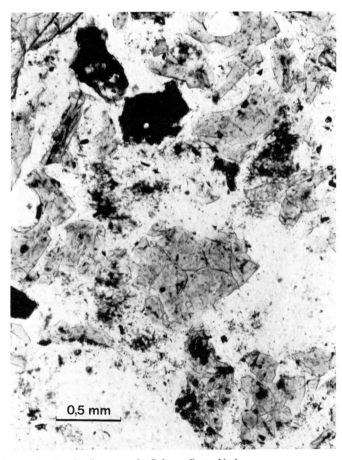

Fig. 9-2. Equant sideromelane grains showing jigsaw cracks. Pahvant Butte, Utah

**Fig. 9-3.** Composite lapillus consisting of two larger and many smaller lapilli set in a dense lava matrix. Deposits formed during transition Vulcanian-Strombolian type eruptions. Prehistoric Marteles Crater (Gran Canaria, Canary Islands) (Schmincke, 1977a)

Blocky mafic shards are common in deposits of maars and tuff rings (Fisher and Waters, 1970; Waters and Fisher, 1971; Walker and Croasdale, 1972), and littoral cones (Fisher, 1968a). Those that are common in seamounts and within the interstices of subaqueously chilled pillow basalt are described in Chapter 10. Honnorez and Kirst (1975) have stressed that blocky, nonvesicular sideromelane shards form during deep water eruptions (below the critical depth of magmatic volatile exsolution) by granulation of extruding lava in addition to spalling of pillow lava rinds. So far, no deposits of significant volume have been described to illustrate this process.

Shards formed by cracking due to thermal shock are typically glassy and nonvesicular. Indeed, these features are a major argument for steam explosions as the main eruption mechanism leading to the formation of maars and tuff rings. However, there are many tuffs associated with maar and tuff ring deposits which are made of shards that are both vitric and slightly to highly vesicular. These shards are apparently formed by a combination of vesiculating magma and quenching by water or steam. Deposits made of such shards – which may show all transitions from blocky through slightly vesicular with scalloped edges to highly vesicular – are characteristic of shallow water eruptions. They may be the most common type

Fig. 9-4. Composite lapilli (cf. Fig. 9-3) formed during transitional stage between Vulcanian (maar-forming) and Strombolian (cone-forming) eruptions. Marteles Caldera (Gran Canaria, Canary Islands) (Schmincke, 1977a)

of ash produced under water and typically occur in seamounts and in the transition from seamount to oceanic island.

Dense to slightly vesicular subspherical lapilli, some consisting of smaller lapilli held together by a lava matrix, occur in many cinder cones and maars within deposits transitional between phreatomagmatic and Strombolian deposits (Schmincke, 1977a) (Figs. 9-3 and 9-4). Layers made largely of such composite lapilli characteristically are inversely graded, caused by rolling down the slopes of the cones. Composite lapilli are similar in size, shape, and structure to "autoliths" described from some kimberlite diatreme breccias and believed by some workers to have formed in the presence of water (Lorenz, 1975; Schmincke, 1977a). Their subspherical shape and internal structure suggest formation and solidification within a vent prior to extrusion. They are interpreted to form when lava droplets are ejected into steam above the level of a magma column; the droplets are quenched and fall back to acquire a rind of new lava, and the process may be repeated several times.

Nakamura and Krämer (1970) first pointed out that surfaces of many lapilli and bombs from hydroclastic eruptions are characterized by a peculiar texture variously described as cauliflower, crackled or bread crust. This texture somewhat resembles that of bread crust bombs (Fig. 9-5), but unlike breadcrust bombs

Fig. 9-5. Dense, quenched basanite cauliflower bomb in Vulcanian deposits rich in accidental clasts. Lummerfeld Maar (East Eifel, Germany) (Schmincke, 1977a)

formed by internal gas expansion, cauliflower bombs commonly have dense or only slightly vesicular interiors (Schmincke, 1977a). Greatly expanded bread crust bombs, such as those formed during the historic eruptions of Vulcano (Walker, 1969), are not found among basaltic ejecta from hydroclastic eruptions because interior residual gas pressures of basaltic bombs are relatively low.

**Accretionary Lapilli**

Accretionary lapilli, which occur in many fine-grained ash layers, were reported from the 1965 phreatomagmatic eruptions at Taal Volcano (Moore et al., 1966), and they have been recognized in many subsequent investigations of hydroclastic deposits (e.g. Fisher and Waters, 1970; Heiken, 1971; Swanson and Christiansen, 1973; Lorenz, 1973, 1974; Schmincke et al., 1973; Self and Sparks, 1978). The occurrence of accretionary lapilli is not conclusive evidence for hydroclastic eruptions, however, because they are common in fine-grained fallout deposits where moisture is supplied by rain that often accompanies pyroclastic eruptions (Moore and Peck, 1962). The abundance of accretionary lapilli in hydroclastic tephra may be due to three factors: (1) abundance of water and steam in the eruption column, (2) production of abundant fine-grained tephra in hydroclastic eruptions, and (3) base surge transport, leading to deposition of fine-grained particles close to the source in contrast to Plinian eruptions where most fine-grained particles are usually deposited far from the vent, out of range of moisture related to the eruption column. One feature of some hydroclastic accretionary lapilli not described from other kinds of accretionary lapilli is the occurrence of vesicles in their outer layer (Lorenz, 1974) and in their core (Schmincke, 1977a).

Armored lapilli (Waters and Fisher, 1971) are a variety of accretionary lapilli containing crystal- or rock-fragment nuclei coated by rinds of fine to coarse ash. They range in diameter from 3 or 4 mm to as much as 10 cm or more, depending to some extent on the size of the nucleus, and have been reported only from hydroclastic deposits. In some cases, flattened lapilli- to bomb-size debris composed entirely of ash without cores are observed in hydroclastic deposits; these were apparently sticky and wet balls of ash when deposited. Armored lapilli apparently develop because the ash cloud contains abundant cohesive ash that sticks to solid particles within it. This mechanism differs from that proposed by Moore and Peck (1962) where moisture or rain drops falling through dry ash in eruption clouds causes agglutination of ash particles. In hydroclastic eruptions, there are probably all transitions between eruptions where large volumes of nearly pure water are initially ejected (Nairn et al., 1979) to blobs of wet ash to individual ash particles coated with moisture within vapor-rich eruption clouds.

**Accidental Clasts**

The form and shape of accidental clasts depends on the type of country rock at the site of fragmentation. Sandstone clasts, for example, are usually angular and blocky, whereas slate clasts are naturally platy. Accidental clasts commonly show little or no signs of thermal metamorphism, suggesting that temperatures are relatively low in much of the hydroclastic eruptive system (Fig. 9-6). Accidental clasts also occur as inclusions in essential lapilli and bombs; most such clasts are only slightly metamorphosed (Schmincke, 1977a) (Fig. 9-7). This suggests that fragmentation and incorporation of the country rock into the magma occurred shortly before or during eruption and thermal quenching. Some maar deposits contain abundant cobbles and pebbles derived from alluvial gravels of an underlying aquifer. Indeed, the presence of such material provides suggestive evidence for steam explosions within buried alluvial gravels, although it is possible that gravel could fall from surface levels into the zone of explosions.

Accidental rock fragments in maar and tuff ring deposits provide additional important information: (1) their maximum size allows estimation of explosion energy; (2) the type of crustal rocks present permits inferences about explosion depths if crustal stratigraphy is known; and (3) mantle-derived ultramafic xenoliths are common in some maar deposits.

Maximum Size of Fragments Related to Energetics

In the Nanwaksjiak Maar, Alaska (200 m deep; 600 × 1000 m in diameter), maximum diameters of blocks decrease from 3.3 m at the rim to 70 cm at 2000 m from the rim. From size relationships, Rohlof (1969; see also McGetchin and Ullrich, 1973) estimated that the eruptive fluid in Nanwaksjiak Maar had a density of about 0.01 g cm$^3$ and a surface velocity of about 500 m/s assuming an ejection angle of 65°. At Hole-in-the-Ground Maar, Oregon, Lorenz (1970) reports similar maximum diameters around the rim, but compared to Nanwaksjiak, the decrease in size away from the rim is less pronounced (Fig. 9-8). Lorenz calculated that pres-

**Fig. 9-6.** Xenolith-rich base surge deposits from 1949 eruption of Duraznero Crater (La Palma, Canary Islands). Note debarked but not burned pine trees rooted on pre-eruption surface

**Fig. 9-7.** Bomb from Quaternary phreatomagmatic deposits (Eifel) showing abundance of Devonian siltstone inclusions that have not been melted or strongly metamorphosed

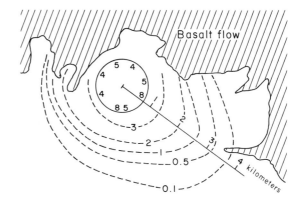

**Fig. 9-8.** Contours of maximum fragment sizes (in m). Hole-in-the-Ground (Oregon). (After Lorenz, 1971)

sures in the vent were over 500 b, ejection velocities of the largest blocks were 90–120 m/s and the fluid density was 0.04 g/cm$^3$. Similar ejection velocities of largest blocks are reported from the 1977 eruption of Ukinrek Maars, Alaska (Self et al., 1980).

Ultramafic Xenoliths

Many maar and tuff ring deposits contain abundant ultramafic rock fragments of different kinds thought to be derived from the mantle. These xenoliths have been used as evidence that maars result from explosions of magma rich in volatiles, especially $CO_2$ exsolved at great depth (McGetchin and Ullrich, 1973; Ringwood, 1975). However, if this were so, several conditions would have to be fulfilled. For example, there should be a close relationship between maar deposits and ultramafic nodules – which has not yet been demonstrated. Lava flows choked with ultramafic nodules are not uncommon, but many eruption centers in monogenetic volcano fields have not emitted lava flows. Secondly, magmas of maar-forming eruptions encompass a wide compositional range, while ultramafic nodules are typically restricted to alkalic mafic magmas, including kimberlites. Thirdly, little is known about the mechanism by which peridotite or other xenoliths become torn off conduit walls and incorporated into ascending magma. Even though mafic alkalic magmas are now thought to be generated at high $CO_2$ contents and high $CO_2/H_2O$ ratios (Wyllie, 1979), and even though $CO_2$ exsolves at much greater depths than $H_2O$ because of its higher partial pressure, no convincing cases have been made so far for fluidized magma particle-gas systems to have developed at depths exceeding a few kilometers.

In some cases, however, magma composition and maar-forming eruptions appear to be related. If so, it might be due to the rate of ascent of magma. Ascent is probably much faster for low viscosity, highly alkalic silica-undersaturated magma than for subalkalic tholeiitic magmas. Fast ascent would favor both nodule transport and, perhaps, the probability of explosive magma-water interactions (Schmincke, 1977a; Delaney, 1982). Kimberlite breccias which have been postulated to form by phreatomagmatic processes (Lorenz, 1975, 1980; Dawson, 1980), and which are known for their abundance of ultramafic nodules, may thus constitute an extreme end member in which very fast ascent rates favor both high transport capacities for nodules as well as high probability for magma-groundwater interaction in the upper crust. Only minor vesiculation of the magma would increase its surface area available for enhanced phreatomagmatic interactions.

## Structures

Deposits of hydroclastic origin are characterized by well-developed beds ranging in thickness from a few millimeters to several tens of centimeters; most are less than about 10 cm thick. The abundance of thin beds presumably results from the large number of short eruptive pulses characteristic of hydroclastic eruptions (Figs. 9-9, 9-10). Layers vary from plane parallel beds to cross-bedded, lenticular beds that show scouring features, giving the misleading impression that they are reworked. However, transport directions radially outward from the crater are shown unambiguously by imbrication of platy fragments, cross-bedding geometry and isopachs. Cross bedding is discussed in the section on base surge deposits.

**Penecontemporaneous Soft Sediment Deformation**

Soft sediment deformation structures have been reported from hydroclastic deposits by several workers (e.g. Fisher and Waters, 1970; Lorenz, 1970, 1974; Schmincke, 1970; Heiken, 1971; Crowe and Fisher, 1973; Schmincke et al., 1973). The most common type resembles "convolute lamination" (Potter and Pettijohn, 1963), consisting of folded beds sandwiched between undeformed layers; deformed layers are several centimeters thick and may extend laterally for several meters. Two main explanations for the development of convolute laminations are: (1) gravity sliding of sloping water-saturated tephra (Heiken, 1971) and (2) shear-deformation caused by an overriding base surge flow (Fisher and Waters, 1970; Schmincke, 1970). A special type of convolute structure is the asymmetric "gravity" or "shear ripples", with wave lengths of 5–10 cm and amplitudes of 1-3 cm, that are inclined downslope in beds with initial dips of 5°–20° (Lorenz, 1974). Spectacular decollement folds (Fig. 9-11) have been observed 5 km from the source of tephra (Laacher See Volcano, Germany); downward sliding of tephra deposited on the slope of an older cone apparently produced the folds.

**Vesicles (Gas Bubbles)**

Vesicles, common in hydroclastic tuff beds of maar volcanoes (Lorenz, 1974), occur as subspherical voids, generally less than 1 mm but rarely exceeding 1 cm in diameter (Fig. 9-12). Most have smooth outlines and are coated by very fine-grained ash. Large vesicles are more irregular in shape than small vesicles and may consist of several coalesced bubbles. Vesicles are most common in beds showing soft sediment deformation but also occur in lahars, in tuff beds with mud cracks, and even in tuff plastered on vertical surfaces. Many tuff beds with vesicles contain accretionary lapilli (Self et al., 1980) which themselves may contain vesicles in their outer fine-grained layer (Lorenz, 1974, Fig. 7) or in their center. Vesicular tuffs

**Fig. 9-9.** Thin-bedded, horizontal base surge layers on rim of Salt Lake Crater (maar), Zuni Salt Lake (New Mexico) (Fisher and Waters, 1970)

**Fig. 9-10.** Well-bedded lithic-rich tuffs formed by Vulcanian eruptions overlying slightly welded scoria deposits formed by Strombolian eruptions. Some bombs were erupted during the Vulcanian phase. Rothenberg volcano (Eifel, Germany). Large bombs 0.5 to 1 m in diameter

**Fig. 9-11.** Soft sediment deformation structure in base surge tuffs, many of which are vesiculated. Late Quaternary upper Laacher See tephra (Eifel, Germany)

**Fig. 9-12.** Vesiculated tuff, base surge deposits, Vulcano (Eolian Islands) (Schmincke, 1977a)

rarely occur more than 2–3 km from their source, although exceptionally they occur as far as 6 km (Laacher See, Germany). Vesicular tuffs are commonly more indurated than overlying and underlying vesicle-free beds.

Several sources of gas can account for vesicles in tuff: (1) gas within the fluidizing phase of the depositing system (derived from the eruptive center, from air incorporated during transit, or from rain falling during movement of the system), (2) gas given off by hot pyroclasts, (3) air rising from the underlying ground, or (4) water evaporating from snow or watersoaked soil beneath a layer of hot ash. Still another source may be (5) rain that turns to steam as it percolates downward into hot ash.

Several features prove that vesicle-bearing ash was water- or vapor-rich at the time of deposition: coating of vesicles with a film of clay or silt, association with soft sediment deformation structures, and preferential lithification of beds containing vesicles compared with associated nonvesicular beds. The amount of water necessary for soft sediment deformation in fine-grained sediment is about 15–20% (Heiken, 1971), far greater than saturation values for magmatic gases in basaltic magmas; thus the water or vapor phase must be mostly or entirely nonmagmatic. Moreover, vesicles can occur in tuff with mostly accidental clasts, thereby excluding them as a source of gas, although vesicles may form locally near large hot fragments exsolving gas or generating steam within a water-rich tuff matrix. Also, vesicles commonly occur several centimeters above the base of a bed, suggesting that air or gas did not rise from the ground below. Similar vesicles also may form in mudflow deposits (Sharp and Nobles, 1953; Bull, 1964).

Together with other criteria, the most common depositional mechanism of beds containing vesicles is probably a base surge resulting from phreatomagmatic eruptions (Lorenz, 1974; Self et al., 1980). However, at Augustine Volcano, Alaska vesicles also occur in fallout deposits. This fallout was probably in the form of "mud rain," a commonly reported event during volcanic eruptions (e.g. Macdonald, 1972). Tephra deposited from mud rains can develop vesicles from air or vapor trapped as bubbles within deposits of wet cohesive clasts.

**Bedding Sags**

Bedding sags, also known as "bomb sags" (Wentworth, 1926), form by the impact of ballistically ejected bombs, blocks, and lapilli upon beds capable of being plastically deformed (Fig. 9-13). They are characteristic of hydroclastic deposits and have been described from the deposits of many maar volcanoes, tiff rings, and tuff cones. Beds beneath the fragments may be completely penetrated, dragged down and thinned, folded, or show microfaulting (Heiken, 1971). Deformation is commonly asymmetrical, showing the angle and direction of impact if three-dimensional exposures are available.

Disruption of bedding also results when fragments fall into dry, noncohesive material. Dry material splashes out radially from the hole in rays or tongues or may construct a raised rim of ejecta around the depression with steep inward and gentle outward slopes (Hartmann, 1967). Because the particles do not stick together they are only affected by compressive forces. Unlike plastic beds, the noncohesive layers have no tensile strength. Thus, fragments beneath and in front of a projectile may be compressed together somewhat, but the force is rapidly dissipated by inter-

**Fig. 9-13.** Ballistically emplaced xenolith (transport direction from left to right) in thinly bedded base surge deposits, Duraznero crater (La Palma, Canary Islands)

granular movement; beds do not deform plastically into folds or become stretched by flow.

The width and depth of disturbance due to an impact is in part a function of the momentum of the projectile, plasticity of the sediments and angle of impact (Fig. 9-14). Compaction of tuff also occurs as some water is forced out during impact and is more slowly displaced by the weight of the block. A few preliminary (unpublished) measurements on bedding sags at Prineville tuff cone in eastern Oregon suggest that the amount of deformation is a direct function of fragment mass. Width-depth ratios of pyroclastic bedding sags produced subaerially should greatly differ from ratios produced by dropstones in water, but such studies have not been published. Knowledge of the ratios, however, should aid interpretations of depositional environments.

**Mudcracks**

Penecontemporaneous mudcracks are observed in places on the surfaces of fine-grained hydroclastic deposits. This feature is reported by Lorenz (1974) at the Hverfjall tuff ring (Iceland) in two vesiculated tuff beds. We have also observed them in tuff cone deposits at Cerro Colorado, New Mexico, Koko Crater, Hawaii and Marteles Caldera, Gran Canaria.

**Fig. 9-14. A** Deformation parameters and terminology for bedding sags. **B** Width-depth data and fragment volume for five bedding sags, Prineville Tuff Cone (Oregon)

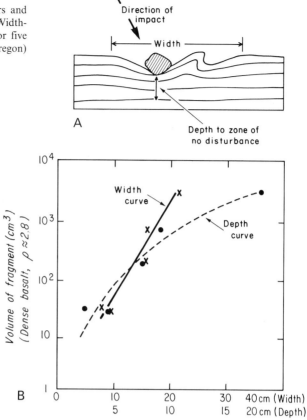

## Base Surge Deposits

Base surges are a type of pyroclastic surge (Chap. 8) that form at the base of eruption columns and travel outward during some hydroclastic eruptions (Fig. 9-15). The name was originally applied to a surge which developed during the 1947 underwater nuclear test at Bikini Atoll (South Pacific) (Brinkley et al., 1950, p. 103–106), and subsequently was recognized in hydroclastic eruptions (Moore, 1967) following the 1965 phreatomagmatic eruption of Taal Volcano (Philippines) (Moore et al., 1966; Nakamura, 1966). Such flows appear to develop mainly by the collapse of vertical eruption columns as detailed in Waters and Fisher (1971). Condensed steam, an integral part of volcanic base surges, becomes thoroughly mixed with particles during flow. Water is trapped by surface tension as thin films around grains causing newly deposited material to be cohesive and behave plastically when deformed.

Single-stage fallback of eruption columns may occur in many cases, but the processes may be more complicated in others. During the 1975 eruption within Lake Ruapehu (New Zealand), for example, an initial base surge apparently developed from the pre-existing crater lake by spillout of water jets and expanding steam

**Fig. 9-15.** Eruption column and base surge (diameter 270 m). Capelinhos (Azores). (Waters and Fisher, 1971)

(Nairn et al., 1979). Drainback of water into the lake was accompanied by collapse of the vertical eruption column to produce a "secondary" base surge composed of a dense aerosol of water droplets and debris. This surge moved at high enough velocities to surmount the rim of the crater 500 m above the lake and leave deposits on the outer slopes.

Eruptions that produce base surges involve release of large volumes of steam capable of supporting or fluidizing many of the particles in the surge.

Base surge deposits are poorly sorted and have an overall wedge-shaped geometry, decreasing logarithmically in thickness away from the source with local thickness controlled by topography (Wohletz and Sheridan, 1979). Distinct changes in facies and bed forms believed to be related to transport mechanism, load, and velocity of the surges are associated with decreasing thickness (distance). In places, the maximum radial distance attained by recognizable base surge deposits is about the same as the diameter of the crater (Table 9-3, Fig. 9-16), but at others, such as Laacher See, Germany (diameter = 2 km), base surge deposits occur 5 km or

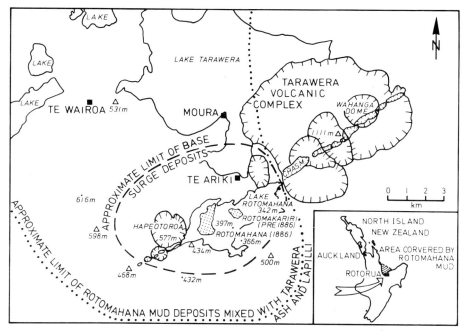

Fig. 9-16. Pre-eruption lakes Rotomahana and Rotomakariri (*stippled*) in Tarawera-Rotomahana-Waimangu area. Approximate limits of fallout tephra and surge deposits of Rotomahana are shown. (After Nairn, 1979)

Table 9-3. Crater size versus spread of surge deposit (Wohletz and Sheridan, 1979)

| Deposit | Diameter of crater rim (km) | Maximum distance from crater rim (km) |
|---|---|---|
| Coronado Mesa, Arizona | 0.5 | 1.1 |
| Peridot Mesa, Arizona | 0.55 | 1.25 |
| Ubehebe Crater, California | 0.80 | 0.82 |
| Sugarloaf Mountain, Arizona | 1.0 | 1.2 |
| Elegante Crater, Mexico | 1.6 | > 1.1 |
| Bishop Tuff, California | 16 × 29 | >15 |

more from source (Bogaard, 1978). Halemaumau, Hawaii, with a present-day diameter of about 1 km, was the site of phreatomagmatic eruptions in 1790 producing base surge deposits possibly as far as 10 km from the crater rim (Swanson and Christiansen, 1973).

## Bed Forms from Base Surges

Bed forms occur as three main kinds – sandwave, massive and planar (plane parallel) beds (Schmincke et al., 1973; Sheridan and Updike, 1975), and are grouped into three facies types (Wohletz and Sheridan, 1979) related to a fluidization model of transport and deposition. Fisher and Waters (1970), Crowe and Fisher (1973) and Schmincke et al. (1973) have emphasized bed forms in terms of the flow regime concept (Chap. 5). These different approaches are treated separately although they are not mutually exclusive.

Sandwave Beds

The term sandwave bed (Sheridan and Updike, 1975) is applied to beds with undulating surfaces or surfaces inclined to the depositional substrate and includes a variety of bed forms such as surface dunes, antidunes and ripples, and internal cross laminations that make up dunes and ripples.

Sandwaves deposited by base surges have a wide range of characteristics believed to be related to the flow regime in which they were deposited (Fisher and Waters, 1970). Most workers believe that base surge bed forms develop within the upper flow regime, but lower flow regime forms might be present (Stuart and Brenner, 1979) (Fig. 9-17). As shown in Figure 9-18, these different types of sandwaves occur at Hunt's Hole (New Mexico) and are developed in a sequence suggesting an upward and lateral decrease in flow regime.

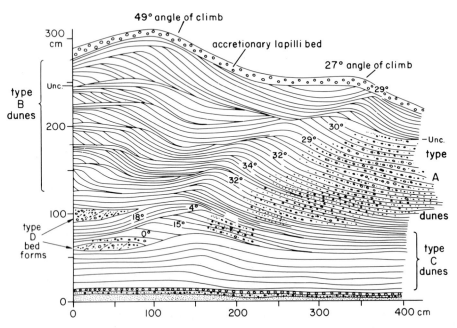

**Fig. 9-17.** Flow regime and bedforms in base surge deposits. *A* Lower-regime ripple drift dunes; foreset laminations approach angle of repose; backsets dip 15° or less toward source. *B* Lower-regime ripple drift dune forms with backset laminations horizontal to about 5° toward the source and foresets dipping up to 15°. *C* Upper regime antidunes. Backset laminations dip 5°–20° toward source, foresets are less than 15°. Laterally, *C* and *D* bedforms may be massive or occur as plane parallel beds. *D* Upper regime chute and pool bedforms. Backset (stoss) beds dip toward source at angles up to 55°; foreset (lee) strata are generally <15°. (After Stuart, unpublished)

250

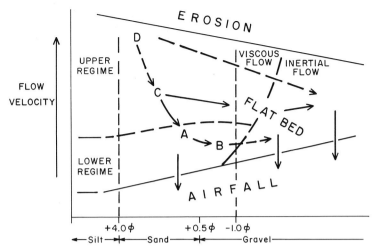

**Fig. 9-18.** Diagrammatic representation of bedform relationships. Bedforms *A, B* and *C* as explained in Fig. 9-17. (After Stuart, unpublished)

At other localities, lateral decreases in dune sizes and particle sizes and increases in sorting suggest decreasing velocities and probably flow regime, although lower-regime bed forms have heretofore gone unrecognized. At Taal Volcano (Philippines), base surge deposits within about 3 km of the vent are characterized by dunes oriented at right angles to movement directions of the base surges. The orientation of internal laminations show that the dunes migrated away from the explosion center. Wave lengths near the explosion center attained 19 m, systematically decreasing to about 4 m at 2.5 km from center. Northeast of the crater, the base surges were slowed by an uphill gradient, and wave lengths of dunes decreased rapidly (Moore, 1967). These relations suggest a direct relationship between wave length and velocity. The low dips of foreset (lee-side) laminations (10°–15°) indicate that the dunes did not advance by gravitational rolling of loose debris down advancing lee slopes, as is the case for desert sand dunes or low flow regime dunes formed in alluvial channels. Wave lengths of the Taal dunes vary directly with total thickness of the deposit, bedding thickness, size parameters of ash and the distance from the source (Table 9-4), suggesting that the carrying capacity of the Taal base surges decreased progressively with distance as velocity slowed.

Cross beds which occur in some flow units at Laacher See (Germany) also progressively change laterally from large dune-like structures characterized as chute and pool structures (Figs. 9-19, 9-20) (Schmincke et al., 1973) near the source to smaller more subdued antidunes farther from the source such as those at Ubehebe, California (Crowe and Fisher, 1973) (Fig. 9-21); the antidunes grade laterally into transitional low-amplitude structures that become plane-parallel beds about 5 km from Laacher See, and which appear to continue another 3–4 km from source. This progressive lateral change in bed forms, together with decreasing size and thickness parameters, are interpreted as reflecting a decrease in flow regime. Wave length probably depends on flow power, and wave height on grain size and volume of bed load. Thus, the downstream change suggests that the bed load was dropped rapidly

Table 9-4. Size parameters of samples from backset and foreset beds of dunes of base surge deposits. Taal volcano, Philippines (Waters and Fisher, 1971)

|  | Median diameter ($Md_\phi$) | Sorting coefficient ($\sigma_\phi$) | Distance from vent (km) | Approx. wavelength (m) | Approx. thickness of bedding set (m) | Approx. total thickness of deposits (m) |
|---|---|---|---|---|---|---|
| 1. Backset | 0.9 | 2.01 | | | | |
| Foreset | 1.3 | 1.50 | 0.75 | 11.5 | 2.0 | >4.0 |
| 2. Backset | 2.4 | 1.65 | | | | |
| Foreset | 2.9 | 1.24 | 1.0 | 9 | 1.0 | ~2.5 |
| 3. Backset | 3.3 | 1.28 | | | | |
| Foreset | 3.0 | 1.20 | 1.5 | 5.5 | 0.7 | 1.0 |
| 4. Backset | 3.4 | 1.16 | | | | |
| Foreset | 3.6 | 1.21 | 2.0 | 4 | 0.5 | 0.5 |

near the source and the energy of transport or capacity to carry a load then decreased more slowly.

Allen (1982) has criticized the hydrodynamic interpretation of sandwave bed forms in surge deposits as antidunes and chute and pool. In Allen's interpretation, these bedforms "record an unstable interaction between the moistened debris driven by the surge and a particle-capturing cohesive bed, that may have been independent of the Froude number" (Allen, 1982, p. 430). Allen subdivided bedforms and internal sedimentary structures in base surge deposits into (a) progressive bedforms – thought to be characteristic of relatively dry and/or hot flows, (b) stationary bedforms, and (c) regressive bedforms – with crests migrating upstream – thought to be deposited from relatively wet and cool flows. Previous authors have noted the problems of interpreting bedforms in systems characterized by cohesiveness. However, the correlation between the assumed wet regressive bedforms thought by Allen to be associated with accretionary lapilli and vesicle tuffs and the absence of these structures in progressive types is not supported by the evidence. At Laacher See, e.g., type C bedforms of Allen – the chute and pool structures of Schmincke et al. (1973) – occur only in the proximal base surge facies in coarse-grained relatively well-sorted deposits with very large wave lengths and amplitudes.

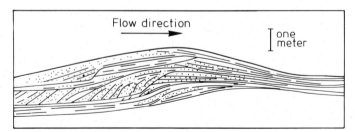

Fig. 9-19. Internal bedding of a chute-and-pool structure, Laacher See (Eifel, Germany). (After Schmincke et al., 1973)

**Fig. 9-20.** Chute-and-pool structures in phonolitic base surge deposits representing a late eruptive phase of the Laacher See Volcano (Eifel, Germany). Scale in 10 cm intervals (Schmincke et al., 1973). Flow direction from *left to right*

**Fig. 9-21.** Dune-like structure, Ubehebe Crater (California) showing build-up and migration from nearly flat beds. Flow direction from *left to right* (Fisher and Waters, 1970)

Downstream, in the more distal facies, the same beds develop progressive bedforms, the sediment being finer-grained and associated with accretionary lapilli and vesiculated ashes.

Although the hydrodynamic interpretation of the bedforms and internal structures of surge deposits is an open problem and incompletely studied as yet, we see no problem in supercritical flow being reached by surges.

Plane-Parallel Beds

Plane-parallel beds have upper and lower contacts which are generally planar and parallel to one another. Such beds in base surge deposits may be concordant with contiguous layers and normally or reversely graded, but unlike fallout layers, may erode into underlying beds. Sorting coefficients may be similar to fallout layers (Crowe and Fisher, 1973) but most surge deposits are more poorly sorted. In places, plane-parallel beds grade laterally from planar conformable sequences into zones of cross bedding, where they steepen into backset laminations (Schmincke et al., 1973). Plane-parallel beds tend to thicken within gentle lows and become thin and finer-grained over crestal parts of undulations, as do their internal laminae (Schmincke et al., 1973), rather than evenly mantling irregular underlying surfaces as is more common for fallout tephra. Platy fragments are imbricated or aligned roughly parallel to bedding surfaces. Internal laminae are commonly very subtly cross-bedded or lenticular over short distances (Fig. 9-22). Inversely graded plane-parallel beds suggest transport and deposition by flow, but are not unequivocal evidence inasmuch as some fallout beds are inversely graded (Chap. 6). Large blocks that rest on lower contacts without deformation are another indication of emplacement by flow, not fall.

Massive Beds

Massive beds usually are thicker, and more poorly sorted than plane-parallel beds or beds within sandwaves. They tend to be internally massive, but usually have pebble trains or vague internal textural variations giving a crude internal stratification that is either planar or wavelike (Fig. 9-23), and many massive beds have inversely graded basal zones. Sheridan and Updike (1975) and Wohletz and Sheridan (1979) postulate that massive beds are transported by a dense-phase fluidized surge and are transitional between sandwave and planar beds.

**Bed Form Facies**

The three facies defined by Wohletz and Sheridan (1979) are (1) sandwave facies (sandwave and massive beds), (2) massive facies (planar, massive and sandwave beds), and (3) planar facies (planar and massive beds). These facies systematically change laterally, with the sandwave facies dominating nearest the vent, massive facies at intermediate distances and the planar facies farthest from the vent (Fig. 9-24) at four volcanic sources described by them. Facies analyses of this kind may provide statistical summations of the dominant flow processes of many different flows through time at a particular locality, but we do not agree that they apply to the processes believed to occur laterally within a single flow.

**Fig. 9-22.** Thin bedded base surge deposits, Ubehebe Crater (California). Tip of hammer handle is 4 cm long. Flow direction from *left to right*

**Fig. 9-23.** Base surge bedforms at Laacher See (Eifel, Germany) (Schmincke et al., 1973). Flow direction from *right to left*

255

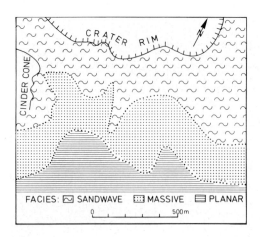

Fig. 9-24. Pyroclastic surge facies relationships at Crater Elegante (Sonora, Mexico). (After Wohletz and Sheridan, 1979)

Our alternative explanation for the three facies is (1) that massive and sandwave bed forms (i.e. sandwave facies) occur closest to the source because this is where the flows begin to separate by gravitational segregation into a laminar-flowing bedload and an overlying turbulent flow (Fisher et al., 1980; Fisher and Heiken, 1982); (2) planar beds from smaller flows begin to occur within the stratigraphic sequence at intermediate distances to give sequences with all three bed forms (i.e., the massive facies); and (3) massive beds within the distal planar facies might be thick planar beds. At Mukaiyama Volcano, Japan (Yokoyama and Tokunaga, 1978), plane-parallel to wavy beds (massive and sandwave bed forms?) occur closer to the source than large antidunes; farther from source the antidunes decrease in size. These relationships do not confirm or deny our alternative explanation for the development of the facies relationships shown by Wohletz and Sheridan (1979), but do point out that additional research is needed to resolve the many problems of facies.

**U-Shaped Channels**

U-shaped channels in base surge deposits, described by several authors (Losacco and Parea, 1969; Fisher and Waters, 1970; Mattson and Alvarez, 1973; Heiken, 1971; Schmincke, 1977b; Fisher, 1977; Nairn, 1979), are symmetrical in cross section, with curving bottoms that clearly cut underlying layers (Fig. 9-25). Most range from about 0.3 m to 7 m across and are a few centimeters to 3 m deep, but unusually large channels (30 m broad, 20 m deep) are reported by Losacco and Parea (1969), Mattson and Alvarez (1973) and Heiken (1971) and occur at Laacher See, Eifel, Germany (Fig. 9-25). The curving bottoms are best described as U-shaped, not parabolic or catenary curves, even though some are very broad in cross section. Infilling beds reflect the shape of the channels, but the curvature of individual beds decreases upward, and the final fill extends uniformly across the channel and is conformable with the sequence outside the channel. Thus, beds thicken toward the centers of channels and therefore do not resemble draped fall-out layers.

Fisher (1977) argues that the shape of the advancing head of a base surge and concentration of particles within the head is responsible for their U-shape. Rather

Fig. 9-25. Light-colored highly differentiated phonolitic fallout and ash flow tuffs with two minor unconformities formed by pyroclastic flows overlain by dark mafic phonolite Vulcanian base surge and pyroclastic flow deposits, the first base surges having eroded a major U-shaped channel. Laacher See (Eifel, Germany)

than having smooth, even fronts, base surges (as do nuées ardentes) develop secondary knuckle-like clefts and lobes that spread outward from the source, each lobe possibly being a separate complexly turbulent cell that joins the main body of the flow behind the advancing front. Moving down a widening slope of a volcano, individual lobes diverge to follow independent paths and carve diverging furrows straight down the slope. The concentration of particles within the turbulent cells is probably greatest along their central axes, where boundary effects are least and forward velocity is greatest. If pre-existing channels are present, the debris becomes more concentrated in the channels to increase the erosive capacity of the currents.

## Maar Volcanoes

Maar volcanoes are low volcanic cones with bowl-shaped craters that are wide relative to rim height (Fig. 9-26). They were originally recognized as small subcircular crater lakes in the Quaternary volcanic district of the Eifel (Germany), the term being derived from the Latin "mare" for sea (Steininger, 1819). Classification, definition, and theories of maar origins are discussed by Noll (1967), Ollier (1967), Waters and Fisher (1970), Lorenz et al. (1971), Lorenz (1973; 1975), Pike (1974), and Wohletz (1980).

### Classification

As modified from Lorenz (1973), we define the various kinds of maar volcanoes as follows:

**Fig. 9-26.** Little Hebe Crater, tuff cone (Death Valley, California). Diameter about 100 m

*Maar (sensu stricto):* a volcanic crater cut into country rock *below* general ground level and possessing a low rim composed of coarse- to fine-grained tephra. They range from about 100 to 3000 m wide, about 10 to more than 500 m deep, and have a rim height of from a few meters to nearly 100 m above general ground level.

*Tuff ring:* a large volcanic crater at or *above* general ground level surrounded by a rim of pyroclastic debris (tuff or lapilli tuff), similar in diameter to maars. *Tuff cones* have higher rims, attaining heights of up to 300 m (Koko Crater, Hawaii), and are essentially tuff rings where volcanic activity was of longer duration. The distinction between tuff cones and tuff rings, however, becomes arbitrary where one side of a crater stands high and another side low. Aliamanu Crater, Hawaii, for example, would be classified as a tuff cone if viewed from the north and a tuff ring if viewed from the south, where it shares its rim with Salt Lake Crater, a low-standing tuff ring. Figure 9-27 depicts a hypothetical volcano showing the transition from a maar to a tuff cone based upon morphology.

### Origin

Most maars result from hydroclastic eruptions (Lorenz, 1973; Kienle et al., 1980); wide craters develop from shallow explosions (Fisher and Waters, 1970), subsidence (Frechen, 1971; Noll, 1967) or a combination of both (Lorenz, 1973). Convincing evidence of a hydroclastic origin is that, in groups of nearly synchronous eruptive centers, those erupting on high ground form spatter or cinder cones, whereas associated eruption centers in valleys, depressions, on alluvial gravels or in coastal regions form maars, tuff rings or tuff cones (Lorenz, 1973). Juvenile clasts within their deposits are glassy, nonvesiculated, and have blocky shapes (Heiken, 1974), suggesting that magma was quenched prior to exsolution of vola-

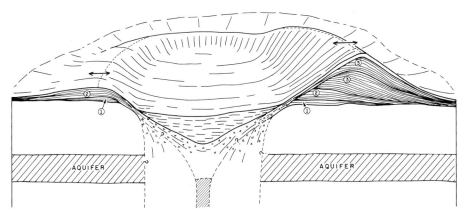

Fig. 9-27. Diagrammatic section of an asymmetrical maar volcano showing rim beds from successive stages of construction from explosion crater (*1*), tuff ring (*2*), and tuff cone (*3*). Anticlinal-form of rim beds is shown. Crater is filled with fall-back tephra and by lake and fluvial sediments. Conduit is shown as either the upper end of a diatreme formed by progressive collapse and partial blow-out of debris (Lorenz, 1973) or else is a funnel-shaped structure formed by explosion due to continuous mixing of magma and water at shallow depth. Unconformities on inner crater wall are from blast-erosion or collapse of steepened slopes

tiles, that breakage of glass resulted from thermal shock and (steam) explosions, and that the vapor and steam phase in the eruption column was partly or largely vapor from external water.

Wohletz and Sheridan (1983) conclude that tuff cones and tuff rings are distinct land forms that result from slightly different types of hydroclastic activity and they present a "hydroclastic continuum" of landforms from cinder cones to pillow lavas relating environments of eruption and mechanical energy of eruptions. According to them, tuff rings evolve through a stage of explosion breccia emplacement to a stage dominated by base surges which deposit thinly bedded layers. Tuff cones may be built when continuing activity evolves into a third stage, characterized by rocks emplaced by poorly inflated base surges and ballistic fallout. They relate these differences to water: melt ratios (Sheridan and Wohletz, 1983) based upon experiments with thermite-water systems (Wohletz, 1980). Fragmentation of melt attains maximum explosive energy when the water:melt ratio is about 0.5 for basaltic compositions. Initial ("vent-coring") eruptions with small ratios result in the formation of breccia with abundant cognate and accidental fragments. Increasing ratios cause development of expanded dilute surges which deposit thin-bedded layers, hence tuff rings. Still higher ratios produce "wetter" and denser eruption columns giving rise to poorly expanded surges, hence dominantly massive beds and tuff cones. The rates of magma and water influx controls the process, therefore such "cycles" may be interrupted, reversed or alternate. We have observed scoria blanketed by phreatomagmatic breccias from the same vent (Figs. 9-6, 9-10), but most commonly, tuff cones may have craters filled or partly filled with lava (Womer et al., 1980), agglutinated spatter and cinders (Prineville, Oregon). In some volcanic fields, many of the scoria cones contain deposits of phreatomagmatic origin commonly developed during their initial eruptive stages (Schmincke, 1977a).

Table 9-5. 20th century maar-forming eruptions (Kienle et al., 1980)

| Maar | Maximum diameter (km) | Year | Source |
|---|---|---|---|
| Corral Quemado, Chile | ≃1 | 1907 | Illies (1959) |
| Stübel crater, Kamchatka[b] | 1.5 | 1907 | Vlodavetz and Piip (1959) |
| Falcon Is., Tonga Islands | 1.5 | 1927 | Hoffmeister et al. (1929) |
| Pematang Bata, Sumatra[c] | 2.1, 1.0 | 1933 | Stehn (1934) |
| Nilahue, Chile | 0.8–1.4[a] | 1955 | Müller and Veyl (1957), Zuniga (1956), Illies (1959) |
| Iwo Jima[c, d] | 0.035 | 1956 | Corwin and Foster (1959) |
| Deception Island (5 maars)[c], Antarctica | 0.3–0.5 | 1967 | Schultz (1972) |
| Vlodavets, Radkevich (Tyatya volcano, Kuriles)[c] | 0.3, 0.3 | 1973 | Markhinin et al. (1974) |
| Taal | 0.2 | 1965–66 | Moore et al. (1966), Moore (1967) |
| Ukinrek (West and East Maars)[c] | 0.17, 0.3 | 1977 | Kienle et al. (1980) |

[a] Varies with different author estimates
[b] This is proposed to be a maar by C. A. Wood (personal communication)
[c] Two or more craters formed during eruptive event
[d] One crater is proposed to be a maar; a second crater is a collapse pit

Traditionally, maars were thought to have originated by the explosive discharge of mantle-derived $CO_2$, as discussed in Chapter 4. However, even carbonatite maars, formed from magmas rich in $CO_2$, appear to occur only in lowland regions of the African Rift Valley where ground water is available, and are therefore of probable hydroclastic origin (Dawson, 1964a,b).

Further evidence for the central role of external water comes from observations of historic maar-forming eruptions (Table 9-5), especially the 1977 Ukinrek Maars, Alaska (Kienle et al., 1980; Self et al., 1980). Characteristically, maar-forming eruptions are accompanied by great volumes of steam and repeated short-interval blasts. Often maars occur in groups of two or more. Indeed some large tuff rings have scalloped shapes that may be caused by several closely spaced eruption centers and/or inward slumping of rims into repeated explosively evacuated central craters.

**Dimensions**

Areal Extent and Geometry

Compared to cinder scoria of similar volume, maar and tuff ring deposits usually extend farther from the eruption center. Scoria cones are built from vertical eruption columns composed mainly of juvenile bombs, lapilli, and ash that are deposited as spatter (agglutinate and agglomerate) and lapilli and ash layers rich in tachylite and scoria; fragments tend to follow ballistic paths, and the bulk of the material falls back near the vent. In maar volcanoes, much of the ejecta is finer-grained than in scoria cones and much may be transported by base surges. The depth of explosions is usually shallow, so that ejection angles are commonly lower than from scoria cones. The contrast between scoria cones and maar volcanoes is well seen when profiles of both are compared (Heiken, 1971). Abundant fines are carried far beyond the sites of eruption but are quickly dispersed and eroded. During the March 30, 1977 eruption of Ukinrek Maars, Alaska, for example, fine ash fell over 20,000–25,000 km² but significant ash accumulation was restricted to a radius of only about 3 km.

Volume

Maar craters are often thought to have larger volumes than the material ejected. However, volume estimates commonly have underestimated amounts eroded from the rim deposits and especially far-distant fallout deposits.

The main problem in determining the volume of magma represented by maar deposits is a realistic estimate of distant fallout material and of essential material hidden in the diatreme beneath the crater floor. Using a formula modified from Lorenz (1971), Mertes (1983) determined the volume of material ejected in three Eifel maars using the term

$$V_D = V_E(\varrho_E - \gamma \varrho_j) - V_c \varrho_B / \varrho_B - \varrho_D + \gamma \varrho_j$$

$V_D$ = volume of essential material in vent
$\varrho_E$ = density of ejecta
$\varrho_j$ = density of essential material
$\varrho_B$ = density of basement rocks
$\varrho_D$ = density of vent filling
$V_C$ = volume of crater in basement and
$\gamma$ = amount of essential material given as percentage of total volume of ejecta $V_E$.

In this calculation, intrusives and larger blocks of country rock are neglected. The relationship

$$V_T = b \cdot R3,$$

where the coefficient b (varying between 0.109 and 0.074) decreases with increasing radius of maar (R), was used to estimate the total volume of ejecta, $V_T$, from all Eifel maars.

Maars and tuff rings encompass eruption centers with volumes up to $12 \times 10^6$ m³ – the range for scoria rings and cones – as well as large centers with volumes between 15 and $30 \times 10^6$ m³ (Mertes, 1983) (Fig. 9-28). Magma volumes may be related to composition. Large maars in the Eifel district, for example, are mostly composed of melilite nephelinite. The low viscosity and, at low pressure, the high volatile content of these magmas could retard freezing of the feeder dikes resulting in especially efficient discharge (Mertes, 1983). Volumes can be estimated from rim diameters as shown in Fig. 9-29.

Taking all factors into account, the total volume of ejecta from Ukinrek Maars is estimated at $10 \times 10^6$ m³ (dense rock equivalent), substantially greater than the combined volume of the two fresh Ukinrek craters calculated at $4.3 \times 10^6$ m³. Kienle et al. (1980) account for the excess ejecta volume ($5.7 \times 10^6$ m³) as juvenile fallout material – that part of the ejecta generally unaccounted for in pre-historic deposits of maars and tuff rings.

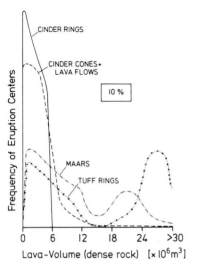

Fig. 9-28. Percentage of four major types of basaltic constructional volcanoes in Quaternary West Eifel volcanic field (about 240 eruptive centers) in relation to magma volume erupted (Mertes, 1983)

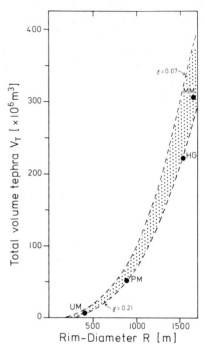

Fig. 9-29. Relationship between total tephra volume and rim diameter of maars. Bounding curves for Gamma = amount of essential clasts in fraction of total volume. *UM* Ulmener Maar, *PM* Pulvermaar, *HG* Hole-in-the-Ground, *MM* Meerfelder Maar (Mertes, 1983)

**Chemical Composition**

The composition of juvenile ejecta from maar volcanoes ranges widely, most being basaltic. In Iceland, for example, where about 20 maars were described by Noll (1967), most are tholeiitic, some with slightly alkalic basalt affinities. In central Oregon, USA, there are about 40 maars and tuff rings, mostly of high-alumina basalt (Heiken, 1971). In the Quaternary West Eifel volcanic field (Germany), most of the 70 maars, tuff rings, and tuff pipes are of melilite nephelinite and sodalite foidite composition, contrasting with the composition of the other 165 eruptive centers in the area (Mertes, 1983). The maars in East Africa are made up mostly of alkali basalt to nephelinite, some even of carbonatite (Dawson, 1964a, b). Maars of phonolitic (Schmincke et al., 1973; Schmincke, 1977a) and rhyolitic (Sheridan and Updike, 1975; Yokoyama and Tokunaga, 1978) composition have also been described.

# Littoral Cones

Littoral cones are mounds of hyaloclastic debris constructed by hydroclastic explosions at the point where lava enters the sea. Littoral cones belong to a group of craters that lack feeding vents connected to subsurface magma supplies (i.e., they are "rootless") and form where lava or pyroclastic flows move over small ponds of water, swamps, springs or streams as, for example, the pseudocraters in Iceland (e.g. Rittmann, 1962), and the phreatic explosion pits in pyroclastic flow deposits at Mount St. Helens (Rowley et al., 1981). Littoral cones commonly occur as crescent-shaped ridges (Fig. 9-30), breached by the source lava or more rarely as complete cones with craters occurring above lava tubes. Explosion centers are near or at the shore line, therefore about half of the radially exploded material falls into the sea, leaving a half-cone on land. A typical littoral cone is characterized by: (1) a wide crater (if the part missing at sea is reconstructed) and low rims; (2) steep inner slopes exposing truncated strata unconformably mantled by in-dipping strata; and (3) gentle outer slopes merging with the slope of the underlying terrain.

About 50 pre-historic littoral cones are known along the shores of Mauna Loa and Kilauea, Hawaii (Stearns and Macdonald, 1946). Twenty-one lava flows have entered the ocean along the shore of the Island of Hawaii between about 1800 and 1973 (Peterson, 1976, Table 1), but only four of the flows have developed littoral cones (Moore and Ault, 1965; Fisher, 1968a; Peterson, 1976).

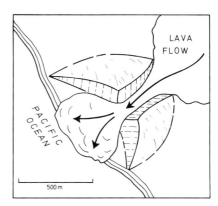

Fig. 9-30. Diagram of typical littoral cone

## Deposits

Littoral cones are typically composed of hundreds of very poorly sorted, poorly defined beds ranging from a few centimeters to over 10 cm thick. They consist of fine- to coarse-grained ash, lapilli, and angular blocks up to 1.5 m and bombs to 1 m in longest dimension. Ash $<0.062$ mm ($>4_\phi$) diameter, however, is commonly no more than 5% of the total ash content. The ash is composed of sideromelane, tachylite, microcrystalline basalt and broken phenocrysts. Sideromelane fragments are predominantly broken fragments indicative of hydroclastic explosions (Heiken, 1974). Some layers contain accretionary lapilli and bedding sags, suggestive of abundant water vapor in the explosion clouds (Fisher, 1975).

**Origin**

The main conditions necessary for the construction of littoral cones appear to be a rapid delivery of large volumes of lava to the water (Moore and Ault, 1965; Fisher, 1968a) and confining conditions where water and lava can become repeatedly mixed. The abundance of beds within littoral cones and the height to which they can be built, about 100 m, suggests that conditions of confinement and water-lava mixing repeatedly occur. Lava tubes are absent at Puu Hou littoral cone, Hawaii, thus explosions probably occurred beneath confining lava crusts and rubble that continued to form as lava was fed to the ocean (Fisher, 1968a). Absence of lava tubes is characteristic of many pre-historic littoral cones on Hawaii's southeast coast, but some occur on top of lava tubes, a likely environment for confining conditions to occur. Indeed, the only observation of a littoral explosion not obscured by steam clouds (Peterson, 1976) indicates that lava tubes are important in the formation of some littoral cones.

Explosions that produce littoral cones may be caused by auto-catalysis (Fisher, 1975). The energy released by a given volume of water and lava during initial explosions may be great enough to cause further mixing and subsequent energy release of a somewhat larger volume of the two liquids in an exponential type of reaction (Colgate and Sigurgeirsson, 1973; see Chap. 4). Explosions subside as available lava is depleted by division into small droplets and expulsion from the mixing site, but take place again as lava continues to be delivered rapidly to the place of confinement under lava crust beneath water, or where water enters and comes in contact with molten lava within the confines of a lava tube at or below sea level.

## Peperites

Peperites are a poorly defined group of volcaniclastic rocks that we include in the broad spectrum of hydroclastic rocks. The name peperite was used by Scrope (1862) for basaltic tuffs and breccias of the Limagne region of Central France. The rocks are composed of dark "basaltic" clasts in a light-colored marly to limy matrix – hence the name "pepper rocks." Scrope believed that they originated by explosions of Oligocene volcanoes, whose erupted material fell into lake water and mixed with sediment. Subsequently the peperites of Limagne were ascribed to explosive volcanic eruption or to the intrusion of basalt into wet sediment (Michel-Levy, 1890; Michel, 1953). Jones (1969b) suggested that peperites form by simultaneous sedimentation of epiclastic volcanic material together with lake sediments. The main argument against entirely epiclastic processes is the discrepancy between the fine-grained lake sediments and the coarse volcanic clasts. Moreover, the commonly glassy nature of the clasts is strong evidence for lava-water contact. The term peperite has been extended to other volcaniclastic breccias formed by mixing of hot lava and wet sediment during surface invasion unrelated to intrusive centers (Schmincke, 1967a). There are clear gradations from surface flows, through hyaloclastites, into peperites and finally "invasive flows" as lava flowed into sedimentary basins.

## *Chapter 10* Submarine Volcaniclastic Rocks

In this chapter we are concerned with underwater volcaniclastic eruptions and products, as well as pyroclastic flows and debris generated on land and transported by mass flowage into the sea. Submarine fallout deposits derived from land-based eruptions are treated in Chapter 7. Clastic materials redistributed as mass flows or turbidity currents originating from the rapid build-up of tephra and epiclastic volcanic debris along coastal parts of volcanoes are also discussed in various places throughout the text (e.g. Chap. 13).

The nature of submarine volcanic explosions is governed by (1) water depth (pressure) at which eruptions take place; (2) composition of magma, especially its volatile content and viscosity, and (3) the dynamics of the interaction of lava and water which are dependent upon many factors such as physical properties of the magma, eruptive rate, vent geometry, degree of magmatic fragmentation and others. Present-day submarine pyroclastic rocks are formed in three distinct settings: (1) at mid-ocean ridges, (2) around oceanic intraplate central volcanoes, and (3) above subduction zones and in back arc basins.

The first half of this chapter is divided into three parts based chiefly on decreasing water depth (pressure) regimes. We begin first with volcaniclastic rocks formed along mid-ocean ridges and during deep water seamount stages by clastic processes below the level at which exsolving magmatic volatiles contribute to clastic processes. This depth, here called volatile fragmentation depth (VFD), depends chiefly on the type and amount of dissolved volatiles and thus on the magma composition. It is generally shallower than 500 m.

Secondly, we discuss degassing of magma at shallower depth which increasingly accompanies fragmentation processes that are not depth-dependent. As an actively building submarine volcano approaches the water surface, pyroclastic eruptions become important. At this stage, the formation of volcaniclastic aprons is initiated by transport of loose unstable volcanic debris toward the lower slope of volcanoes and into surrounding sedimentary basins.

Thirdly, during the emergent stage, lava flows and pyroclastic flows entering the sea contribute large volumes of volcanic debris into the sea. To these are added increasing amounts of eroded epiclastic volcanic materials as an island grows and finally decays.

## Deep Water Stage

No deep water eruption has been observed, all evidence for volcaniclastic processes below the VFD coming from analyses of the clastic products, from studies of uplifted extrusive oceanic crust, basement drilling by the Glomar Challenger and submersible studies. The relative percentage of volcaniclastic rocks formed in deep water is small compared to pillow and sheet flow lavas. But it is larger than indicated by drilling in young ocean floor, because low core recovery of unconsolidated material tends to overemphasize massive lavas and pillow lavas compared to unconsolidated clastic rocks. Drill core data from the Atlantic (Hall and Robinson, 1979) and mapping of the extrusive section of the Troodos ophiolite (Schmincke et al., 1983) suggest that 8–10 percent of the extrusive section are various kinds of volcaniclastic rocks, mostly breccias (Fig. 10-1). Judging from other studies, these figures appear realistic and we suspect that volcaniclastic rocks will generally constitute between about 5 and 15 percent of submarine extrusive crustal sections. Studies of these rocks, which range widely in grain size, structures, textures and origin, are rare and their classification and nomenclature are not generally agreed upon. For discussions of various aspects see Rittmann (1958), Hentschel (1963), Carlisle (1963), Furnes (1972), Furnes and Fridleifsson (1979), Jones (1970), Moore (1975), Moore et al. (1973), Sigvaldason (1968), Dimroth et al. (1978), Lonsdale and Batiza (1980), Staudigel and Schmincke (1984), van Andel and Ballard (1979), and others.

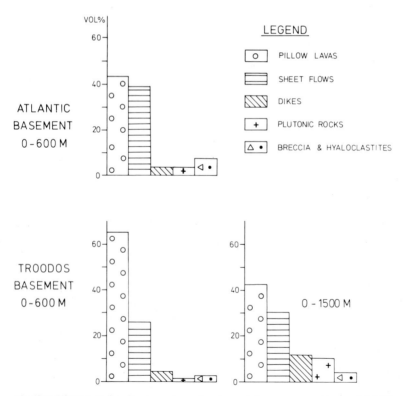

Fig. 10-1. Histogram showing percentage of major rock types recovered by crustal drilling in the North Atlantic Ocean and by mapping the Troodos Ophiolite (Cyprus). (Data from Hall and Robinson, 1979; Schmincke et al., 1983)

## Pillow Breccias

The most abundant fragmental volcanic rocks in the deep sea are various kinds of coarse breccias composed of lapilli to block size material. At one end of the breccia spectrum are broken pillow lavas with segments of individual pillows slightly offset against one another. The joints between fragments are either healed during cooling of the lava (Fig. 10-2) or, if the fragmentation occurred under subsolidus conditions or after complete cooling to ambient temperature, by secondary minerals such as carbonate or smectite. Such breccias are monolithologic with little or no matrix. They are *in situ* breccias that appear to form, for example, when tube lava extrudes on steep flanks or when flanks become oversteepened, unstable and partially collapse during growth of pillow volcanoes. Their origin is similar to autoclastic subaerial lava flows where slowly moving viscous lava is broken during flowage. Partially brecciated sheet lava flows of basaltic composition occur in some deep water submarine sequences such as in the Troodos ophiolite (Cyprus). However, criteria for recognizing such autobrecciated lavas from drill cores and in older lava sequences have not yet been developed.

Also, directly associated with closely packed pillow assemblages are breccias interpreted to have formed by implosions of tube lava due to water pressures on the brittle crust, the pressure difference being caused by cooling contraction of a gas phase (Moore, 1975). Such breccias were described from Archean pillow lavas (Cousineau and Dimroth, 1982) but are probably restricted to shallower water.

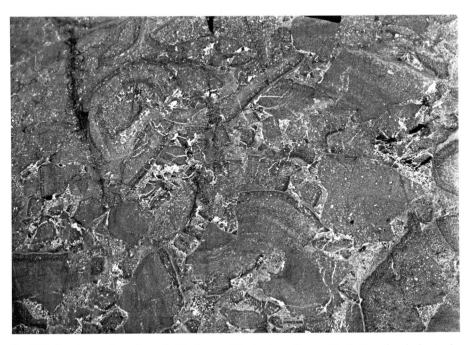

Fig. 10-2. Breccia showing pillows slightly fractured during formation and healed together during cooling. Photograph 1 m across. La Palma seamount section (Canary Islands)

A "mechanical unrimming" process (Carlisle, 1963) has long been recognized as a mechanism of fragmentation characteristic of lava erupted underwater. Rocks so generated were called hyaloclastite by Rittmann (1958). The mechanical contrast between the brittle glassy crust of tube or sheet lava flows commonly results in spalling off, especially during expansion of tubes, a process that can be repeated several times, thereby generating multiple rinds on the lava flows. The curved fragments of such glassy lava rinds are common in pillow lava piles, especially in their upper parts, where pillows are not closely packed (Fig. 10-3).

Perhaps the most widespread kind of coarse breccia associated with pillow volcanoes has little or no matrix but may have more than one type of clast (heterolithologic), some of which may show slight rounding. Pie-segment-shaped pillow fragments, the curved side being fine-grained to glassy, are characteristic of these breccias (Fig. 10-4), but sorting and bedding structures vary widely. Many form tabular units 5–20 m thick. Partly broken pillows with well-fitted pieces (Fig. 10-5) are transitional between pillow lavas and pillow-fragment breccias.

There appear to be two major groups of pillow-fragment breccias: those that form on the steep slopes of pillow volcanoes or larger seamounts, and talus breccias

**Fig. 10-3.** Spalled-off glassy rinds forming matrix between pillow and sheet flow lobes. White material carbonate cement. Photograph 3 m across. Carboniferous basalts Dillenburg (Germany)

**Fig. 10-4.** Pillow fragment breccia showing heterolithologic assemblage of nonvesicular to vesicular blocks of basalt, some representing pillow segments. La Palma seamount section (Canary Islands)

**Fig. 10-5.** Pillow fragment breccia traversed by basaltic dikes. Quaternary seamount sequence La Palma (Canary Islands)

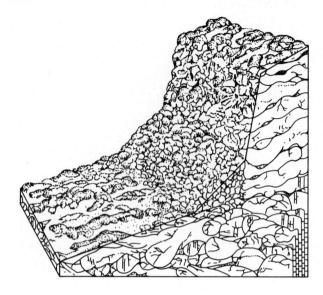

Fig. 10-6. Diagrammatic composite representation of a constructional seamount escarpment. Steep escarpment is composed of broken ends of pillows and tubes with talus at base composed of broken pillow fragments (After Fornari et al., 1979)

accumulated at the foot of submarine fault scarps which are common in faulted mid-ocean ridge environments and also on seamounts (Fig. 10-6). In drill cores, they are especially difficult to distinguish from breccias formed along the slopes of pillow volcanoes. Some pillow-fragment breccias are emplaced as debris flows as indicated by reverse grading, presence of a lapilli and tuff matrix and slight rounding of fragments. Some units grade laterally or vertically into poorly sorted tuff breccias consisting of ash- to lapilli-sized clasts with interspersed larger pillow fragments. Block fields on the slopes of seamounts represent this type of breccia.

Another kind of breccia ("isolated-pillow breccia" of Carlisle, 1963) is composed of irregular pillows, lava stringers and isolated "mini pillows" or droplets ranging from a few millimeters to a few decimeters in diameter. These breccias may contain a tuffaceous matrix and broken pillow fragments (Fig. 10-7). The spherical, ovoidal to teardrop shape of the droplets and their common occurrence at the top of pillow sequences suggest processes such as leakage from larger pillows by budding (Moore, 1975) or submarine lava fountaining (Carlisle, 1963; Schmincke et al., 1983).

**Fine-grained Hyaloclastites**

The term hyaloclastite has been used in the last few years mostly for fine-grained deposits of sideromelane shards. Characteristic are curved or blocky, vesicle-free sideromelane shards (Figs. 10-8, 10-9). These may occur together with detached glassy or formerly glassy pillow selvages discussed above to deposits several meters thick composed dominantly or entirely of shards (Dick et al., 1978; Schmincke et al., 1978). The shards appear to have originated mainly by spalling of glassy crusts of pillows of sheet lavas during cooling contraction of the flow interior or expansion of growing pillow tubes and were therefore called spallation or spalling shards (Schmincke et al., 1978; Schmincke, 1983). Some authors suggest that such shards form by cracking and granulation of lava or lava globules extruded beneath water (Fuller, 1931; Carlisle, 1963; Honnorez, 1972; Dick et al., 1978).

**Fig. 10-7.** Breccia composed of irregular large and globular minipillows. Lower part of extrusive section at Klirou. Troodos Ophiolite (Cyprus)

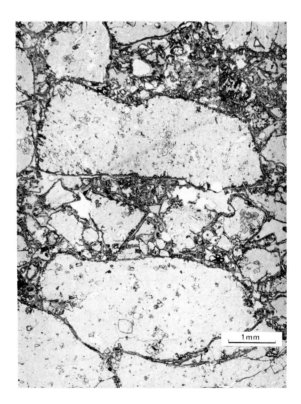

**Fig. 10-8.** Photomicrograph of glassy basaltic pillow rim with olivine and plagioclase microphenocrysts showing transition from pillow rind (lower part of photomicrograph) to broken off and detached angular pieces of sideromelane showing dark marginal palagonite rims. Palagonia (Sicily)

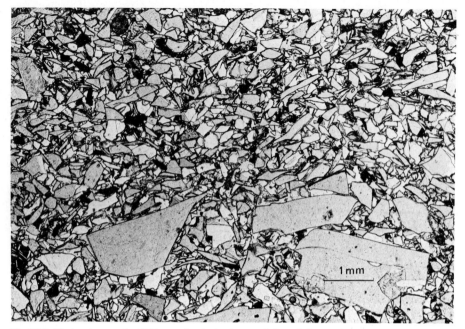

**Fig. 10-9.** Photomicrograph of Miocene hyaloclastite consisting of slivers of sideromelane and plagioclase, presumably formed by spalling of glassy crusts of pillows or lava flows extruded at the Mid-Atlantic Ridge at 22°N (Schmincke et al. 1978)

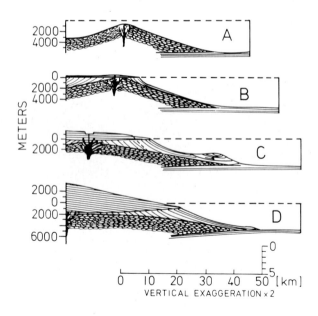

**Fig. 10-10.** Inferred basement structure of Kilauea and Mauna Loa Volcanoes, Hawaii. *Oval pattern:* pillow lavas, fragments and sediment. *Sloping lines:* clastic rocks derived from hydroclastic eruptions, littoral cones, pillowed flows and pillow breccias. Slides, slumps, mass flows and turbidity currents carry sediments to deep levels. *Horizontal lines:* subaerial lava lows. A–D shows successive stages of volcano island growth. (After Moore and Fiske, 1969)

Sideromelane shards not uncommonly occur as distinct ash layers or are dispersed within pelagic sediments in the Pacific Ocean (Fox and Heezen, 1965; Scheidegger and Kulm, 1975; Schmincke, 1983). Absence of vesicles, and their cuniform shapes, suggest an origin by spalling, the chemical composition indicating ocean floor tholeiites as source rocks. However, questions concerning the process of transport and transport distances for these ashes are unsolved. Local seamount eruptions penecontemporaneous with sediment deposition are a plausible origin. Vallier et al. (1977) describe basaltic ash consisting of spherical to teardrop-shaped glassy vesicular shards mixed with basaltic rock fragments within pelagic sediments. They suggest high Strombolian subaerial eruptions from a seamount within 100 km of the site which must have been of Fe-, Ti-rich MORB-composition. These magmas must have been unusually volatile-rich, however, to produce high eruption columns. The origin and transport of basaltic ashes in pelagic sediments clearly remains a problem.

Observations by Lonsdale and Spiess (1979) suggest that the outer slopes of some young seamounts are lava-flow aprons rather than the sediment aprons that fringe emerged volcanic islands (Lonsdale, 1975) (Figs. 10-6, 10-10). The basal slopes of East and West Seamounts, a pair of cratered volcanoes on the East Pacific Rise 1500 m below sea level, are 1°–2° and these slopes steepen upward into concave slopes averaging 25° and even as high as 30° before they flatten onto the crests which have caldera-like craters (Lonsdale and Spiess, 1979). The sides of East Seamount descend from the summit to the sea floor in a single concave slope consisting of bedrock lava flows strewn with 10–20 cm diameter angular slabs of rock, a few large boulders and a light dusting of sediment. One side of West Seamount, however, is a succession of benches similar to Gilliss Seamount (Stanley and Tay-

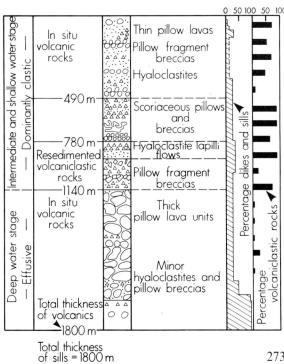

Fig. 10-11. Stratigraphic section through the La Palma uplifted sea mount formation. See text for discussion. (After Staudigel and Schmincke, 1984)

lor, 1977) and Tortuga and Tutu Seamounts (Fornari et al., 1979). The benches are nearly horizontal and are covered by coarse sediment and volcanic gravel-size fragments. The outer limits of the benches are interpreted as steep (60°–70°) flow fronts of tubular and pillow basalts at the base of which are short talus slopes of pillow basalt joint blocks. Such a succession of benches, flow fronts and talus may be typical of volcano slopes built mainly by successive flows of pillow basalts (Fornari et al., 1979; Lonsdale and Spiess, 1979) (Fig. 10-6).

Staudigel and Schmincke (1984) divided a continuously exposed 3600-m-thick cross section through the uplifted submarine part of La Palma into a lower 2500-m-thick section composed of pillow lavas and sills with rare pillow breccias (Fig. 10-2). Pillow fragments breccias form tabular bodies 5–10 m thick in the lower 250 m of the 1150 m thick upper section (Fig. 10-4, 10-5). The rocks become increasingly heterolithologic upwards and are interpreted to record the increasing steepness of the submarine volcano leading to the formation of a clastic sheet enveloping the core of the volcano consisting of lavas and intrusions (Fig. 10-11).

## Shoaling Submarine Volcano

As a volcano grows to within a few hundred meters of sea level, explosively derived hydroclastic materials can be formed due to decreased pressure of the water column. This enables both intense exsolution of magmatic volatiles and increased conversion of sea water to steam. From geologic evidence, it has been suggested (Moore and Fiske, 1969; Jones, 1970) that explosions can occur at about 100 m below water level for basaltic volcanoes. Largely on theoretical grounds, Allen (1980) concludes that for subglacial basaltic eruptions in Iceland the transition from effusive to explosive activity involves magma vesiculation combined with rapid chilling by melt water between water depths of 100 to 200 m. Moore and Schilling (1973) suggested that explosive volcanic activity begins about 200 m below sea level in tholeiitic islands. Silicic eruptions in water shallower than 500 m may give rise to floating pumice, ejecta, and steam rising to considerable heights above sea level. Most seamounts are made of more alkalic basalts – at least during later stages – than the ocean crust and thus were volatile-rich.

Breccia units of highly scoriaceous lapilli in the alkalic La Palma seamount section apparently formed *in situ* about 780 m below the top of the section (Staudigel and Schmincke, 1984) (Fig. 10-11), thus the critical depth for explosive volcanism for these more volatile magmas was about 700–800 m. Volcanic sequences produced under "shallow" subaqueous conditions tend to be complex, commonly consisting of localized lava complexes interfingering with *in situ* breccias and varying amounts of resedimented highly vesicular, very shallow water volcanic debris.

In the present oceans, volcaniclastic sediments are interbedded with (1) uppermost lava flows comprising the sea floor basement, (2) lavas just above basement rocks, and (3) intercalated in the nonvolcanic sedimentary sequences along most aseismic ridges. Aseismic ridges are sites of high magmatic production rate formed especially at ridge-transform sections. Activity can persist for many tens of millions of years, leading to aseismic ridges several hundreds to thousands of kilometers long. The Ninetyeast Ridge in the Indian Ocean (Pimm, 1974; Vallier and Kidd,

1977; Fleet and McKelvey, 1978) and the Walvis and Rio Grande Rises in the South Atlantic (Simon and Schmincke, 1983) have been drilled in most detail. The rocks in these areas are almost exclusively basaltic, and the string of seamount and broad volcanic swells grew to within the sea surface and locally formed emergent islands. At least 388 m of layered, Middle Eocene basaltic pyroclastic sediment occur at DSDP Site 253 on the Ninetyeast Ridge above basement (Fleet and McKelvey, 1978). On the Walvis Ridge, volcaniclastic deposits are dominantly vesicular basaltic lapilli and shards totalling up to about 50 m thick. They are intercalated with and directly overlie basement lavas within some 50 km of the presumed source (Simon and Schmincke, 1983). At the most distant sites drilled, about 100–150 km from the source, volcaniclastic sediments are <20 m thick, are interbedded with pelagic sediments and are dominantly silicic, presumably generated during a later emergent island growth stage on top of an aseismic ridge.

## Transition Submarine – Subaerial

A common question asked by geologists studying submarine volcaniclastic rocks evidently deposited in shallow water concerns the subaqueous *vs* subaerial nature of the volcanic eruptions, a problem confronting all students of older, partly volcaniclastic marine sequences whose source areas are not preserved or not exposed in the geologic record. The following criteria suggest subaerial volcanic activity: (1) the presence of vesicular tachylitic pyroclasts indicates slower cooling in subaerial compared to subaqueous eruption columns; (2) iddingsitized olivine is a useful although not unequivocal criterion as is (3) the presence of rounded epiclastic fragments of many types (heterolithologic) indicating subaerial erosion of a multicomponent volcanic terrain (Schmincke and von Rad, 1979; Simon and Schmincke, 1983).

Transition from submarine to emergent stages of volcanism has been documented by several studies. At drill sites close to presumed sources, subaerial stages are indicated by the occurrence of conglomerates and lignite beds (Pimm, 1974; Fleet and McKelvey, 1978). Farther from sources, the transition is reflected in contrasting types of submarine volcaniclastic debris flows (Schmincke and von Rad, 1979) (Fig. 10-12). Hyaloclastite flows containing angular, blocky, and only slightly vesiculated sideromelane shards – now replaced by smectite and phillipsite – are a submarine growth stage perhaps at water depth below the VFD. Stratigraphically higher, volcaniclastic debris flows are composed mainly of tachylitic, vesicular basalt, pyroxene, plagioclase and altered olivine crystals, and a variety of epiclastic rock fragments ranging from dolerite to trachyte (Fig. 10-12). Large volumes of tachylitic to crystalline basaltic debris are probably generated at the emergent shield stage of an island volcano where lava flows enter the sea. Interpretation of emergence and near-shore volcanic activity and clastic processes is consistent with the fact that similar rocks are lacking higher in the stratigraphic sequence. The nature of clastic particles produced in littoral cones and during entry of lava flow in shallow water has been described from historic and prehistoric eruptions in Hawaii (Fisher, 1968a; Moore et al., 1973; Peterson, 1976).

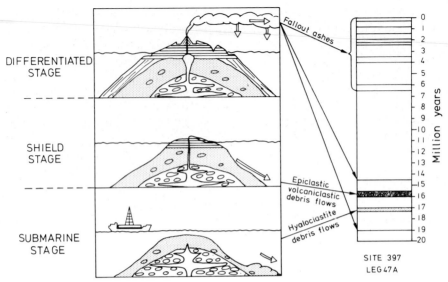

**Fig. 10-12.** Evolution of an ocean island (Canary Islands) as reflected by different types of volcaniclastic rocks in DSDP site 397 drill cores. (After Schmincke, 1982a)

Tuffs composed of highly vesicular, glassy – or once glassy – shards and small lapilli are the dominant volcaniclastic rocks (1) in the upper few hundred meters of exposed seamounts sections (Staudigel and Schmincke, 1984), (2) in very thick volcaniclastic sequences formed in shallow water along aseismic ridges (Fleet and McKelvey, 1978) and (3) in dredge hauls from the upper slope of volcanic islands (Moore and Fiske, 1969). Significant features of fragments in the deposits are a combination of small grain size and high vesicularity yet vitric nature, indicating near-synchronous pyroclastic disruption and quenching by water or water vapor (Fig. 10-13). Such glassy particles form in shallow water, but also during emergent stages if abundant water can interact with the explosively fragmented lava column. Moreover, as shown by recent deposits of Surtsey (Fig. 10-14), quickly accumulated subaerial or shallow-water tephra can be remobilized and transported into deep water, resulting in further problems of interpretation. The shape and vesicularity of shallowly formed particles can vary widely, and can include blocky and slightly to moderately vesicular shards formed by thermal stress and steam eruptions (Walker and Croasdale, 1972). In our experience, highly vesicular shards predominate.

## Volcaniclastic Aprons

Submarine and subaerial clastic volcanic activity and the redistribution processes mentioned above produce large volumes of clastic material that form volcaniclastic aprons around seamounts and volcanic islands. Transporting agents are sediment gravity flows which include turbidity currents (Menard, 1956). Thus, aprons may

Fig. 10-13. Bedded trachytic hyaloclastites formed in shallow water during shoaling stage of seamount. Porto Santo (Madeira)

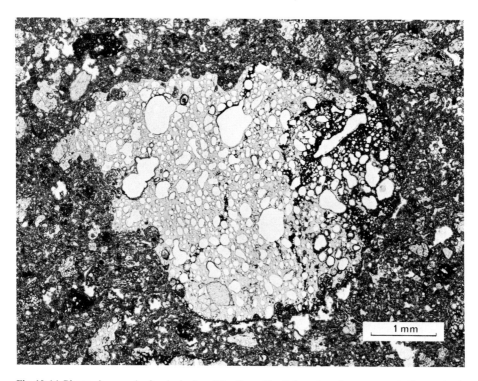

Fig. 10-14. Photomicrograph of vesicular basalt lapillus with olivine microphenocryst set in fine-grained matrix of vesicular basalt shards. 1971 Surtsey eruption (Iceland)

extend for more than 100 km from an island. Submarine pyroclastic flows of such aprons provide excellent seismic reflectors. South of the Canary Islands, for example, such a horizon traced by seismic profiling over an area of 3000 km² was correlated with a volcaniclastic grain flow deposit drilled at site 397, its volume amounting to about 80 km³ (Schmincke and von Rad, 1979). The sedimentary apron around Samoa, built during the last 5 m.y. is estimated to have a volume of $5 \times 10^{13}$ m³ of volcaniclastic sediment mixed with some pelagic materials and is up to 1 km thick (Lonsdale, 1975).

Subglacial volcanoes (Mathews, 1947; Jones, 1966, 1969a, 1970; Allen, 1980) have morphologies and structural successions similar to those of seamounts as envisioned by Moore and Fiske (1969) (Fig. 10-15). Pillow lavas become increasingly vesicular and interbedded with hyaloclastic layers upward (Fig. 10-16). These in turn are overlain by lava flows that formed clastic debris where they entered water (flow foot breccias) and contributed talus to their flanks. The "pillow-palagonite tuff" complexes formed where lavas have entered rivers and lakes (Fuller, 1931) in many ways resemble pillow-hyaloclastite assemblages formed under shallow marine conditions (see Sigvaldasson, 1968).

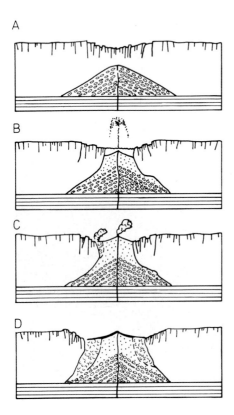

**Fig. 10-15 A–D.** Growth stages of a subglacial volcano, Iceland. **A** Extrusion beneath ice in meltwater builds a pillow pile. **B** At depth of <200 m, hydroclastic activity commences resulting in vitric ash above the pillows. **C** After build-up above melt-water level, extrusion of lava is resumed. **D** Flows move into ponded water resulting in pillow and pillow breccia complexes. Melting of ice leaves steep-sided breccia-mantled and essentially flat-topped volcano. (After Jones, 1970)

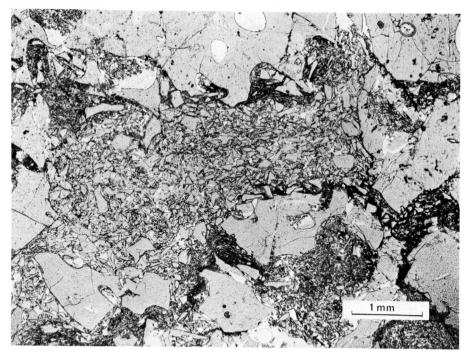

Fig. 10-16. Photomicrograph of slightly vesiculated basaltic sideromelane lapilli set in a matrix of curved sideromelane spalling shards. Subglacial hyaloclastites (Reykjanes, Iceland)

## Silicic Submarine Eruptions

The substructures of silicic volcanoes are probably similar to basaltic central volcanoes on oceanic crust except that exsolution of volatiles and explosive disruption may be initiated at deeper levels (e.g. Bevins and Roach, 1979; Dimroth et al., 1979; de Rosen-Spence et al., 1980; Furnes et al., 1980). The main overall difference is that the subaerial parts of intermediate to silicic composition volcanoes contain a greater abundance of pyroclastic materials than do basaltic volcanoes. Compared with basaltic hyaloclastite sequences (pillow lavas, pillow breccias, sideromelane fragments, sometimes in vertical sequence), silicic hyaloclastites are more diverse, consisting of lithic blocks, lapilli and ash, structureless to flow-banded fragments of pumice and obsidian in intimate association with larger irregular to subspherical bodies of vesicular to nonvesicular rhyolite (possibly analogous to pillows) enclosed within obsidian layers (Furnes et al., 1980) (Fig. 10-17).

Dacite-andesite and rhyodacite lava flows 100–400 m thick and up to 2000 km² in areal extent may occur within a deep marine environment as inferred by their conformable position within a succession of flysch-like Paleozoic sedimentary rocks (Cas, 1978). Their tops and bases are coherent and sharp or else fragmental and irregular. Geological evidence rules out an origin as sill injections, or as pyroclastic flows. Rather, the lava flows apparently formed from eruptions at water depths (pressure) sufficient to prevent a rapid release of most volatiles. The flows remained mobile because they retained their volatiles, and the rate of increase in viscosity due to cooling was far less than the rate of emplacement.

Furnes et al. (1980) describe two kinds of hyaloclastite formed in water beneath ice; one type, formed by vesiculation and explosive disruption, consists largely of pumice, the other type is characterized by obsidian, flow-banded pumice and lithic rhyolite fragments formed by shattered rhyolite flows or outer margins of intrusions into water-logged pumice-bearing hyaloclastite (Fig. 10-17). Pillows are absent, one possible reason being that the pressure of the water column – estimated to have been 200–300 m (Grönvold, 1972; Saemundsson, 1972) – was not great enough to inhibit explosive expansion of the volatile-rich rhyolitic magmas. Pillowed silicic lavas evidently form at deeper levels or from less volatile-rich flows.

De Rosen-Spence et al. (1980) describe Archean subaqueous felsic flows and their lateral facies within the Rouyn-Noranda area (Quebec) alternating with pillow basalts and massive basalt flows with pillow breccias at their tops – evidence of subaqueous emplacement. The proximal facies of the felsic rocks generally consists of massive flows; some of the flows have a thin veneer of flow breccia at the

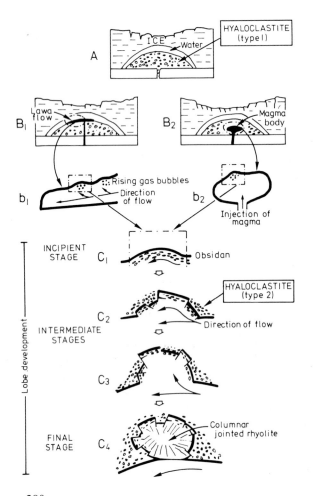

Fig. 10-17. Diagrammatic representation showing development stages of two types of subglacial acid hyaloclastites. See text. (After Furnes et al., 1980)

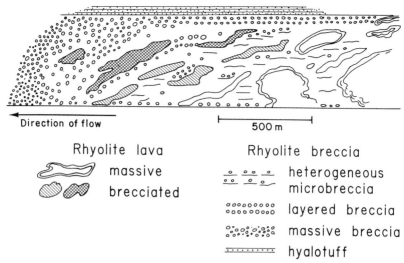

Fig. 10-18. Diagram showing lateral changes and facies of a felsic lava flow interpreted as subaqueous. (After Provost, 1978, in de Rosen-Spence et al., 1980)

top (Fig. 10-18). These grade laterally to the median facies composed of irregular ellipsoidal masses of rhyolite lava (pods), folded sheets (lobes) and finger-shaped bodies (tongues), all of which are enveloped by microbreccia. Additionally, there are associated unbedded to more rarely bedded flow breccias which commonly occur at the tops and locally at the bases of the rhyolite flows. The distal facies resembles the median facies except that flow breccia surrounds pods, lobes and tongues, and resedimented planar- to cross-bedded tuff overlies parts or all of some flows. Some of these tuff units are turbidite sequences.

Despite the tectonic and metamorphic overprint, features in the Archean rocks similar to modern and Quaternary rhyolitic and basaltic pillow associations can be recognized. The rhyolite flows differ from mafic flows by being (1) thicker and less extensive, (2) lava pods, lobes and tongues are larger than typical basalt pillows, (3) the flow-laminated skin on pods, lobes and tongues is much thicker than on basalt and (4) vesicles are larger (Dimroth et al., 1979).

## Subaqueous Pyroclastic Flows

Most examples of sediment gravity flow deposits discussed in the first part of this chapter are from seamounts and oceanic islands, and are of basaltic composition. While such material is more common in the sedimentary column than generally realized (Vallier and Kidd, 1977), it is much less voluminous than the vast amount of volcanogenic debris shed into fore-arc and back-arc basins next to island arcs and continental margin volcanic chains, with volcanoes being mostly calcalkalic in composition and highly explosive. Fallout ash from such volcanoes has been treated in previous chapters and some general stratigraphic and tectonic implications are discussed in Chapters 13 and 14. Here we focus upon subaqueous sediment

**Table 10-1.** Subaqueous pyroclastic flow occurrences

| Environment | Age | Location | Welding temperature |
|---|---|---|---|
| Marine | 1883 | Krakatau, Java | Unwelded |
| | ~28,000 yr. B.P. | Grenada Basin, Lesser Antilles | Unwelded. Resemble debris flow deposits |
| | Late Quaternary | Subaqueous west flank, Dominica, Lesser Antilles | Deposits not examined |
| | Middle Miocene | Santa Cruz Island, Calif., USA | Unwelded. Pumice-rich. Massive beds, some with pinkish oxidized tops |
| | Miocene | Tokiwa Formation, Japan | Unwelded. Pumice-rich |
| | Oligocene | Island of Rhodos, Greece | Unwelded |
| | Oligocene-Eocene | Mt. Rainier Nat'l. Park, Wash., USA | Unwelded. Massive lower to bedded upper division |
| | Early Oligocene-Late Eocene | Philippines | Unwelded. Pyroclastic turbidites |
| | Paleogene to Cretaceous | Philippines | Unwelded to welded (?) tuff |
| | Lower Mesozoic | Sierra Nevada, California | Metamorphosed |
| | Permian | Alaska, USA | Unwelded. Massive lower to bedded upper divisions |
| | Mississippian | Oklahoma and Arkansas, USA | Unwelded. Turbidite structures |
| | Devonian | Rhenish Massif, Germany | Welded |
| | Ordovician | Ireland | Welded |
| | Ordovician | Wales, U.K. | Welded |
| | Lower Ordovocian | South Wales, U.K. | Welded |
| | Archean | Rouyn-Noranda, Canada | Unwelded |
| Lacustrine | Plio-Pleistocene | Japan | Unwelded. Resemble turbidites |
| | Mio-Pliocene | Japan | Unwelded. Thermoremanent magnetism suggests emplacement at T 500 °C |

gravity flow deposits, which have become increasingly recognized in the geologic record following the pioneering studies of Fiske (1963) and Fiske and Matsuda (1964).

Pyroclastic flows are known to have entered the sea following subaerial eruptions in several cases (e.g. 1902 Mt. Pelée, Lacroix, 1904; Augustine Volcano, 1976), and Self and Rampino (1981) suggest that pyroclastic flows generated dur-

| Remarks | Reference |
|---|---|
| Generated destructive tsunamis | Self and Rampino (1981) |
| Derived from subaerial hot pyroclastic flows on Dominica | Carey and Sigurdsson (1980) |
| Deposits traceable to subaerial flows (block and ash; welded ignimbrite) | Sparks et al. (1980a) |
| Within a marine sequence | Fisher and Charleton (1976) |
| Interbedded conformably with fossil-rich beds | Fiske and Matsuda (1964) |
| Deep water, marine | Wright and Mutti (1981) |
| Marine shelf deposition inferred | Fiske (1963); Fiske et al. (1963) |
| Interbedded with marine limestone | Garrison et al. (1979) |
| Within a marine sequence | Fernandez (1969) |
| Ponded in shallow marine calderas | Busby-Spera (1981a, b) |
| Structures and sequence similar to Tokiwa formation, Japan | Bond (1973) |
| Interbedded within marine flysch sequence | Niem (1977) |
| Tidal flat environment | Scherp and Grabert (1983) |
| Associated with shallow water deltaic sediments | Stanton (1960), Dewey (1963) |
| In marine sequence. Traceable to subaerial pyroclastic flows | Francis and Howells (1973), Howells et al. (1979), Howells and Leveridge (1980) |
| Within marine sequence | Lowman and Bloxam (1981) |
| Structures and sequence similar to Tokiwa Formation, Japan | Dimroth and Demarcke (1978), Tassé et al (1978) |
| Interbedded with lacustrine rocks. Flow from land into water | Yamada (1973) |
| Massive units resembling debris flows | Kato et al. (1971), Yamazaki et al. (1973) |

ing the 1883 Krakatau (Java) eruption entered the sea and caused destructive tsunamis. Subaqueous pyroclastic flow deposits have been reported from marine or lacustrine rock sequences from the Archean of Canada (Dimroth and Demarcke, 1978) to the Recent in the Lesser Antilles (Carey and Sigurdsson, 1980) (Table 10-1). Attention was focused on the subject by a Penrose Conference in 1977 (Fisher and Dimroth, 1978).

It has been postulated that pyroclastic flows can form entirely from underwater eruptions (Fiske, 1963; Fiske and Matsuda, 1964). Most workers hold that it is not possible for hot pyroclastic flows to originate by eruptions occurring beneath water, or to flow from land into water, without mixing with the water in sufficient quantities to decrease the temperature below the minimum needed for welding. Kato et al. (1971) and Yamazaki et al. (1973), however, present evidence from natural remanent magnetism studies indicating that the subaqueous deposits studied by them came to rest at about 500 °C, and Sparks et al. (1980b) present theoretical arguments showing that welding is likely to occur in subaqueous environments once a subaerially produced hot flow has penetrated the air-water interface. A vertical eruption column entirely beneath water probably cannot develop into a hot pyroclastic flow because turbulent mixing with water would tend to destroy the column. There is the possibility, however, that a voluminous "boiling-over" type of eruption may be extruded at rates great enough to produce a flow protected from the water by a carapace of steam and retain enough heat to become welded.

Detailed studies of subaqueous flow deposits are not numerous even though such rocks are widespread. We suspect that closer analyses of the several 1000 m of volcaniclastic sediments drilled by the Glomar Challenger during the last 5 years in the South Pacific will provide some answers to the many questions posed by these rocks.

These questions are centered around four main problem areas. The first point to be clarified is whether a subaqueous pyroclastic flow was indeed deposited under water or on land. Interpretations of a subaqueous depositional environment of ancient deposits are based largely upon interbedding with marine or lacustrine rocks. This is suggestive of subaqueous deposition but is not conclusive because marine transgression or a rise in lake level may follow emplacement. In many instances, however, special pleading is required to make a case for transgression. Even more difficult to decide is the question of whether a mass flow originated from a subaqueous or subaerial eruption. Moreover, how reliable are the criteria that distinguish between flows that are the underwater continuation of a subaerial pyroclastic flow or the direct result of an eruption and flows that are generated by slumping and resedimentation of unstable volcanic debris that had accumulated on the subaerial or submarine slope of a volcano?

Many subaqueous pyroclastic flow deposits resemble turbidites and thus may be merely remobilized pyroclastic debris originally deposited by fallout or other processes along the shoreline or shallow-water flanks of an active volcano. At present, there is little to distinguish flows that originate in this manner from flows initiated by underwater eruptions. In either case, however, it is likely that such deposits come to rest at low temperatures.

Finally there is the tantalizing question of whether a primary flow was hot during underwater emplacement or whether textural criteria suggesting welding are really due to diagenetic compaction.

## Terminology

Because structures and textures to distinguish hot from cold pyroclastic flows are poorly understood, we use the term subaqueous pyroclastic flow in a broad sense (see Fiske, 1963). Kato et al. (1971) and Yamazaki et al. (1973), restrict the term to subaqueous pyroclastic flows emplaced at high temperatures.

## Nonwelded Deposits

### Environment of Deposition

Deposition beneath water in ancient deposits can only be determined indirectly on stratigraphic grounds and this is an interpretation requiring careful documentation. One of the most convincing examples of submarine deposition of pyroclastic flows are those in the Tokiwa Formation, Japan (Fiske and Matsuda, 1964). The deposits occur within a conformable 1500-m sequence of fossiliferous marine mudstones and small amounts of nonvolcanic graywacke, suggesting nearly continuous marine sedimentation during early and middle Miocene time. The Wadera Tuff Member of the Tokiwa Formation is a conformable, richly fossiliferous mudstone sequence containing five main separate subaqueous pyroclastic flow sequences. Foraminifera within the mudstone indicate deposition in the open sea at depths of 150 to 500 m, corresponding to the lower part of the continental shelf or upper part of the continental slope.

Subaqueous pyroclastic flow deposits described by Niem (1977) occur within a 4000-m-thick sequence of deep marine shales and subordinate nontuffaceous fine-grained quartz wackes and feldspathic sandstones. Conodonts from turbidite sandstones and shales, and trace fossils, suggest bathyal depths.

Subaqueous pumice flow deposits resulting from subaerial eruptions occur within a lacustrine environment from the Onikobe Caldera, Japan (Yamada, 1973). One of the pumice flow sheets which can be correlated throughout the caldera basin is intercalated with thin-bedded lacustrine siltstone in the northern and southwestern parts of the caldera, but toward the inferred eruption center in the subaerial southeastern part, where volcanic rocks exceed 900 m in thickness, the pumice flow sheet thickens, becomes coarser-grained, and overlaps the volcanic pile.

### Components

Subaqueous pyroclastic flow deposits in the broad sense are composed of lithic, crystal, and pumice clasts in variable proportions and sizes (Figs. 10-19, 10-20). Shale rip-ups and blocks measuring up to several meters across derived from underlying beds, foreign rock fragments and fossils may be incorporated near the base of a sequence (Mutti, 1965; Yamada, 1973; Niem, 1977; Schmincke and von Rad, 1979; Carey and Sigurdsson, 1980; Garrison et al., 1979; Wright and Mutti, 1981). Broken crystals are common in subaqueous pyroclastic flows (Fiske and Matsuda, 1964; Fernandez, 1969; Yamada, 1973; Niem, 1977).

**Fig. 10-19.** Pumice lapilli set in dark shale matrix, part of 1-m-thick subaqueous pyroclastic flow deposit believed to have resulted from underwater eruption in a poorly oxygenated shallow basin. Carboniferous (Bestwig, Germany)

**Fig. 10-20.** Thinly bedded graded submarine tuffs of basaltic composition believed to have resulted from underwater eruptions. Devonian (Madfeld, Germany)

Based upon lithologic grounds, Fiske (1963) recognized three main types of subaqueous pyroclastic flow deposits: (1) the most common kind is crowded with a great variety of lithic fragments, with the larger fragments chaotically mixed with pumice lapilli, whole and broken crystals and 20 to 70 percent glassy ash matrix; (2) a less common variety contains abundant pumice and delicate glass shards with up to 70 percent fine ash matrix; (3) the least common kind contains abundant lava fragments, little pumice, minor amounts of crystals, and commonly less than 50 percent tuffaceous matrix. He interpreted the first variety as the product of powerful phreatic eruptions, reshattering parts of an already rubbly volcano. The second variety was interpreted as resulting from large underwater eruptions of vesiculating magma, and the third variety as derived from explosive disintegration of fairly homogeneous bodies such as domes, spines, and lava flows as they enter water.

### Grain Size, Sorting, and Fabric

Poor sorting is characteristic of subaqueous pyroclastic flow deposits, particularly in the massive lower division. In the sequence studied by Fiske and Matsuda (1964), however, massive beds contain only 10 to 25 percent pyroclastic material finer than 2 mm, mostly sand-size ash, in contrast to most subaerial pyroclastic flow deposits which commonly contain more than 50 percent ash-size clasts (Smith, 1960a; Murai, 1961).

Large pumice fragments in the massive parts of submarine pyroclastic flow deposits tend to be randomly oriented except near their bases, where elongation parallel to contacts is common. In some deposits, pumice with tube-like vesicles is randomly oriented, indicating that attenuation occurred during eruption rather than by compaction. Niem (1977) reports elongate pumice in a massive division and attributes it to later compaction or to shearing during tectonism, but also notes that elongate crystals are locally imbricated and aligned parallel to bedding, indicating that orientation took place during flowage.

### Bedding and Grading

Unwelded subaqueous pyroclastic flow deposits characteristically have a massive to poorly bedded and poorly sorted lower division and an upper thinly bedded division (Fig. 10-21). Unlike most turbidite sequences, the lower massive division forms 50 percent or more of the total sequence (Fiske and Matsuda, 1964; Bond, 1973; Niem, 1977), and massive sequences of up to 300 m thick are reported by Bond (1973). Also, unlike turbidite sequences believed to form by the passage of a single flow, the two-division sequence is interpreted in terms of a waning, initially voluminous eruption (Fiske and Matsuda, 1973). On the other hand, Yamada (1973) recognizes five divisions (Fig. 10-22) similar to a Bouma sequence (Bouma, 1962). We have recognized a two-fold sequence capped by a thin, very fine-grained bioturbated pelagic layer in California (USA) (Fig. 10-23) where basal contacts may be sharp on undisturbed strata, but locally there may be flute marks, grooves and load casts. Even where bottom contacts are plane and sharp, however, missing underlying strata may indicate erosion (Yamada, 1973).

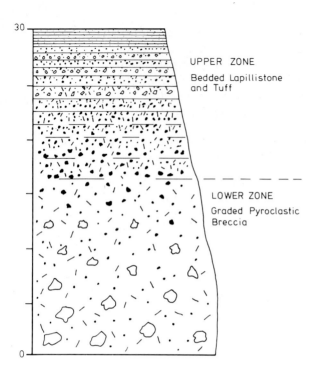

Fig. 10-21. Generalized sequence of massive to thin-bedded units of subaqueous pyroclastic flow. (After Bond, 1973)

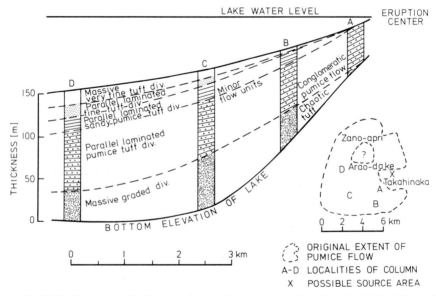

Fig. 10-22. Facies and distribution of unwelded pyroclastic flow deposit within Onikobe Caldera (Japan). (After Yamada, 1973)

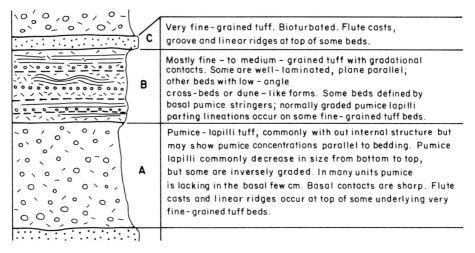

Fig. 10-23. Idealized depositional unit of unwelded subaqueous pyroclastic flow near Santa Maria (California). Miocene Obispo Tuff Formation

The Massive Lower Division

The lower division of a subaqueous pyroclastic flow unit is a single coarse-grained bed lacking internal structures or laminations (Fiske and Matsuda, 1964; Bond, 1973), although it may have incipient cross laminations outlined by a crude orientation of coarse fragments (Kato et al., 1971; Yamazaki et al., 1973). Coarser and more dense fragments generally occur near the base of the massive divisions, with crystals, glass shards and pumice becoming more abundant toward the top (Schmincke and von Rad, 1979).

Within the lower massive division described by Yamada (1973), upward density (inverse) grading of pumice occurs, but dense andesite fragments at its base are also inversely graded according to size. Maximum diameters of the andesite fragments (up to 3 cm) occur a few meters above the base, but become smaller upward, and therefore show first an inverse grading that reverts upward to normal grading; pumice fragments in the layer continue to increase in size upward, and attain maximum diameters of up to 25 cm.

The massive coarser-grained lower unit shows inverse grading similar to that at the bases of many debris flow and lahar deposits (Chap. 11), as well as in subaqueous mass flows and coarse-grained turbidites (Schmincke, 1967b; Fisher, 1971; Walker, 1975). Similar grading relationships of lithics and pumice are reported by Yamazaki et al. (1973).

A widespread, coarse-grained massive pyroclastic flow unit in the Grenada Basin, Lesser Antilles (Fig. 10-24) exceeds 4.5 m in thickness, covers a minimum area of $1.4 \times 10^4$ km$^2$ (volume 30 km$^3$) and extends up to 250 km from its source on the island of Dominica (Carey and Sigurdsson, 1980). The deposit, composed principally of rhyolitic glass shards and pumice (fine ash to pumice clasts to 6.5

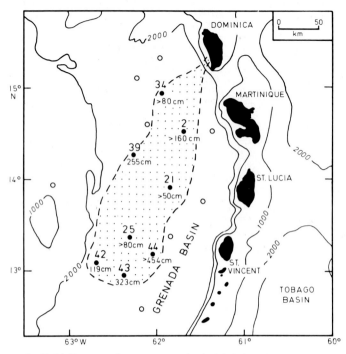

**Fig. 10-24.** Roseau subaqueous pyroclastic debris flow in Grenada Basin. *Solid circles* indicate cores containing flow deposits; *open circles* are cores without deposits. *Upper number* is core number; *lower number* is thickness of deposit. Contours in meters. (After Carey and Sigurdsson, 1980)

cm), forms a single massive unit lacking an upper laminated division in most cores (Fig. 10-25).

Upper Division

The upper parts of many subaqueous pyroclastic flow deposits are characteristically composed of many thin, fine- to coarse-grained ash beds. In a 15 m-thick sequence described by Fiske and Matsuda (1964), graded beds at its base contain ash-size particles ranging up to 2 mm in diameter, whereas beds at the top are composed of particles 0.5 mm or less. The sequence contains about 200 very evenly bedded tuff units from 3 mm to 9 cm thick. The term double-grading is used to note that (1) each individual bed is graded, and (2) the entire sequence of beds is graded, becoming finer-grained upward (Fig. 10-26). The lower part of each bed is rich in crystal fragments and pieces of glassy dacite, and grades upward to a top rich in pumice shreds and glass shards. Double grading is also described by Yamada (1973).

The upper division described by Niem (1977), however, is not double-graded. The beds, which lie above massive unstratified lower divisions, are composed of alternating pumice-rich layers (7–10 cm thick) and pumice-poor, vitric-crystal layers (5–7 cm thick). Generally, the largest pumice fragments are concentrated

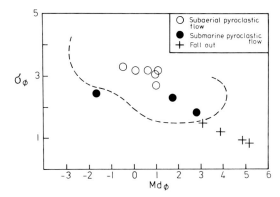

Fig. 10-25. Median diameter ($Md_\phi$) versus Inman Sorting Coefficient ($\sigma_\phi$) for Roseau subaqueous (*solid circles*) and subaerial (*circles*) pyroclastic flow deposits. *Crosses:* submarine fallout tephra. *Dashed line* is boundary between pyroclastic flows and fallout tephra (Walker, 1971). (After Carey and Sigurdsson, 1980)

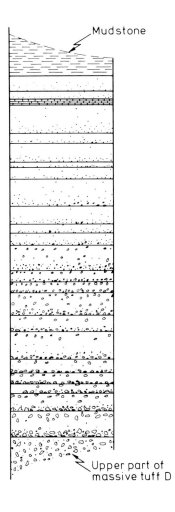

Fig. 10-26. Diagrammatic sketch of doubly graded tuff beds in upper part of one pyroclastic flow unit (Wadera Tuff D), Japan. The total sequence, a maximum of 15 m, becomes finer-grained upward and each bed is normally graded from coarse to fine. (After Fiske and Matsuda, 1964)

near the central parts of the pumice-rich layers. Subaqueous pyroclastic flow deposits lacking double-grading are interpreted by Fiske (1963) to indicate deposition in water shallower than indicated by double-grading, but the units described by Niem (1977) contain trace fossils suggesting bathyal depths. Presence or absence of double-grading, or indeed, the upper division itself, is most likely related to how the flows originate and to flow processes, rather than to water depths.

Within the Tertiary marine Aksitero Formation, Philippines, there are thin (3–15 cm thick) graded beds of tuff composed mostly of crystals, bubble wall and pumice shards and glassy lithic clasts, some with complete Bouma (Bouma, 1962) sequences, interbedded within a sequence of marine pelagic limestone of turbidite origin (Garrison et al., 1979). Even though the tuff sequences lack the thick massive lower divisions described by Fiske and Matsuda (1964) and others, Garrison et al. (1979) suggest that the pyroclastic turbidites were fed by underwater eruption columns because the beds lack constituents that indicate a shallow water or subaerial source.

## Relationship to Eruptions and Eruptive Centers

Double grading is interpreted by Fiske and Matsuda (1964) to signify contemporaneous subaqueous volcanism with deposition from waning thin turbidity flows following deposition of the massive bed. Deposits without double grading are taken to signify sloughing off oversteepened pyroclastic slopes from the edge of a volcano.

Fiske and Matsuda (1964) present the following model for underwater eruptions (Fig. 10-27): A large eruption column emerges with great force from an underwater vent. Large amounts of dacite fragments torn from the vent are violently thrown upward in suspension at first, but later are replaced by increasing amounts of juvenile pumiceous debris. Progressive sorting begins within the eruption column, and the first fragments to settle back are mostly dense lithic fragments plus feldspar and quartz crystals. The debris from the dispersing eruption column forms a dense slurry that flows along the sea floor into deeper parts of the basin where it deposits a thick, massive layer. In the closing stages of eruption, small intermittent turbidity flows entrain pumice lapilli and crystals of plagioclase and quartz, which are deposited in thinly graded beds. Ash from a submarine eruption cloud that breaks the water surface, or that has spread widely away from the vent within the water body, continues to rain downward after the eruption ends, but in decreasing amounts and becoming finer-grained.

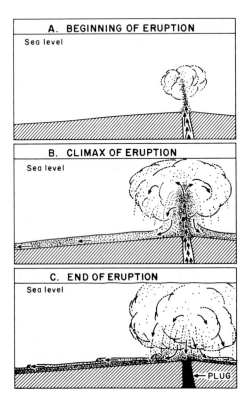

**Fig. 10-27 A–C.** Sequence of events of eruption giving rise to subaqueous pyroclastic flow units shown in Fig. 10-26. (After Fiske and Matsuda, 1964)

The pyroclastic flow deposit described by Carey and Sigurdsson (1980) from the Grenada Basin is interpreted to be a debris flow(s) initiated by the entry of hot subaerial pyroclastic flows into the sea at the mouth of the Roseau Valley, Dominica as indicated by (1) tracing the subaqueous deposit to a subaerial pyroclastic flow deposit at the mouth of the Roseau Valley, (2) the similarity in composition and grain-size characteristics of the subaerial and subaqueous deposits, and (3) the occurrence of charcoal in the subaqueous deposit. Mixing with pelagic sediment and sea water occurred downslope, resulting in an interstitial matrix of high strength and considerable mobility, able to suspend large fragments, but unable to produce a turbidite, as indicated by lack of Bouma sequence features and the generally massive and coarse-grained character throughout its extent. Plio-Miocene rocks interpreted by Kato et al. (1971) to have come to rest at about 500 °C based on thermal remanent magnetism data have similar features.

Subaqueous pyroclastic flows interbedded with lacustrine sediments of Okinobe Caldera, Japan (Yamada, 1973) are interpreted as having originated from subaerial eruptions at the margin of the caldera. There is little evidence, however, to indicate whether hot pyroclastic flows entered the water from land, or if slumping of unstable slopes on the margins of the lake initiated the flows. Normal grading of the 150-m-thick pyroclastic unit and the density grading of the five subdivisions are interpreted by Yamada to have developed from the passage of a single, large pyroclastic turbidity current.

Thus far, there is only reported occurrence of a subaqueous caldera associated with subaqueous ignimbrites. This is an early Mesozoic caldera complex in the southern Sierra Nevada, California, showing evidence of four collapse events and four ponded rhyolite ignimbrite eruptions (Busby-Spera, 1981a, b). The ponded ignimbrite units are 1000 to 1500 m thick and each has a volume of at least 40 km$^3$. The ignimbrites are massive, poorly sorted, monolithologic, and pumice-rich; systematic changes in pumice-flattening are related to cooling history rather than to tectonic deformation. The ignimbrites are interstratified with shallow marine quartzites and carbonates and differ from subaerial ignimbrites by passing gradationally upward into sequences of well-sorted ash, lapilli, and block layers. Caldera floor and wall rocks consist of marine sandstones showing soft-sediment deformation, indicating incomplete lithification during volcanism.

## Welded Deposits

Welded pyroclastic flow deposits interpreted to be subaqueous because of the marine nature of the enclosed sediments are controversial. The best-documented example is from the Ordovician of Snowdon in Wales. There, a welded tuff interpreted to be submarine is correlated with welded tuff of probable subaerial origin within the same formation (Capel Curig Formation) (Francis and Howells, 1973; Howells et al., 1973; Howells et al., 1979). The subaerial tuff consists of an uninterrupted sequence of two or more flow units of greater combined thickness than correlative subaqueous units. The base of the subaerial tuff is planar, unwelded, contains fiamme, and overlies a conglomerate with a reddened top. The subaqueous pyroclastic flow deposits are commonly less than 50 m thick, welded to their

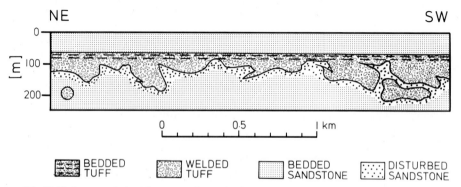

Fig. 10-28. Large-scale load features at base of subaqueous welded tuff (Wales, U.K.). (After Francis and Howells, 1973)

base, are interbedded with marine rocks and in places are underlain by sandstones containing brachiopods. Subaqueous welded tuffs are also reported by Lowman and Bloxam (1981) in South Wales.

Although the subaqueous pyroclastic flows are intensely welded, most of them lack fiamme. Basal contacts tend to be irregular with zones resembling load structures of very large dimensions and flame structures (Fig. 10-28). These flow deposits are massive in the lower and central portions and grade imperceptibly through an ill-defined zone with faint bedding upward to evenly bedded tuff which locally has broad, low-angle cross bedding (reworked?). Their tops are in sharp contact with overlying sandstone with similar broad cross-bedded structures. The sediments beneath the welded tuff are highly disturbed where the base is irregular. There is some question as to whether the welded tuffs were emplaced beneath the water, or else flowed onto soft-mud tidal flats with marine sediments above the tuffs being transgressive (Francis and Howells, 1973, p. 641). The thickness of the welded tuff units (up to 30 m), however, are greater than the estimated Ordovician tidal range.

In northern Snowdon, welded tuffs have diapiric structures interpreted to have formed during deposition and welding of pyroclastic flows in shallow marine water (Wright and Coward, 1977). They are nearly circular in plan and funnel-shaped in cross section and are interpreted to be rootless vents formed from secondary small steam eruptions. Quartz nodules in the welded tuffs represent original gas vesicles that have since filled with drusy quartz, and are interpreted to indicate deposition of a pyroclastic flow in water or water-saturated sediments (Francis and Howells, 1973; Howells et al., 1979), but quartz "nodules" (lithophysae) are abundant in subaerial welded tuffs elsewhere (see Ross and Smith, 1961).

**Discussion**

Kato et al. (1971) and Yamazaki et al. (1973) suggest that the subaqueous hot pyroclastic flows studied by them were emplaced as laminar, high-concentration dispersions. The lower parts of their flow units have incipient cross laminations outlined by a crude orientation of coarse debris interpreted to have formed by overlapping lobes of subflows on an irregular front. Because such flows probably were not turbulent, mixing with water in significant amounts would be minimized. They suggest that water surrounding the hot flows become vaporized, creating a thin layer of water vapor around the flow which helps to prevent mixing.

It is difficult to imagine the process by which hot pyroclastic flows penetrate the air-water interface without explosively mixing with the water and destroying the character of the pyroclastic flow. Indeed, Walker (1979) suggests that a major rhyolitic ignimbrite in New Zealand entered the sea, causing immense explosions resulting in a widespread fallout layer of ash and probable destruction of the pyroclastic flow. Evidence from Dominica (Sparks et al. 1980a), however, suggests that pyroclastic flows can enter the water, and, at Mt. Pelée in 1902, the high-concentration underflow parts of the nuées ardentes flowed into the sea as the overriding low-concentration ash clouds flowed across the water (Anderson and Flett, 1903; Lacroix, 1904). As Sparks et al. (1980b) emphasize, many pyroclastic flows, including even pumice and ash flows, are denser than water ($>1.0 \text{ g/cm}^3$).

Initial explosive interaction at the front of a flow should result in complex internal textures and structures, and appreciable thickening of the deposits might occur where the flow decreases velocity upon entering water. The nature of such a transition zone remains problematical and its preservation in the littoral zone doubtful. One of the members of the pyroclastic flow deposit of the Capel Curig Volcanic Formation has a local pyroclastic breccia at its base which Howells et al. (1979) ascribe to initial quenching of the early stages of the flow at the shoreline which was then overridden by later stages of the same flow as it continued beneath the water.

Apart from the question of what would happen to a hot pyroclastic flow as it passes through the air-water interface, if conditions permit penetration of a flow with temperatures of several hundred degrees without destruction, welding is theoretically possible (Sparks et al., 1980b). As the flow moves into water, hydrostatic pressures increase by 1 b for every 10 m water depth and water drawn into the hot pyroclastic flow can be absorbed into glass shards. At increased pressures, water content of glass can be higher, resulting in lower viscosities. These conditions favor

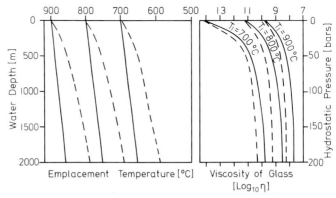

**Fig. 10-29.** Diagram on *left* shows temperature change of a pyroclastic flow deposit as a function of water depth. *Solid* and *dashed lines* are extreme limits. Diagram on *right* shows effect of absorption of water into glass shards on glass viscosity as a function of water depth. (After Sparks et al., 1980b)

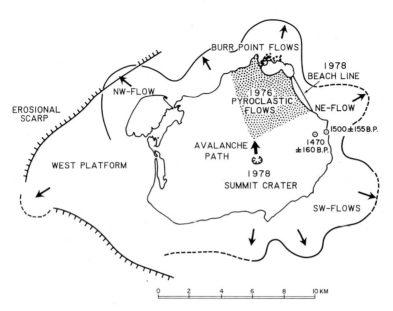

**Fig. 10-30.** Probable limits of historic pyroclastic flow deposits offshore from Augustine Island Volcano. (After Kienle and Swanson, 1980)

welding at elevated temperatures (Smith, 1960a; Riehle, 1973). Calculations of viscosity of glass shards as a function of water depth (Fig. 10-29) suggest that the viscosity of glass decreases substantially with depth despite cooling of the flow. For subaerial pyroclastic flows at 800 °C, the viscosity of rhyolite glass is about $2 \times 10^{11}$ poise, but under 500 m of water at 755°C, viscosity of glass is reduced by dissolution of water to an estimated $2 \times 10^9$ poise and thus would compact much faster than a subaerial flow under the same uniaxial stress (Sparks et al., 1980b).

This theoretical permissibility of subaqueous welding lends some support to the interpretation that the Capel Curig rocks are subaqueous welded tuffs (Francis and Howells, 1973), but more unequivocal examples are needed. Inescapable, however, is that the cumulative volume of pyroclastic flows should be very large in seas adjacent to explosive volcanoes. At Augustine Volcano, Alaska, for example, where eruptions generally accompanied by pyroclastic flows occur about every forty years (Kienle and Swanson, 1980), enlargement of the island's circumference is probably dominated by pyroclastic flows that have entered the sea (Fig. 10-30). The small volume of individual flows at Augustine makes it unlikely that they can remain hot very long under water, but they undoubtedly pile up very quickly. Further, such unstable wedges of loose sediment augmented by laharic and alluvial material derived from rapid subaerial reworking of unconsolidated pyroclastic flows are ideal source areas for volcanic turbidites generated by periodic slumping.

## Chapter 11  Lahars

The name lahar is Indonesian for volcanic breccia transported by water (van Bemmelen, 1949, p. 191) but has come to be synonymous in geological literature with volcanic debris flow, a mass of flowing volcanic debris intimately mixed with water. The term lahar refers both to the flowing debris-water mixture, and also to the deposit thus formed. A classic review of the various origins of lahars is that of Anderson (1933). A more recent discussion of lahar deposits by Parsons (1969) is included in a review of volcanic breccias. Crandell (1971) gives an account of the origin and characteristics of post-glacial lahars from the slopes of Mount Rainier volcano (Washington), and Neall (1976) has prepared a bibliography of their global occurrences. The recent eruptive phases of Mount St. Helens (Washington) produced lahars in a variety of ways (Christiansen, 1980; Janda et al., 1980; Janda et al., 1981; Harrison and Fritz, 1982) but investigations of these have not been completed to date.

Many lahars are associated with stratovolcanoes of which they may comprise significant volumes of the volcanoes' bulk. Most stratovolcanoes are of andesitic to dacitic composition, and hence most lahars have been reported from Indonesia, the western U.S. (Cascade volcanic chain), Japan, New Zealand, and Central and South America, but they are also associated with stratovolcanos of other compositions such as Vesuvius (Italy) and Hekla (Iceland). Lahars of much smaller dimensions occur during many phreatomagmatic eruptions of diverse chemical composition.

Most modern, Holocene, or Pleistocene lahars are relatively limited in extent and occur in valleys or on alluvial aprons or lowland areas immediately surrounding volcanoes, but in the geologic record there are extensive areas of laharic accumulations where volcanic edifices no longer exist. These can cover thousands of square kilometers and span several million years in time. For example, in the Absaroka Mountains (Parsons, 1969), extensive early Eocene to early Oligocene laharic breccias associated with lava flows and related volcaniclastic sediments once covered about 10,350 km$^2$. In the central and northern Sierra Nevada, late Miocene and early Pliocene lahars (Curtis, 1954) extended over 31,080 km$^2$, and in the southern Cascade Range, late Pliocene lahars covered 5180 km$^2$ (Anderson, 1933; Lydon, 1969). Although the rocks of these large volcanic tracts are of diverse origin, lahars are dominant. However, there are no comprehensive studies that treat the facies of such large accumulations or their relationships to the evolution and possible periodicity of the growth of a volcano.

## Debris Flows as Fluids

Many lahars are initiated directly by volcanic eruption, whereas others originate in ways similar to nonvolcanic debris flows, but once flow begins their fluid characteristics appear to be similar or identical. Thus, studies of all kinds of debris flows contribute to our understanding of the fluid properties of lahars. Debris flows are non-Newtonian fluids that have a yield strength. They behave like plastic materials similar to wet concrete, have a high bulk density, and exhibit the property of strength which greatly influences the final textures and structures of the deposits. The Newtonian properties of water (i.e. lacking in yield strength) begin to be modified by particle interference when the volume of solids exceeds 9 percent (Bagnold, 1954b, 1955). Estimates vary, but at volume concentrations of about 20 or 30 percent particle interactions almost completely dominate flow behavior (Middleton, 1967).

Beverage and Culbertson (1964) defined hyperconcentrated streams as those with 40 to 80 percent by weight of solids, and mudflows (debris flows) as containing 80 percent by weight or more (about 60 percent by volume) of solids. Debris flows, however, differ from hyperconcentrated streams (streams in flood) in flow behavior and depositional mechanisms, but the concentrations that determine behavior depend also upon the grain sizes and their size distribution. In stream flow, including that of hyperconcentrated streams, large and small particles are carried in the water by turbulence and traction processes; as velocity decreases, progressively smaller fragments settle out of the water. On the other hand, debris flows are fluids in which the water and solids form an intimate mixture that flow with laminar motion. As velocity decreases, the entire flow stops rather abruptly, after which water separates from the granular material by percolation or evaporation. On steep slopes, velocities may be rapid enough (internal shear stress high enough) to keep the entire mass in motion, but as slope decreases, internal shear stresses fall below the critical yield stress, so that the mass will congeal unless it is thick enough to maintain a high shear stress at the base of the flow. If so, the basal part of the flow will continue to move in laminar fashion and carry the rigid plug above it. As the gradient decreases, velocities decrease, and the flow thins, shear stresses increase until the flow congeals to its very base and deposition is complete (Johnson, 1970). Observations by Broscoe and Thomson (1969) on a debris flow in the Yukon showed that the newly deposited debris remained in a quasi-fluid state for many days, and 2 weeks after coming to rest, a thin crust developed over still-fluid material beneath.

If the concentration of solids in a debris flow is taken to be 80 percent or more by weight, then many flows called lahars or mudflows with lower concentrations actually are water floods (hyperconcentrated streams). Lahars described by Waldron (1967), for example, were mostly floods because they varied in concentration from 20 percent to about 80 percent. Maximum concentration values greater than 90 percent for nonvolcanic debris flows have been reported (Curry, 1966); a commonly reported lower limit is 70 percent (Sharp and Nobles, 1953). The distinction between debris flow and hyperconcentrated stream deposits, however, is poorly defined and it is not certain if their deposits can be separated on field criteria.

A useful concept for the theoretical and practical treatment of debris flows is to consider them to be composed of two phases: (1) a continuous phase (matrix or fluid phase) consisting of an intimate mixture of water with particles <2 mm; and (2) a dispersed phase consisting of particles >2 mm (Fisher, 1971). Thus, even though there may be a continuum of grain sizes from clay to boulders, it is possible to conceptually consider viscosity, density, strength, etc. of high concentration dispersions without regard to the individual properties or behavior of single particles: the continuous phase is the fluid that transports the large fragments. Moreover, treatment of the continuous (matrix) phase separately from the dispersed phase would be useful for standardizing size limits used by various authors to characterize and compare different debris flow deposits.

## Distribution and Thickness

Lahars follow pre-existing valleys and may be interstratified with alluvium, colluvium, pyroclastic rocks of diverse origin and lava flows derived from the same source area. They may leave thin deposits on steep slopes and in the headwaters of valleys, but become thicker in valley bottoms and form fans that coalesce or else form broad individual digitate lobes in lowland areas on very low slopes somewhat similar in distribution to pyroclastic flow deposits (Figs. 11-1, 11-2). The movement of lahars down valleys generally occurs in surges, or peaks of flow. During their course down a valley, lahars tend to leave thin "high water" marks (veneers) where a constriction momentarily causes a large debris flow to pond up to several tens of meters above the valley bottom and then drain away. Also, their momentum may carry them farther up the outer part of a bend in a stream curve. Veneers of over 150 m above present valley floors are reported by Crandell (1971).

Lahar assemblages at Nevado de Toluca volcano, Mexico (Bloomfield and Valastro, 1977) occur as an older series of overlapping and coalescing sheets and fans that give rise to a smoothly-rounded undulating topography. They lie upon a sloping (6°–8°) piedmont area surrounding the volcano. Farther up on the volcano, the rugged, forested slopes are underlain by lava flows. Younger lahars radiate outward from the volcano and occupy valleys cut in the older lahar assemblages and lava flows. It is possible, however, that some of the lithic-rich deposits described as lahars by Bloomfield and Valastro (1977) are pyroclastic flows. As evidence for a lahar origin they cite absence of bread-crust blocks and carbonized wood and small content of pumice and glass (Bloomfield and Valastro, 1977). However, the May 8 and May 20, 1902 nuée ardente deposits from Mt. Pelée (Fisher et al., 1980) fit this description, as do some of the pre-1980 deposits at Mount St. Helens Volcano, Washington (Crandell, personal commun., 1978). Thermoremanent magnetization of some of the lithic-rich Mount St. Helens deposits (Hoblitt and Kellogg, 1979) indicates that they were emplaced above their Curie temperatures, and thereby suggests that water was not the mobilizing agent.

Lahars vary greatly in thickness. They tend to maintain a relatively constant average thickness on relatively low slopes but locally vary depending upon the configuration of underlying topography. Lahars and other debris flows come to rest with steep sloping lobate fronts (Johnson, 1970). Most lahars are probably less than 5 m thick (Mullineaux and Crandell, 1962; Schmincke, 1967b; Crandell,

**Fig. 11-1.** Peripheral laharic fans formed a few days to weeks after the January to April 1976 eruption of dacitic Augustine Volcano (Alaska). Deposits are covered with pumice pebbles and are cut by stream valleys. (Photograph taken August, 1976)

**Fig. 11-2.** Lahars of tephritic composition of Pliocene Roque Nublo Formation interbedded with 5-m-thick lava flow in lower part. Gran Canaria (Canary Islands) (Schmincke, 1976)

Table 11-1. Dimensions of some lahars

| Name of lahar, volcano or formation | Date of eruption | Distance travelled (km) | Thickness (m) | Area (km²) | Volume (km³) |
|---|---|---|---|---|---|
| Mount St. Helens, N. Fork Toutle River, Washington, USA (Janda et al., 1981) | May 18, 1980 | >120 | 1–2 | | >0.36 |
| Santa Maria, Guatemala (Anderson, 1933) | 1929 | 100 | | 15 | |
| Kelut, Java (Anderson, 1933) | 1919 | 40 | 50 (max.) | 130 | |
| Mt. Lassen, USA (Macdonald, 1972) | 1915 | 46 | | | |
| Cotopaxi, Ecuador (Anderson, 1933) | 1877 | >240 | | | |
| Galungung (Macdonald, 1972) | 1822 | 65 | | | 0.03 |
| Electron, Mt. Rainier, USA (Crandell, 1971) | 600 y. BP | 50 | 4.5 (av.) 8 (max.) | 36 | 0.15 |
| Mount St. Helens, USA (Mullineaux and Crandell, 1962) | 2000 y. BP | 65 | | | |
| Osceola, Mt. Rainier, USA (Crandell, 1971) | 5700 y. BP | 110 | 6 (av.) 60 (max.) | 260 | >2.0 |
| Paradise, Mt. Rainier, USA (Crandell, 1971) | 6000 y. BP | 30 | 4.5 (max.) | 34 | 0.1 |
| Raung, Java (Macdonald, 1972) | Prehistoric | 56 | | | |
| Yatsuga-dake, Japan (Mason and Foster, 1956) | Pleistocene | 24 | | | 9.6 |
| Ellensburg Formation, USA (Schmincke, 1967b) | Miocene | 60 | | | |

1971), but some are more than 200 m thick (Bloomfield and Valastro, 1977) and may be as thin as 0.5–1 m (Curtis, 1954). Despite their importance as common products of stratovolcanoes and as one of the most dangerous of volcanic hazards, there are few detailed sedimentological studies of either fossil or historic lahars (Table 11-1).

## Surface of Lahars

Lahar surfaces tend to be remarkably flat over wide areas but in detail contain local swells and depressions interpreted to be caused by differential compaction over an irregular underlying surface (Crandell and Waldron, 1956). The form, shape, and size of irregularities, however, depend upon the viscous properties of the flows and the number and characteristics of multiple lobes. In the past, deposits interpreted

Fig. 11-3. Dark, hummocky landslide-debris flow of May 18, 1980 Mount St. Helens eruption, Washington (USA). Hummocks about 30 m high. Light colored area with planar surface is underlain by pumice flow deposits. (Photograph taken September 1980)

as lahars with unusually hummocky surfaces have been reported by Escher (1920), Grange (1931), Mason and Foster (1956), Aramaki (1963), Gorshkov and Dubik (1970) and others. However, the collapse and avalanching of the north side of Mount St. Helens volcano on May 18, 1980 produced a hummocky deposit much like those previously described as lahars and cast doubts upon their interpretation as lahars (Voight et al., 1981). The Mount St. Helens rockslide-avalanche deposit, with a volume of 2.8 km$^3$, has hummocks that are as much as 170 m wide and protrude about 30 m above the mean elevation of the surface of the deposit (Fig. 11-3). The material was emplaced at a temperature that approached boiling water. It was unsaturated by water during emplacement, but its momentum imparted to it an enormous mobility. The hummocks consist of huge brecciated chunks of the mountainside set in a poorly sorted "matrix" ($S_0 = 2.9$ to 13.0; average $= 7.1 : S_0 = Q_{75}/Q_{25}$, where Q is the size measure on a cumulative curve at the indicated percentages 25 and 75 and $S_0$ is a sorting measure; Table 5-6). There is no systematic down-valley change in sorting values of the matrix. The question of how the matrix developed from the original solid rock of the mountainside remains unsolved.

## Basal Contact of Lahars

Although lahars and other debris flows may be very thick and carry large boulders, they commonly do not erode the surfaces on which they flow except on very steep slopes. Curry (1966) reports that talus was incorporated by a bouldery debris flow observed by him moving on slopes of 35° to 41°, but on slopes of 7° to 10°, where velocities were low, the flow did little harm to meadow grass despite the fact that large boulders were abundant. The 1941 Wrightwood, California, debris flow rests in places on a carpet of pine needles covering low slopes (Sharp and Nobles, 1953), and Crandell (1957, 1971) notes that debris flow deposits conformably overlie soft soil profiles, peat deposits and thin layers of sand and volcanic ash on slopes of up to 7.5°. Molds of inclined grass were noted at the base of several Miocene lahars of the Ellensburg Formation, Washington (Schmincke, 1967b). Lahars can pick up loose materials from surfaces on steep slopes or where local turbulence develops within the flow owing to highly irregular channels. Some Pleistocene lahars in the southern part of the Puget Sound lowland, however, have traveled 60 to 80 km from their source without picking up appreciable debris from the surface on which they flowed (Crandell, 1963).

## Components of Lahars

Depending upon their origin, lahars may be monolithologic or heterolithologic. Monolithologic varieties are likely to be derived directly by eruption, whereas collapse of crater walls or avalanching of rain-soaked debris covering steep volcanic slopes are more likely to give rise to heterolithologic types. Pumice-rich lahars are described (Bond and Sparks, 1976; Wright, 1978), which resemble pumice-rich deposits of hot, dry pyroclastic flows (Mullineaux and Crandell, 1962), but are distinguished from hot flows mainly by thermal analysis of the magnetism (Aramaki and Akimoto, 1957).

Lahars characteristically contain dense angular to subangular rock of dominantly andesitic to dacitic composition mixed with ash-sized minerals and lithic particles.

Many lahar deposits contain charred wood (Crandell and Waldron, 1956; Fisher, 1960; Mullineaux and Crandell, 1962; Schmincke, 1967b; Crandell, 1971), indicating that they were initiated as hot pyroclastic flows then cooled down during transport. Analysis of fragments from one lahar containing charcoal showed clustered rather than random orientation of north-seeking poles, suggesting that parts of the deposit were above the Curie point when the deposit came to rest (Mullineaux and Crandell, 1962). Emplacement temperatures of various deposits are discussed by Hoblitt and Kellog (1979).

## Grain-Size Distribution

Particles carried by lahars range from clay- to boulder-size, but the percentages of each size fraction vary enormously from deposit to deposit and also within a single deposit. In general, lahars are coarser-grained and more poorly sorted than pyroclastic flow deposits, although there are many exceptions. The block-and-ash flows from the ill-famed 1902 eruptions of Mt. Pelée, for example, are coarser-grained than many lahars that originate from loose ash on the steep slopes of volcanoes.

Grain-size parameters reported by various authors (Table 11-2) show the obvious fact that lahars and nonvolcanic debris flows have a wide range in grain size and are coarse-grained and poorly sorted, but the data are not strictly comparable

**Table 11-2.** Grain-size parameters of some lahars compared with nonvolcanic debris flows

| Locality | Md$_\phi$ | $\sigma_\phi$ $(=\phi_{16}-\phi_{84}/2)$ |
|---|---|---|
| Mt. Rainier, Washington (Crandell, 1971) Lahar Deposits | Range: 3.4 to $-3.7$ Av.: $-1.7$ (30 samples) | Range: 2.78 to 5.79 Av.: 4.44 (38 samples) |
| Irazu Volcano, Costa Rica (Waldron, 1967) Flowing Lahars | Range: 3.87 to 0.75 Av.: 1.88 (10 samples) | Range: 2.62 to 4.04 Av.: 3.12 (10 samples) |
| Tokachi-Dake Volcano, Japan (Murai, 1960) Lahar Deposits | Range: 0.20 to 1.23 Av.: 0.58 (4 samples) | Range: 3.07 to 5.43 Av.: 4.06 (4 samples) |
| Non-volcanic debris flows from an alluvial fan (Bull, 1964) | Range: 0.2 to 10.0 Av.: 2.9 (48 samples) | Range: 4.0 to 6.2 Av.: 4.7 (27 samples) |

because of different sampling procedures, laboratory techniques, and total number of samples analyzed by individual authors. Also, because lahar deposits tend to contain abundant coarse-grained fragments, fine-grained lahars or the finer-grained matrix of coarse-grained lahars are more apt to be analyzed granulometrically for technical convenience. In outcrop (Fig. 11-4), however, many lahar deposits appear to be coarser-grained than shown by granulometric analysis because the statistically few boulders that might be present are visually more impressive than the smaller particles and thereby give a false impression of true size values. The presence of large boulders, commonly exceeding 1 m in diameter, is one of the most characteristic features of lahars except perhaps, in their terminal zones (Crandell and Waldron, 1956; Crandell, 1971; Curtis, 1954; Schmincke, 1967b).

A study by Sharp and Nobles (1953) of the 1941 Wrightwood debris flow showed lateral changes in grain size of boulders. The large fragments progressively decreased in number and size away from the source, although the finer constituents (matrix) did not show corresponding changes. Erratic fluctuations in median diameter were attributed to the longitudinal inhomogeneity of the flow caused by deposition from individual debris tongues that differed in grain size. The flow occurred as a succession of many debris flow surges per day over a period of 10 days; the longest of the surges travelled a maximum distance of 26 km. The total deposit is a sequence of overlapping tongues of variable length. One study of a lahar in Japan (Murai, 1960) showed that median diameters did not vary systematically over a distance of 3 km, but only four samples were analyzed.

Because boulders cannot be included in standard size analyses and therefore lahars cannot be completely characterized granulometrically, we compare (Fig. 11-5) matrix phases (sand/silt/clay recalculated to 100%) of different debris flows and also the May 18, 1980 Mount St. Helens rockslide avalanche and blast deposit (Voight et al., 1981). As shown, lahars tend to contain less clay-size material than nonvolcanic debris flows. A possible reason is that fragments in volcanic deposits on the whole may be diagenetically less mature than nonvolcanic debris which is derived by weathering rather than explosive or other volcanic processes. The abun-

| $S_0 = \sqrt{Q_{75}/Q_{25}}$ | Comments |
|---|---|
| Range: 3.41 to 17.0<br>Av.: 10.37<br>(39 samples) | $Md_\phi$ figures converted from mm units using graph |
| Range: 2.58 to 7.01<br>Av.: 4.61<br>(10 samples) | $Md_\phi = 0.75$ converted from mm units given by Waldron. Published figure is wrongly given as $Md_\phi = 3.35$ |
| Range: 1.81 to 3.72<br>Av.: 2.70<br>(4 samples) | The 4 samples reported are from the lahar of May 24, 1926 |
| Range: 5.1 to 25<br>Av.: 9.7<br>(46 samples) | |

dance of clay that occurs in the matrix of a few lahars has been a matter of some debate, but Crandell (1971) convincingly shows that the clay in Mount Rainier lahars is derived from a source area where marked hydrothermal alteration had occurred. This kind of plot does not distinguish the Mount St. Helens rockslide avalanche matrix from lahars.

**Fig. 11-4.** Dacitic lahar ($\sim 4$ m thick) in late Miocene Ellensburg Formation (Washington, USA), showing 10-cm-thick fine-grained base and concentration of larger boulders in lower third (Schmincke, 1974a)

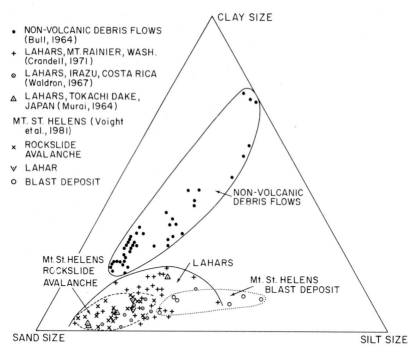

Fig. 11-5. Grain size of matrix of volcanic and nonvolcanic debris flows, and Mount St. Helens rockslide avalanche deposit (Washington, USA)

## Vesicles

Air spaces that we call vesicles occur in lahars (Crandell and Waldron, 1956; Crandell, 1971), base surges and other hydroclastic deposits (Chap. 9). Vesicles also have been reported by Sharp and Nobles (1953) and Bull (1964) in nonvolcanic debris flows. In fine-grained deposits air spaces tend to be spherical whereas in coarse-grained deposits air spaces are irregular in shape and therefore may be overlooked. Vesicle diameters range from nearly a millimeter up to a centimeter or more and may by scattered or concentrated adjacent to large particles or impermeable clastic horizons.

Vesicles in lahars have been explained as trapped air bubbles (Crandell and Waldron, 1956; Crandell, 1971) rather than by the draining away of free water after the lahar came to rest. The best evidence of air bubble origin is the occurrence of spherical cavities. Steam cavities also may form in some hot lahars, similar to those in tuffs formed by phreatomagmatic eruptions. We observed that small cavities are common in debris flow deposits at Wrightwood, California and elsewhere, but nearly all such cavities are irregular in shape; only rarely can spherical cavities be found, and these are confined to muddy parts of the deposit.

# Grading

Many lahar deposits show a subtle grading of the coarse-grained (>2 mm) dispersed phase, but it may not be evident in the matrix phase (see Crandell, 1971, Table 2). Single depositional units generally have an irregular but slightly more concentrated arrangement of large fragments a short distance above the base of the lahar (Schmincke, 1967b); such layers are reversely graded (Fig. 11-4). The large fragments in a lahar rarely rest directly upon the depositional surface. However, reverse grading with the coarse fraction becoming progressively larger to the top of a deposit is very rare unless low density pumice is abundant. Photographs in Crandell (1971, Figs. 10, 21, 27, 29, 33), Macdonald (1972, Plate 8-7), Parsons (1969, Plate 3) and examples of many other lahars, e.g. in the Canary Islands and Eifel, Germany, observed by us also show reverse-to-normal grading within the same bed (i.e., large boulders tend to be more common in the lower central zones). A relatively fine-grained basal layer from a few to several centimeters thick is indeed a common feature of lahars as well as pyroclastic flows (Sparks, 1976; Fisher et al., 1980) and also of nonvolcanic debris flows (Fisher, 1971).

Understanding how grading (or its absence) is developed is aided by observations of moving debris flows and by laboratory experiments. Reports of boulders bobbing along on the surface of flowing debris are common (Blackwelder, 1928), but whether the large fragments are actually floating at the top of a flow, tumbling within it, or saltate in slow fashion and bob to the surface occasionally, depends upon the settling velocity of the large fragments relative to the density and the plastic strength of the fluid. Some workers have suggested that large boulders are suspended by turbulence. Johnson (1970), however, convincingly shows that debris flows move in laminar fashion; therefore, large boulders are suspended by a combination of high density (buoyancy) and high strength of the matrix. His conclusion is based in part upon laboratory experiments with kaolin-water slurries that tended to move in laminar fashion when the clay content was greater than 10 percent by weight, and in part by observation of moving debris flows in the field. Field observations showed an essentially smooth-flowing surface indicative of laminar flow rather than a choppy surface characteristic of turbulent flow. Evidence from deposits that indicates gentle handling of debris, hence the absence of turbulence, includes unmodified fragile fragments such as tin cans, large blocks of brittle shale and wood fragments, but most convincing is the presence of unweathered fractured boulders that are still coherent or else so slightly scattered that the fragments can be fitted together like jig-saw puzzles. Johnson (1970, p. 513) attributes the gentle handling to plug flow (see section on fabric).

The mechanisms by which reverse grading develops are not well understood. According to Bagnold (1954b, 1955) dispersive forces act normally to flow boundaries during movement of concentrated dispersions. The transfer of momentum from grain to grain or from close grain encounters during flow supports individual grains throughout the flowing bed. Bagnold's equations show that the dispersive force acting upon a particle is proportional to the rate of shear, suggesting that when particles are sheared together, the larger particles will drift toward the zone with the least rate of shear (Johnson, 1970, p. 462). Sanders (1965, p. 202), Schmincke (1967b) and others have used this concept to explain reverse grading. Middleton (1970), however, suggests that reverse grading develops by smaller clasts falling downward between the larger clasts during movement, thereby preventing the larger ones from moving downward; hence the larger fragments would progressively work themselves relatively upward. The difference between these two ideas, however, appears to be one of *how* the process occurs rather than of different causes. Fisher and Mattison (1968) and Mattinson and Fisher (1970) attempt to explain reverse grading in terms of lift forces supplied to individual large particles resulting from lower pressures at the top than at the bottom of large particles due to different velocity gradients within the flow. Experiments by Southard (1970), however, suggest that such lift forces are very small although the sediments used by him were fine-grained.

Differences in grading, whether it be absent, weakly or strongly developed, normal or reverse, appear to be related to the relative concentration of solids and fluid; the lower the concentration of solids, the more likely it is that normal grading can develop because viscosity, density, and strength of the fluid are less able to support large dense particles as velocity decreases. Where concentration values and therefore viscosity, density, and strength are high, reverse grading is more likely to develop especially if the density of fragments is relatively low. Inasmuch as there may be a wide range in concentration values in different flows, from hyperconcentrated streams to debris flows of Beverage and Culbertson (1964), it is expected that all gradations between different kinds of grading will occur.

## Fabric

The fabric of lahars, and indeed most debris flows, is commonly regarded as isotropic, but in some lahars disc-shaped pebbles and uncharred twigs and tree trunks concentrated low in the central parts of the deposit are oriented subparallel to the base (Schmincke, 1967b).

The development, or lack thereof, of clast fabric in debris flows depends upon the mechanism of movement and deposition, and is a matter of some debate. Convincing arguments by Johnson (1970) and Hampton (1972), however, suggest that matrix strength in debris flows may produce a rigid plug where shear stress is below the yield threshold throughout (Johnson, 1970), and this plug rides on a zone of laminar flow within which the shear stress is greater than the yield threshold. Flow stops when the plug expands to the base of the flow at the expense of the zone of laminar flow, thus fabrics in the shearing region adjacent to the base become frozen in place during the last stages of flow and preserve the clast orientations, textures and structures of the debris flow.

In modern debris flow deposits, preferred orientations of platy or elongate fragments are reported as strongly aligned approximately parallel to flow surfaces (Fisher, 1971, Fig. 1), or random, parallel, and nearly perpendicular to channel axes within a single debris flow deposit (Johnson, 1965, p. 24, 31). Random orientation may be expected within the rigid plug if shearing does not occur, but within the basal zone of flow, movement is probably laminar and should leave its imprint with fragments either parallel to flow, inclined to flow, or imbricated (Enos, 1977). Fabric in debris flow is discussed by Lindsay (1966, 1968) and Enos (1977) and in other kinds of mass flow deposits (subaqueous) by Davies and Walker (1974), Hubert et al., (1975), Hendry (1976) and others.

# Comparison of Lahars with Other Kinds of Coarse-Grained Deposits

Other coarse-grained deposits that have characteristics similar to lahars and may be difficult to distinguish from them if the source rock is largely volcanic include till and tillite, fluvial gravels (flood deposits) and pyroclastic flow deposits. These deposits have no single unique feature that separates them, but several features taken together may help to discriminate between them (Table 11-3).

Lahars may be distinguished from volcaniclastic fluvial deposits by a greater abundance of clay-size particles and presence of extremely large boulders, that is, their extremely poor sorting, general absence of internal layering and channeling, greater thickness, distribution as flat-topped lobate deposits outside valleys, nonerosive basal contacts and presence of charred wood. Poor sorting and large boulders are also characteristic of till, but till lacks charred wood and commonly rests on striated bedrock.

The presence of striated fragments within coarse-grained deposits is regarded as evidence of a glacial origin, but as has been stated many times in the past, they also occur in lahars (Anderson, 1933; Cotton, 1944, p. 239–247; Crandell and Waldron, 1956; Curtis, 1954; Mason and Foster, 1956; and others). Grooves on underlying surfaces generally occur beneath glacial deposits but this also may occur on the surface beneath some lahars (Bloomfield and Valastro, 1977) and pyroclastic flow deposits (Brey and Schmincke, 1980).

The presence of abundant pumice may distinguish unwelded pyroclastic flow deposits from lahars, but where lahars have originated from hot pyroclastic flows that enter streams and become mixed with water, they may be difficult to identify. However, a coarse-grained poorly sorted deposit with individual rock fragments that have random directions of remanent magnetism is probably a lahar, and a deposit containing large groups of fragments having a preferred orientation is inferred to have been formed as a hot pyroclastic flow (Aramaki and Akimoto, 1957; Crandell, 1971; Crandell and Mullineaux, 1973; Hoblitt and Kellogg, 1979). Hot pyroclastic flow deposits may be oxidized by hot gases to pale red in their upper few meters. Some lahars derived from hot pyroclastic flows that become mixed with water and carry hot debris may confound all attempts to determine origin until detailed field mapping is done.

# Origin

Macdonald (1972) lists 12 different ways that lahars can originate, and these can be grouped into three major categories (Crandell, 1971):

1. Those that are the direct and immediate result of eruptions: eruptions through lakes, snow or ice; heavy rains falling during or immediately after an eruption; flowage of pyroclastic flows into rivers, or onto snow or ice.
2. Those that are indirectly related to an eruption or occur shortly after an eruption: triggering of lahars by earthquake or expansion of a volcano causing the rapid drainage of lakes or the avalanching of loose debris or altered rock.

**Table 11-3.** Comparison of coarse-grained deposits with lahars

|  | Lahars | Till (excluding water-laid till) | Unwelded ignimbrite | Fluvial deposits |
|---|---|---|---|---|
| Large fragments (>2 mm) | May have boulders weighing many tons | May have boulders weighing many tons | Extremely large boulders absent | Extremely large boulders rare |
| Sorting | Poor. May contain abundant clay-size material | Poor. May contain abundant clay-size material | Poor. Clay-size material rare or absent | Poor to good. Clay-size material sparse |
| Grading | Commonly reverse. May be normal or absent | Commonly absent | Commonly absent, but may be normal or reverse | Commonly normal |
| Bedding and thickness | Commonly very thick with vague internal bedding | Very thick. Bedding poor or absent | Commonly very thick with vague internal layering | Thin with channels and cross beds. Shingled gravels |
| Composition | Commonly 100% volcanic. May be pyroclastic or mixed with epiclastic materials. May contain bread crust bombs | Commonly hetero-lithologic with admixtures from many sources. Plutonic, metamorphic and sedimentary clasts commonly more abundant than pyroclasts | Pyroclastic. May contain abundant bread crust bombs | Material usually 100% epiclastic except in areas of active volcanism |
| Rounding of large fragments | Commonly angular to subangular | Commonly faceted subangular to subrounded. May be faceted with striations and chatter marks | Commonly sub-angular | Commonly sub-rounded to rounded |
| Carbonaceous matter | Uncharred to charred | Uncharred | Charred | Uncharred if present |
| Pumice | Common in some lahars | Not present except on active volcanoes | Common | Not present except in areas of active volcanism |
| Distribution | In valleys spreading onto flat piedmont surfaces | Plains and valleys. May mantle all surfaces. Moraines with steep fronts | Lower parts of valleys and flat piedmont surfaces | Confined to valleys |
| Lower surfaces | Commonly not erosional | Erosional. Commonly rests on striated bedrock | Commonly not erosional | Erosional |

[a] Except close to caldera walls and in very proximal facies

3. Those that are not related in any way to contemporaneous volcanic activity: mobilization of loose tephra by heavy rain or meltwater; collapse of unstable slopes (in particular of diagenetically or hydrothermally altered clay-rich and water-soaked rocks); bursting of dams due to overloading; lahars that originate on the steep slopes of volcanoes of other volcanic terrane undergoing active weathering and erosion; sudden collapse of frozen ground during the spring thaw.

Perhaps the most common type of lahar forms during the waning stages of an eruption when large amounts of loose pyroclastic fall or flow deposits on the slopes become soaked by heavy rains that commonly occur during this stage of an eruption. Several workers have presented maps showing that with increasing distance from the center of an eruption, nuée ardente deposits are succeeded laterally by lahars on lower slopes of volcanoes (Wolf, 1878; Zen and Hadikusumo, 1965; Moore and Melson, 1969; Lipman and Mullineaux, 1981).

The water for some lahars may be from stores of snow and ice within the crater or locked up inside the porous superstructure of a volcano and driven out by an advancing heat wave (Roobol and Smith, 1975, p. 14) but much is meteoric water vapor drawn into the eruption plume and condensed upon contact with the cold atmosphere aloft. In other cases, the rain may be completely unrelated to an eruption. Other sources of water are melted snow or ice on the slopes of a volcano, rivers invaded by hot avalanches or pyroclastic flows, or crater lakes or dammed-up slope basins whose dams are broken by lava flows or other extruded products. Earthquakes may also trigger lahars, either during an eruption, or later.

Great floods formed at the beginning of many subglacial volcanic eruptions may be associated with lahars. They are especially common in Iceland where they are known as "jökulhaups" (Kjartansson, 1951).

## *Chapter 12*  Alteration of Volcanic Glass

There are few rock types that offer better opportunities for alteration studies (weathering, diagenesis, hydrothermal) than glassy volcanic and volcaniclastic rocks. The main reason is that volcanic glass is thermodynamically unstable and decomposes more readily than nearly all associated mineral phases. Volcanic glass is a super-cooled silicate liquid with a poorly ordered internal structure consisting of loosely linked $SiO_4$ tetrahedra with considerable intermolecular space. Hydration and concomitant breakdown of glass results in fluxes of some elements out of the glass into interstitial pore waters. Precipitation of secondary (authigenic) minerals from such solutions, replacement of glass shards by new minerals and filling of pore space created by dissolution of glassy particles during alteration are some of the most rapid low temperature lithification processes known. Moreover, changes in composition of pore solutions and elevation in temperatures during hydration and burial result in variable mineral compositions and rapidly changing mineral assemblages because of the restricted P/T stabilities of some of the alteration products.

Alteration of volcanic glass is receiving attention for a variety of reasons. Hydrothermal alteration of basaltic glass along mid-ocean ridges may contribute in a major way to the formation of submarine massive sulfide ore bodies (Wolery and Sleep, 1976; Spooner, 1976). Zeolite and bentonite deposits, largely formed by reaction of pyroclastic material with pore water, have become of great economic importance (Mumpton, 1978; Grim and Güven, 1978). The mobility of elements of economic interest such as U during glass alteration is being studied in some detail (Zielinski, 1982). Waste disposal site stability in tuffs and characterization of alteration environment at waste sites is a topic of much present concern (Zielinski, 1980).

## Diagenesis

Diagenetic processes are part of a continuum between weathering (subaerial or submarine) and metamorphic changes which occur beneath the surface at elevated temperatures and pressures. Diagenetic changes are both physical and chemical. They include (1) compaction and reduction in porosity, (2) dissolution of components, (3) cementation by precipitation of new minerals and (4) recrystallization in response to P/T changes. Diagenesis, therefore, causes both mineralogical and textural changes within a rock. However, the boundary between diagenesis and metamorphism is not easily drawn (e.g., Pettijohn, et al., 1972; Blatt et al., 1972;

Winkler, 1979). For many sedimentologists, reduction of pore space to <5 percent roughly forms the boundary, but the formation of metamorphic mineral assemblages, defined as those not stable in a sedimentary environment, often does not coincide with the low-porosity boundary (Winkler, 1979).

Even the most cursory examination of volcaniclastic rocks illustrates the difficulty in drawing the boundary between diagenesis and metamorphism. Two main reasons for this are (1) the instability of the most common constituent – glass, and (2) the wide spectrum of conditions under which the most common diagenetic products are formed – the smectites, zeolites, and varied polymorphs of silica.

Most zeolites are stable from near-surface temperatures up to about 100 °C. Laumontite (Ca-rich), however, forms in the range 150°–200 °C (<5 kb). It and pumpellyite commonly have been used by students of metamorphic rocks as index minerals for the beginning of metamorphism, zeolites being regarded as diagenetic (Winkler, 1979). One of the most significant advances during the past 15 years is the recognition that the chemical activity of various pore fluid components is a major control on the composition of zeolites and other diagenetic minerals. It is now known that laumontite can crystallize at rather low load pressures and temperatures in ancient rocks as well as in deep sea sediments (Hay, 1966; Surdam, 1973; Boles and Coombs, 1975; Viereck et al., 1982) and thus cannot be used as a metamorphic index mineral.

Many lithified volcaniclastic rocks are difficult to analyze microscopically because of the bewildering textural varieties resulting from the extensive dissolution of glass, the precipitation of diagenetic minerals and because of the small grain size of volcanic dust and some of the authigenic mineral phases. Prior to diagenesis, glass and mineral components of pyroclastic rocks may also change in texture and composition during initial weathering and transport, especially in moist and warm climates; such changes may be impossible to separate from subsequent diagenetic processes. Weathering of pyroclastic rocks has been studied by Hay (1959b), Hay and Jones (1972), Kirkman (1976, 1980), Nagasawa (1978), and others. Antweiler and Drever (1983) have stressed the importance of organic compounds during weathering of volcanic glass. These control the release and transport of solutes by complexing Al and Fe, and by causing low pH-values.

The alteration environment in glassy rocks may be viewed as a dynamic system which consists mainly of three components:

a) The glassy starting material with compositions spanning the range of natural magmas from basalt to rhyolite. In predominantly glassy rocks, whole rock composition can be relatively homogeneous. Phenocrysts can range widely in composition and stability.
b) Physical conditions such as temperature, grain size, porosity, and permeability: They are highly variable initially, but become more uniform with time.
c) The pore solutions, whose composition initially ranges from marine to meteoric water, but which changes drastically during the course of alteration.

At low temperatures, composition of original material is probably the most important factor in determining the type of alteration. We will thus discuss alteration of basaltic and silicic glass separately, with the understanding that many processes and alteration products are quite similar. This approach seems justified, as the geologic literature indicates the bimodal occurrence of mafic and silicic tephra. The effect of pore water chemistry will be reviewed in the section on alteration of silicic glass. The third section deals with alteration of tephra at higher temperature.

## Alteration of Basaltic Glass

### Palagonite

Sartorius von Waltershausen (1845) described a yellow to brown wax-like substance in marine basaltic tuffs in the Iblean Mountains near Palagonia (Sicily) which he named palagonite (Fig. 12-1). He found the same substance to be widespread in Iceland. Bunsen (1847) recognized the similarity of Icelandic palagonite with basalt except for the addition of water and loss of Ca and Na. Penck (1879) showed that palagonite is not a mineral nor a mineral family; since then, palagonite has been used as a general term for any hydrous, altered, basaltic glass. Opinions continue to differ, however, on the exact structural and chemical nature of palagonite as well as on the conditions of its formation. Palagonite is extremely widespread in basaltic tuffs, especially those of subaquatic origin because of the much greater proportion of true basaltic glass, sideromelane, as compared to tachylite which is full of skeletal crystals and grades into fully crystallized basalt. Honnorez (1972) has reviewed the literature up to 1969 in detail.

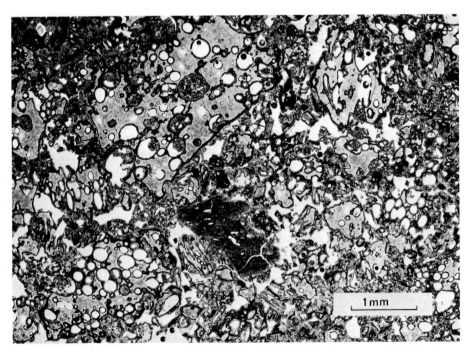

**Fig. 12-1.** Photomicrograph of slightly palagonitized sideromelane tuff (dark rims around light-colored, slightly vesicular shards) from the type locality at Palagonia (Sicily)

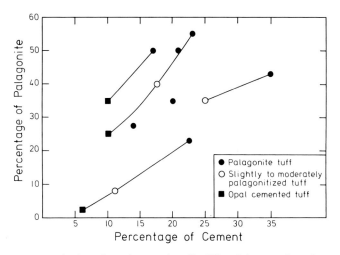

Fig. 12-2. Relationship between amount of palagonite and cement in tuffs of Honolulu group from Oahu. Lines connect rocks from the same deposit. (After Hay and Iijima, 1968a)

Physical Properties

In hand specimen, palagonite is a dull, resinous or waxlike substance in various shades of brown, yellow, orange or gray. In younger deposits, irregular patches of black fresh glassy sideromelane occur as "islands" coated by brownish palagonite envelopes of varying thickness. Palagonite tuff is commonly massive and breaks with conchoidal fracture. Both density and refractive index are significantly lower than in sideromelane for the entire range of basaltic compositions (Table 12-1). Density of palagonite increases with age due to its transformation into smectite (see below). In palagonite tuff, there is a drastic loss in porosity and increase in bulk density by about 0.3–0.4 gcm$^{-3}$ due to deposition of secondary gels and minerals in the pore spaces. Concurrently, grain density decreases significantly because the glass is more dense than palagonite and most of the secondary mineral phases, chiefly zeolites, clays, and opal whose densities vary from about 2.1–2.3 gcm$^{-3}$ (Fig. 12-2).

Textural Changes

The initial textural changes from fresh sideromelane into its first-stage major alteration product, palagonite, are striking. On a microscopic scale, there is commonly a sharp boundary between the transparent light yellow sideromelane and the more strongly colored yellow to brown palagonite (Fig. 12-1). Dark-red to red-brown palagonite is more rare, forms in tropical climates, and may have formed by fumarolic palagonitization at moderately elevated temperatures (Jakobsson, 1978). Palagonite typically forms around the edges of a sideromelane grain, along cracks in the glass, and from the glass walls of vesicles. Very fine-grained shards may become completely palagonitized at an early stage, while fresh sideromelane cores of palagonite tuffs have been found in rocks as old as Cretaceous (Byerly and Sinton, 1979; Staudigel et al., 1979, 1981), Triassic (Muffler et al., 1969) and Carboniferous

Table 12-1. Density and refractive index of sideromelane, palagonite and palagonite tuff from different source basalts

| | Sideromelane | | Crystallized basalt | Palagonite | | Palagonite Tuff | | | Reference |
|---|---|---|---|---|---|---|---|---|---|
| | $d$ (g cm$^{-3}$) | n | $d$ (g cm$^{-3}$) | $d$ (g cm$^{-3}$) | n | $d_1$ (bulk) (g cm$^{-3}$) | $d_2$ (grain) (g cm$^{-3}$) | Porosity (%) | |
| Mid-ocean ridge basalt | 2.7–2.8 | 1.594–1.611 | | 1.82–2.50 | 1.533–1.583 | | | | Noack and Crovisier (1980) |
| Alkali basalt | | | 2.89 | 2.39–2.57 | | 2.32–2.43 | | | Furnes (1980) |
| Ti-, Fe-rich primitive nephelinite | | | 3.0–3.2 | 2.47–2.85 | | 2.63–2.70 | | | Furnes and El-Anbaawy (1980) |
| Nephelinite | 2.56–2.73 | ~1.6 | | 1.90–2.1 | 1.52–1.61 | 1.82–2.22 | 2.21–2.49 | 6.1–11.5 | Hay and Ijima (1968b) |
| Tholeiite | | | | | 1.47–1.50 | | | | Peacock (1926), Peacock and Fuller (1928), Hoppe (1940) |

(Schmincke and Pritchard, 1981). Peacock (1926) distinguished two main varieties which he termed gel-palagonite and fibrous palagonite. Gel-palagonite is isotropic and commonly banded, forming the alteration zone next to the unaltered glass. It is separated sharply from the glass by a zone, about 100 μm thick, containing opaque material and, in cases, microtubules, penetrating into the glass (Peacock, 1926; Melson and Thompson, 1973). Morgenstein and Riley (1975) called this transition zone the immobile product layer. With advanced palagonitization, an outer rim of more intensely colored fibrous, slightly birefringent material develops next to gel-palagonite. These palagonite zones may be accentuated by zones of finely divided Fe and Ti-oxides. Single or multiple zones of such opaques which parallel the outline of, or vesicle inside, a shard survive more advanced stages of alteration and nonpenetrative metamorphism and are a characteristic texture to identify formerly palagonitized sideromelane shards, such as in Archean hyaloclastites described by Dimroth and Lichtblau (1979). Refractive indices of palagonite vary widely (Table 12-1).

Mineralogical Changes

X-ray diffraction studies of *isotropic* palagonite show weak reflections at 3.0, 3.21 and 3.9 Å (Stokes, 1971; Furnes, 1978). Some of the "isotropic" palagonite shows peaks of poorly crystalline montmorillonite and, more rarely, mixed layer mica-montmorillonite (Hay and Iijima, 1968b). Illite was noted by Furnes (1978). Palagonitization in open marine systems appears to lead to nontronitic clays while in more closed systems, trioctahedral smectites are formed initially, followed by ferripotassic dioctahedral smectites at more advanced stages of palagonitization, associated with crystallization of phillipsite (Noack, 1981).

Eggleton and Keller (1982) using high resolution transmission electron microscopy noted that palagonite forming from basanite glass by hydration consists of spherical structures with diameters ranging from 200 Å to 600 Å. Exfoliation of 10 Å (2:1) clay layers allows the development of thin (30–60 Å) crystals of dioctahedral smectite with significant Mg in the octahedral site, which ultimately forms a tangled network of submicron sized bent flakes. Eggleton and Keller (1982) noted the similarity of average palagonite to smectite when recalculated to a cation charge of $+22$ (Table 12-2). They found that most palagonites contain more Mg in octahedral and Al in tetrahedral sites than "average smectite" and that increasing crystallinity of palagonite causes a decrease in Ca content and in tetrahedral Al.

Basaltic, nearly isotropic brown to tan "glass" that lacks textural evidence of having gone through gel- or fibro-palagonite, occurs in some pyroclastic rocks that have been subjected to more advanced diagenesis and burial metamorphism. These altered glass fragments show strong 14 Å X-ray peaks, splitting into a 14 Å and 16 Å peaks after glycolation, indicating complex mixtures of chlorite and interlayered chlorite and montmorillonite (Surdam, 1973). Chlorite is the principal "end product" of low grade alteration of palagonite and basaltic glass, and is stable over a wide range of conditions (Smith, 1967; Kuniyoshi and Liou, 1976; Viereck et al., 1982) (Fig. 12-3). Secondary minerals other than layer silicates can be grouped into zeolites, opal, carbonate, Fe-Mn oxides or carbonates, and, rarely, gypsum.

**Table 12-2.** Chemical analysis of palagonite (1) and structural formula of average dioctahedral smectite (3) and average palagonite (2) based on cation charge of +22 (Eggleton and Keller, 1982)

|  | 1 |  | 2 | 3 |
|---|---|---|---|---|
| $SiO_2$ | 50.0 | Si | 3.26 | 3.72 |
| $TiO_2$ | 3.0 | Al | 0.73 | 0.24 |
| $Al_2O_3$ | 14.7 | Ti | 0.15 | 0.02 |
| FeO | – | Al | 0.39 | 1.26 |
| $Fe_2O_3$ | 18.2 | $Fe^{3+}$ | 0.89 | 0.43 |
| MgO | 7.7 | Mg | 0.69 | 0.29 |
| CaO | 3.7 | Mg | 0.05 | 0.02 |
| $Na_2O$ | 1.5 | Ca | 0.26 | 0.09 |
| $K_2O$ | 1.3 | Na | 0.17 | 0.11 |
|  |  | K | 0.13 | 0.14 |

## Zeolites

Next to smectite, zeolites are the most common secondary minerals formed from volcanic glass. Some recent reviews on zeolites in volcaniclastic rocks are by Deffeyes (1959), Blatt et al. (1972), Stonecipher (1978) and Kastner (1979). Hay (1966) discusses the occurrence and origin of zeolites in some detail.

**Fig. 12-3.** Former basaltic (sideromelane) glass shard completely altered to chlorite. Miocene interbasalt tuff (Reydarfjordur, Iceland)

Zeolites are hydrated aluminosilicates of alkali and alkaline earth cations with framework structures. Their general formula is $X_u Y_r Z_n O_{2n} \times mH_2O$; X stands for Na and K, Y for Ca, Sr, Ba, and Mg, and Z for Al and Si. The ratio $(Si+Al):O$ is 1:2. Channels and large cavities are ubiquitous elements of these structures and loosely bound cations and water can be easily removed or replaced without destroying the aluminosilicate framework. Zeolites are characterized by reversible dehydration, ion exchange, catalysis and molecular sieve properties.

Zeolite species common in pyroclastic rocks are briefly discussed below. Some alkalic silicic zeolites such as clinoptilolite, are generally not associated with mafic glass and palagonite but with silicic glass precursors (Table 12-3).

*Phillipsite*, the most common zeolite in palagonite tuffs, is monoclinic and occurs chiefly in colorless to yellowish euhedral to subhedral prismatic crystals up to 250 μm long and in spherulitic aggregates. Its mean refractive index varies from 1.477 to 1.486. Rapid growth rates are indicated by numerous inclusions of Fe-oxyhydroxides and smectite. Etched crystal faces are common, indicating rapid dissolution, evidence that phillipsite like other zeolites are metastable reaction products. The composition of phillipsite is strongly controlled by the composition of both host rock and solution (Fig. 12-4). For example, the Si/Al-ratios in phillipsite in mafic igneous rocks varies from 1.1 to 2.4, those from marine sediments 2.3 to 2.8 and those formed from silicic glass shards in saline alkaline lakes 2.6 to 3.4. Harmotome, isostructural with phillipsite, contains Ba as a major cation. Ca is a significant cation in terrestrial phillipsite, but only occurs in small amounts in marine sediments and is higher in Ca-rich lavas than in Ca-poor pyroclastic host composition (Iijima and Harada, 1968; Brey and Schmincke, 1980) (Fig. 12-4). It is still lower in silicic tuffs and is almost absent in marine phillipsite. On the other hand, marine phillipsites are much richer in K than terrestrial phillipsites (Honnorez, 1978).

*Clinoptilolite* and *heulandite* are the high and low silica members of the heulandite group and must be heat-treated to enable their distinction by X-ray diffraction.

**Table 12-3.** Formulas of common zeolites (Deer et al., 1963)

(General formula: $(Na_2, K_2, Ca, Ba)[(Al, Si)O_2]_n \cdot XH_2O$)

| Group | Name | Formula |
|---|---|---|
| K ~ Na | Phillipsite | $(\tfrac{1}{2}Ca, Na, K)_3[Al_3Si_5O_{16}] \cdot 6H_2O$ |
| | Harmotome | $Ba(Al_2Si_6O_{16}) \cdot 6H_2O$ |
| | Erionite | $(Na_2, K_2, Ca, Mg)_{4.5}(Al_9Si_{27}O_{72}) \cdot 27H_2O$ |
| | Heulandite | $(Ca, Na_2)(Al_2Si_7O_{18}) \cdot 6H_2O$ |
| | Clinoptilolite | $(Na, K)_4 CaAl_6 Si_{30} O_{72} \cdot H_2O$ |
| Na-rich | Analcime | $NaAlSi_2O_6 \cdot H_2O$ |
| | Natrolite | $Na_2Al_2Si_3O_{10} \cdot 2H_2O$ |
| Na Ca | Chabazite | $Ca(Al_2Si_4O_{12}) \cdot 6H_2O$ |
| | Mordenite | $(Ca, Na_2, K_2)(Al_2Si_{10}O_{24}) \cdot 7H_2O$ |
| | Gmelinite | $(Na_2, Ca)(Al_2Si_4O_{12}) \cdot 6H_2O$ |
| Ca-rich | Wairakite | $CaAl_2Si_4O_{12} \cdot H_2O$ |
| | Stilbite | $(Ca, Na_2, K_2)(Al_2Si_7O_{18}) \cdot 7H_2O$ |
| | Laumontite | $Ca(Al_2Si_4O_{12}) \cdot 4H_2O$ |

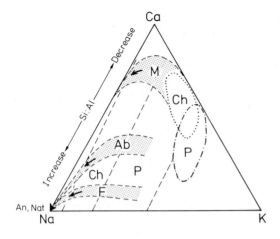

Fig. 12-4. Dependence of trend of chemical composition of zeolites on chemical composition of precursor glass. After Iijima and Harada (1968). *F* Felsic tuff; *Ab* alkali basalts; *M* melilite nephelinite. Main fields of chabazite and phillipsite between *dashed lines*. Chabazite (*dotted*) and phillipsite (*dash-dotted*) fields from basanitic to tephritic precursor glass (Brey and Schmincke, 1980). *An* analcime; *Nat* natrolite; *Ch* chabazite; *P* phillipsite

Clinoptilolite is by far the most common zeolite in marine sediments. It is monoclinic, has a mean refractive index of 1.484 and forms small euhedral platy crystals generally <45 µm long. Si/Al-ratios range from 4.2 to 5.2. In contrast to phillipsite, inclusions are rare, probably indicating slow growth rates. Moreover, unlike phillipsite, clinoptilolite is not always associated with volcanic detritus. The occurrence of clinoptilolite is discussed more fully in a later section on marine sediments.

*Analcime,* the Na-dominated zeolite, is isometric and occurs in discrete clear crystals up to 50 µm in diameter or as a late coating. Its refractive index varies from 1.489 to 1.494. The Si/Al ratios in analcime from nephelinite tuff are <2.0, but from siliceous tuffs range from 2.0 to 2.9. Analcime is the third most common zeolite, occurs in calcareous volcaniclastic sediments where it replaces earlier alkali zeolite, is commonly associated with basaltic glass and phillipsite, and is precipitated in cavities. The transformation of alkali zeolites into analcime is discussed at the end of this chapter.

The most common zeolites in palagonite tuffs are phillipsite, chabazite, analcime and members of the natrolite-gonnardite-thomsonite family (Table 12-2). Of these, the prismatic to spherulitic phillipsite and the rhomb-shaped chabazite are usually the first to be formed. The fibrous to fanlike natrolite-thomsonite zeolites also may form easily in pores and veins. Most zeolites vary widely in their chemical composition. The K:Ca:Na and Si:Al ratios and the general sequence, K-zeolites, Ca-Na-zeolites and Na-zeolites, depend upon the composition of the host rock (Iijima and Harada, 1968).

Opal and carbonate are generally late-stage cements in palagonite tuffs and may extensively replace primary and secondary components, although opal appears early in some cases (Jakobsson, 1972).

Chemical Changes

Following von Waltershausen (1845) and Bunsen's (1847) early work, Peacock (1926), in his classic study of palagonite from Iceland, showed that conversion of sideromelane to palagonite involves addition of about 18 to 30 weight percent

$H_2O$, oxidation of Fe, and loss of Ca and Na. Hoppe (1940) in a detailed analytical and experimental investigation confirmed the earlier results, and also noted that many elements are lost during the process, especially the alkalis, but also Ca, Mg and other elements, with Fe being relatively constant. Based on these studies, it was believed some 20 years ago that "the chemistry of the material (palagonite) and the process of alteration can be said to be fairly completely elucidated" (Jonsson, 1961).

Present ideas on chemical changes of major and minor elements during conversion of glass to palagonite are largely based on microprobe analyses of glass and juxtaposed palagonite (Hay and Iijima, 1968a,b; Jakobsson, 1972, 1978; Melson and Thompson, 1973; Andrews, 1977; Baragar et al., 1977; Brey and Schmincke, 1980; Staudigel and Hart, 1983). Changes in major and a few trace elements determined in separated glass and palagonite have been studied by Honnorez (1972), Furnes (1978, 1980), Furnes and El-Anbaawy (1980), Ailin-Pyzik and Sommer (1981) and Staudigel and Hart (1983). The data demonstrate that most elements are redistributed during palagonitization. Data are still sparse, however, and in part contradictory with respect to relative mobility of different elements.

The most drastic change from sideromelane to palagonite is a pronounced uptake of $H_2O$, generally 10 to 20%, this being the reason why palagonite was once thought to be hydrated glass. This hydration is accompanied by conversion of FeO to $Fe_2O_3$ and relative gains and losses of other elements. A positive correlation between the amount of $H_2O$ and the degree of alteration has been noted by several workers.

Several procedures have been employed to monitor the relative mobility, i.e. the losses or gains of elements. Some authors have assumed some elements to be constant, most commonly Ti or Fe, and have reported relative changes of other elements compared to "immobile" elements. Others have simply compared analyses of sideromelane and palagonite recalculated on a $H_2O$-free basis. Furnes (1978, 1980) following earlier suggestions (Hay and Iijima, 1968a) that palagonitization is an isovolumetric process calculated percent changes as $100 \cdot (A \cdot B - C \cdot D)/C \cdot D$ where A = density of palagonite, B = weight percent oxide of palagonite, C = density of sideromelane and D = weight percent oxide sideromelane.

In a recent study of element migration during submarine palagonitization, relative losses and gains of eight major elements are compared at different degrees of Ti accumulation, reflecting approximately increasing degrees of palagonitization (Staudigel and Hart, 1983) (Fig. 12-5). In Fig. 12-5 concentrations along the solid horizontal line ($SiO_2$, $Al_2O_3$, FeO) indicate passive accumulation similar to that of Ti. Only Fe behaves similarly to Ti but shows more scatter. The broken line in Fig. 12-5 indicates loss of an element corresponding to the measured gain in $TiO_2$. The network forming cations Si and Al approximately follow this line, although Al can be retained depending on the pH of the solution. MgO and $Na_2O$ are lost in "moderate" amounts, although Na appears to be removed much more effectively during alteration by meteoric waters.

Calcium is especially mobile but commonly is in part reprecipitated by late carbonates. In some cases, the amount of cement is proportional to the amount of palagonite, indicating that the cement is produced by palagonitization of the glass (Fig. 12-2). This is also shown by a similar bulk composition (except for Na and Ca) of fresh and thoroughly palagonitized tuff from the same bed (Hay and Iijima,

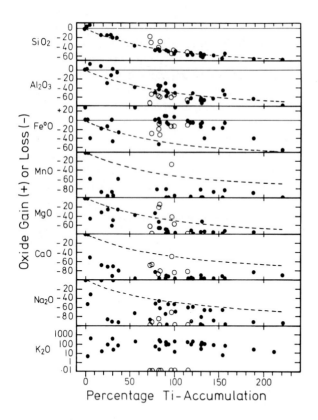

Fig. 12-5. Relative gain (+) and loss (−) of elements in % in various palagonites compared at different degrees of Ti-accumulation, increasing accumulation approximately reflecting increasing degrees of palagonitization. *Dots* are marine, *circles* terrestrial palagonites. (After Staudigel and Hart, 1983)

1968b). Ti may be fixed in palagonite in the form of poorly defined hydrates (leucoxene) while it occurs in sphene in more strongly altered rocks (Fig. 12-29).

It is now recognized that the entire volume of sea water of the oceans circulates through the oceanic crust once every 8 to 10 m.y. (Wolery and Sleep, 1976). Interaction of sea water with the volcanic rocks of the oceanic crust, especially the glass component, affects the composition of sea water, buffering it with respect to some elemental concentrations. Some elements released from the glass are reprecipitated at the sediment/water interface by oxidation of pore fluids. For example, Correns (1930), Hoppe (1940), Bonatti (1965), Moore (1966) and others made the important observation that Mn released from basaltic glass weathered on the sea floor could result in the formation of manganese crusts and nodules. Dissolved potassium is taken up from the sea water by palagonite and by authigenic clay minerals during low temperature alteration (Bloch and Bischoff, 1979). However, large amounts of potassium are released again during high temperature alteration in lower parts of the oceanic crust, a process confirmed by analysis of submarine hot springs. This illustrates a complex set of temperature-dependent redistribution processes which in the case of K may partially compensate each other.

Trace element mobility during palagonitization has been studied recently by Furnes (1978, 1980), Furnes and El-Anbaawy (1980), Ailin-Pyzik and Sommer (1981) and Staudigel and Hart (1983). A detailed study of Sr concentrations and $^{87}Sr/^{86}Sr$ ratios has demonstrated fluxes out of and into palagonite, the final concentration of Sr reflecting the total budget of these exchange reactions (Staudigel and Hart, 1983). Approximately 60% of the REE are removed during palagonitization but relative abundances stay constant due to the very low concentrations of these elements in sea water. Relative increase in light REE (Juteau et al., 1979) is explained by prolonged submarine weathering, i.e. very high water/rock ratios (Staudigel and Hart, 1983). Also lost during alteration are Zn, Cu, Ni, Mn, Sc, Co, and Hf and, to a lesser degree, Cr (Ailin-Pyzik and Sommer, 1981). These elements may be available for redistribution and reprecipitation in hydrothermal ore deposits.

Trace elements that are enriched in palagonite relative to sideromelane are Ba, Cs, Rb and U. Staudigel and Hart (1983) noted early uptake of alkalis on the surface of altered glass in proportions close to their ratio in sea water while at a later stage, perhaps during the transition of palagonite to smectite, Rb and Cs are taken up preferentially in the alteration product. This latter process will increase the K/Rb and K/Cs ratios in sea water.

## Process of Palagonite Formation

Most early workers interpreted palagonite as an alteration product of sideromelane, the quenched basaltic glass, with alteration occurring mostly at ambient temperatures but also around hot springs (von Waltershausen, 1845; Bunsen, 1851; Penck, 1879; Peacock, 1926; Noe-Nygaard, 1940; Jonsson, 1961). Following Fuller (1932), however, many workers believed that some palagonite forms at high temperature during and shortly after eruption of hot lava into the sea (Fuller, 1932; Wentworth, 1938; Nayudu, 1964a; Bonatti, 1965, 1967). Correns (1930) and Hoppe (1940), while emphasizing the importance of elevated temperatures (about 200 °C) of water vapor, also recognized the importance of sea floor weathering. Mathews (1962) assumed gel-palagonite to be formed at high and fibro-palagonite at low temperature. Moreover, palagonitization proceeds rapidly at temperatures of about 50°–100 °C, but much slower at temperatures <50 °C (Jakobsson, 1972, 1978) (Figs. 12-6, 12-7) on the young volcanic island of Surtsey, around a hydrothermal anomaly caused by near-surface intrusions. This palagonitization occurs in the presence of hydrothermal vapor which is of mixed sea water and meteoric water origin and which produces alteration to dense palagonite within one to two years. Such high rates of palagonitization are now known to be restricted to recently active near-vent areas and to sites of fumarolic and hot spring activity. Experimental studies clearly demonstrate a marked increase in the rate of palagonitization with temperature (Hoppe, 1940; Surdam, 1973; Furnes, 1975) (Fig. 12-8).

Most palagonitization occurs during low temperature weathering and diagenesis (Hay and Iijima, 1968a, b; Honnorez, 1972; Heiken, 1972; Furnes, 1974, 1975, 1978; Hekinian and Hoffert, 1975; Brey and Schmincke, 1980). This is indicated by two fundamental observations: One is that palagonite rinds on the glassy rim of submarine pillows systematically grow in thickness with time (Moore, 1966). The other is that sideromelane is increasingly depleted in many elements during palagonitization.

Changes in element concentrations can at present be interpreted only in a most general way. A better understanding of reaction processes requires finer analytical

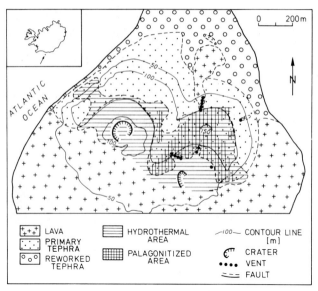

**Fig. 12-6.** Central part of volcanic island of Surtsey formed off Iceland in 1963–1967 showing areal extent of palagonitization in thermal area of central part of the island. (After Jakobsson, 1978)

tools, more precise sampling of different stages of the alteration process, as well as better discrimination of the variables noted above.

The influence of solution composition on co-existing palagonite is suggested by the contrasting concentration of potassium in terrestrially derived and marine palagonite. All studies of palagonitization under entirely or partly terrestrial conditions indicate a loss in K-concentrations, while submarine palagonitization universally shows drastic gains in K. This is interpreted to be the result of a higher concentration of K in sea water compared to fresh water but there are compli-

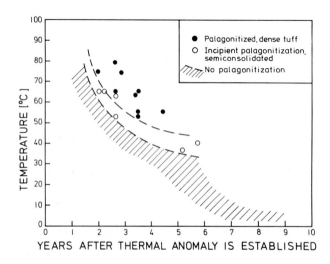

**Fig. 12-7.** Rate of palagonitization and consolidation in Surtsey tephra as function of temperature and time. (After Jakobsson, 1978)

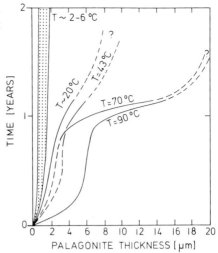

Fig. 12-8. Growth rate of palagonite at different experimental temperatures compared to growth rate at sea floor temperatures (*stippled* from Moore, 1966). (After Furnes, 1975)

cations. For one, K is adsorbed in surface layers in both environments initially (Jezek and Noble, 1978; Staudigel and Hart, 1983), although the terrestrial samples are rhyolitic obsidian which may preferentially accumulate K due to ion exchange (see below). Secondly, most terrestrial basalts studied have much higher initial K-concentrations than ocean floor tholeiites. This and the poorly understood difference in glass structure between tholeiitic and alkalic basalts suggests that K budgets are not simply controlled by a single environmental factor.

A second major aspect concerns the initial water/rock ratio during alteration and the change in this ratio during the process of alteration. That is, palagonitization will proceed differently when glass is exposed to sea water of constant composition for a long time, such as during prolonged submarine weathering, than when hyaloclastite is rapidly buried and the elemental flux out of the glass is increasingly restricted to the immediate environment of the glass. This leads to precipitation of secondary phases and drastic changes in the composition of the pore fluids, especially a change from acid to neutral. Both processes are highly interdependent.

Pore solution pH is a major factor in dissolving glass (Hoppe, 1940; Hay, 1963, 1966; Hay and Iijima, 1968a, b; Furnes, 1978, 1980; Furnes and El-Anbaawy, 1980). There are several lines of evidence for a change in pH with time. Furnes noted a change from depletion to enrichment for most trace elements during palagonitization and argued that trace elements such as REE are initially leached by acid solutions, but later absorbed on clay minerals under neutral and alkaline conditions. The increasing amounts of zeolites formed after initial alteration stages to palagonite suggest removal of Al from the glass, Al-solubility and zeolite formation being favored by pH > 9 (Hay and Iijima, 1968b). Furnes and El-Anbaawy (1980) interpreted a change in clay mineralogy from early kaolinite through mixed layer illite-montmorillonite to hisingerite and illite in palagonitized tuffs on Gran Canaria as reflecting a change in solution pH from mildly acidic to alkalic environments, with subsequent formation of zeolites and still later carbonate being favored by increasingly alkaline pore water compositions. Kaolinite formation does not require high acidity, however, only a low K/H activity ratio (Zielinski, personal communication). High leach ratios may be sufficient.

Experimental studies of the interaction of basalt and sea water are generally perfomed at temperatures exceeding 70 °C (see Seyfried and Bischoff, 1979 and literature therein) and thus result in the formation of higher temperature layer silicates rather than palagonite. However, they do demonstrate an effect of water/rock ratios on solution pH. High effective water/rock ratios during the initial stage of alteration resulted in precipitation of $Mg(OH)_2$ and accompanying acid pH, whereas a decreasing water flux and more effective pH buffering by glass hydrolysis and phases such as montmorillonite produced a rise in pH.

**Rate of Palagonitization**

The main factors controlling the rate of palagonitization are surface area of sideromelane particles, permeability and porosity, temperature, and chemistry of pore solutions.

Palagonitization at ambient surface temperatures develops within a few thousand years, as reported from studies of historically erupted tholeiitic tuffs on Iceland (Jonsson, 1961; Furnes, 1978), and nephelinitic tuffs at Oahu (Hay and Iijima, 1968a, b).

For low temperature palagonitization, Moore (1966), following Friedman and Smith (1960), used the rate

$$T = kt^{1/2}, \tag{1}$$

where T is the thickness of the surface layer (palagonite or hydrated obsidian), k is a constant depending on glass composition (480 to 2000 $\mu m^2/10^3$ yr for

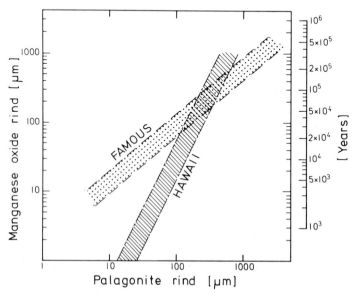

Fig. 12-9. Growth rate of palagonite rinds from Atlantic mid-ocean ridge tholeiites (FAMOUS area) and ocean island tholeiites from Hawaii (Moore, 1966). (After Hekinian and Hoffert, 1975)

sideromelane from Hawaii, and 0.36–14 µm²/10³ yr for obsidian) and t is the time since the process began. Moore suggested that the rate of palagonitization is diffusion-controlled and that palagonization proceeds faster in sea water than in fresh water (see also Hawkins and Rustum, 1963). However, Furnes (1975) found no difference in the rate of palagonitization in sea water $vs$ fresh water. Hekinian and Hoffert (1975) (Fig. 12-9) measured a rate of palagonite formation of 3 µm²/10³ yr in the Mid-Atlantic rift valley near 36°50'N. The rate formula for palagonitization ($R_p$), according to Morgenstein and Riley (1975) is

$$R_p = (N-2) Q/T, \qquad (2)$$

where N equals the number of palagonite bands, Q equals the thickness of each band and T equals time. The rate of hydration of basaltic glass increases with temperature. The fact that fresh sideromelane occurs in some pre-Tertiary rocks suggests that it can be sufficiently isolated from altering solutions or elevated temperatures (Muffler et al., 1969; Furnes, 1974; Schmincke et al., 1978; Schmincke and Pritchard, 1981).

## Alteration of Silicic Glass

The alteration of basaltic glass and that of silicic ($SiO_2 > 65$ wt.%) glass differs in many ways. Simply put, the alteration of sideromelane appears to proceed mostly in the solid state, and commonly involves the formation of the intermediate but chemically different product palagonite which is eventually transformed into thermodynamically stable crystalline phases, chiefly smectite. In contrast, alteration of silicic glass appears to involve an initial stage of diffusion-controlled hydration and alkali ion exchange but minor overall chemical changes. This is commonly followed by a stage of glass matrix destruction and precipitation of secondary phases in the pore space created by the dissolution of the glass. The reasons for these differences are poorly understood. Differences in the glass structure are probably important, the stronger resistance of the silicic glasses during the initial stages of alteration being due to much higher concentrations of the network-forming elements Si and Al. The higher initial alkali content of silicic glasses and their early release into pore solution will generate high pH conditions under relatively closed system conditions. This leads to rapid dissolution of $SiO_2$ at pH > 9. Experimental alteration studies under elevated temperatures indicate greater solution and reactivity of basaltic compared to silicic compositions (Hawkins and Rustum, 1963; Khitarov et al., 1970).

### Hydration and Ion Exchange

Perlite is hydrated obsidian. It is characterized by curved concentric "perlitic" cracks that give an onion-skin appearance and result from hydration and thus expansion of the glass (Friedman and Smith, 1958) and not from cooling. Refractive indices of perlite decrease with increasing $H_2O$-content (Ross and Smith, 1955) (Fig. 12-10). The boundary between nonhydrated obsidian and perlite is generally sharp. Perlite is also distinguished from coexisting obsidian by its characteristic strain birefringence (Friedman and Smith, 1960).

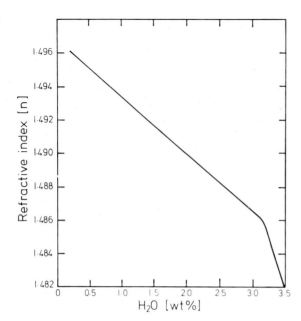

Fig. 12-10. Relationship between water content and refractive indices in rhyolitic perlite determined by successively removing the $H_2O$ through heating. (After Ross and Smith, 1955)

Ross and Smith (1955) pointed out that hydrated obsidians (perlite) differ chemically from their nonhydrated equivalents and cautioned against petrogenetic interpretations based on analyses of perlite, the common form of "glass" in older rocks where glasses are preserved. Numerous subsequent studies have confirmed and expanded these fundamental observations (Friedman and Smith, 1958, 1960; Friedman et al., 1966; Lipman, 1965; Noble, 1967, 1968; Truesdell, 1966; Rosholt et al., 1971; Steen-McIntyre, 1975; Zielinski et al., 1977; Jezek and Noble, 1978).

Obsidian takes up $H_2O$ to a saturation value of about 3%. This water appears to be present in the glass as $H_2O$-molecules rather than $H^+$ and/or $OH^-$ ions because accompanying ion exchange is limited. The most notable chemical change is a decrease in Na and an increase in K by ion exchange with groundwater, with potassium also being depleted during more advanced alteration. Fe, which does not change much in concentration, is strongly oxidized during hydration. Si can also be lost to a minor degree. Volatile components such as $Cl_2$ and $F_2$ may be lost from the glass during hydration while U and Th are relatively immobile. Ion exchange is more intense, however, along the fine cracks of the perlite (Jezek and Noble, 1978).

The reaction by which water hydrates natural obsidian has recently been studied by measuring hydrogen concentration profiles using the $^{15}N$ resonance nuclear reaction method (Laursen and Lanford, 1978). It appears that the hydration involves the interdiffusion of hydronium ions ($H_3O^+$) with mobile alkali ions in the glassy phase of the obsidian following the formation of hydronium ions at the surface (Doremus, 1975). This ionic interdiffusion model also explains the drastic changes in alkali contents of natural glasses because in obsidian the only major mobile charge carriers are the monovalent alkali ions $Na^+$ and $K^+$. Exchange of these alkali ions with positive hydronium ($H_3O^+$) ions proceeds according to the following equation:

$$2 H_2O + Na^+ \text{ (glass)} \rightarrow H_3O^+ \text{ (glass)} + NaOH. \tag{3}$$

Only part of the alkali ions take part in this reaction, however, possibly because of incipient crystallization or devitrification that is not detectable by optical methods. Tsong et al., (1978), studying chemical variation in altered surfaces of obsidian by sputter-induced optical emission, showed constant concentrations for network-forming cations such as Si and Al along depth profiles but strongly varying concentrations for network modifying elements (K, Na, Li, Mg). The deepest penetration was noted for H, indicating that alteration is initiated by hydration.

The hydration rate of obsidian has been developed as a powerful dating method in archeology (Friedman and Smith, 1960; Friedman et al., 1966; Friedman and Long, 1976; Friedman and Trembour, 1978). Hydration is considered a two-stage process: rapid adsorption of water on the surface followed by slow diffusion of water into the obsidian. Hydration depends exponentially on both time and temperature as well as on chemical composition, hydration rates increasing with temperatures but decreasing with decreasing silica and increasing CaO and MgO in the range of silicic glasses ($> 72$ wt.% $SiO_2$) studied (Friedman and Long, 1976). A monolayer of $H_2O$ is sufficient to start hydration (Friedman and Trembour, 1978).

The proposed encapsulation of radioactive waste into synthetic glass for long-term underground storage requires assessment of current knowledge on the durability of natural glasses in geologic environments. A succinct discussion of important factors is given by Zielinski (1980), with considerable data derived from studies of altered tephra. While breakdown of glass in the absence of water is extremely slow, contact between glass and hot aqueous solutions greatly accelerates both hydration and dissolution rates. At low temperatures, important factors in glass destruction are solution pH (highly acid or alkaline), solid surface area and composition, solution Eh, salinity and the presence of complexing agents.

In continental settings, alteration processes depend very much on dynamic changes in the glass-water system (Zielinski, 1982). Alkaline solutions are generated during glass hydrolysis and dissolution by exchanges of $H_3O^+$ for $Na^+$ and $K^+$ ions as discussed above. If the rate of dilution of these solutions by fresh water is low, redistribution of leached elements such as U is on a short range as discussed below for saline alkaline lake environments. However, in well-flushed fluvial sediments and in lake sediments bordering highland areas, interstitial solutions are frequently replenished with fresh oxygenated water, which is under-saturated with respect to glass components. Leaching of glass is rapid and distant transport of leached elements such as uranium is more probable. Clay minerals are the dominant alteration product under these relatively open system conditions. Because of its economic importance, the leaching of U from silicic tephra and its redistribution has recently been studied in detail (Zielinski, 1982; Zielinski et al., 1980; Goodell and Waters, 1981).

**Advanced Stages of Alteration**

The process by which perlite transforms to a rock containing significant amounts of diagenetic mineral phases is poorly known. Boundaries between volcaniclastic

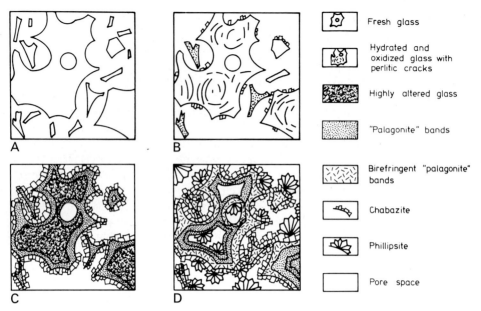

**Fig. 12-11 A–D.** Sequence of three alteration stages (**B** to **D**) of tephritic phonolite glass from the Pliocene Roque Nublo Formation (Gran Canaria, Canary Islands). (After Brey and Schmincke, 1980)

rocks consisting essentially of "fresh" (hydrated) glass and those dominated by zeolites and/or montmorillonite are sometimes sharp, the transition occurring within a few centimeters to decimeters. Hay (1963) recognized three principal but overlapping stages within the transition zone: (a) formation of clay (commonly montmorillonite), perhaps representing the outer leached skin of shards or the more intensely altered fractures in perlite noted by Jezek and Noble (1978), (b) partial to complete dissolution of glass shards and (c) precipitation of authigenic minerals, especially zeolites, in the new cavities as well as original pore space. A succession of alteration stages in glass of intermediate alkaline composition is shown in Fig. 12-11.

**Saline Alkaline Lake Environment**

Many of the economically significant zeolite deposits are produced from the alteration of silicic tephra in saline alkaline lakes. Modern studies of these deposits include Deffeyes (1959), Hay (1966), Sheppard (1969, 1971, 1973), Sheppard and Gude (1968) and Surdam (1972). The following section is mostly based on the excellent review by Surdam and Sheppard (1978).

Saline alkaline lakes form in closed hydrographic basins in areas where evaporation exceeds precipitation. Many are in rift zones or in areas of block faulting, a type of tectonic deformation commonly accompanied by volcanism (Fig. 12-12). Glassy fallout tephra deposited in a playa-lake complex or a rift system where influx of clastic debris by fluvial processes is minimal provides the setting and ideal parent material for accumulation of rather pure zeolite deposits.

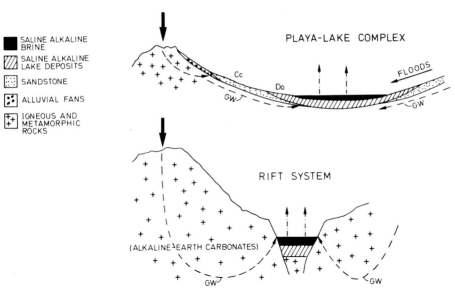

Fig. 12-12. Hydrology and brine evolution in playa-lake complex and rift system. *Solid arrows* rain water, *GW* ground water circulation path, *Cc* calcite precipitation, *Do* dolomite precipitation; *vertical dashed arrows* indicate evaporation. (After Eugster and Surdam, 1973)

In a playa-lake complex, saline and alkaline brines develop gradually. Alkalinity is defined as

$$[HCO_3^-]_{total} + 2[CO_3^=]_{total} + [OH^-] - [H^+]$$

(Stumm and Morgan, 1981, p. 130). Groundwater which is $CO_2$-charged becomes soft by reacting with igneous and metamorphic country rocks. After it emerges at the foot of an alluvial fan, calcite forms in the capillary zone during evaporation of the groundwater flowing at shallow levels. pH and the Mg:Ca ratio increase as the groundwater moves toward the center of the lake; trona, a sodium carbonate, forms as the end result of extreme evaporation. In a rift system, groundwater circulation is deep, Ca and Mg have precipitated at depth and lakes are fed by alkaline hot springs. Two main evolutionary paths of water chemistry are distinguished (Fig. 12-13). If alkalinity/$2(Ca^{+2} + Mg^{+2}) > 1$, the solution will be depleted in $Ca^{+2}$ by calcite precipitation and with further evaporation be enriched in $CO_3^{-2}$, resulting in a saline and alkaline brine. If alkalinity/$2(Ca^{+2} + Mg^{+2}) < 1$, the solution will be depleted in $CO_3^{-2}$ by calcite precipitation. Enrichment of $Ca^{+2}$ due to further evaporation will lead to saline but not alkaline brines.

Lateral zonation in the distribution of authigenic minerals is characteristic of saline lake deposits and is well illustrated by the Pleistocene Lake Tecopa deposits (Fig. 12-14) (Chap. 13). Fresh glass along the margins of the lake can be traced basinward to a zone of zeolites (in this case chiefly phillipsite, erionite, and clinoptilolite). In the central part of the lake basin, potassium feldspar is deposited. In many similar cases of lateral zonation of mineral facies a zone of analcime separates the other zeolites from the zone of potassium feldspar (Fig. 12-15).

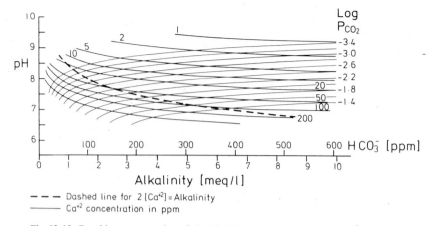

– – – Dashed line for $2[Ca^{+2}]$ = Alkalinity
——— $Ca^{+2}$ concentration in ppm

Fig. 12-13. Graphic representation of chemical divide in brine evolution. $Ca^{2+}$ balances alkalinity *along dashed line*. Water, upon further concentration, will precipitate calcite and evolve into an alkaline brine above and into a saline but not alkaline brine *below dashed line*. (After Drever, 1972)

The mechanism by which zeolites form from volcanic glass is poorly understood. In the presence of saline and/or alkaline brines, erionite can form directly from trachytic glass by the addition of only $H_2O$ as in the Lake Magadi region (Surdam and Eugster, 1976). An alumino-silicate gel may first precipitate from the solution of the glass, and then zeolites grow from the gel (Mariner and Surdam, 1970). Deffeyes (1959) and Hay (1963, 1966), however, emphasized solution of the glass shards and subsequent precipitation of zeolites from the solution (Fig. 12-16).

The important compositional solution parameters in zeolite genesis are alkaline or alkaline earth cation/$H^+$ ratios, and $SiO_2$ and $H_2O$-activity. The rate of solution

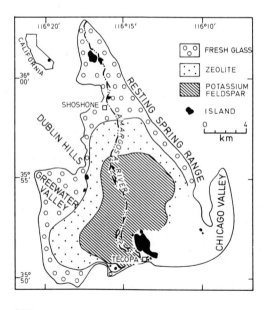

Fig. 12-14. Map of ancient Lake Tecopa near Shoshone (California) showing concentric facies changes from the margin toward the center of the lake. (After Sheppard and Gude, 1968)

Fig. 12-15. Reaction sequence of lateral mineral changes typical of saline, alkaline lake deposits. (After Sheppard and Gude, 1973)

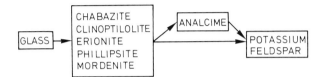

of glass increases with both salinity and alkalinity (Hay, 1966; Mariner, 1971). Boles and Surdam (1979) infer that diagenesis of volcaniclastic sediments in the Tertiary Wagon Bed Formation (Wyoming) occurred in moderately saline but not highly alkaline pore fluids (pH probably < 9), as evidenced by the absence of saline minerals such as sodium carbonates and the presence of zeolites such as clinoptilolite and heulandite.

## Marine Environment

The occurrence of zeolites such as phillipsite on the sea floor and their association with basaltic materials, first recognized by Murray and Renard (1891), has been repeatedly emphasized by students of marine geology (Bramlette and Posnjak, 1933; Nayudu, 1962, 1964a; Arrhenius, 1963; Bonatti, 1965). In recent years, the Glomar Challenger has recovered tens of thousands of meters of cores from Recent to Mesozoic rocks which allow a more systematic study of diagenetic alteration of

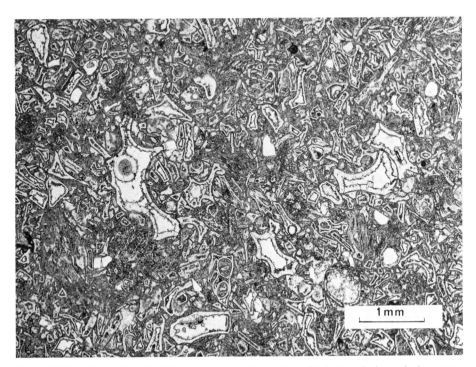

Fig. 12-16. Rhyolitic zeolitized tuff in which glass shards have been filled along the inner rim by a coating of clinoptilolite subsequent to their dissolution. Miocene Obispo Formation (California, USA)

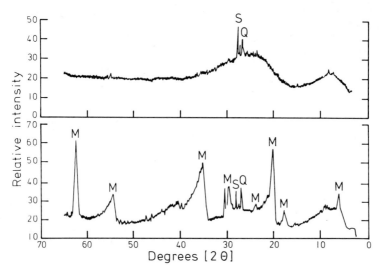

Fig. 12-17. X-ray diffractogram of fresh Miocene volcanic glass shards with minor quartz (Q) and sanidine (S) phenocrysts and of some ash altered to clay (M montmorillonite). (After Zielinski, 1982)

volcanic material to begin. Many studies have concentrated on the identification of diagenetic minerals (especially zeolites), their distribution in time and space and their relationship to parent material which commonly, although not exclusively, is volcanic glass. These data have been recently reviewed by Iijima (1978), Kastner and Stonecipher (1978), Stonecipher (1978), and Boles and Wise (1978). Other investigations have been concerned with the identification of ash layers and dispersed volcanic components and their alteration to diagenetic minerals such as smectite (e.g. Hein and Scholl, 1978).

The most common types of diagenetic silicates formed from volcanic precursors (Fig. 12-17) are zeolites and clay minerals. Authigenic feldspars are much less commonly observed (Kastner and Siever, 1979). The importance of volcanogenic source rocks for the authigenic silica in marine sediments is uncertain. Phillipsite occurs in many types of deep sea sediments, especially in the Pacific. It occurs preferentially in areas of low sedimentation rate with the following mineralogical associations: Clay-rich > clay-volcanic > calcareous > siliceous (Stonecipher, 1978). Its derivation from basaltic parent material is commonly assumed because of its association with sideromelane and/or palagonite and smectite. It may form from volcanogenic smectite or its precursor of palagonite, with some additional dissolved silica. Phillipsite may also form from more silicic parent material in deep sea environments.

Clinoptilolite occurs in many types of sediments. Unlike phillipsite, it is more abundant in more carbonate-rich sediments in the Atlantic and shows the following preference in mineralogical associations: Calcareous > clayey > calcareous-siliceous (opal-CT) > volcanic > siliceous (opal-A). Thus, the higher sedimentation rate appears to favor the formation of clinoptilolite.

The formation of clinoptilolite is more difficult to analyze than phillipsite because crystals are small, as are the most common precursors, silicic glass shards.

**Fig. 12-18.** Precursor materials of diagenetic phillipsite and clinoptilolite in deep sea sediments. (After Iijima, 1978)

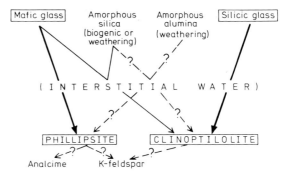

Clinoptilolite is commonly associated with siliceous sediments. Thus, it may form either via the dissolution of volcanic glass (or phillipsite) or via diagenesis of biogenic siliceous skeletons (Berger and von Rad, 1972; Boles and Wise, 1978; Kastner and Stonecipher, 1978) (Fig. 12-18).

The contrasting distribution of clinoptilolite and phillipsite in Tertiary-Mesozoic sediment columns provides important evidence for the conditions of their formation (Boles and Wise, 1978; Kastner and Stonecipher, 1978): Phillipsite occurs in the upper 500 m of sediments younger than Eocene, while clinoptilolite is most abundant in Eocene and Cretaceous sediments (Fig. 12-19). Assuming conservation of alumina: phillipsite + dissolved silica (mainly biogenic) + $H_2O \rightarrow$ clinoptilolite. Thus, phillipsite may be a metastable, silica-deficient phase in marine pore fluids, which is eventually replaced by clinoptilolite. Both phillipsite (Si/Al-ratios between 2.5 and 2.8) and clinoptilolite (Si/Al-ratios 4.5-5.0) show a more restricted composition in marine than in other environments, reflecting the more restricted pore fluid composition (Kastner and Stonecipher, 1978).

Analcime is much less common than either phillipsite or clinoptilolite in marine sediments. The mode of its formation is poorly understood although basaltic parent materials ± zeolite intermediates are generally assumed. Other observed zeolites are chabazite, erionite, gmelinite, harmotome, mordenite, laumontite, thomsonite, thaumasite, and natrolite, but they are all rare. Authigenic K-feldspar of

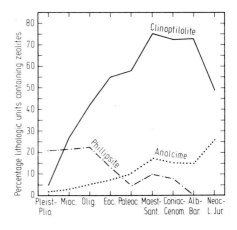

**Fig. 12-19.** Relationship between relative abundance of the three main marine sedimentary zeolites clinoptilolite, phillipsite, and analcime and age showing the decrease in phillipsite in pre-Tertiary and of clinoptilolite in pre-Cretaceous rocks. (After Kastner, 1979)

low temperature origin has been found in association with altered volcaniclastic sediments from both the Atlantic and Pacific Ocean basins (e.g. Kastner and Gieskes, 1976).

Although most clay minerals within marine sediments are detrital, smectites appear to be predominantly formed *in situ*, as shown by their occurrence as delicate rims in detrital grains in volcaniclastic rocks, association with authigenic zeolites and a Ce-anomaly typical of sea water (Piper, 1974). While Fe-montmorillonite appears to form from basaltic glass, smectite with higher Mg and Al appears to form from nonbasaltic volcanic glass (Kastner, 1976; Kastner and Gieskes, 1976). Jeans et al. (1982) interpret widespread Cretaceous glauconites in England and Ireland as argillized basaltic pyroclasts, and smectite-rich clays to be derived from silicic alkaline ash.

The alteration of volcanic glass significantly influences the pore water chemistry of marine sediments as shown by numerous studies of Glomar Challenger cores. Pore fluids from sediments (Layer 1) overlying older oceanic crust commonly show increases in $Ca^{2+}$, decreases in $Mg^{2+}$ and $K^+$ and decreases in $^{18}O$ concentrations (see summary in Gieskes and Lawrence, 1981). These gradients are believed to be due to uptake of $^{18}O$, $Mg^{2+}$ and $K^+$ and release of $Ca^{2+}$ during low temperature alteration of volcanic glass and its transformation into smectite and zeolites. Although most of these reactions occur in the extrusive basalts of Layer 2 – part of which is hyaloclastite – volcanic ash in the sediment column is also a chemical source and a sink. For example, assuming that the sediments of Layer 1 contain 10 volume percent volcanic ash, and upon subduction are 60 m.y. old and 400 m thick, then about 16 percent of the $^{18}O$ flux into the oceanic sediments is due to ash alteration. The remaining 84 percent comes from the underlying basalts (Gieskes and Lawrence, 1981).

Relatively pure ash layers that are completely altered to smectite or, in some cases, to kaolinite, are called bentonites and tonsteins respectively and have been the subject of many special studies.

## Bentonites and Tonsteins

Thin, laterally persistent clay-rich layers interbedded with marls, shales and especially coal-bearing strata are generally called *tonsteins* (German for "claystones") in the European literature following Bischof (1847-55) (see Williamson, 1970; Burger, 1980). The term *bentonite* was first applied to clay-rich rock near Fort Benton (Wyoming) (Knight, 1898), later recognized as altered volcanic ash by Hewitt (1917).

There is some argument concerning usage of the terms tonstein and bentonite. Bentonite was used originally for deposits of smectite-dominated expanding clays having colloidal properties, mostly, but not necessarily, derived from tephra deposits (Grim and Güven, 1978; Gary et al., 1974; Hein and Scholl, 1978). This definition is based on the clay mineral species and not on the precursor material, and this is the way the term bentonite is used in industry. However, for many decades the term bentonite has also been used to describe thin, widespread, clay-altered ash layers that are commonly but not necessarily dominated by smectite (e.g. Weaver, 1963; Ross, 1955; Smith, 1967) (Fig. 12-20). Bohor et al. (1979, p. 16) use bentonite in this general sense to include all altered volcanic ashes that "fulfill the

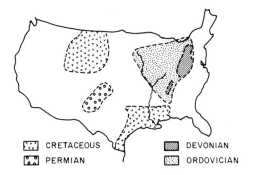

**Fig. 12-20.** Areal distribution of Mesozoic and Paleozoic bentonites in North America. (After Ross, 1955)

requirements for an isochronous, datable marker horizon useful for correlation over broad areas." Roen and Hosterman (1982), working on Devonian ash-derived clay-rich beds in the Appalachian basin, have recently advocated restriction of the term bentonite to layers consisting essentially of smectite regardless of the precursor material and to use the term ash beds for those older layers consisting largely of nonsmectite clays such as kaolinite, illite, and mixed layer minerals (Fig. 12-21).

The term ash bed, however, is in general use for fallout beds containing recognizable shards. Moreover, the term bentonite has become so firmly entrenched as a general term for thin, laterally widespread clay-rich beds that are of probable volcanic origin, that restriction to the type of clay mineral species seems less practical. We therefore use the term bentonite without connotation of geologic age or composition of the clay mineral species.

The term tonstein is also used in different ways. Strictly speaking, it refers to kaolinite-dominated, thin claystone beds in coal measures. However, the clay mineral species changes to smectite or illite when the tonstein layer passes from the coal

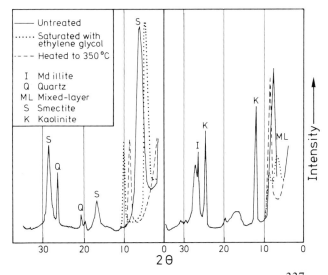

**Fig. 12-21.** X-ray diffractogram traces of bentonite from the type locality (Wyoming) (*left*) and Devonian altered ash bed ("bentonite") (Appalachian Basin) (*right*). (After Roen and Hosterman, 1982)

337

measure into the country rock. To circumvent nomenclatorial problems, we suggest that the term tonstein could be eliminated and that bentonite could be used in a broad sense to be specified once the dominant clay mineral species has been identified as follows: *smectite(s)-bentonite, illite(i)-bentonite* or *kaolinite (k)-bentonite*. The term K-bentonite has previously been used for illite-rich bentonites (Weaver, 1953).

There is firm evidence that most bentonites and tonsteins result from alteration of vitric fallout ash. The main lines of evidence are:

a) Thinness (generally < 10 cm, but with some major exceptions) combined with wide lateral extents exceeding tens to sometimes hundreds of kilometers. Such beds commonly have sharp upper and lower contacts with the enclosing marine or nonmarine sedimentary rocks. Pyroclastic fallout seems the most reasonable geologic mechanism for producing the observed distribution.
b) Vitroclastic textures have been found in some tonsteins and in most rocks called bentonites. Some bentonites are interlayered with unaltered or less altered vitric tuffs. A lateral gradation from unaltered tuff to bentonite also has been recorded from several localities.
c) Euhedral and/or high-temperature mineral phases such as sanidine, high-temperature quartz, biotite, zircon, allanite, apatite, sphene, rutile, and others are common and may dominate the relic mineralogy of these claystones. These volcanic minerals differ in composition and morphology from the abraded detrital mineral suite in adjacent sediments.
d) The chemical composition, although strongly affected by alteration, contains elemental abundances and ratios of relatively immobile elements that are more characteristic of vitric tuff than of nonvolcanic sediments.

Most bentonites are interlayered with shallow marine sediments, but they also occur in terrestrial sequences. Smectites comprising members of the montmorillonite-beidellite series are the dominant mineral phase in geologically young bentonites, Na and Ca being the most common extractable interlayer cations. With progressive alteration (diagenesis), chlorites and illites increase in abundance relative to smectites and mixed-layer illite/smectite. Cristobalite and tridymite occur in some bentonites. Chemically, most bentonites consist of 55 to 75% $SiO_2$, 15–25% $Al_2O_3$, up to about 5% $Fe_2O_3$ and 3% MgO, $K_2O$ and $Na_2O$ generally <1% with $K_2O > Na_2O$. Alteration of ash to bentonite is generally thought to occur during early diagenesis.

The original composition of altered glass is hard to determine (Fig. 12-22). Where coexisting glass survives, rhyolitic to dacitic compositions are most common. High silica content (>70%) does not appear to favor rapid bentonite formation, whereas a moderate MgO-content (5 – 10%) does (Grim and Güven, 1978, p. 128). It appears that alteration of silicic ash to bentonite requires significant incorporation of MgO, apparently from sea water; leaching removes most of the alkalis, calcium, and also some $SiO_2$ which may be redeposited within or adjacent to bentonite deposits. Smectite, as discussed earlier, is the main alteration product of basaltic glass.

Kaolinite-rich bentonites ("tonsteins") have been studied in some detail in the Ruhr coal measures (Central Europe) where they generally occur as 0.5–2 cm thick

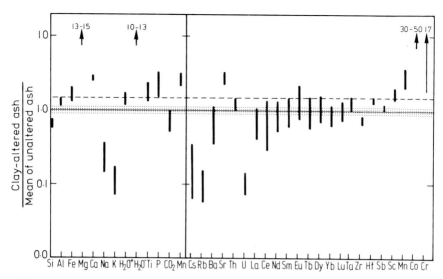

Fig. 12-22. Major and trace element concentrations in Miocene rhyolitic clay-altered ash normalized to concentration in fresh glass. Values that plot outside the *stippled zone* are regarded as analytically significant. The *horizontal dashed line* represents the postulated limit of enrichment produced by concentration of immobile elements in an insoluble residuum. (After Zielinski, 1982)

layers (rarely up to 20 cm). The main mineral component, kaolinite, occurs as several textural varieties (Burger, 1980; Williamson, 1970). Phosphates, pyrite, and siderite occur in most tonsteins in addition to several igneous minerals listed earlier (e.g. quartz, sanidine [rare], zircon). Chemically, nearly all Ruhr tonsteins contain less than 50% $SiO_2$, 20–40% $Al_2O_3$, <1% MgO, 0.5–4% $Fe_2O_3$, <0.5% $Na_2O$, 0.2–2% $K_2O$ and highly variable amounts of CaO and $P_2O_5$ owing to local concentrations of phosphates. Tonsteins and bentonites generally contain higher concentrations of Y, Zr, U, Th, Rb than of Cr and Ni and thus differ strongly from other clay-rich sediments.

Recently, attempts have been made to chemically identify the parent material of tonsteins. Among the major elements, $TiO_2$ and the $TiO_2/Al_2O_3$ ratio have been used to try to distinguish tonsteins derived from silicic and mafic tephra, and from normal coal measure sediments. Ratios ranging from about 0.14 to 0.18 are thought to reflect mafic, and ratios <0.02 silicic igneous compositions (Price and Duff, 1969; Spears and Kanaris-Sotiriou, 1979). However, the $TiO_2/Al_2O_3$ ratio may change during alteration of ash. Published chemical data for tonsteins show clearly that the $Al_2O_3$ content has increased by a factor of 1.5 to 2 over that of mafic to silicic volcanic rocks, while $TiO_2$ values are more variable. For example, in one of the most detailed discussions of the chemistry of tonsteins, Spears and Kanaris-Sotiriou (1979) list four tonstein analyses (3–6 in their Table 1), with $TiO_2$ ranging from 1.73–4.15, that they consider to have mafic precursors. For other tonsteins with mafic sources $TiO_2$ is invariably less than 1%. Apart from uncertainties introduced by mixing of mafic and silicic tephra and nonvolcanic debris as well as by alteration, major differences in primary $TiO_2$ concentrations also must be considered. For example, alkali basalts generally have $TiO_2$ concentrations of more than 2.5%, mid-ocean ridge basalts 1–1.5%, and many arc type basalts <1%, whereas the alumina contents in these major basalt groups are similar.

In an attempt to identify the igneous precursor of European tonsteins, Spears and Kanaris-Sotiriou (1979, Table 3) used Ti, Cr, Ni, and Zr. Some problems arise if diagnostic trace elements are highly concentrated in trace phases. For example, Zr is highly concentrated in zircon, and contamination of ash by unrecognized detrital zircon can compromise conclusions based on bulk chemical data (Schmincke et al., 1982). Multi-elemental analyses of fresh glass and adjacent completely altered ash (Fig. 12-22) are needed in order to better define the chemical changes accompanying tonstein formation.

The origin of kaolinite-rich bentonites has been the subject of lively debate for decades. Most authors favor an origin by alteration of tephra layers in the acid environment typical of coal swamps. One theory calls for solution of the glass, possible formation of a gel, and crystallization of the kaolinite from the gel. Low salinity also appears to be a major factor leading to alteration of vitric ash to kaolinite. Complete alteration of ash to clay is aided by thinness of the ash layer and slow burial (Spears and Rice, 1973). This would keep the environment within the kaolinite stability field. Local availability of organic matter and the common occurrence of sulfides in the rocks are evidence for low Eh values during diagenesis.

The importance of tonsteins lies in their use as stratigraphic marker beds (Chap. 13), especially within coal-bearing sequences that show rapid facies changes. Correlation of individual bentonite and tonstein beds is greatly facilitated by distinct mineral suites and trace element compositions (Weaver, 1963; Slaughter and Early, 1965; Smith, 1967; Winter, 1981).

**Burial Diagenesis and Metamorphism**

While Glomar Challenger cores have provided excellent material to study early diagenetic changes in volcaniclastic sequences, these drill holes rarely penetrate more than 1000 m. In contrast, observations of tectonically uplifted regions allow study of diagenetic alterations at depths which exceed 10,000 m. Assuming average geothermal gradients of 30 °C/1000 m depth, it is clear that the transition from diagenesis to metamorphism (temperatures exceeding about 200 °C) should have occurred within these thick sequences. Lithostatic load pressure appears to be less important than temperature, pore fluid, and initial glass and mineral composition during early and low grade diagenesis, but becomes more important when "truly" metamorphic conditions are approached. Following the classic study of Coombs (1954), who introduced the concept of the zeolite facies, most work has been done in andesitic to rhyolitic volcaniclastic series in New Zealand (Coombs, 1954; Boles, 1974; Boles and Coombs, 1975, 1977); Chile (Levi, 1970), USA (Dickinson, 1962; Sheppard and Gude, 1973) and Japan (Utada, 1970, 1971; Iijima and Utada, 1972; Iijima, 1978). The transition from diagenesis to metamorphism has also been studied in volcaniclastic rocks of basaltic (Surdam, 1973; Kuniyoshi and Liou, 1976) and mixed basaltic and rhyolitic composition (Viereck et al., 1982).

These studies demonstrate a wide range of PT stability for some authigenic minerals. For example, laumontite, heulandite and albite occur throughout the 10.4-km-thick sequence of Mesozoic sediments in the Southland syncline (New Zealand) (Fig. 12-23). The effect of parent composition on alteration assemblage

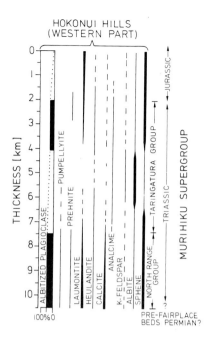

Fig. 12-23. Mineral distribution in thick sequence of volcaniclastic sandstones of Murihiku Supergroup (New Zealand) developed during burial diagenesis and metamorphism. (After Boles and Coombs, 1977)

is shown by the association of laumontite, heulandite, prehnite, pumpellyite, calcite, interlayered chloritic minerals, and sphene with andesitic clasts and of quartz, albite, and K-feldspar with dacitic to rhyolitic clasts (Boles and Coombs, 1977). Other factors of major importance in burial diagenesis are incomplete reactions, permeability, ionic activity in pore fluids, $P_{CO_2}$, contrast of $P_{fluid}$ and $P_{total}$ in sediments and fractures and the effect of rising temperature following burial (Boles and Coombs, 1977).

Burial diagenesis of thick volcaniclastic rock series in Japan, studied in detail by Utada and Iijima, is reviewed by Iijima (1978). A downward succession of four zones is distinguished, each dominated by specific mineral assemblages of the reaction series silicic glass – alkali zeolites – albite, representing dehydration with depth (Figs. 12-24, 12-25). Silicic glass, partly altered to montmorillonite and opal-A or opal-CT, is characteristic of zone I. In zone II, reaction of silicic glass with interstitial water has led to the formation of the alkali zeolites clinoptilolite and mordenite, opal-CT and montmorillonite. These alkali zeolites are transformed into analcime in zone III, where two subzones are distinguished: IIIa where clinoptilolite changes partly to heulandite and IIIb where heulandite is transformed into laumontite. In zone IV, analcime is transformed into albite. Relics of precursor zeolites persist in the succeeding zones.

Some of these zonal successions were studied in deep boreholes. In these, the transition between the zones occurs over a rather restricted temperature interval, regardless of $P_{total}$ and $P_{H_2O}$. Thus, in four deep oil field wells, the boundary zone II/III occurs at 84°–91 °C at depths from 1700–3500 m and that between zone III and IV at 120°–124 °C at 2500–4500 m depth. The importance of temperature and the independence of the reaction

$$\text{analcime} + \text{quartz} = \text{albite} + H_2O$$

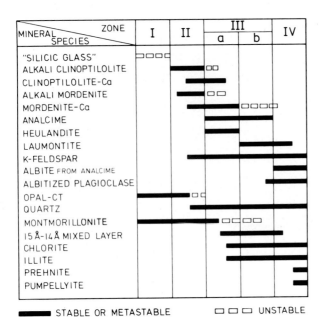

Fig. 12-24. Zones of authigenic zeolites and other silicates developed during burial diagenesis in a thick section of marine silicic volcaniclastics. Zone IV grades into metamorphic regime. (After Iijima, 1978)

on pressure is also shown by experiments and thermodynamic calculations by Campbell and Fyfe (1965) and Thompson (1971). At $P_{H_2O} <$ ca. 2 kb, the reaction temperature is constant at 180–190 °C. While this pressure is consistent with the conditions of burial diagenesis discussed above, temperatures are significantly higher. Similarly, the reaction clinoptilolite-analcime is known to occur at surface temperatures in alkaline saline lake deposits while experimentally the same transformation in a $Na_2CO_3$-NaOH solution has only been observed at temperatures above 100 °C (Boles, 1974).

Obviously, reaction temperatures are influenced by other factors, especially time and kinetics. The one environmental change that may lower the reaction temperature is an increase in $Na^+$-concentration (Fig. 12-26).

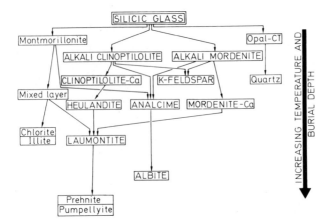

Fig. 12-25. Flow diagram showing development of authigenic zeolites and silicates from silicic glass in thick marine deposits during burial diagenesis and metamorphism (prehnite-pumpellyite). (After Iijima, 1978)

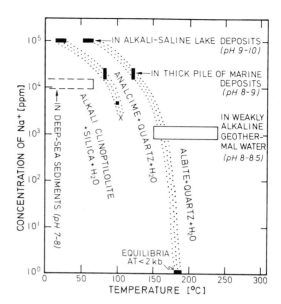

Fig. 12-26. Equilibria of reaction series alkali clinoptilolite – analcime – albite at $P_{H_2O} < 2$ kb in silica-saturated environment with respect to temperature and Na-concentration of interstitial water. (After Iijima, 1978)

Viereck et al. (1982) studied alteration of basaltic and rhyolitic volcaniclastic rocks in a 2-km borehole in Eastern Iceland. Alteration temperatures are believed to have ranged from about 100 (top) to about 330 °C (base of the section) based on mineral equilibria. Three alteration stages were recognized (Figs. 12-27, 12-28). Stage I is characterized by filling of open spaces by alkali zeolites and replacement of primary components by illite, smectite, and Si-rich chlorite. Earlier secondary phases are replaced during stage II, during which phases such as Al-rich chlorite,

| SITE OF SECONDARY MINERALS IN VOLCANICLASTIC ROCKS | EXOTIC ROCK CLAST | PHENOCRYST | SHARD | PORE/MATRIX | VESICLE | VEIN |
|---|---|---|---|---|---|---|
| SMECTITE |  |  |  | ● | ● | ● |
| CHLORITE |  | ● | ● | ● | ● | ● |
| ILLITE-CHLOR.MIX.M. |  | ● | ● | ● | ● | ● |
| ILLITE |  | ● | ● | ● | ● | ● |
| MORDENITE |  |  |  | ● | ● | ● |
| HEULANDITE |  | ● | · | ● | ● | ● |
| LAUMONTITE |  | ● | · | ● | ● | ● |
| WAIRAKITE |  |  |  | ● | ● | ● |
| ALBITE |  | ● | · | · | · | · |
| ADULARIA | ● | · |  | · |  | · |
| QUARTZ |  | · | · | ● | ● | ● |
| SPHENE |  | ● | · | ● | · | ● |
| EPIDOTE |  | ● | · | ● | · | ● |
| PUMPELLYITE | ● | ● |  | ● |  |  |
| PREHNITE |  | ● |  |  |  | ● |
| CARBONATE |  | · | · | ● | ● | ● |
| ANHYDRITE | ● |  |  | ● |  |  |
| GARNET |  | · | · |  |  |  |
| PYRITE |  |  |  | ● |  | ● |
| FE-OXIDES | ● | · | · | ● | ● | ● |

Fig. 12-27. Site of occurrence of secondary minerals in tuffs interbedded with basalt flows in 1,920 m thick Miocene volcanic series (Reydarfjordur, Iceland). (After Viereck et al., 1982)

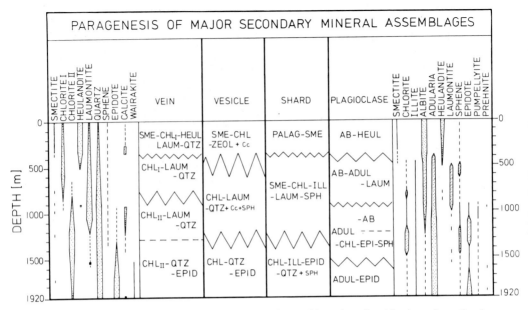

Fig. 12-28. Paragenesis of major secondary mineral assemblages in volcaniclastic rocks at Reydarfjordur (Iceland) (see Fig. 12-27). (After Viereck et al., 1982)

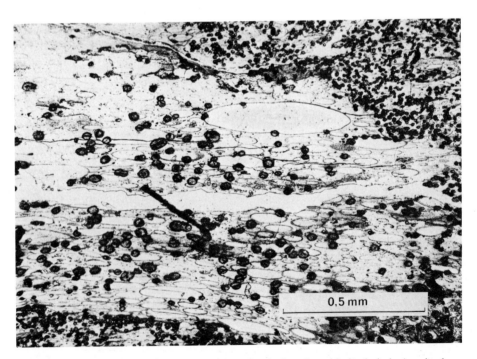

Fig. 12-29. Former basaltic vesicular sideromelane shards altered to chlorite (colorless) and sphene (dark round blebs). Miocene interbasalt tuffs (Reydarfjordur, Iceland)

laumontite, and calcite crystallized. Stage III is characterized by crystallization of quartz and epidote with sphene crystallizing during stages II and III (Fig. 12-29). While the more pervasive stage III alteration (below 1400 m) is similar for all components, different sites such as vitric shards, crystals, vesicles, and fractures respond very differently to low grade alteration (Fig. 12-27). Apart from the important influence of texture and type of clast, primary compositions have significantly influenced the composition of secondary phases. For example, the abundance of illite and adularia is clearly related to the abundance of rhyolitic components. Moreover, volcaniclastic rocks are more sensitive to alteration compared to the interlayered lava flows owing to the abundance of vitric fragments and high initial porosity and permeability. The alteration stages occur stratigraphically higher in volcaniclastic rocks compared to lava flows.

# Chapter 13    Stratigraphic Problems of Pyroclastic Rocks

Stratigraphic methods used to study volcanic rocks are similar to those used to study sedimentary rocks and have similar purposes: establishing correlations, vertical time sequences, determining facies changes and the like. Additional aid in pyroclastic stratigraphy comes from igneous petrology, studies of magma evolution, and studies of the growth history of volcanoes. We stress that stratigraphic analysis – (1) mapping and subdivision of volcanic sequences into members, formations and groups, (2) determining vertical and lateral facies changes and (3) interpreting the types of eruptions, and the origin and manner of transport of rock types and the environments of deposition – provides the necessary framework for petrological and geochemical work, and that pyroclastic and epiclastic volcanic rocks often provide the most important parameters for establishing the stratigraphic framework in volcanic areas. Because of the many ways that volcanic sequences originate, however, different stratigraphic and petrologic approaches may be necessary in different situations such as, for example, in areas of plateau basalts, ignimbrite plateaus, clusters of scoria cones, composite andesite volcanoes, deep-sea ash layers or thick nonmarine tuff accumulations. Moreover, stratigraphic problems on active or dormant volcanoes commonly differ from those that must be solved in ancient volcanic regions, such as greenstone belts, or zeolitized volcaniclastic sediments, where the volcanic record may be best preserved in sedimentary accumulations derived by erosion of primary volcanic deposits. Volcanic stratigraphy can be studied on many overlapping levels depending upon the purposes of study (Table 13-1).

Most of this chapter is concerned with the stratigraphy of subaerial pyroclastic materials, but as shown in Chapters 7, 10 and 12, pyroclastic debris is abundant and especially well-preserved in the geologic record in marine environments. Ma-

Table 13-1. Volcanic stratigraphic studies

| Purpose of study | Field sources of information |
| --- | --- |
| Correlation, tephrochronology, emplacement mechanisms, types of eruptions, volcanic energy | Individual beds, bedding sets of layered sequences |
| Paleotopography, environment of emplacement, basin analysis and tectonic history | Facies associations and depositional history of layered sequences |
| Magma composition and evolution, tectonic setting and volcanism, regional stratigraphy | Rock sequences, composition and tectonic history in volcanic regions, provinces, fields and centers (see Table 13-5) |

rine tuff layers commonly altered to various diagenetic mineral phases are found in many parts of the world, but systematic stratigraphic studies of them are few. In the United States, there are large tracts of Ordovician, Devonian, Permian and Cretaceous bentonites (Fig. 12-20). They are common in Mesozoic rocks of central Oregon (Dickinson and Vigrass, 1964) and many other parts of the Cordillera (Dott and Shaver, 1974). Miocene bentonite has been valuable in correlating rocks from separate Tertiary basins of California and for dovetailing physical age determinations with the microfaunal and megafaunal time scales (Turner, 1970). Subaqueous tuff beds are reported from the Precambrian of Sweden, Devonian of Germany (Winter, 1981), Carboniferous of Scotland (Francis, 1967), England and Wales (Francis, 1970b); tonsteins (kaolinite bentonites) (Chap. 12) have been very important as stratigraphic marker beds in many parts of the world. Packham (1968) gives an excellent account of marine Ordovician, Silurian, and Devonian volcaniclastic rocks in New South Wales, Australia. As volcaniclastic rocks are better understood, their importance in the stratigraphic record is appreciated in many sedimentary sections where they were formerly overlooked or misinterpreted.

In some ways, pyroclastic sequences are unique within the fields of sedimentology and stratigraphy. Most importantly, a set of primary pyroclastic fragments is formed almost instantly, and, unlike epiclastic fragments, is given an initial lift of kinetic energy that is independent of slope or base level of deposition. Thus, some pyroclastic ejecta from a given explosive eruption are able to bypass or overtop physiographic barriers; and, in a minor way, deposition may defy the Law of Original Horizontality. As an example, base surges may deposit initially cohesive layers of graded ash on vertical or even overhanging surfaces (Chap. 9).

# Relation of Volcanic Activity to Rock Stratigraphy

### Volcanic Activity Units

Volcanic phenomena are diverse, ranging from seismic activity, fumarole action, high rates of heat flow, and explosive ejection of tephra to quiet effusion of lava. We define volcanic activity broadly to include any or all of these phenomena, but relate it in this chapter to the explosive ejection or quiet outpouring of material from a volcanic vent which produces deposits of volcanic rock. A volcanic eruption can be subdivided into *volcanic activity units* (Fig. 13-1) which include: (1) an *eruptive pulse* that may last a few seconds to minutes, (2) an *eruptive phase* that may last a few hours to days and consist of numerous eruptive pulses, and (3) a single *eruption,* composed of several phases, that may last a few days to months, or in some basaltic volcanoes, for a few years. We define the eruption as the basic unit of volcanic activity for use in studying deposits. For convenience, Simkin et al. (1981) define eruptions in terms of quiet (inactive) periods. For example, an eruption which follows its predecessor by less than 3 months is considered to be a phase of the earlier eruption unless it is distinctly different in some way. Clearly, we cannot pre-empt the term eruption for use in describing witnessed events, but definition of eruptions in terms of deposits they may produce is geologically helpful as discussed in a following section.

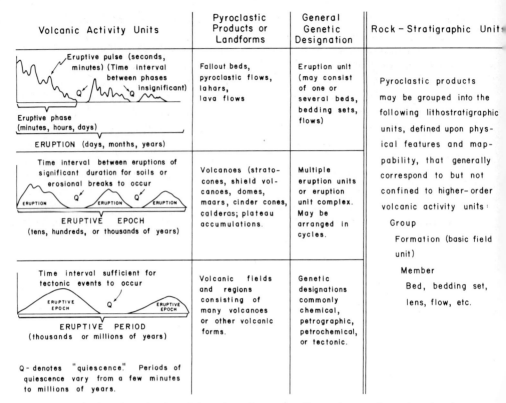

Fig. 13-1. Terminology for volcanic activity units and corresponding rocks, eruption unit and rock-stratigraphic terms

Some volcanoes (e.g., domes and basaltic scoria cones) may form completely within the time span of an eruption as defined above. Others, such as shield volcanoes, stratovolcanoes or volcanic islands, may show high-order discontinuities such as major chemical changes, volcano-tectonic events like caldera collapse, or long erosional intervals, and may last 10 m.y. or more before volcanism completely dies out. On a still higher order, several volcanoes may form volcanic chains along which volcanic activity can migrate.

Volcanic activity terms at orders higher than eruptions are more conceptual and therefore are of somewhat less practical use. Nevertheless, it is sometimes useful to conceptualize longer units of volcanic activity when considering volcanoes and other volcanic forms. For example, large andesitic stratovolcanoes are constructed by many eruptions over thousands of years, with large intervening intervals of quiescence between, and the whole may be considered as constructed during an *eruptive epoch*. Likewise, volcanic fields or regions composed of products from many volcanic centers active over tens of thousands or millions of years could be considered to have been formed during an *eruptive period*. Eruptive periods may themselves be parts of larger cycles during which volcanic centers migrate. The combined products produced during eruptive periods may form a petrogenetically related chemical series developed in the magmatic system. The terms epoch and period are *sensu stricto,* time-stratigraphic terms of long standing; distinction must be made between them and volcanic activity terms.

Volcanic activity as defined above must be carefully distinguished from the deposits (rock-stratigraphic units) derived from such activity. A great deal of information about volcanic activity can be derived by the study of rock units, but to interpret rocks in terms of volcanic activity requires a conceptual bridge between the two; thus we use the genetic concept of "eruption unit" discussed below.

**Eruption Unit**

An eruption unit is a deposit defined as *a thickness of volcanic material deposited from an eruptive pulse, an eruptive phase or an eruption.* Thus, an eruption unit is neither a volcanic activity unit nor a rock-stratigraphic unit, but rather it is a conceptual entity that relates the two. Depending upon recognized manner of emplacement, different kinds of eruption units are: *Pyroclastic fallout unit, pyroclastic flow unit, lava flow unit, lahar unit, etc.*

An accumulation of volcanic material comprising a mappable unit and defined as a stratigraphic formation may include several kinds of eruption units. In most places, individual lava flows can be referred to safely as single eruption units. However, the mere presence of a pyroclastic bed does not warrant the interpretation that its constituents are from a single eruptive pulse, phase, or eruption, although, of course, it does not preclude that likely possibility. A pyroclastic fallout unit may form only one small part of a bedded tephra sequence classed as a member or formation. On the other hand, a widespread and distinctive mappable unit consisting of a single bed may be defined as a member or formation even though, in fact, it consists of only a single eruption unit. More difficult to decipher is the stratigraphy of pyroclastic flow units. An individual pyroclastic flow may be derived from a single eruptive phase, and thereby be a single eruption unit. Commonly, however, pyroclastic flow sheets are composed of several flow units that are difficult to distinguish.

It is useful to compare volcanic activity units with rock-stratigraphic units to provide information about the type of eruptions, volcanic energy involved, possible cyclicity of eruptions and the chemical and physical evolution of the original magma. Experience shows that the interpretation of volcanic activity can be more easily done for an active or very young volcano, but becomes increasingly difficult for older volcanic sequences. Thus, in young volcanoes, a member could be a single bed derived from an eruptive phase, whereas a member of a formation in older terranes generally consists of the products from many phases or eruptions. However, the May 18, 1980 Mount St. Helens eruptive phase instructively shows the stratigraphic complexities of fallout units that can arise within a period of a few hours. Within a few kilometers of the volcano, there are four principal fallout units (Chap. 5). East of the volcano these laterally change to three units. Beyond 200 km much of the deposit is only two units, and at several hundred kilometers just one unit occurs (Sarna-Wojcicki and Shipley et al., 1981; Waitt and Dzurisin, 1981).

## Stratigraphic Problems in Young Volcanic Terranes

In young volcanic terranes, there is a common tendency to define members and formations in terms of volcanic activity, but here we will adhere to the accepted principle that rock-stratigraphic units are defined strictly on the basis of physical criteria. The following definition can be used for either young or old volcanic terranes: A formation is a *mappable* bed, bedding set or sequence of beds of any thickness set apart from rock units above and below by distinctive physical criteria such as texture, color, lithologic or mineralogic characteristics, or by weathered zones or erosional unconformities; a member is a convenient subdivision of a formation. A formation might, for example, be a single bed, two beds (each designated as a member if desirable) or a sequence of many well- to poorly bedded tuffs 1000 m or more thick that could even include lava flows, etc. Physical characteristics are commonly related to composition of magma or style of eruption and emplacement. Therefore, most rocks within formations are genetically related in some way.

Defining members and formations by physical criteria does not prohibit them from being *interpreted* as being derived from an eruptive pulse, phase or an eruption. This is more commonly done in younger volcanic terranes; but in ancient terranes, distinctive units such as ash flow layers may be interpreted in terms of volcanic activity.

The usefulness of relating volcanic deposits to eruptions which produced them is well illustrated by Nakamura's (1964) classic volcanic stratigraphic study of Oshima Volcano, Japan (see later section).

In discussing stratigraphic problems of the young Taupo volcanic region, New Zealand, Healy et al. (1964) try to relate the volcanic activity units, such as eruptive phases, to rock units. They state that in geological mapping the basic field unit is the formation, and that it would be convenient if a formation were to represent an eruptive phase separated by appreciable time breaks above and below; in practice, however, they recognize that this is not always possible. Beds near the source can be very thick, but farther away they commonly become very thin and may merge by soil-mixing processes with others. Therefore they confine the term formation to single or multiple units that can be readily identified over wide areas.

A useful concept for defining formations in active volcanic regions is that of *optimum thickness* as used by New Zealand workers (Healy et al., 1964). Near eruption centers, formations are generally too thick to be defined adequately and their tops or bottoms may not be exposed. In areas distant from the sources, such formations wedge out and in places merge into soil zones. But in intermediate areas, it is common to find an optimum thickness (or distance) where the top and bottom are exposed, and beds which characterize the formation can be described. Moreover, the facies changes, both toward the vent and away from it, are more easily described by relating them to the optimum thickness.

## Stratigraphic Nomenclature in Older Volcanic Terranes

The standard rules of stratigraphic nomenclature may be difficult to apply to large volcanic accumulations. In addition to rapid facies changes in near source areas, products from adjacent volcanoes or even from nonvolcanic sources may interfin-

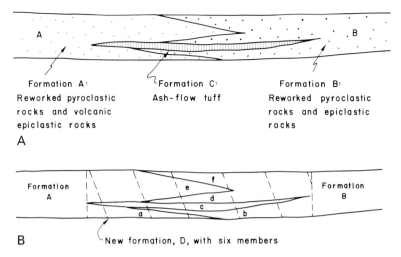

**Fig. 13-2 A, B.** Diagrammatic stratigraphic relationship of **A** three intertonguing formations and **B** one possible solution for redefining the rocks in the area of intertonguing by using the arbitrary cut-off. See text for further explanation

ger in an exceedingly complex manner resulting in a complex assemblage of local to widespread mappable units. This may result in the unacceptable situation where members, formations, etc. may succeed each other in vertical sequence, thereby violating the Law of Superposition.

A simplified hypothetical example of local formations that interfinger with more distal formations is shown in Fig. 13-2. Formational names (A, B, and C) can be applied where they are easily identifiable, but where they interfinger, formations A and B succeed each other in vertical sequence. One solution is to apply vertical cut-offs (arbitrary cut-off of Wheeler and Mallory, 1953) and define a third independent formation composed of six members (see North American Commission on Stratigraphic Nomenclature, 1983, Article 23b). Such a solution, however, requires a detailed knowledge of local and regional relationships acquired through many field seasons, usually by several geologists working in different areas that eventually converge. During the course of long-term investigations, formations, members, and other rock stratigraphic units are defined, commonly resulting in a proliferation of names in different local areas. Thus, the same rock body may receive different names, eventually requiring redefinition and renaming of one or several units, elevation of members to formations, formations to groups, etc. Given the complex nature of rock bodies, the human propensity to view problems in a provincial manner, and the fact that mapping of regional areas may take place over decades by different generations of geologists working on different problems from different points of view, and during which there are changes in ideas and improvement of technology, the problem of increasing nomenclatural confusion is virtually inevitable.

In the San Juan Mountains, an area of complexly interfingering Tertiary volcaniclastic deposits, flows, and intrusive rocks, Lipman and Steven (1970) set aside the names that had been applied earlier to different parts of the coalescing complex. In their place, they devised a surrogate name for Group, the "early intermediate composition volcanics," based in large part upon general composition. This permitted them to define clearly identifiable formations where necessary, and to deal with individual stratigraphic problems without the burden of excessive formal nomenclature or of knowing the details of the stratigraphic relationships within the entire volcanic assemblage.

In many circumstances, however, members, formations and groups in older volcanic terrane may be treated in accordance with standard stratigraphic procedures used in nonvolcanic rocks, and the volcanologist needs to become familiar with standard stratigraphic texts. A point of difference between nonvolcanic sedimentary rocks and volcanic rocks, in addition to the complexities of multiple sources mentioned above, lies in the mechanisms of emplacement and the importance of vertical chemical compositional progressions in volcanic rocks for interpreting magma genesis. Some of these problems are dealt with in the sections on facies and in the stratigraphic examples. Another problem is the absence of fossils in most, but not all, volcanic rocks. Age dating by radiometric methods can successfully overcome this void.

Using standard stratigraphic procedures and information from deep sea cores, Cook (1975) defines four lithologically distinct "oceanic formations" of Tertiary age which occur on the Pacific Ocean floor off the coast of Central America. The formations represent two broad sedimentation regimes; an easterly volcanogenic regime and a westerly pelagic regime. His isopach reconstruction and facies analysis suggest that they are diachronous, and show a close association of the main volcanogenic formation (San Blas "Oceanic Formation") to the East Pacific Rise and the coast of Central America during Tertiary time.

## Tephrochronology

Tephra layers are unique time-stratigraphic markers especially useful in Quaternary sequences. These layers are emplaced geologically instantaneously over broad regions. Under favorable circumstances, the precise time of formation of clasts and deposition of tephra layers can be determined by radiometric dating of unaltered glass and phenocrysts. The layers may be chemically and petrographically characterized by a variety of techniques. Thus, tephra layers are excellent key horizons for wide correlations in different facies environments and physiographic provinces, and for dating of associated nonvolcanic sedimentary sequences. If their sources and distribution patterns are known, paleowind directions at the time of eruption are easily determined.

Chronological (and correlation) studies using tephra layers are known as *tephrochronology,* a term coined by Thorarinsson (1944, see Thorarinsson, 1981) from detailed stratigraphic work on post-glacial deposits in Iceland. He originally included only fallout pyroclastics in the definition of tephra, but later (1974) expanded the meaning to include pyroclastic material emplaced by flowage. At present the term tephra has become essentially the equivalent of "pyroclastic material" without reference to fragment sizes, and it also includes all subaqueous varieties, which eliminates the necessity of interpreting the environment of deposition before using the term. Thus "pyroclastic material" and "tephra" are now used interchangeably.

The identification and correlation of widespread tephra layers have elucidated many problems in Quaternary and Holocene geology, archeology, paleopedology, glaciology, palynology, and, last but not least, volcanology. Two world bibliographies on Quaternary tephrochronology (Westgate and Gold, 1974; Vitaliano, 1982) testify to its importance, and Blong (1982) has written an informative book combining archeology, mythology, and tephrochronology. Methods of sampling, laboratory preparation, petrographic examination, as well as tephrochronological dating, are given by Steen-McIntyre (1977). Several excellent up-to-date summaries of many aspects of tephrochronology are contained in a recent book on tephra studies (Self and Sparks, 1981).

Many ash beds are contaminated with extraneous material and by chemical precipitates incorporated during or after deposition, or by biologic reworking. In lakes or in the sea, for example, glass and mineral particles may co-mingle with concurrently deposited clay, organic material or other detritus. Layers range from nearly pure ash to those containing a small but recognizable amount of volcanic material. Unless diagenesis has completely obliterated all vestiges of the original ash, scattered pyroclastic grains within a single layer can be concentrated in the laboratory and analyzed chemically and mineralogically for purposes of correlation.

The overall petrologic and chemical features of ash layers derived from two separate volcanoes or even from several widespread centers within a petrographic province can be nearly identical. Therefore, ash layers must be carefully distinguished, as pointed out in an important review by Wilcox (1965). Precise petrochemical, petrographic and paleomagnetic studies are commonly necessary to identify ash in distal regions (Izett et al., 1972; Izett, 1981; Westgate and Gorton, 1981).

Environments favorable for the burial and preservation of a newly deposited ash layer are lakes, bogs, valley bottoms, and the sea floor. Locally, a deposit may be reworked by creep or running water, but for most stratigraphic purposes the age of the reworked ash can be regarded as the same as the original ash layer. Ash is easily stripped from sloping upland areas, therefore cores from peat bogs, lake bottoms, and other sedimentary basins may contain duplicates of a single eruption – the upper one washed in from adjacent hills. Despite such complications, the stratigraphy of many local areas can be pieced together and the lateral extent of the original layer determined.

The physical and chemical description of volcanic ash is determined in several ways (Westgate and Gorton, 1981). These include field criteria such as color, mineralogy, stratigraphic relations, sedimentary structures, thickness and distribution (Mullineaux, 1974) but detailed chemical analyses of individual phases – glass shards and primary phenocrysts – are of particular importance. Individual particle chemistry is more useful than bulk chemistry of layers because of eolian fractionation (Chap. 5). Similarities in total mineral assemblages are somewhat less useful as correlation parameters although Westgate and Gorton (1981) have found that all mineral species within a given tephra bed persist to the most distal end of a recognizable fallout unit.

**Table 13-2.** Relative value of analytical techniques for the geochemically important groups of elements. (Westgate and Gorton, 1981)

|  | Probe | XRF | AA | INAA |
|---|---|---|---|---|
| Major elements | 1 | 1 | 2 | 3 |
| Transition/heavy metals | 2–3 | 2 | 1 | 2 |
| Large cations | 3 | 1 | 3 | 3 |
| Rare earth elements | 3 | 2–3 | 3 | 1 |
| High atomic number metals | 3 | 2–3 | 3 | 1 |

Notes: 1 = Excellent. 2 = Satisfactory. 3 = Unsatisfactory

Major and minor element analysis of glass shards and pyrogenic minerals are accomplished by microprobe, instrumental neutron activation analysis (INAA), X-ray fluorescence (XRF) and atomic absorption (AA). The relative value of these analytical techniques are compared in Table 13-2.

One standard technique is determining the range and modal values of the refractive index of glass (Ewart, 1963; Wilcox, 1965; Steen and Fryxell, 1965). However, this is equivocal because the refractive index of glass increases with hydration, and the amount of hydration varies in different environments. Therefore it is best to combine this technique with other parameters such as the phenocryst suite and chemical and petrographic properties of glass shards and individual phenocrysts (Izett, 1981; Westgate and Gorton, 1981).

Tephrochronological studies of 68 different fallout layers (0.1 m.y.–4.0 m.y.) in the western United States by Izett (1981) have yielded important results regarding the classification of silicic ash. Izett (1981) divides silicic ashes into W-type rhyolite, G-type rhyolite and dacite according to the iron and calcium content in glass shards (Fig. 13-3). Each kind of ash also shows other distinctive characteristics (Table 13-3).

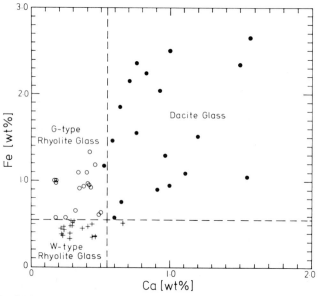

**Fig. 13-3.** Classification of some silicic fallout ashes in the western United States. (After Izett, 1981)

**Table 13-3.** Some features of silicic fallout ashes from the western United States. (After Izett, 1981)

| Type of glass shard | Color | Characteristic phenocrysts | Chemical features (in wt.%) | Characteristics of glass shards | Comment |
|---|---|---|---|---|---|
| W-type rhyolite | Chalky white megascopically | Always biotite, but no fayalite. Also quartz, Na-poor sanidine, oligoclase, clinopyroxene, hornblende, magnetite, ilmenite, apatite and allanite | Fe <0.55<br>Ca <0.55<br>$SiO_2$ = 76–79 | Large percentage of micropumice shards. R.I.[a] = 1.494–1.497. Shards have few microlites and a narrow range of major element composition | Commonly generated in regions underlain by granitic crust undergoing extension |
| G-type rhyolite | Light to medium gray megascopically | No biotite, but contains fayalite. Also Na-rich sanidine, oligoclase, clinopyroxene, hornblende (may be iron-rich), magnetite, ilmenite, allanite and chevkinite | Fe = 0.55–2.0<br>Ca <0.55<br>$SiO_2$ = 72–79 | Mainly colorless platy bubble wall and bubble junction shards. R.I. = 1.497–1.520. Shards have few microlites and a narrow range of major element composition | Same as above |
| Dacite | White to light gray or light grayish brown megascopically | Frequently lack quartz, sanidine and biotite. May have andesine, clinopyroxene, orthopyroxene, hornblende and cummingtonite, magnetite, ilmenite and apatite | Fe >0.55<br>Ca >0.55<br>$SiO_2$ = 67–77 | May have complex mixture of shard shapes including micropumice, fibrous, chunky and bubble wall and bubble junction types. R.I. = 1.496–1.530. Microlites common. Shards may have wide range of major element composition in a single bed | Generated chiefly in areas of continental-oceanic plate convergence |

[a] R.I. = Refractive Index

Cerling et al. (1975) used oxygen isotope ratios to correlate tuffs in northern Kenya, but the ratios are primarily a result of alteration history of ash and therefore not as discriminant as trace element concentrations Ash layers have been identified from thermomagnetic properties of ferromagnetic minerals by Momose et al. (1968), and remanent paleomagnetism has been used by Dalrymple et al. (1965), Izett et al. (1972), and Reynolds (1975) to determine the polarity epoch in which ash was deposited.

The three important radiometric methods for dating Quaternary tephra – fission-track, K-Ar and radiocarbon ($^{14}$C) are reviewed by Naeser et al. (1981). The $^{14}$C method is only applicable to samples less than about 50,000 years old, and zircon and glass shards older than about 100,000 years can be routinely dated by fission track. Sanidines as young as 70,000 years can be dated by K-Ar while the lower limit of plagioclase K-Ar dates is about 200,000 years old.

Progressive hydration and superhydration (filling of closed vesicles with water) may provide the means to determine roughly approximate ages of Holocene, Pleistocene, and pre-Pleistocene volcanic ash beds (Steen-McIntyre, 1975). Complete superhydration may take as long as 10 m.y. (Roedder and Smith, 1965).

## Volcanic Facies

The term facies refers to physical, chemical, and biologic variations of rock bodies deposited within a specific interval of geologic time. The term has been used in several ways (see discussion by Dunbar and Rogers, 1957), but a facies is usually named according to that aspect emphasized by the particular study being made – tectonic, paleogeographic, paleoecologic, depositional environment and so forth. General treatments of sedimentary facies models may be found in Selley (1978), Matthews (1974), Reading (1978) and R. Walker (1979).

Without distinctive tephra horizons, tracing and correlating time-equivalent volcanic facies can be difficult because of the variety of products (lava flows and pyroclastics) that can be erupted concurrently from different vents, with synchronous rapid erosion in near-source regions and deposition in distant areas.

### Facies Based upon Position Relative to Source

The products of ancient as well as recent volcanoes within nonmarine or marine environments may be divided into a *near-source facies, intermediate-source facies* and *distant facies,* and each of these may be divided into many different subfacies depending upon desired emphasis, such as lahar or pyroclastic flow subfacies (mechanism of transport), lacustrine or alluvial subfacies (environment of deposition), etc. As we will point out, near-source, intermediate-source, and distant facies are relative terms chosen to fit different situations. Moreover, there are special categories such as intracaldera facies and caldera outflow facies depending on position relative to source. Parsons (1969) applied "cone-complex facies" and "alluvial facies" to special varieties of near- and intermediate-source facies of large subaerial volcanoes. Studies by Nakamura (1964), Healy et al (1964), Gwinn and Mutch (1965), Lipman (1968) and Steven et al. (1979) illustrate facies studies of ancient subaerial volcanic accumulations. A study describing the details of near-source ("vent") facies and intermediate-source ("alluvial") facies of single and coalescing subaerial volcanoes is given by Smedes and Prostka (1972) for the Absaroka volcanics (Eocene) in northwestern Wyoming and southwestern Montana.

Nonmarine volcanic deposits from active, pyroclastic flow-producing volcanoes in the fore-arc area of Guatemala are divided into four facies by Vessel and Davies (1981) (Fig. 13-4). The vent facies ("core facies" of Vessel and Davies) consists of interbedded lavas, coarse- to fine-grained fallout tephra and breccias caused by erosion of the steep volcano flanks. The near-source facies ("proximal and medial facies" of Vessel and Davies) consists of breccias deposited from pyroclastic flows and eroded debris within valleys on the volcano flanks, and fallout tephra, grading to alluvial fans at the base of the volcano composed of lahars, fluvial debris and tephra beds. The intermediate-source facies ("distal facies" of Vessel and Davies) consists of fluvial deposits of braided streams interbedded with thin (a few centimeters to millimeters thick) fallout ash beds. Offshore are small wave-dominated deltas fringed by well-developed beaches (Kuenzi et al., 1979). Kuenzi et al. (1979) and Vessel and Davies (1981) are the first to describe nonmarine facies associations of an active volcanic region based upon a depositional facies model approach. Similar facies are to be expected in nonmarine back-arc regions (Chap. 14).

Sedimentation in and adjacent to volcanic island arcs is dominated by pyroclastic and other volcaniclastic debris, and three genetic types of facies can be recognized in arc assemblages (Dickinson, 1974b): (1) a central (near-source) facies of eruptive centers flanked by (2) subaerial and/or subaqueous fans, aprons or shelves (intermediate-source facies) and (3) basinal facies (intermediate-source to distant facies) within arc structures (trench, force-arc, and back-arc basins; Chap. 14).

Research on subaqueously deposited volcanic rocks in the Archean of Canada, using a facies approach, is yielding important results (Ayres, 1977, 1983; Tassé et

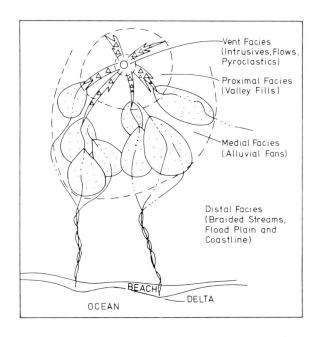

Fig. 13-4. Sedimentary facies, of modern volcaniclastic debris on an active volcano. Unornamented areas are older rocks of similar kinds. (After Vessel and Davies, 1981). The vent, proximal, and medial facies are included in the near-source facies as used in this text, and the distal facies is equivalent to the intermediate-source facies. Distant facies (fallout tephra far from source) not shown

al., 1978; Lichtblau and Dimroth, 1980). For example, facies assemblages of an Archean caldera complex are outlined by Lichtblau and Dimroth (1980). In attempts at paleogeographic reconstruction including the environments of deposition, they recognize seven pyroclastic facies defined on the basis of geometry and sequence of sedimentary structures. These facies are interpreted as (1) coarse-grained lag deposits derived from a collapsing submarine eruption column, (2) filling of pit craters by steam explosion breccias, (3) pyroclastic debris flow, (4) pyroclastic turbidite, (5) subaerial pyroclastic fallout tephra settled through water, (6) coarse-grained subaerial fallout tephra redeposited in a shallow marine environment by offshore currents and (7) fine-grained fallout tephra deposited by shallow marine currents. In the interpretation of the evolution of an Archean island volcano, Ayres (personal communication), following Smedes and Prostka (1972), divides the lateral sequence into (1) a vent facies consisting of flows and primary pyroclastic rocks, (2) a proximal alluvial facies consisting of secondary (reworked) pyroclastic rocks and (3) a distal alluvial facies of epiclastic rocks.

Near-source, intermediate-source and distant facies are relative terms used in reference to the source. Except in rare specific cases, absolute distance cannot be defined. A large and highly explosive volcano, for example, will distribute coarse-grained and fine-grained debris to greater distances than a less energetic source. Also, the eruptive behavior of a volcano may change. Ancestral Mount Mazama, Oregon, for example, was a large stratovolcano whose deposits have so far been recognized only close to the source. Late cataclysmic eruptions accompanied by formation of a caldera (now Crater Lake), spread great volumes of pumice flows on its flanks and into surrounding valleys to as far as 60 km from the original volcano. Fallout tephra was blown easterly in a wide fan whose distal edges are more than 1000 km to the northeast (David, 1970). Therefore, near-source, intermediate-source, and distant facies of the final eruptive stage of Mount Mazama have different meanings in terms of absolute distance from its earlier volcanic products.

Near-Source Facies

The near-source facies of active or deeply dissected volcanoes or volcanic complexes consists of lava flows and pyroclastic debris produced directly by volcanism or as a direct consequence of erosional and gravitational processes acting on the steep slopes of a volcano. In areas where erosion has removed all but the volcanic roots, the near-source facies consists mainly of a complex of stocks, sills, and dikes and some intrusive or extrusive breccia and tuff. Where volcanic cones are so deeply dissected that the cone form is completely destroyed, a *vent facies* may be identified by poorly bedded or unbedded pyroclastic breccias and collapse breccias in intrusive relation with the basement rocks. Widespread fumarolic alteration may appear, and one or more small, roughly circular to elongate plutons may have been unroofed. Rock types arranged in a roughly circular configuration help to identify a near-vent facies. Dikes radiating outward may target the central vent source.

Initial fragmentation of material in the near-source facies is accomplished by explosive eruptions. Pyroclastic accumulations are typically thick, wedge-shaped and discontinuous volcanic breccias which are poorly sorted and poorly bedded. Deposition is by slump and flow within valleys, or by fallout mantling ridges and valleys alike. Lenticularity, erosional unconformities, and deposition on highly irregular surfaces make it difficult to establish lateral continuity. Interbedded breccias and lava flows may originate from different places and at different times from the volcano. Breccias may interfinger with tuff from different volcanoes within the region. In some instances, stratigraphic continuity of contemporaneous but isolated sections can be established by relating them to distinctive fallout tephra layers. More commonly, it can only be done by the piecing together of several interfingering units that link the local sections to known stratigraphy – or, in many cases, not at all (see Smedes and Prostka, 1972).

Intermediate-Source Facies

The intermediate-source facies includes rocks around a volcanic center and its various constructional cones. The facies includes rocks deposited from pyroclastic flows, lava flows, fallout processes, and their reworked products. As distance from the source increases, this facies will include an increasing amount of resedimented pyroclastic and epiclastic volcanic debris. Still farther away, nonvolcanic clastic material is increasingly mixed with the volcanic debris, or it may form discrete layers interbedded with reworked volcanic materials. Particular facies characteristics depend upon the topographic, structural, and environmental factors affecting the volcanic edifice during its formation, as well as the type of volcanic activity.

Distant-Source Facies

The distant facies consists of fallout tephra in areas far from the source, much farther than lava or pyroclastic flows can travel. Its transition with the intermediate-source facies is generally lost by erosion. Isolated ash deposits may occur as one or more thin individual layers in marine, lacustrine or subaerial deposits hundreds of kilometers from the source. Layers of the distant-source facies are generally thin, well-sorted, and differ markedly in composition from interbedded nonvolcanic sediments. Such widespread tephra layers are the concern of many tephrochronology studies. However, if deposition of ash occurs slowly and relatively continuously for a long period of time in an area that is not receiving other kinds of sediment, ash sequences may become very thick. The Oligocene-Miocene John Day Formation of eastern Oregon, for example, is composed predominantly of fine-grained tuff in sections hundreds of meters thick (Fisher and Rensberger, 1973). This formation was deposited chiefly from many volcanic sources about 160 km to the west within the Cascade Mountains (Peck et al., 1964), but also received contributions from a few local sources. Thus, most of the John Day Formation is the distant-source facies of the source Oligocene-Miocene Cascade volcanoes, but the coarse-grained tuff and ignimbrite units (Fisher, 1966a) are intermediate-source facies from vents within the area of the John Day Formation.

Caldera Facies

Caldera collapse resulting from large-scale pyroclastic eruptions (Smith and Bailey, 1968) gives rise to special varieties of near-source and intermediate-source facies rocks that accumulate both within the caldera and outside the collapsed part (Fig. 13-5). Cunningham and Steven (1979) have called these the *intracaldera fa-*

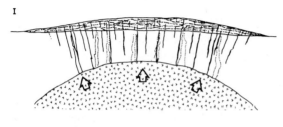

I. Regional tumescence, propagation of ring and radial fractures with possible apical graben subsidence. Eruptions from radial or ring fractures. Erosion of the volcanic highland with sediments deposited on flanks and surrounding lowland areas.

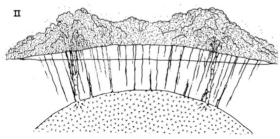

II. Major ash-flow eruptions with flows extending tens of miles beyond the volcano.

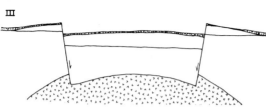

III. Caldera collapse. Ash-flow deposits within the caldera form part of the intra-caldera facies along with avalanches and slides from caldera wall. Ash-flow deposits beyond the caldera walls form the dominant rock of the outflow facies.

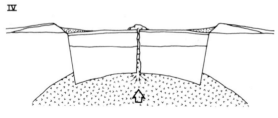

IV. Minor pyroclastic eruptions and lava flows occur on caldera floor of some calderas and occur with continuing deposition of slides, fans, and lake deposits.

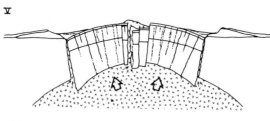

V. Resurgent doming and possible ring-fracture volcanism and/or eruption or intrusion in dome fractures. Caldera fill from non-volcanic processes continues.

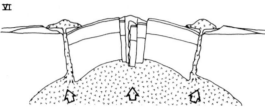

VI. Possible regional tumescence and reopening of ring fractures with ring-fracture volcanism and development of domes along margin. Caldera filling continues (lake sediments, marginal erosional, volcanic products).

**Fig. 13-5.** Evolution of caldera collapse, resurgence and related sediments. (After Smith and Bailey, 1968)

*cies* and *caldera outflow facies*. These facies are related to eruption and outflow of pyroclastic material with concurrent collapse of the source area. Such collapse may be followed by development of a resurgent cauldron as shown in Fig. 13-5.

The intra-caldera facies within the subsided area may include ignimbrite deposits measuring hundreds of meters in thickness. If resurgence occurs, the resulting moat may be filled by pyroclastic rocks, lava flows, lake sediments, epiclastic volcanic sediments, and particularly by landslide or talus breccias from the caldera wall. The caldera-outflow facies is characterized by ignimbrite sheets that may extend for many tens of kilometers outside the caldera. These are at least partly time-correlative with the thick ignimbrite-filling of the intra-caldera facies. Resurgence may occur without filling, and filling may occur without resurgence. Moreover, fills within calderas can be derived from younger ash flows from any nearby younger source.

**Facies Based upon Environment of Deposition**

In the broadest sense, an environment of deposition includes all physical, chemical, biological, and geological features that affect sedimentation within a specific area. Pettijohn (1975) holds that we should discriminate between local environment, best defined in geomorphic terms (alluvial fan, delta, lacustrine, etc.) and the tectonic environment, which includes the relation of sediment accumulation to larger tectonic elements such as cratons and geosynclines. We discuss here only the local environments.

Physical, chemical and/or biological parameters may be investigated to determine salinity (marine or nonmarine environments), agent of transportation (fluvial, glacial, eolian) and others, but more commonly, studies are directed toward determining the total environment of deposition, thereby integrating several parameters. Wolf and Ellison (1971) recognize four facies types of volcaniclastic rocks derived from contemporaneous volcanism and erosion of older rocks (Fig. 13-6). Notable is a comprehensive study by Hay (1973, 1976) of the sediments (volcaniclastic) in a semiarid basin at Olduvai Gorge, Africa. Schmincke et al. (1973) and Schmincke (1977b) illustrate channel-facies, fan-facies, and overbank-facies of pyroclastic flows and pyroclastic fallout deposits at Laacher See, Germany.

An important direction in stratigraphic environmental analysis is the facies model concept which is the conceptual scheme linking a particular vertical sequence to a particular sedimentary environment (see Selley, 1978). Various facies models developed by stratigraphers include rock sequence characteristics of alluvial fans, deltas, deep sea fans and many of their subfacies. Development of such facies models has yet to be fully exploited in studies of volcaniclastic rocks, although an instructive summary of developing facies models for volcaniclastic rocks is given by Lajoie (1979), who describes some of the vertical and lateral variations and the kinds of volcanic assemblage characteristic of flow breccias, hyaloclastites, pyroclastic fall, and pyroclastic flow deposits (see also Tassé, et al., 1978). Many of these facies aspects are discussed in our chapters on individual types of deposits, and others are summarized below.

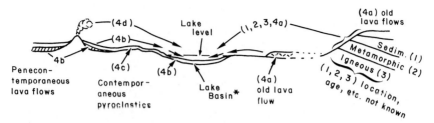

Provenance of sedimentary rocks in lake basin (Pliocene Lake Beds, Oregon)

Key to origin of sediments is as follows:

| Lithology of source | Age of source | Location of source | Mode of origin of grains |
|---|---|---|---|
| (1) Sedimentary | old | extrabasinal | epiclastic |
| (2) Metamorphic | old | extrabasinal | epiclastic |
| (3) Plutonic | old | extrabasinal | epiclastic |
| (4) Volcanic | (a) old | extrabasinal | epiclastic |
|  | (b) penecontemporaneous | extrabasinal | epiclastic |
|  | (c) contemporaneous | extrabasinal | pyroclastic: extensive reworking and transportation into Rome basin |
|  | (d) contemporaneous | intrabasinal | pyroclastic: (*i*) no reworking (*ii*) slight reworking, (*iii*) extensive reworking |

*Lake basins facies in western part of lake are organized as follows:

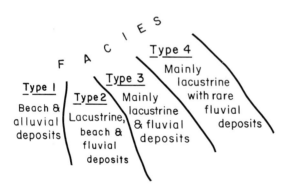

Fig. 13-6. Lacustrine volcaniclastic rock facies in an eastern Oregon basin. (After Wolf and Ellison, 1971)

## Facies Based upon Primary Composition

The highly complex stratigraphic rock associations near a volcanic source generally present facies problems that differ from those farther from the source where reworked material from other sources may be interbedded or intermixed. But in both facies, a petrologic-chemical approach is generally used to separate rock assemblages into broad compositional categories which minimize the need to determine the precise stratigraphic relations of individual beds. These facies categories are discussed below under the headings of (1) compositional facies and (2) petrofacies.

## Compositional Facies

Compositional facies consist of a petrologically and chemically related association of volcanic rocks that are essentially time-equivalent but may be separated in space. Lateral discontinuities frequently render them unsuitable to map as rock-stratigraphic units of formation or other ranks. A single compositional facies can consist of a set of highly lenticular rock masses composed of lahars, fallout layers, ash-flow deposits and associated erosional products, deposited in different environments by different processes. To give a general example, andesite lava flows and tuff may occur on different sides of a volcano, but may have been emplaced from the same eruption or at different times from separate andesitic eruptions. Later erosion can completely destroy evidence of either age relation. However, if a previous cycle of eruptions were dominantly dacitic and a later cycle were dominantly basaltic, the products at different localities of the volcanic pile would show the same vertical compositional succession, but not necessarily all parts of the succession at every locality.

A more specific example illustrating compositional facies in older volcanic terranes is Independence Volcano, an Eocene stratocone in the Absaroka Volcanic Field, Wyoming, eroded nearly to its roots (Rubel, 1971). Erosion of the former cone contributed thick volcaniclastic accumulations to the surrounding low-lying areas. At the volcanic center within the cone complex, abundant minor unconformities prohibit subdivision of individual beds or flows into correlative units and,

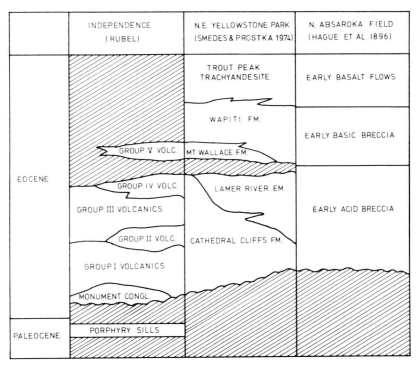

**Fig. 13-7.** Correlation chart for volcanic formations, Northern Absaroka volcanic field. (After Rubel, 1971)

Table 13-4. Compositional facies of Independence volcano. (After Rubel, 1971)

| | |
|---|---|
| Facies V | Trachybasalt flows; minor felsic tuffs |
| Facies IV | Conglomerate, and sandstone reworked from Facies I through III |
| Facies III | Sequence of volcaniclastic debris of dacitic, quartz latitic, and trachyandesitic composition |
| Facies II | Hornblende-quartz latite flows |
| Facies I | Sequence of lava flows, tuffs, pyroclastic breccias composed of rhyodacite, quartz latite, trachyandesite, and basalt |

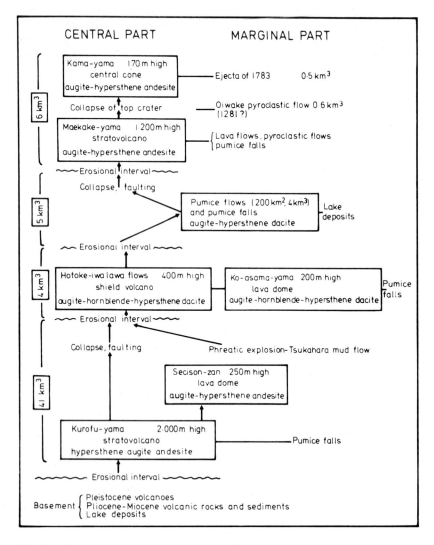

Fig. 13-8. Block diagram showing development of Asama Volcano. *Rectangular frames* represent times of construction of volcanic edifices. *Arrows* show confirmed stratigraphic succession. Volumes shown at *left*. (After Aramaki, 1963)

as should be expected, no single mappable unit can be traced completely around the periphery of the present-day intrusive complex identified as the source. Eruption, deposition, erosion, and redeposition occurred together, and their relative importance varied unsystematically in time and space.

Because of these stratigraphic complexities, the varied rock types of Independence Volcano are subdivided into five compositional facies. Correlation of these facies with Eocene volcanic rocks in Yellowstone Park and in the northern part of the Absaroka volcanic field are given in Fig. 13-7. These facies cannot be defined as formations. Rock types within these various facies, from youngest to oldest, are shown in Table 13-4.

An example of compositional facies in a young volcanic terrane is given by Aramaki (1963) for Asama Volcano, Japan. Asama Volcano had a long and complex growth history which included the construction of several edifices consisting of mafic andesite to silicic dacite erupted and deposited in a variety of ways. The stratigraphic subdivision of Asama rocks is somewhat similar to the compositional facies of Independence Volcano, but differs in detail because the volcanic edifice (Asama) still exists and the rocks mapped can be related to different phases of cone building (Fig. 13-8).

Compositional facies also may be characterized in terms of magma suites such as the calc-alkalic suite, etc., but the concept of magma suites far transcends local volcanic centers, and has more significance in broad-scale tectonic interpretations (Chap. 14).

Petrofacies

Sandstone sequences with a stratigraphic and geographic distribution of distinct detrital mineral populations constitute sandstone petrofacies (Dickinson and Rich, 1972; Ingersoll, 1978). Petrofacies studies can be used to establish gross stratigraphic relationships in areas of complex structure and are especially applicable to pyroclastic or epiclastic volcanic rocks, as shown, for instance, by Stanley (1976) in the Cenozoic High Plains sequence of eastern Wyoming and Nebraska. Sandstone detritus may come from tephra that falls directly into a basin and mixes with sediments from other sources, or be derived from the resedimentation of volcanic material that has changed in composition through time, or from erosion that cuts downward through successively older rocks of different composition. Petrofacies may be more persistent laterally than lithofacies based upon grain size, bedding style, mode of emplacement or environment of deposition. Such parameters change laterally, but the composition of the clastic debris that is supplied from a particular source will be recognizable.

Parameters of petrofacies distinguished by Dickinson and Rich (1972) include modal percentages of quartzose grains (Q), feldspar grains (F), and unstable lithic grains (L) expressed as percentages of the Q-F-L population recalculated to 100 percent, as well as the ratios (1) plagioclase to total feldspar grains, P/F, and (2) volcanic rock fragments to total lithics, V/L. By using these parameters, five stratigraphically controlled petrofacies were recognized in correlative petrologic intervals within the Great Valley Sequence of California and correlated with intrusive and possibly coeval volcanic episodes in the Sierra Nevada and Klamath Mountains, California and northern Nevada (Fig. 13-9).

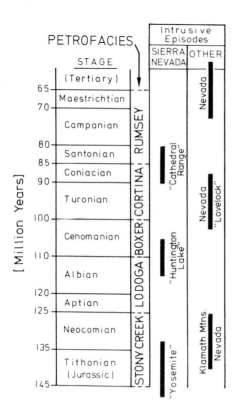

Fig. 13-9. Correlation of clastic petrologic intervals of Great Valley Sequence, biostratigraphic stages and various intrusive episodes. (After Dickinson and Rich, 1972)

## Diagenetic Rock Facies

Diagenesis, a subject explored in Chapter 12, is controlled by the composition of the pore fluids, pH, Eh, and temperature conditions, as well as by initial composition of the rocks. Thus, the resultant diagenetic minerals may be used to interpret the physicochemical conditions of the environment of deposition or chemical environment within a sequence of already deposited rock containing either circulating or stagnant groundwater.

An illustration of diagenesis of dacitic tuff is the Late Oligocene-Early Miocene John Day Formation (750 to 1000 m thick). Alteration took place shortly following deposition within different physiographic environments (lakes, hilltops) (Fisher, 1968b) as well as following relatively deep burial, by circulating groundwater (Hay, 1963).

Present-day color imparted to the John Day Formation by diagenesis has influenced its subdivision into members. Early in the study of the formation, for example, color was the main criterion for its subdivision (Merriam, 1901), but it is now known that diagenetic alteration broadly cuts across lithologic boundaries (Fisher and Rensberger, 1973). This early subdivision based upon color led to problems in correlating the abundant vertebrate fauna from place to place within the formation, thereby causing incorrect conclusions about the stratigraphic placement of the fossils and their evolutionary sequence. From the base, the formation at many places is essentially red overlain by green and then topped by white tuff beds (Fig. 13-10).

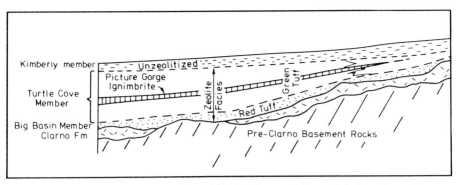

Fig. 13-10. Diagrammatic facies relationships within the Oligocene-Miocene John Day Formation, eastern Oregon. (After Fisher and Rensberger, 1973)

Hay (1963) divided the John Day Formation into two main diagenetic facies: the fresh glass facies (white upper part) and the clinoptilolite facies divisible into a montmorillonite subfacies (drab or reddish) and a celadonite subfacies (green rocks). These facies formed prior to, during, and after mild folding.

Groundwater responsible for zeolite genesis originated as meteoric water, which increased in pH and ion concentration downward through the beds (Hay, 1963). Inferred Eh and pH conditions for the environment of deposition of the drab and red rocks of the lower part of the formation are given by Fisher (1968b). The initial "weathering" color (red beds) of the lower part of the formation apparently was not significantly altered by groundwater circulating through the formation after burial.

## Stratigraphic Examples

Volcanism takes place over vast areas which include smaller subdivisions such as regions, fields, and centers that may overlap in time and space (Table 13-5). A model study of a large region that synthesizes abundant age and petrochemical data and its implications in the plate tectonic evolution of the western United States is given by Lipman et al. (1972) and Christiansen and Lipman (1972). While the object of their study is not stratigraphic, it is based upon stratigraphic-petrologic-chemical studies of many volcanic fields and centers, and so illustrates the ultimate end toward which some studies of volcanic rocks may be directed.

Within this section, out of numerous excellent examples we select and briefly outline three such studies, arranging them in order of increasing age of the rocks: the first, a single volcano (Oshima Volcanic Center, Japan); the second, a volcanic field (San Juan Mountains, USA.) whose products were vented from many volcanic centers; and a third which illustrates Archean greenstone-belt volcanoes in Canada. These studies illustrate different approaches, but show the kinds of information that students of volcanic rocks in general, and of pyroclastic rocks in particular, can obtain.

**Table 13-5.** Geographic-volcanic subdivisions

*Volcanic region:*

Geographic area of large size which includes more than one volcanic province in which volcanism spanned eras of geologic time. Most are linear and these are called volcanic belts or volcanic chains. Example: Cordilleran Volcanic Region which includes volcanic provinces within Alaska, the Cascade Mountains. Sierra Nevada. Rocky Mountains. Mexico and western South America.

*Volcanic province:*

Geographic area of volcanic rocks which includes more than one volcanic field (or district) of the same or differing time spans. Rocks within a province may overlap in time and space, form a continuous zone or be erosionally disconnected. A volcanic province may be either elongate or irregular in map view.
Examples: Cascade Volcanic Province divisible into two overlapping fields or districts: (1) the Western Cascade Range (late Eocene to late Miocene) and (2) the High Cascade Range (Pliocene to present). Various Pacific island arcs.

*Volcanic district or field:*

An association of consanguineous volcanic rocks that includes more than one volcanic center. Volcanism may span long periods of time. Centers may be arranged in volcanic chains.
Example: San Juan Volcanic Field in the Rocky Mountains of Colorado and northern New Mexico.

*Volcanic center:*

Area that includes one source or many closely spaced sources of volcanic rocks. May consist only of central vents or dikes if deeply eroded. Land forms generally include stratovolcanoes, cinder cones, and domes. The products of a volcanic center also may form volcanic plains or plateaus.

*Volcano:*

Pyroclastic debris or lava flows may pile up around a vent or series of vents to produce a constructional landform (hill, mound, mountain) called a volcano, measuring from several tens of meters at its base to several kilometers: commonly with a centrally located crater or caldera.

*Volcanic vent:*

A relatively small, generally roughly circular or elongate area from which volcanic products have erupted. May be filled with pyroclastic breccias, landslide blocks and/or talus from the crater rims. Dikes or dike swarms radiate from the vents of some volcanoes. Erosion progressively exposes solid plugs or more rarely pipe breccia complexes, dike complexes and finally transitions to plutonic rocks such as granite.

## Oshima Volcano, Japan

Nakamura's (1964) study of Oshima Volcano was one of the first to deal specifically with the pyroclastic stratigraphy of a single active volcano. The principles developed by Nakamura are applicable to all volcanoes that contain significant portions of pyroclastic debris. Oshima is particularly ideal for pyroclastic stratigraphic studies because its form is simple, the ejecta are fresh, and stratigraphic units are thinner and therefore easier to work with than most andesitic or dacitic volcanoes. Because the island has been continuously inhabited for the last 5000 years, archeological remains are important in determining the chronology of the younger deposits.

Oshima is part of the Izu volcanic chain extending southward from central Honshu, Japan and is built upon three underlying basaltic stratovolcanoes. Nearly all of the island was built by ejecta from the central vent area, but the nearly 40 parasitic volcanoes located on the flanks of Oshima Volcano have also contributed volcanic products to the island's growth.

During late Pliocene time, three basaltic stratovolcanoes were constructed and then beheaded by marine erosion, leaving only a group of small islands prior to the birth of Oshima Volcano. Several thousand years ago, submarine phreatomagmatic activity began on the shallow sea floor at the northwestern part of the present-day island. The products of these eruptions soon blanketed several dissected islands and formed a new volcanic island. Then the subaerial volcanism of Oshima Volcano proper began, and has continued recurrently to the present, producing a cone of pyroclastics and lavas totalling about 45 km$^3$ in volume. Nakamura divides the Oshima volcanic products into an Older Oshima Group, and a Younger Oshima Group, with no great time break between them. The younger group was deposited over the last 1500 years, and is coincident with, or younger than, the formation of a summit caldera on Oshima Volcano.

The Younger Oshima Group (3.6 km$^3$ in volume) forms the basis for the stratigraphic study. Because of the petrologic similarities of the erupted products of this later volcanic activity, there are no readily identifiable marker beds around the volcano, requiring the stratigraphic history to be pieced together from many individual sections.

The smallest subdivision which Nakamura defines to aid in reconstructing the eruptive history of Oshima is the fallout unit, which is a single bed with distinctive physical properties that decreases in grain size and thickness away from the source and covers an elliptical area within which the vent is located at one focus. Because a single fallout unit is difficult to trace laterally, Nakamura defined larger units, called members, as the basic stratigraphic subdivision for island-wide mapping and correlation. On Oshima, a member consists of several fallout units, in places containing intercalated lava flows, which show a weathered zone at their top. Each member is composed of a different succession of beds. Members are distinguished from one another by position in sequence, the color of the upper weathered zone and their internal stratigraphic sequence (Fig. 13-11). Their thickness distribution is given in Fig. 13-12. Toward the summit of the volcano, fallout units within members increase in number, thickness, and grain size. Unconformities become more pronounced but weathering profiles are more poorly developed than on the volcano's flanks.

With a few exceptions, the common succession within a member (see Fig. 13-12) is, from youngest to oldest:

e) Brown loamy layer (weathered horizon).
d) Explosion tuff breccia, in places interbedded with layers of accretionary lapilli tuff.
c) Alternating ash fall beds, in places with minor amounts of scoria-fall and accretionary lapilli.
b) Lava flows.
a) Scoria beds.

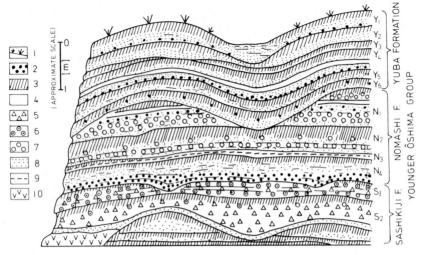

**Fig. 13-11.** Idealized sketch of road-cut on mid-slope of Oshima volcano. *1* Present ground surface, *2* weathered ash or soil, *3* fine volcanic ash, *4* tuff breccia, *5* accretionary lapilli tuff, *6* rounded lithic lapilli, *7* coarse volcanic ash, *8* rhyolite ash in the $N_3$, *9* lava flow, *10* agglutinated driblet. (After Nakamura, 1964)

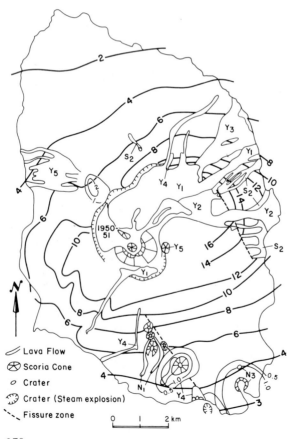

**Fig. 13-12.** Distribution of Younger Oshima Group. Isopachs of fallout deposits in meters. Designated stratigraphic units ($S_2$, etc.) as shown in Fig. 13-11. (After Nakamura, 1964)

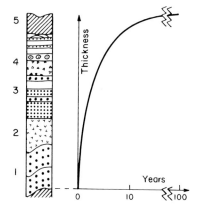

Fig. 13-13. Idealized succession of deposits in a member, and a time versus thickness curve. Note that each member is defined as a cycle beginning with rapid extrusion of explosive (and some effusive) products then decreasing in rate of extrusion and ending with a period of weathering and reworking. Each member is lithologically distinct but is not mappable in the field. Each succession may be lithologically defined but is given a cyclic genetic interpretation. (After Nakamura, 1964)

5. Weathered ash
4. Reworked ash and phreatic deposits
3. Ash, interbedded with phreatic breccia and accretionary lapilli
2. Lava flow
1. Fallout scoria

Not all of these components are present in every member, and some members are interbedded with products from the parasitic vents.

Each member is defined as an eruptive cycle with high initial but declining later eruptive rates (Fig. 13-13). In our terminology, however, each "cycle" represents an eruption consisting of several phases, but rather than call the products a member, as Nakamura does, in this discussion we call them an eruption unit. Each eruption begins with an initial explosive phase (scoria falls) that grades to the next phase when lava is extruded. The explosion products represent the first release of volatiles, whereas the lava flows are extruded as the head of the magma column rises and gradually loses most of its volatiles. Following lava eruption is a phase when the ejection of ash predominates. The intermittent ash falls suggests that the magma level fluctuates as inferred from the following: coarse breccias at the uppermost horizons in a few members suggest that, in the final stages of eruptions, the magma column lowers to below the groundwater table as probably occurred at Kilauea Volcano, Hawaii, in 1924 (Stearns, 1925). Such hydroclastic events are also suggested by the occurrence of accretionary lapilli in the ash of several members.

Unconformities between formations as defined by Nakamura do not represent any more time than those between members, although some are separated by fairly distinct regional unconformities that can be recognized in several stratigraphic sections. However, Nakamura does not treat the formations as mappable units in the formal sense required by the North American Commission on Stratigraphic Nomenclature, but each has volcanological significance. The Yuba and Nomashi Formations (Fig. 13-11) show a decrease in volume percent of scoria and an increase in amount of ash upward in the section. This is interpreted by Nakamura to indicate that the internal pressure of the magma gradually decreased through time. Deposits of the Sashikiji Formation differ from the later formations by their greater abundance of breccia.

### San Juan Volcanic Field, USA

The San Juan volcanic field, Colorado, USA, is a volcanic complex containing many separate centers (Fig. 13-14). Covering about 25,000 km² in Oligocene and

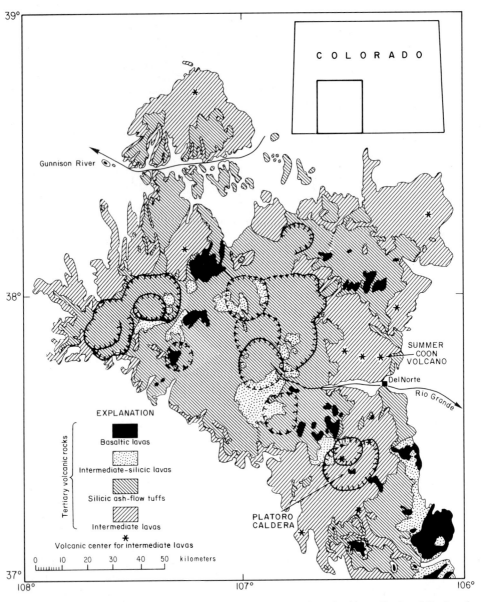

Fig. 13-14. San Juan volcanic field in Colorado showing location of calderas (*hachured lines*) and stratovolcanoes. (After Lipman, 1968; Lipman et al., 1970; Steven and Lipman, 1976)

later time (Steven, 1975), it was a nearly continuous volcanic field that extended over most of the southern Rocky Mountains and adjacent parts of the Colorado Plateau. Late Cretaceous and Paleocene volcanism which occurred earlier in the western part of the field is not treated here.

Larsen and Cross (1956) had described the San Juan volcanic field earlier but without the aid of modern isotopic age-dating techniques and the modern concepts

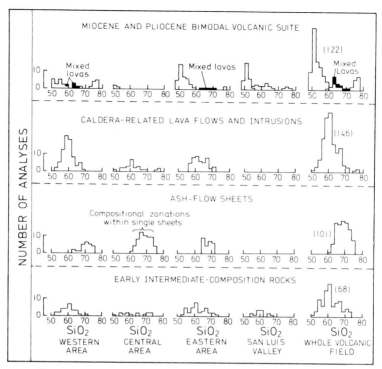

**Fig. 13-15.** SiO$_2$ content of igneous rocks, San Juan volcanic field. (After Lipman et al., 1978)

of volcanic processes and caldera mechanics available to, and in part pioneered by, later workers (Steven et al., 1967; Lipman et al., 1970; Lipman, 1975; Lipman et al., 1978). Other important studies of the San Juan field include Luedke and Burbank (1963, 1968), Steven and Ratté (1964, 1965), Steven and Lipman (1968, 1976), and Lipman et al. (1973).

The San Juan volcanic field evolved in essentially three major volcanic episodes, each of which consists of two distinct suites of igneous rocks (Fig. 13-15). From oldest to youngest these are an Oligocene calc-alkalic suite including: (1) intermediate-composition lavas and breccias of the San Juan, Lake Fork and West Elk Formations in the western San Juan Mountains, and of the Conejos Formation in the eastern San Juan Mountains and (2) silicic ash-flow tuffs erupted from numerous calderas with penecontemporaneous eruption of intermediate to silicic lavas from many centers both closely associated with and independent of the calderas; and (3) a Miocene-Pliocene bimodal suite of basaltic lavas and silicic alkalic rhyolite flows and tuffs. The radiometric ages and stratigraphic position of these rocks are given in Fig. 13-16.

The voluminous calc-alkalic suite, interpreted by Lipman et al. (1978) to be related to subduction along the western margin of the American plate, has chemical and isotopic characteristics which suggest complex interactions with a Precambrian cratonic lithosphere and appears to record the rise, differentiation, and crystallization of a large composite batholith beneath the volcanic field (Fig. 13-17).

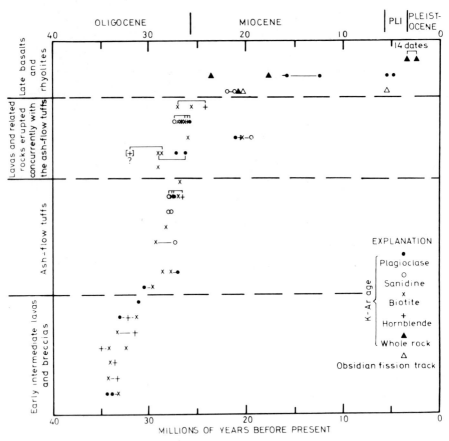

**Fig. 13-16.** Ages of minerals and rocks arranged in stratigraphic sequence, San Juan volcanic field. (After Lipman et al., 1970)

The Miocene-Pliocene bimodal suite originated in an extensional tectonic environment with the basalts thought to represent magma of upper mantle origin and associated rhyolites which may be partial melts of lower crust generated by the thermal effects of basaltic magma emplacement.

Intermediate-composition volcanism (35–25 m.y.) occurred at many volcanic centers that constructed large andesitic to rhyodacitic stratovolcanoes (Steven et al., 1974), one of which (Summer Coon Volcano) is described below. The rocks derived from the many centers can be divided into a near-source facies and an intermediate-source facies. The near-source facies consists of lava flows, flow breccias, and explosion breccias of the early volcanic activity; the intermediate-source facies consists of lahars, conglomerates, and other sedimentary deposits derived from the volcanic centers, and fills topographic lows between volcanoes.

Summer Coon Volcano (Fig. 13-18), an instructive example of a cone eroded to its roots, is one of the early intermediate-composition centers (Lipman, 1968). Rocks of the roots of this volcano occur in three facies: (1) a central nearly circular complex of diorite to quartz monzonite porphyry; (2) lava flows and breccias com-

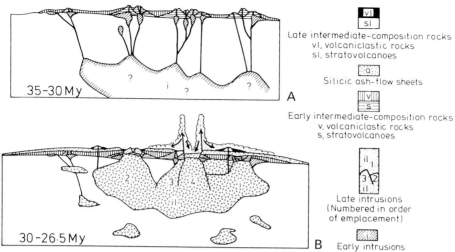

Fig. 13-17 A, B. Diagrammatic model of evolution of Oligocene batholith, San Juan Mountains. A During time of early intermediate volcanoes, B during time of pyroclastic flow eruptions and formation of calderas. (After Lipman et al., 1978)

prising the dissected remains of the cone and dipping 15° to 20° away from the central complex; and (3) hundreds of radial dikes composed of mafic, intermediate, and silicic composition. The outer part of this root zone of the eroded cone is covered by younger lava flows and pyroclastic flow deposits from other centers. Although the radial dikes do not intersect one another, their genetic associations are inferred from the stratigraphic sequence of petrographically similar flows and breccias that dip outward from the core. From oldest to youngest, this sequence is: (1) a thick stratified unit of mafic breccias of probable debris flow and explosive origin with subordinate lava flows; (2) a small rhyolite flow; and (3) intermediate-composition lava flows and breccias. This stratigraphic stacking-order of rocks appears to preclude differentiation of a single parent magma by simple fractional crystallization.

In the San Juan field as a whole, major volcanic activity changed from intermediate-composition eruptions to explosive ashflow eruptions of quartz latite and low-silica rhyolite about 30 m.y. ago. More than 15 major sheets of ash-flow tuff were then erupted from source areas now marked by major calderas in the western, central, southeastern, and northeastern parts of the field (Fig. 13-14). The ash-flow sheets extend over 25,000 km$^2$ and have an estimated total volume of 20,000 km$^3$. Some individual sheets, with a volume as great as 3000 km$^3$, once covered as much as 15,000 km$^2$.

The Platoro caldera complex is one of the many calderas in the region (Lipman, 1975). It is a composite structure about 20 km in diameter that collapsed coincident with the emplacement of a 29 to 30 m.y. old composite ash-flow sheet known as the Treasure Mountain Tuff which itself consists of a coextensive assemblage of ash-flow sheets and associated rocks divided into six members. Figure 13-19 shows the distribution of one of the more extensive members. Similar chemistry, pheno-

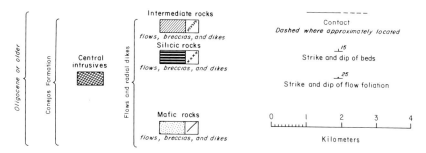

**Fig. 13-18.** Geology of Summer Coon Volcano, San Juan Volcanic field (see Fig. 13-14 for location), illustrating volcaniclastic facies of an eroded stratovolcano. (After Lipman, 1968)

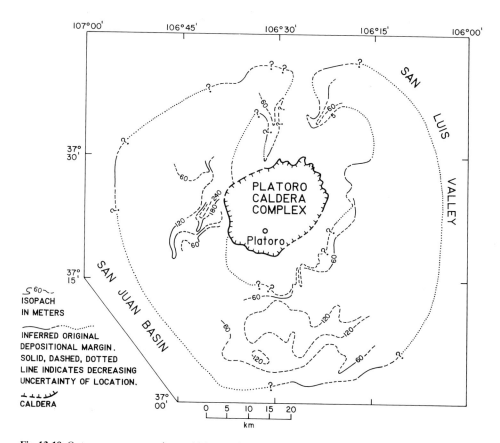

**Fig. 13-19.** Outcrop area, approximate thickness of selected sections, and inferred original depositional extent of the La Jara Canyon Member, Treasure Mountain Tuff, Platoro Caldera, Colorado. The La Jara Canyon Member is a multiple-flow compound cooling unit that makes up the first extensive ash-flow sheet from the caldera. The outflow sheet is up to 250 m thick, extending as far as 30 km; in the caldera it is > 800 m thick where it ponded concurrently with subsidence. (After Lipman, 1975)

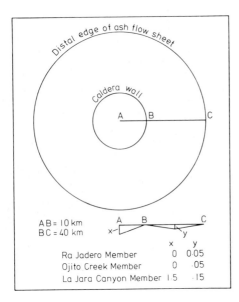

**Fig. 13-20.** Geometric model for volume calculations of three ignimbrite sheets of the Treasure Mountain Tuff. $x$ and $y$ are approximate thicknesses in kilometers. Platoro Caldera (Colorado). See Fig. 13-14 for location. (After Lipman, 1975)

cryst mineralogy and areal distribution of the ash-flow members of the Treasure Mountain Tuff indicate that they are closely related, but despite their similarities, the units are in part distinguished by different phenocryst proportions and $SiO_2$ content and by different natural remanent magnetic polarities.

Three of the Treasure Mountain Tuff members are ash-flow sheets each of which extend over about 5000 km². Mapping of these three members suggests a simple geometric model that illustrates how their original volumes can be estimated (Fig. 13-20).

Interlayered with and overlying the major ash-flow sequence are intermediate to silicic lava flows and breccias with a probable aggregate volume of 5000 to 10,000 km³. Some of these rocks are clearly related petrographically and chemically to the ash-flow tuff sequence and closely related in time to the caldera collapse structures. Other interbedded rocks are similar to the early intermediate-composition rocks that preceded caldera development, and may represent a waning continuation of the early intermediate-composition eruptions.

**Archean Greenstone-Belt Volcanoes, Canada**

The volcanic heritage of Archean greenstone belts in Canada has long been known, but only in the last decade have detailed studies of volcanostratigraphy in these regions been started (Buck, 1976; Ayres, 1977; Dimroth and Demarcke, 1978; Tassé et al., 1978). Fragmental rocks of greenstone belt systems commonly contain Cu-Zn-Ag and Au-Ag deposits, and their exploitation has spurred modern research. A totally different approach to volcanostratigraphy and paleovolcano reconstruction is required in regions where volcanism is recorded only in metamorphosed but layered stratigraphic sequences formerly deposited on volcano flanks and depositional basins. Such an approach relies heavily upon unmetamorphosed volcano analogs to develop constructional and depositional models, and mimics the mod-

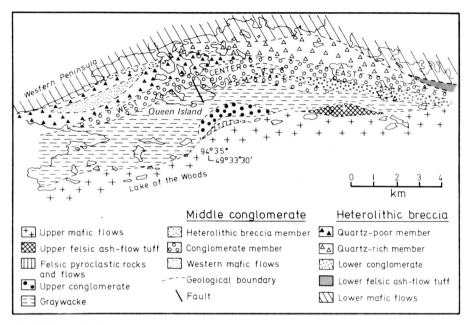

Fig. 13-21. Isoclinally folded Precambrian volcanic sequence within a greenstone belt, Western Peninsula of Lake of the Woods (Ontario, Canada). (After Ayres, personal communication)

Fig. 13-22. A Idealized model of development of a felsic to intermediate-composition pyroclastic cone on top of a mafic shield volcano. Calculations are based upon assumption that initial felsic to intermediate pyroclastic volcanism began at a water depth of 100 m. B Idealized model of a second-generation cone developed upon a wave-eroded first-generation pyroclastic cone assuming no isostatic downsinking. (After Ayres, 1983)

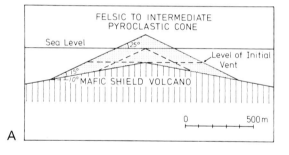

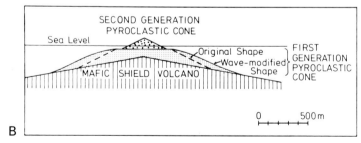

ern sedimentary facies approach for linking vertical sequences to sedimentary environments (Selley, 1978). Rock types described by Ayres (personal communication) include many of those discussed in this book, and the study presents a model of volcano-edifice reconstruction based upon facies analysis and interpretation of

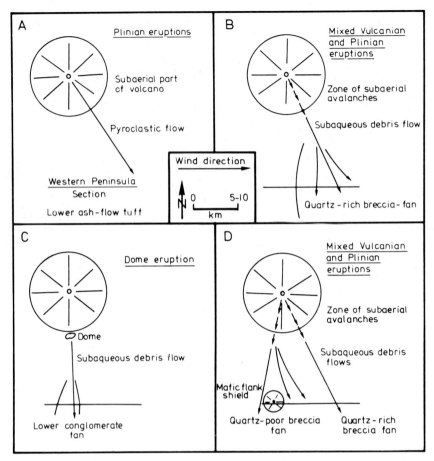

Fig. 13-23 A–D. Inferred paleogeography and events giving rise to deposits of the lower part of the Archean Western Peninsula fragmental sequence. **A** Lower pyroclastic flow tuff formation; **B** quartz-rich member of the heterolithologic breccia formation; **C** Lower conglomerate formation; **D** quartz-poor member of heterolithologic breccia and a western mafic formation. (After Car, 1979)

the origin and emplacement mechanisms of the rocks. Interpretation of recovered deep sea cores also requires use of this technique (e.g. Schmincke and von Rad, 1979) (Chap. 10).

Western Peninsula of Lake of the Woods, Ontario, Canada shows an isoclinally folded sequence which includes a 4000-m-thick rock sequence, within which is a group of dominantly dacitic fragmental rocks bounded at the top and bottom by basaltic pillow lavas and massive flows (Fig. 13-21). Despite a metamorphic grade of lower to middle greenschist facies and a variable metamorphic foliation overprint, many primary textures and structures of the volcanic rocks are preserved.

The dacitic fragmental rocks have been divided into nine formations with a total of up to 2500 m in thickness. The principal rock-stratigraphic units consist of heterolithologic breccia, monolithic conglomerate, graywacke, minor ash-flow tuff, monolithologic breccia and lava flows. The over- and underlying basaltic pillow and massive flows are interpreted to be of subaqueous origin.

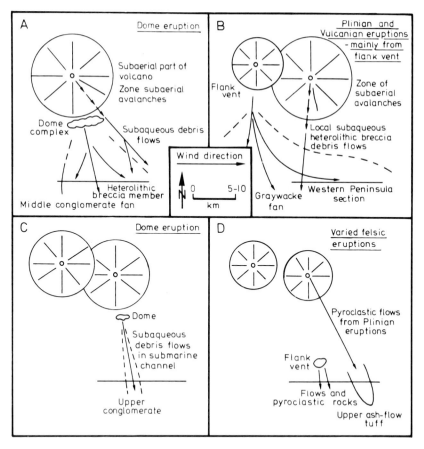

Fig. 13-24 A–D. Inferred paleogeography and events giving rise to deposits of the upper part of the Western Peninsula fragmental sequence. **A** Middle conglomerate formation; **B** graywacke formation; **C** upper conglomerate formation; **D** upper pyroclastic flow tuff and felsic formations. (After Car, 1979)

Ayres (personal communication, 1982) presents a paleovolcano model (Fig. 13-22), based upon evidence from the fragmental stratigraphic sequence and with the following assumptions: (1) the eruption of large volumes of pyroclastic ejecta commonly results in construction of a subaerial cone or stratovolcano, (2) most of the subaerial volcanic debris will be transported down the flanks of the volcano by several transporting processes and ultimately be deposited on the subaqueous flanks as heterolithic, remobilized pyroclasic ejecta or erosional epiclastic material, (3) primary pyroclastic material deposited subaerially will be preserved only rarely because of erosion, (4) a major mode of transport of reworked pyroclastic debris is by avalanching or debris flows either subaerially or subaqueously and (5) structures in the deposits record only the final transport stage, but earlier transport modes can be inferred from clast shape, abundance, composition, sorting, and other features.

The sequence described by Ayres is interpreted to be part of the subaqueous flank of a large felsic to intermediate-composition stratovolcano constructed on a platform of a large mafic subaqueous to locally subaerial shield volcano. The lower part of the fragmental sequence rests on pillow basalt and consists of two felsic porphyritic tuffs interpreted as pyroclastic flows produced by Plinian eruptions because: (1) pumice is present in the rocks, (2) crystals are abundant and many are broken, (3) small monolithologic lapilli similar in composition to the rest of the rock are present, and (4) layering can be defined by variations in crystal content but well-defined bedding is absent. The very fine-grained and strongly foliated matrix is interpreted to represent recrystallized glass shards.

At three stratigraphic levels monolithologic conglomerates occur that show features characteristic of debris flow deposits. These are considered to be parts of submarine fans. The conglomerates are thick-bedded with diffuse bedding contacts, and are poorly sorted with large rounded felsic clasts in a fine- to medium-grained matrix. Because the conglomerates do not have admixtures of compositionally different rock types, they are interpreted to be derived from possible dacite domes located between the main vent of the volcano and the depositional area. Rounding of dacite clasts may have occurred in a beach environment. The conglomerate formations are separated by heterolithologic breccia and graywacke formations.

Some heterolithologic breccias are quartz-rich, others are quartz-poor. The quartz-rich heterolithologic breccias, which overlie the porphyritic pyroclastic flows within the lower part of the sequence, are interpreted to be subaqueous debris flows because the breccias are poorly sorted and have a moderately high matrix content which supports large angular clasts. The clasts are intermediate volcanics (40%), felsic volcanics (25%) and mafic volcanics (5%). Ayres further suggests that because of their large size and angularity, the lithic clasts were derived initially from Vulcanian eruptions, rather than erosion of flows. The quartz crystals, on the other hand, are believed to be quartz phenocrysts that were liberated from dacitic magma by Plinian eruptions. To produce this mixed clast population from essentially monolithologic ejecta without rounding of particles, Ayres suggests transport by avalanches and debris flows that incorporated previously erupted debris from underlying pyroclastic deposits.

The final major stage of fragmental volcanism is recorded in thick deposits of graywacke of turbidite origin. The distribution pattern and the composition of clasts in the graywacke sequence suggest that turbidity currents were fed by Plinian eruptions from more westerly sources than the source of the breccias. Penecontemporaneity of graywacke deposition with volcanism is indicated by interbedded volcanic breccia, lithic clasts in the graywacke of dominantly felsic to intermediate volcanic composition, euhedral and subhedral shapes of feldspar clasts, and the greater abundance of plagioclase relative to quartz. The change to turbidite deposition of graywacke indicates an abrupt shift in depositional regime. This shift could reflect one or all of the following: (1) lowered relief of the source volcano, (2) change of eruptive character from Vulcanian and dome extrusions to Plinian eruptions, and (3) development of new and more distant source vents.

Reconstructed paleogeography and volcanic events based upon the stratigraphic analysis of the Western Peninsular section are illustrated in Figs. 13-23, 13-24. Ayres' study well illustrates the importance of stratigraphic analysis in areas where volcanic sources have been erased, and the value of modeling based upon careful description and interpretation of the rocks.

# Chapter 14  Pyroclastic Rocks and Tectonic Environment

The close proximity of continental borderlands, active and ancient tectonic regions, and modern and ancient volcanoes has long been known. Pre-plate tectonic ideas about this association culminated in Kay's (1951) masterful synthesis showing the relationship between volcanic island arcs and ancient eugeosynclinal sedimentary basins containing abundant volcaniclastic rocks. This association is now considered to develop within convergent plate margins. Wide acceptance of plate tectonic processes began during the late-1960's, and has resulted in a continuing and accelerating flood of literature concerned with reinterpreting a vast amount of descriptive geology in addition to new studies. Many formerly difficult geologic problems appear amenable to interpretation based on plate tectonic theory, but we add a note of caution: it is far from certain whether or not plate tectonic theory adequately explains tectonism and volcanism in Archean time which includes nearly three-quarters of earth history.

Interpreting geologic history in terms of plate tectonic processes is complex. Acceptable syntheses require bringing together the many facets of the geological sciences...tectonic, petrologic, geochemical, and geophysical. Here we briefly describe pyroclastic and erosional volcanic products and how they fit into the plate tectonic framework. Volcaniclastic rocks are a significant facet of the problem, for, as pointed out in Chapter 1, they were produced in large volumes at many points in geologic time. We hazard a guess that volcanically derived sedimentary material may in fact exceed the nonvolcanic contributions. As with many paleogeographic and paleotectonic reconstructions, most of the record is preserved within basins of sediment accumulation. Our main attention centers upon convergent plate settings associated with magmatic arcs along which large volumes of volcaniclastic sediments of epiclastic and pyroclastic origin occur. Volcanic arcs are arcuate volcanic chains that can occur as islands, hence volcanic island arcs, or else can occur on continental margins. Deep erosion of volcanic arcs exposes their plutonic roots, and such belts can be called magmatic arcs. Our main examples are drawn from the Cordilleran system of America.

## Convergent Margins, Magmatic Arcs, and Sedimentation

Convergent boundaries are characterized by many kinds of rocks and tectonic elements (Fig. 14-1). These various elements, their physiographic-bathymetric configurations and ideas about how they develop have been discussed by many authors (e.g. Burke and Drake, 1974; Dickinson, 1974c; Dott and Shaver, 1974). Several kinds of sedimentary basins form adjacent to magmatic arcs along continental

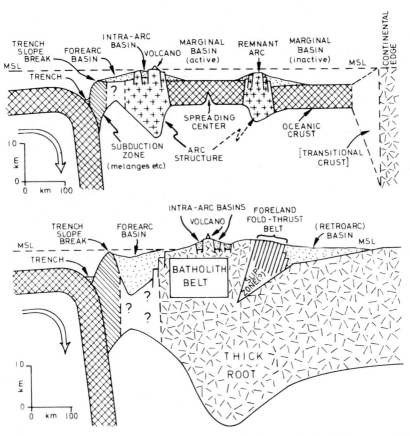

Fig. 14-1. Diagrammatic illustration of sedimentary basins associated with intraoceanic island arcs (*above*) and continental margin magmatic arcs (*below*). (After Dickinson, 1977)

margins (Dickinson, 1974a, b). These include: (1) the trench, (2) fore-arc basin, (3) intra-arc basin, (4) inter-arc (or marginal) basin within an arc-trench gap and (5) a retro-arc basin, the latter two being in back-arc settings. The types of sediments received within these basins (Table 14-1) depend upon the proximity to the source area, depositional environment (subaqueous and subaerial), whether or not the arc is active or inactive, and whether the basin is in front of, or in back of, the arc.

A generalized arc rock-assemblage includes intertonguing effusive volcanic rocks and volcaniclastic strata, intruded at depth by plutonic rocks which are compositionally related to the eruption centers (Dickinson, 1974b). Depending upon (1) the amount of time involved, (2) the possible recurrence of active volcanism, (3) the later structural and metamorphic events that may occur and (4) the rate of erosion, this assemblage may be eroded within the magmatic arc to any level at any stage of development to leave a complex lithologic assemblage. In areas adjacent to arc sources, the volume of volcaniclastic debris dispersed explosively, by reworking of newly erupted pyroclastic deposits or by erosion of solidified volcanic rocks can be far greater than the net volume of volcanoes on the arc at any given time.

Volcaniclastic rocks differ in composition depending upon the type of arc. Mafic varieties are most common in areas where adjacent plates are oceanic; inter-

**Table 14-1.** Possible dominant depositional facies and depositional environments in tectonic settings within and adjacent to magmatic arcs. (After Dickinson, 1974b)

| Setting | Crust is oceanic or semi-oceanic (<15–20 km) | Crust is continental or semicontinental (>15–20 km) |
|---|---|---|
| Volcanic fields | Andesite to mafic lavas and pyroclastics dominant: pillow lavas and pillow breccias common in central facies: biogenic reefs common as proximal facies on flanks of partly submerged cones | Andesite to silicic subaerial pyroclastics dominant: ignimbrites and pyroturbidites common in central facies: fluvial breccias common as fluvial and debris flow aprons on flanks of high cones |
| Intra-arc basins | Volcaniclastic turbidites and marine tuffs in deep troughs flanked by narrow belts of local shelf facies: proximal turbidites on steep submarine slopes | Volcaniclastic red beds and subaerial tuffs in grabens having lacustrine basin fills: also, unstable shelf facies, conglomerates and local unconformities |
| Arc-trench gaps | Turbidites deposited in deep troughs as wedges of subsea fans showing evidence of transverse paleocurrents, and as prisms of fill showing evidence of longitudinal paleocurrents parallel to arc trend | Fluviatile, deltaic, and prodeltaic beds of progradational coastal plain complexes built upon subsiding shelfs open to the sea or within partly enclosed embayments |
| Back-arc basins | Turbidite aprons built into deep interarc basins of marginal seas, or into open oceans | Piedmont or deltaic wedges built into interior seaways or fluvial lowlands of shallow foreland basins |

mediate and silicic varieties usually occur in areas where oceanic slabs slide down beneath continental lithosphere. Apparently reflecting the composition of their source areas, the volcaniclastic rocks become increasingly rich in $K_2O$ and $SiO_2$ (quartz) away from the oceanic margins where plutonic roots have been exposed, and thereby may be broadly used as indicators of paleotectonic environment along with the effusive and plutonic rocks. Effusive rocks are mostly basalt to andesite, but rhyolitic to dacitic ignimbrites are also abundant and widespread.

The kinds of volcaniclastic rocks and sequences also reflect the position of arcs (oceanic, marginal to continents, continental). The proportion of submarine mafic lavas is high in marine island arc areas and includes pillow basalts, hyaloclastites and tuffs associated with turbidites that sweep across narrow shelves around the volcanoes to form submarine canyons, deep-sea fans (Fig. 14-2) and rises that descend into bathyal to abyssal depths. Depending upon water depths and temperatures, fringing reef-forming carbonates may form parts of the stratigraphic sequences, and in the deeper bathyal to abyssal zones, small amounts of cherty nannofossil limestones and bedded cherts along with abundant clastic materials form sequences of up to 15 or 30 km in thickness – much thicker than in trenches where such deposits rarely exceed 3 km in thickness. Locally, silicic subaqueous fallout and even pyroclastic flow tuffs may occur within this rock assemblage. Thicknesses of fallout tephra depend in large part upon prevailing wind directions. On continental margins having large subaerial stratovolcanoes, thick alluvial and laharic piedmont aprons are predominant, commonly containing interbedded pyroclastics and reworked pyroclastics with minor amounts of lava flows. In such areas, deep

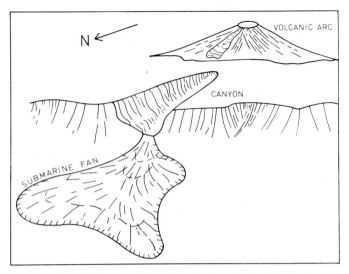

**Fig. 14-2.** Generalized geographic setting showing relationship of volcanic arc, shelf and adjacent deep sea area. (After Grippi and Burke, 1980)

erosion may expose plutonic rocks contemporaneous with the older eruptions (e.g., Mount Rainier National Park, Washington, USA). Thus, many sediments may carry quartz and other "continental-type" mineral assemblages eroded from the plutonic cores along with pyroclastic debris.

**The Trench**

Trenches are deep linear troughs along subducting plates. Many occur on the seaward sides of island arcs (e.g. Japan) or magmatic arcs near continental edges (e.g. the Andes). Others occur between two oceanic plates (e.g. Tonga-Kermadec). Accordingly, the composition and thickness of the sedimentary fill in trenches vary greatly. Because trenches are caused by subduction, plate tectonic theory predicts that they should contain accreted pelagic oceanic sediments rafted to the trench as the oceanic plate moves toward the subduction zone, and such sediments may be largely scraped off the plate as it descends. Other rocks within the trench assemblage include pillow basalts, hyaloclastites, and plutonic ophiolitic rocks tectonically mixed with pelagic sediments, as well as much volcaniclastic debris and even plutonic debris eroded from the magmatic arc. However, the trench may be cut off from the magmatic arc by the growth of an outer arc ridge composed of the offscrapings of the incoming plate and tectonically stacked shelf, slope and trench facies (Seely et al., 1974). Connelly (1978) describes such a Cretaceous subduction complex from the Kodiak Islands, Alaska, but recent investigations of the Deep Sea Drilling Project have cast doubt on the theory of accreting pelagic sediments (e.g. Leg 56, Japan Trench; Leg 60, Mariana arc and trench) because drill holes on the landward side of these trenches gave no evidence of stacked accretions.

One of the many problems concerning modern trench sedimentation is that the thickness of the sedimentary fill does not exceed 3 km and some are considerably less. Explanations of why accumulations are not thicker are: (1) they are transported down the subduction zone, (2) they are cut off from voluminous sediment sources of the arc by development of outer arc ridges, or (3) ocean bottom currents transports the sediments away from the trench.

Stewart (1978) reports as follows on the volcaniclastic sediments within an arc-trench environment from the Atka Basin, Aleutian Ridge on a slope basin near a trench: Neogene sediments recovered at DSDP Site 186 (Fig. 14-3) consist of interbedded diatomaceous silty clay, volcaniclastic turbidite sands, and vitric volcanic ash. The sand layers, up to 4.5 m thick, are medium- to coarse-grained with parallel laminations, size grading and convolute laminations. Some layers are predominantly vitric glass shards. Clasts are angular to subrounded. Most rock fragments are volcanic. Glass shards are followed in abundance by calcic plagioclase (Table 14-2). Quartzose debris consists of chert and bipyramidal volcanic quartz. Plutonic fragments and potassium feldspar along with sedimentary clasts are minor. Clinopyroxene, chiefly augite, forms over half of the heavy mineral component. These sands, therefore, clearly are derived from an andesitic volcanic terrane consisting of both volcanic and plutonic rocks (Fig. 14-4).

The source area for the sands in the central Aleutians (Andreanof Islands, see Fig. 14-3) is composed of an older rock assemblage consisting of keratophyre, spilite, tuff, chert, cherty argillite, and graywacke of Eocene to early (?) Miocene age intruded by sills, dikes and plutonic rocks (gabbro to granite with granodiorite predominating). These older rocks are unconformably overlain by middle Miocene to Holocene sedimentary and volcanic rocks (basalt to rhyolite with andesite predominating) which apparently supplied most of the volcaniclastic debris analyzed from DSDP Site 186. The most likely sources of the tephra layers in the Atka Basin are stratovolcanoes and calderas of the central Aleutian arc.

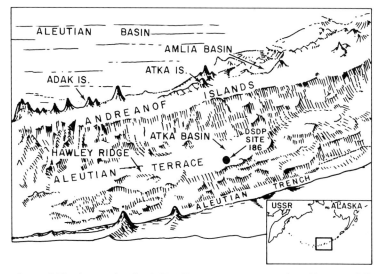

**Fig. 14-3.** Diagram of central Aleutian arc showing DSDP Site 186. Components of sand layers at drill site given in Table 14-2. Vertical exaggeration × 10. (After Stewart, 1978)

**Table 14-2.** Components in sand layers, DSDP site 186, Atka Basin, Aleutian Ridge. (After Stewart, 1978)

| Clast type | Average percent (27 samples) |
|---|---|
| Plagioclase (mostly calcic) | 32 |
| Volcanic rock fragments (chiefly andesite and basalt) | 29 |
| Glass shards (equal amounts of bubble-wall and bubble-pumice shards) | 28 |
| Chert | 3.9 |
| Quartz (bipyramidal) | 2.5 |
| Plutonic rocks | 1.6 |
| Sedimentary rocks | 1.1 |
| Potassium feldspar | 1.0 |
| Clinopyroxene (chiefly augite) | 59 |
| Orthopyroxene | 21 |
| Amphiboles (common hornblende) | 12 |
| Epidote and trace minerals[a] | 6.8 |
| | *Average Ratio* |
| Plagioclase/total feldspar | 0.97 |
| Volcanic clasts/total lithics | 0.96 |
| Chert/total quartzose debris | 0.68 |

[a] Basaltic hornblende, tremolite-actinolite, olivine (up to 1%), calcite, garnet, pumpellyite, sphene, tourmaline, zircon

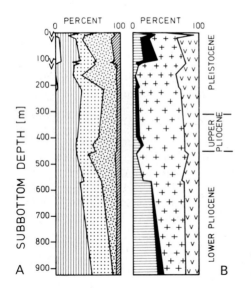
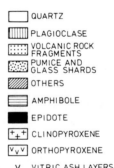

**Fig. 14-4 A, B.** Stratigraphic variation in frequencies of light (**A**) and heavy (**B**) mineral/rock fragment fractions from sand layers, DSDP Site 186, Atka Basin (Alaska). Also see Fig. 14-3 and Table 14-2. (After Stewart, 1978)

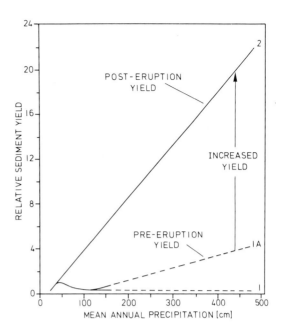

Fig. 14-5. Hypothetical curves showing relationship between precipitation and sediment yield before and after an explosive volcanic eruption. *Curve 1:* normal precipitation. *Curve 1A:* mean annual precipitation >113 cm. *Curve 2:* after violent volcanic eruption

Modern sedimentation within a narrow arc-trench gap has been investigated by Kuenzi et al. (1979) and Vessel and Davies (1981) in Guatemala. The narrow coastal plain in Central America adjacent to the Pacific Ocean forms the subaerial part of the arc-trench gap which is bordered 110 to 150 km offshore by the Middle America Trench, marking the convergent boundary between the Cocos and Caribbean lithospheric plates (Malfait and Dinkelman, 1972). Twenty-five to 60 km inland, the coastal plain is abruptly terminated by the steep slopes of an active Quaternary volcanic arc; volcanic cones lie another 10 to 15 km inland. Within such a narrow shelf-to-trench setting, Kuenzi et al. (1979) show that fluvial-deltaic sedimentation under conditions of torrential rainfall significantly differs in peak sediment loads (Fig. 14-5), delta construction and destruction from fluvial-deltaic sedimentation in nonvolcanic regions. They attribute this difference to periodic volcanic eruptions which locally result in high-relief unvegetated slopes with abundant loose clastic materials that allow extremely high local rates of erosion, fluvial-deltaic sedimentation, and delta extension. As volcanic eruptions diminish and sediment supply decreases, the delta front is quickly cut back (Fig. 14-6). Vessel and Davies (1981) show general subaerial facies associations from the volcanic arc to the shoreline (Chap. 13). Mathisen and Vondra (1983) describe nonmarine volcaniclastics in an interarc setting.

An important summary of an ancient arc-trench-ocean basin complex of rocks in New Zealand is given by Carter et al. (1978). Their synthesis combines the arc-trench model of Karig and Sharman (1975) with the lithofacies model of Mutti and Ricci-Lucchi (1972) and of Ricci-Lucchi (1975) to present an integrated characterization of a well-preserved volcanotectonic-sedimentary environment (Fig. 14-7).

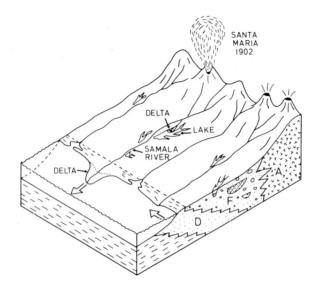

**Fig. 14-6.** Fluvial-deltaic system after 1902 eruption of Santa Maria volcano, Guatemala. Inferred facies resulting from long-term progradation along entire length of coastal plain. *A* arc-derived lava flows, pyroclastic flows, lahars and fallout ash; *F* braided stream, deltaic and lacustrine facies environments consisting of volcaniclastic rocks and sediments; *D* littoral and strand-line volcanic sands composed largely of coalesced deltas. (After Kuenzi et al., 1979)

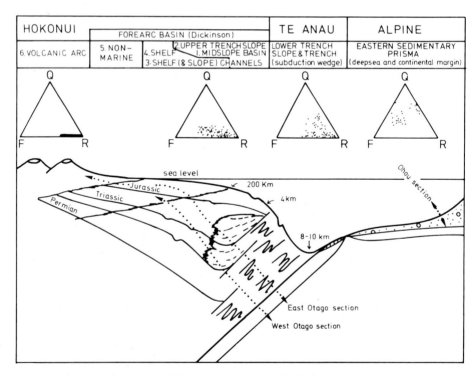

**Fig. 14-7.** Diagrammatic summary of facies environments and sediment composition (*Q* quartz, *F* feldspar, *R* rock fragments) within a subduction plate tectonic setting through time. (After Carter et al., 1978)

## Fore-Arc and Back-Arc Basins

It is within the fore-arc and/or back-arc basins of the magmatic arc (Fig. 14-1) that the greatest volumes of volcaniclastic materials accumulate (Dickinson and Seely, 1979; Mathisen and Vondra, 1983). Fore-arc basins attain widths of as much as 100 to 300 km and receive very large volumes of pyroclastic material, such as silicic subaqueous pyroclastic flows and subaqueous fallout layers, reworked land-derived pyroclastics, and epiclastic volcaniclastic materials which may eventually be reworked into thick graywacke sequences. The width of the fore-arc and back-arc basins and the amount of subsidence that they undergo can accommodate a large variety of potential sedimentary environments such as fluvial-deltaic shorelines, unstable shallow marine shelves, turbidite and grain-flow-covered marine slopes, and rises composed of coalesced active and inactive submarine fans, especially in fore-arc basins (see articles in Burk and Drake, 1974; Dott and Shaver, 1974). The rocks are commonly broadly andesitic with lesser amounts of basaltic and dacitic materials.

In back-arc regions of continental margin arcs, sedimentation occurs in clastic wedges in foreland basins (retro-arc basins, see Fig. 14-1), and may contain either marine sediments or continental sediments such as east of the Andes in South America. The thick continental volcaniclastic sequence in Magdalena Valley, Colombia is an unusually complete record of late Andean orogenic and postorogenic volcanism (Van Houten, 1976). Major volcanic activity and consequent sedimentation occurred from latest Oligocene to latest Miocene time; Pliocene-Pleistocene volcanism produced only minor volumes of volcanic detritus. Andesitic lithic clasts and mineral grain suites predominate in proximal debris flows and fluvial channel deposits, whereas reworked dacitic pyroclastic material is more common in distal debris flows, sheeted overbank sands, and flood plain muds (Table 14-3).

Because belts with thick marine sediments are elongate and commonly adjacent to continental regions, they closely resemble what was formerly called an eu-

Table 14-3. Transport mechanisms and depositional environments of volcaniclastic debris, Magdalena Lowland, Colombia. (After Van Houten, 1976)

| Transport | Depositional environment | Clasts |
|---|---|---|
| Unbedded debris flows | Broad valley fill; proximal alluvial fan | Very coarse, angular, poorly sorted clasts (mostly lava) |
| Unbedded debris | Distal fan; lowland mantle | Unsorted, rounded pumice clasts in volcanic sand |
| Mixed debris flows and torrential deposits | Proximal fan; broad valley trains; confined valley fill | Very coarse, poorly sorted, angular to subrounded lava and nonvolcanic clasts |
| Torrential deposits, minor debris flows | Broad lowland fill | Coarse, poorly sorted, subrounded lava and nonvolcanic clasts |
| Sheet overbank, flood-plain and minor channel deposits | Distal fan; lowland mantle; confined valley fill | Poorly sorted rounded pumice clasts and cross-stratified volcanic sand |

geosynclinal assemblage, characterized by thousands of meters of deep-water turbidites composed of immature mineral suites and lithic epiclastic volcanic sands. The sequences may contain some carbonates and the rocks commonly are poor in quartz. Structures within such assemblages are complex and suggest a long history of sedimentation. For example, early sedimentation is usually in wide belts which tend to become narrower through time. During the course of their history there is internal uplift and folding, therefore parts of the tectonic belt are cannibalized to provide materials for the low foreland or submerged parts of the system formerly known as miogeosynclines.

The deformed and partly metamorphosed circum-Pacific belts of sedimentary basins are dominated by volcanic-derived and terrigenous sediments, and are associated with pelagic sequences of cherty nannofossil limestones. The parallels between the eugeosynclinal belts and the fore- and back-arc elements of the plate tectonics model are unmistakable. The back-arc basin nearest the arc that contains abundant volcaniclastic materials is considered to be part of the eugeosynclinal assemblage, but these rocks become thinner toward the continental interior (the craton), gradually lose their dominant volcaniclastic component which interfingers with materials derived from continents with their fringing shore-line-shelf-slope-rise bathymetric sedimentary environments. Such environments can be considered to be the miogeosynclines within the pre-plate tectonic scheme.

**The Cordilleran System**

During nearly every period since Precambrian time, the western margin of North and South America has undergone volcanism and deformation within various parts of a wide linear belt known as the Cordilleran geosyncline. Its development is far too complex to attempt a review in this short chapter (see Dickinson, 1976). It is sufficient for our purpose to describe some of the volcaniclastic and related rocks from parts of the Cordilleran system, dividing the discussion essentially into (1) sediments deposited in parts of the ancient "eugeosyncline" and (2) those deposited in Cenozoic time which are similar to volcaniclastic rocks in modern volcanic arc-basin regions.

Western North America: Paleozoic Rocks

According to Churkin (1974), the major petrotectonic elements of the Cordilleran geosyncline during Paleozoic time from Alaska to California were similar to the present-day Japanese arc, which includes a trench and fore-arc basin oceanward of the magmatic arc, the magmatic arc, and a marginal sea (inter-arc basin) between the arc and the continent. Thus, the Paleozoic rocks can be divided into four generalized belts (Fig. 14-8), with their stages of development shown in Fig. 14-9. In early Paleozoic time, the volcanic-graywacke belt on the west side of an island arc system was initially deposited on primitive oceanic crust, but through a long period of volcanism, erosion and sedimentation, metamorphism and plutonism, the crust increasingly became more continental in character. Most newly formed sedimentary rocks were derived from volcanic island chains and deposited within fore-arc, inter-arc, and ocean margin basins. Deep-sea pelagic deposits, turbidites, graptolite shales and chert apparently accumulated on basaltic crust between the

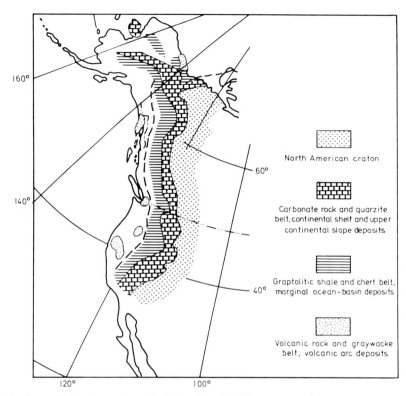

**Fig. 14-8.** Major Lower Paleozoic stratigraphic belts of the Cordilleran geosyncline, western North America. (After Churkin, 1974)

arcs and the continent in places where newly formed marginal ocean basins appeared as the volcanic arc migrated away from the continent.

During middle Paleozoic time, the marginal basins apparently closed (Fig. 14-9) as the ocean floor moved eastward, and the sediments within the basins were thrust eastward over the continental shelf. These rocks were then uplifted and partly eroded to provide thick wedges of coarse-grained clastic rocks in new frontal basins. In late Paleozoic time, the compressed earlier Paleozoic marginal ocean basins were rifted to form a new system of basins probably floored by new oceanic crust, but these basins and their sediments were again closing during late Permian and early Triassic time and the rocks were again thrust eastward. According to Dickinson (1976), through Mesozoic time the complex continental margin was the site of an arc-trench system of the Andean type, and the former marginal basins between the magmatic arc and the continent became sites of back-arc basinal sedimentation and thrusting.

"Eugeosynclinal" rocks deposited in the outer oceanic margin of the late Paleozoic geosyncline are interpreted as the deposits in fore-arc basins and trenches on the oceanic side of active volcanic arcs. The rocks include thick sequences of graywackes, polymict rudites, volcanic rocks (including many volcaniclastic graywackes), subaqueous lava flows and biogenic limestones. Lavas and volcanic rocks

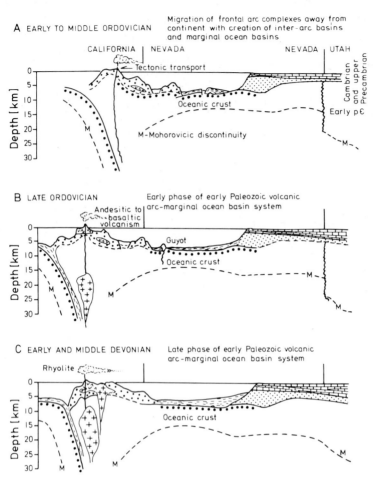

**Fig. 14-9 A–D.** Paleozoic geosyncline reconstructions of western North America in Ordovician to Permian time. (After Churkin, 1974)

are mainly basaltic and andesitic, but range to rhyolitic in composition and include some subaerial welded tuffs. Basaltic lavas commonly are pillowed and are associated with pillow breccias and hyaloclastites. Plutonic igneous rocks occur as cross-cutting deep-seated plutons and shallower porphyry bodies, many of which are contemporaneous with parts of the volcanic sequence. The more coarsely crystalline rocks range in composition from monzonite-syenite to granodiorite-diorite but include gabbro and some ultramafic rocks. The porphyritic bodies are massive and brecciated fine-grained igneous rocks rich in feldspar, pyroxene, and amphibole phenocrysts closely associated with compositionally similar submarine lavas and volcaniclastic rocks. They usually occur as cross-cutting stocks but in some places they are sills within stratified sedimentary rocks.

Basement rocks upon which the more recent volcanic rock sequences rest, include metavolcanics of possible early Paleozoic age in Alaska that are metamor-

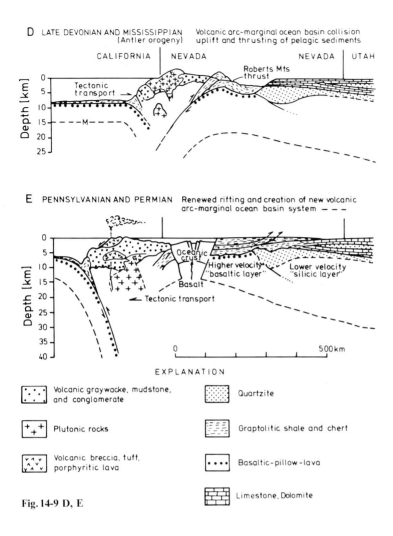

Fig. 14-9 D, E

phosed to greenschist or amphibolites. Metamorphosed plutonic and volcanic rocks of Precambrian and early Paleozoic age form the basement of the San Juan Islands and northern Cascades. Farther south in northern California and southern Oregon, Precambrian basement rocks have not been recognized. Instead, Ordovician and Silurian volcanic flows and volcaniclastic rocks are tectonically bordered by ultramafic rocks, and by melanges of metamorphosed graywackes, blueschists and serpentinites of Mesozoic age.

East of the magmatic arc, in early Paleozoic time, sedimentary rocks were deposited on oceanic crust within a marginal ocean basin (Churkin, 1974; Wrucke et al., 1978). These rocks include graptolitic shale and chert with quartzite forming marker beds, but in places quartzite forms a large part of the section. Greenstone, limestone, sandstone, and pillow basalts also occur. The volcanic component of the sequence may be quite large, as is suggested by the fact that bulk chemical analyses

395

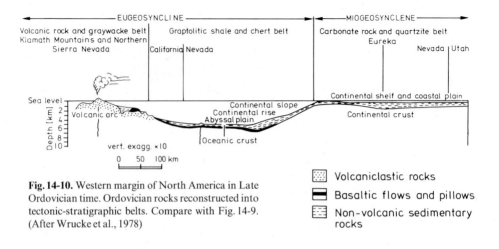

**Fig. 14-10.** Western margin of North America in Late Ordovician time. Ordovician rocks reconstructed into tectonic-stratigraphic belts. Compare with Fig. 14-9. (After Wrucke et al., 1978)

of argillaceous rocks within the graptolite shale and chert belt show 75 to 90 % $SiO_2$, much higher than modern deep-sea clays or radiolarian oozes. In places these rocks also contain relic pumice shards, porcelaneous grains that may be devitrified volcanic glass and silt-size feldspar grains (15–25%), suggesting that fallout ash from active volcanoes in the island arc region spread eastward across the marginal sea. In northern Nevada, thick pillowed greenstone lava sequences are interbedded with pelagic sedimentary rocks of Ordovician age within the graptolite chert and shale belt shown in Fig. 14-8. Wrucke et al. (1978) suggest a Late Ordovician plate tectonic model for western North American as shown in Fig. 14-10.

Southern South America: Upper Mesozoic Flysch

The upper Mesozoic volcaniclastic and related rocks of Tierra del Fuego and South Georgia Island, southern South America, are typical of the rock associations, depositional environments and tectonic processes within an active calc-alkalic arc terrane. Winn (1978) demonstrates that field-oriented sedimentologic studies can be used to decipher the complex lithospheric plate motions responsible for such associations.

South Georgia Island presently lies 2000 km east of Tierra del Fuego and is surrounded by oceanic crust. Yet the island is underlain by a thick sequence of strongly deformed flyschlike upper Mesozoic graywacke, mudstone, and tuff intruded by silicic to intermediate plutonic rocks. These rocks are interpreted as having once been adjacent to Tierra del Fuego. This surprising conclusion comes from study of the rocks – their ages, depositional environments and dispersal patterns, deformation style and metamorphic grades, but chiefly because of their petrographic provenance characteristics. The rocks are remanents of the infill, chiefly interfingering submarine fans from both sides, of a marginal basin which formed between the South American continent and an active calc-alkalic magmatic arc (Fig. 14-11). The basin closed, its rocks were deformed and metamorphosed to the prehnite-pumpellyite grade, and then South Georgia was translated relatively 2000 km eastward during the Late Cretaceous Andean orogeny.

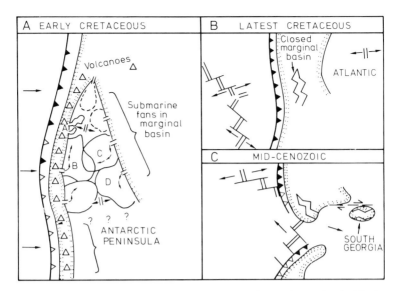

**Fig. 14-11 A–C.** Tectonic evolution of northern Scotia arc, southern South America. **A** Marginal basin opened in Late Jurassic to Early Cretaceous time with deep sea fans filling the gap from both sides. Fans have similar dimensions of several fans of the Oregon and Washington coasts (*A* dimensions of La Jolla-Navy-Coronado fan systems; *B* dimensions of Astoria fan; *C* dimensions of Delgado fan; *D* dimensions of Monterey fan). **B** Atlantic opened in Early Cretaceous time. Marginal basin in South America closed by 80 m.y. ago. About 64 m.y. ago (early Cenozoic time) a new ridge system formed in the eastern Pacific. **C** In mid-Cenozoic time South Georgia was translated 2,000 km relatively eastward following a ridge-continent collision. (After Winn, 1978)

There are two Upper Mesozoic flyschlike sequences on South Georgia Island, each of which closely matches with rocks on Tierra del Fuego. The lower sequence contains abundant quartz and plagioclase, uncommon potassic feldspar and essentially no pyroxene and amphibole grains. These rocks have identical counterparts in Middle to Late Jurassic silicic volcanic and interbedded sedimentary rocks in Tierra del Fuego and are believed to have been tectonically separated from them. Table 14-4 summarizes the clastic components of the sequence. Such rocks are typical of volcaniclastic rocks derived from calc-alkalic volcanic terranes. In the upper sequence, plagioclase is common, whereas quartz, ferromagnesian minerals, and mafic volcanic fragments are sparse.

The lower volcaniclastic flyschlike sequence on South Georgia Island contains sedimentary structures, fabrics, bedding styles and trace fossils, suggesting that it was deposited in submarine fans by sediment gravity flows in a deep water environment. The rocks are evenly stratified, poorly sorted and graded, sandstone and slaty mudstone. Sole marks, flames, mudstone intraclasts, and small scours are common. The only known source for the quartz-rich sandstones are the silicic volcanic and sedimentary rocks of mid- to late-Jurassic age in southern South America. These Jurassic volcanics are porphyritic with phenocrysts of quartz and sodic plagioclase, less common potassic feldspar and rare biotite, pyroxene, and amphibole.

**Table 14-4.** Summary of detrial modes of Yahgan, Cumberland Bay, and Sandebugten sandstones, southern South America. (After Winn, 1978)

|  | Yahgan Formation (Tierra del Fuego) | | |
|---|---|---|---|
|  | Intermediate volcaniclastic rocks | | Quartzose |
|  | Average | Range |  |
| Lithic fragments | 69.7 | 24–95 | 21.9 |
| Plagioclase | 22.9 ($\pm 5.4$) | 1–75 | 17.3 |
| Potassic feldspar | 0.1 | 0–1 | 1.2 |
| Quartz | 3.6 ($\pm 3.1$) | 0–14 | 58.9 |
| Pyroxene | 1.1 | 0–16 | 0 |
| Amphibole | 1.1 | 0–29 | tr. |
| Biotite | tr. | 0–tr. | tr. |
| Muscovite | tr. | 0–tr. | tr. |
| Epidote | 0.1 | 0–1 | 0.1 |
| Opaques | 0.6 | 0–3 | 0.6 |
| Plant and shell debris | tr. | 0–3 | tr. |
| Polycrystalline quartz/total quartz | 0.18 | | |
| Plagioclase/total feldspar | 0.99 | | |
| Volcanic rock fragments/total unstable lithic grains | 0.99 | | |
| Mica (%) | < 1 | | |
| No. of specimens | 55 | … | 8 |

The upper flyschlike volcaniclastic sequence on South Georgia Island and its Tierra del Fuego counterpart is composed of interbedded graywackes and mudstones, diamictites, conglomerates and sedimentary breccias deposited by sediment gravity flows. Thin (a few mm to 20 cm) interbedded quartz-prehnite beds with rare shard and ash fabrics indicate former tuff units. The rocks coarsen upward and in places are interbedded with mafic lava flows. According to Winn (1978), sediment gravity flow fabrics, sedimentary structures, bedding styles, and trace fossils suggest deposition on deep-sea fans similar to the overlying quartz-rich sequence. On Tierra del Fuego, the sequence rests on ophiolites of the marginal basin on the south, and upon silicic volcanic rocks of the cratonic margin to the north which supplied quartz-rich volcaniclastic debris to the lower sequence on South Georgia Island.

Cenozoic Tectonism and Volcanism: Western North America

Throughout Paleozoic and Mesozoic time, the subduction zone, magmatic arc, and flanking basins formed fluctuating but essentially continuous regional strips along the western margin of the North American continent (Churkin, 1974; Dickinson, 1976). During Cenozoic time, however, the continuity was disrupted and parts of the Andean-type margin were converted to transform margins by the evolution and migration of triple junctions that formed when the trench encountered a spreading center (Atwater, 1970; Dickinson, 1976; Lipman et al., 1972; Christiansen and Lipman, 1972).

| Cumberland Bay (S. Georgia Is.) | | Quartzose sandstones | Sandebugten (S. Georgia) | | |
|---|---|---|---|---|---|
| Intermediate volcaniclastic rocks | | | Intermediate volcaniclastic rocks | Quartzose sandstones | |
| Average | Range | | | Average | Range |
| 81.1 (±5.2) | 54–95 | 43.1 | 79.2 | 43.1 (±6.6) | 17–59 |
| 15.1 (±4.7) | 4–37 | 26.1 | 17.2 | 19.4 (±5.2) | 8–39 |
| 0.1 | 0–1 | 1.1 | tr. | 1.8 | 0–1 |
| 2.4 (±1.4) | 0–10 | 28.4 | 3.5 | 35.4 (±6.3) | 23–59 |
| 0.3 | 0–3 | 0 | tr. | tr. | 0–tr. |
| tr. | 0–1 | 0 | tr. | tr. | 0–1 |
| tr. | 0–tr. | 0.2 | 0 | tr. | 0–tr. |
| tr. | 0–tr. | 0.1 | 0 | tr. | 0–1 |
| 0.1 | 0–1 | 0.2 | tr. | 0.1 | 0–2 |
| 0.8 | 0–2 | 0.9 | 0.1 | 0.2 | 0–3 |
| tr. | 0–1 | 0 | 0 | tr. | 0–1 |
| 0.15 | | | | 0.50 | |
| 0.99 | | | | 0.92 | |
| 0.99 | | | | 0.97 | |
| <1 | | | | <1 | |
| 37 | ... | 3 | 3 | 19 | ... |

Lower and Middle Cenozoic intermediate volcanic and associated intrusive rocks of western North America (Fig. 14-12) related by Lipman et al. (1972) to plate convergence, include andesite, dacite, rhyodacite and quartz latite with associated granodiorite, monzonite, and quartz monzonite. Somewhat more silicic ash-flow tuffs of quartz latite composition and low-silica rhyolite rocks that are abundant in some areas are interpreted as high-level differentiates of intermediate-composition magmas. Very mafic and very silicic rocks are much less abundant or are absent. Mafic rocks that show oceanic affinities, however, are abundant at the Pacific margin. The thick sequences of Lower to Middle Eocene basalts and associated marine sedimentary rocks exposed along the coastal regions of Oregon and Washington, atypical of andesitic arc volcanism, are interpreted as oceanic seamounts and volcanic archipelagoes that were rafted against the continental edge as the ocean plate was consumed (Cady, 1975; Dickinson, 1976). The petrochemistry of some of these basalts is similar to oceanic tholeiites and alkalic basalts found in Hawaii (Snavely et al., 1968; Glassley, 1974); moreover, pelagic limestones are associated with some pillow basalts in the sequences.

Andesitic arc magmatism began in Late Eocene time in the Cascades of Washington (Fisher, 1961b) and was fully established through the Cascades and adjacent regions to the east in Oligocene time (Peck et al., 1964) (Fig. 14-13). Rocks within the various volcanic fields of this vast region include the cores of deeply dissected stratovolcanoes, bedded volcanic epiclastic sandstone, breccia and conglomerate, fallout tuff and large areas of ashflow tuff, some of which are described in Chapter 8. In general, the volcanic rocks increase in $K_2O$ content toward the continental interior.

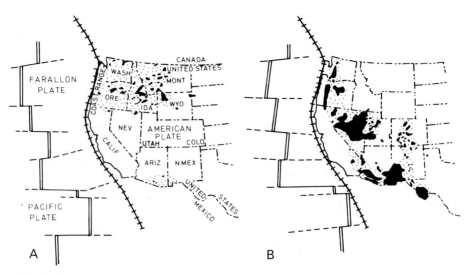

**Fig. 14-12 A, B.** Approximate distribution of lower and middle Cenozoic igneous rocks in the Western United States and their relationship to Pacific plate and subduction boundary (suture pattern). *Dark pattern* present distribution of continental igneous rocks (*stippled pattern* inferred original extent). *Double lines* spreading ridges; *single lines* transform faults. **A** Eocene igneous rocks (40 to 55 m.y.) and plate geometry at about 50 m.y. **B** Oligocene igneous rocks (25 to 40 m.y.) and plate geometry at about 30 m.y. (After Lipman et al., 1972)

## Oceanic Island Arc Settings

Volcaniclastic Rocks and Facies; Cenozoic

Lower and middle Miocene rocks deposited in and around a volcanic chain between subducting oceanic plates are described by Mitchell (1970). Lower Miocene rocks exposed on Malekula Island, New Hebrides (Fig. 14-14) consist of a sequence of quartz-free volcaniclastic rocks, detrital limestones, pelagic sediments, and a few lava flows grouped by Mitchell into several facies (Table 14-5). This succession is cut by andesitic and basaltic dikes. Paleogeographic reconstruction suggests that lower Miocene time in the area was characterized by conditions resembling volcanic chains in present island arcs. Some of the volcaniclastic rocks were derived from subaerial eruptions, whereas others were from subaqueous volcanoes, but many of the sediments were derived from subaqueous mass transport and by turbidity current processes rather than direct deposition from erupting volcanoes. Clasts within the limestones are largely from benthonic foraminifera or algal and coral growths. Above the lower Miocene succession are middle Miocene turbidites composed of volcanic and limestone reef debris derived from erosion of the lower Miocene rocks. Rare fallout tuffs are found in parts of this sequence.

To explain the stratigraphic facies succession at Malekula, Mitchell (1970) infers that initial eruptions in early Miocene time were entirely subaqueous (Fig. 14-15), giving rise to cones consisting of peperites, agglomerates, and tuffs. The submarine volcanoes built to above sea level and became stratovolcanoes of lava flows and tephra. Coastal erosion around the stratovolcanoes maintained

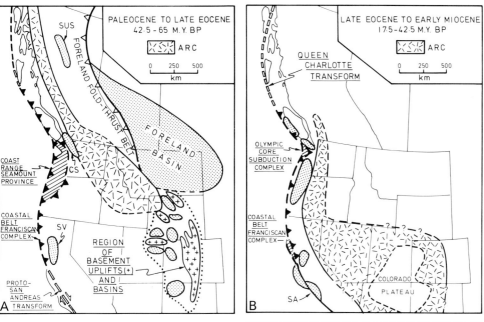

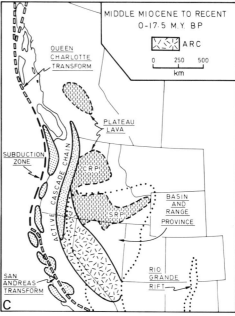

**Fig. 14-13 A–C.** Cenozoic Cordilleran arc-trench system. **A** Paleocene to Late Eocene time: *SUS* Sustat assemblage, British Columbia; *CS* Chuckanut-Swauk sequence in Northern Cascades; *SV* Sacramento Valley, California. Basins shown by *stipple pattern* in all figures. **B** Late Eocene to Early Miocene time: *SA* San Andreas fault which offsets earlier rocks. **C** Middle Miocene to Recent time: *CRP* Columbia River plateau (mainly Mid-Miocene flood basalts); *SRP* Snake River plain (mainly Late Miocene and younger basalts). South-eastern part of arc has been extinguished as San Andreas transform lengthens due to northward migration of Mendocino triple junction (not shown). (After Dickinson, 1976)

steep slopes, and coarse detritus was reworked to form rounded beach boulders, some of which were incorporated in lahars that entered the sea and continued to flow down steep subaqueous slopes into deep water. Sand-size and coarser detritus was carried down the subaqueous slopes as turbidity currents. Reefs that fringed the islands contributed calcareous detritus in slides and talus, as well as constitu-

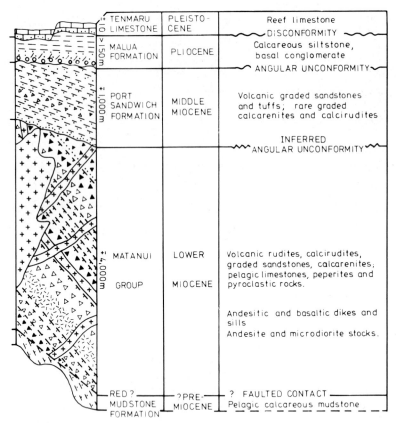

**Fig. 14-14.** Composite stratigraphic column, Malekula Islands, New Hebrides (see Table 14-5). (After Mitchell, 1970)

ents in debris flows and turbidites which came to rest beyond the talus slopes, but in the same environment as the lahars and volcaniclastic turbidites. Pyroclastic eruptions produced fallout tuffs that are interbedded and locally mixed with epiclastic material. As volcanic activity waned, regional subsidence submerged the volcanic island, but in adjacent areas dikes intruded the succession of rocks, causing submarine eruptions which produced hydroclastic and pyroclastic ejecta that overlie epiclastic sedimentary rocks. The eruptions again built volcanoes above sea level, and supplied clastic sediments which buried older submerged island volcanoes. Erosion during middle Miocene time and continued reef-building supplied mostly fine-grained volcanic turbidites and clastic limestones; distant volcanism supplied a few fine-grained fallout tuff layers. These events were followed by block faulting, uplift, and erosion.

In Fiji, Dickinson (1968) describes the lithologic variations within a Pliocene volcanogenic unit (Fig. 14-16) that includes a near-source facies composed of leucocratic and mesocratic monzonite through diorite to gabbro stocks and dikes which are adjacent to stratified pyroclastic breccias and a few lava flow piles that attained 760 m or more in thickness. Distally from this cluster of volcanic centers are volcaniclastic units that came to rest as interbedded fluvial rudites and sand-

**Table 14-5.** Facies and their origin, Malekula Island, New Hebrides. (After Mitchell, 1970)

| Facies | Environment | Source | Mode of transport |
|---|---|---|---|
| A. Gravity-transported facies | | | |
| C. Graded sandstones | Marine; below wave base | Subaerial volcanic rocks: near-shore marine deposits | River-generated turbidity currents |
| F. Graded calcarenites | Marine; below wave base | Fore-reef talus | Turbidity currents |
| E. Structureless rudites | Marine; below wave base | Subaerial volcanic rocks; near-shore marine deposits | Lahars Submarine cold lahars |
| D. Fine-grained rudites with oriented clasts | Marine; below wave base | Subaerial volcanic rocks; near-shore marine deposits | Fluvial Submarine cold lahars |
| H. Structureless calcirudites | Marine; below wave base | Fore-reef talus | Submarine calcareous mudflows |
| G. Fine-grained calcirudites with oriented clasts | Marine; below wave base | Fore-reef talus | Submarine calcareous mudflows |
| J. Poorly-bedded limestones | Marine; reef talus? <2,000 m | Growing reef | Rolling and sliding of individual fragments |
| K. Slumped beds | Marine; below wave base | Poorly lithified turbidites; below wave base | Mass flow and plastic deformation |
| B. Pyroclastic and autoclastic facies | | | |
| M. Tuffs | Marine; below wave base | Subaerial or? submarine explosive eruptions | Wind, ocean currents; gravity settling |
| O. Agglomerates | Marine; below wave base | Submarine or subaerial explosive eruptions | Gravity settling; ? submarine lahars |
| P. Volcanic flow breccias | Marine; below wave base | Basaltic intrusions | In situ breccias |
| Q. Peperites | Marine; below wave base | Submarine intrusions | In situ |
| C. Other facies | | | |
| A. Pelagic limestones | Marine; below wave base | Planktonic foraminifera | Gravity settling |
| B. Red mudstones | Deep marine; ?4,000–6,000 m | Mainly volcanic detritus and planktonic foraminifera | Wind, ocean currents; gravity settling |
| L. Lignite beds | Marine; below wave base | Subaerial | Ocean currents; gravity settling |
| N. Cross-bedded sandstones | Shallow-marine or fluviatile | Subaerial volcanic | Traction currents |

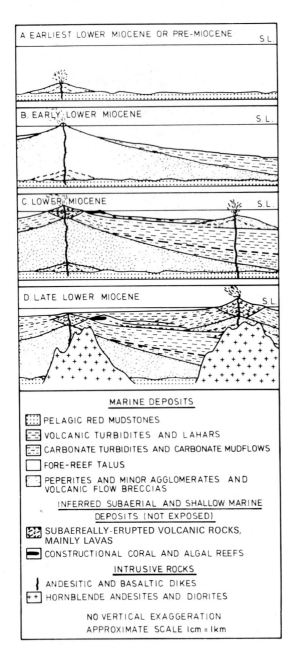

**Fig. 14-15 A–D.** Lower Miocene evolution and volcanic facies of Malekula Islands (New Hebrides). (After Mitchell, 1970)

stones. Still farther away, the volcaniclastic strata interfinger with finer-grained, evenly bedded, well-sorted neritic volcaniclastic sandstone and siltstone. Directional indicators show that the bedded volcaniclastic rocks were carried outward in all directions from the volcanic centers. Correlation of units was chiefly accomplished by petrographic affinities of the volcaniclastic rocks with the plutonic and pyroclastic rocks.

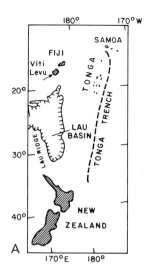
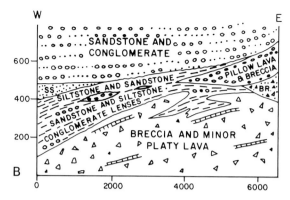

Fig. 14-16 A, B. Facies relationship of volcaniclastic rocks, northwest Viti Levu. A Location of Fiji Islands. B Schematic columnar section. Source toward right. Vertical exaggeration ×6. (After Dickinson, 1968)

Lau Basin and Tonga Arc

The Lau Basin, with a depth of about 2250 m, is flanked on the east by the volcanically active Tonga Ridge, and on the west by the Lau Ridge. The basin is about 700 km long and is a maximum of 400 km wide. Both ridges are largely submerged. Visible parts consist of a basement of eroded volcanics capped by carbonate reefs. Tonga, however, is capped by many volcanoes, both emergent and submergent, ranging in composition from basaltic to dacitic (Karig, 1970; Bryan et al., 1972).

The center of Lau Basin consists of a surface of fractured basalt with a patchy veneer of pelagic carbonate (Karig, 1970; Burns et al., 1973; Hawkins, 1974). The oldest sediments within the basin lie close to the ridges, are latest Miocene or earliest Pliocene, and are correlative with basement rocks on the Lau and Tonga Ridge, which in places may be as old as Eocene (Karig, 1970; Ewart and Bryan, 1973).

Pyroclastic debris from volcanoes of the Tonga Ridge, however, has begun to encroach into the basin. It includes mass flows, turbidites, fallout ash, and drift pumice. Sediments derived from the Lau Ridge consist largely of carbonate debris. Differences in easily erodible source material on the two sides of the Lau Basin result in an asymmetric fill with the thickest wedge of debris adjacent to the Tonga Ridge (Karig, 1970; Burns et al., 1973). The tectonic setting and rock facies of a middle Paleozoic sequence in eastern Australia is similar to that of the Lau Basin and adjacent ridges (Cas and Jones, 1979).

Lesser Antilles Arc

During the last 100,000 years, about 530 km$^3$ of volcanic material has been produced from five active volcanic islands in the Lesser Antilles Arc. More than 80% of this material is composed of volcaniclastic sediments, now distributed over an

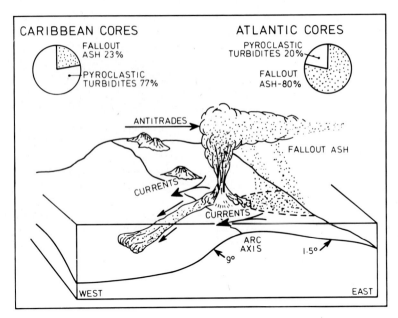

**Fig. 14-17.** Illustration of asymmetry of sedimentation processes and relative percentages of volcaniclastic rock types east and west of the Lesser Antilles volcanic arc. (After Sigurdsson et al., 1980)

area of 0.5 million km$^2$ (Carey and Sigurdsson, 1980; Sigurdsson, et al., 1980). Deep sea piston-coring on both sides of the arc reveals primary pyroclastics, remobilized pyroclastic deposits, and pyroclastically derived volcanic sands distributed asymetrically on both sides of the arc (Fig. 14-17).

Primary pyroclastic material includes fallout tephra, debris flows derived from hot pyroclastic flows (Carey and Sigurdsson, 1980), and primary hot pyroclastic flows transported into the basins from land. Remobilized deposits are mostly turbidites composed of pyroclastic debris.

The transition from land to the marine environment is characterized by nearshore conglomerates, block and ash-flow, debris-flow, and pyroclastic-flow deposits. In places they are interbedded with coral limestones.

Piston and gravity core operations have been unsuccessful on steep submarine flanks, but it is inferred that reworked pyroclastic sands are dominant, and attain their thickest expression, on the west sides of island passages where major lobes of well-sorted volcanic sands have been carried by prevailing marine currents.

The distribution of sediment on either side of the arc is asymmetric (Fig. 14-18). On the Caribbean side (back-arc), the Grenada Basin contains by far the greatest thickness and types of volcaniclastic sediments, mostly debris flow and turbidite deposits transported during or shortly after eruptions. Except for a pyroclastic flow sequence on Dominica (Sparks et al., 1980a), individual pyroclastic flows cannot be traced from land to under the water. The asymmetric distribution of pyroclastic debris flows – they are virtually absent on the Atlantic side east of the arc – is held by Sigurdsson et al. (1980) to be a function of slope. Steep western slopes

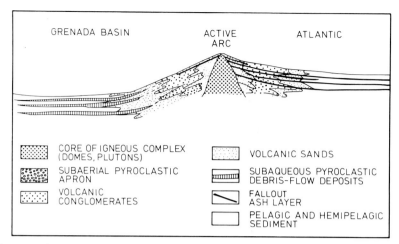

**Fig. 14-18.** Diagrammatic east-west stratigraphic section across the Lesser Antilles island arc showing facies asymmetry in fore-arc (Atlantic side) and back-arc (Grenada Basin) regions. (After Sigurdsson et al., 1980)

promote transport of the flows into the Grenada Basin, whereas low slopes of the eastern flank causes slower flow rates that result in increased interaction of the pyroclastic flows with air and sea water leading to their disintegration. Asymmetry of fallout tephra deposits in the water (Carey and Sigurdsson, 1980) (Fig. 14-19) is caused by transport of airborne ash in westerly antitrade winds at altitudes of 6–17 km.

The three factors that influence distribution of volcaniclastic sediments in the Lesser Antilles, and therefore thickness and volume of deposits, are prevailing winds, slopes off the arc, and marine currents.

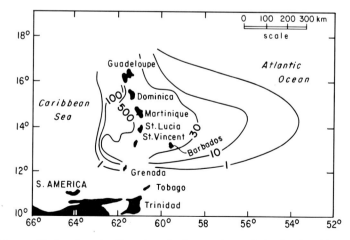

**Fig. 14-19.** Cumulative thickness of tephra (in cm) sediment gravity flows and reworked volcanic sands deposited in the last $10^5$ years adjacent to the Lesser Antilles arc. (After Sigurdsson et al., 1980)

# The Pre-Cambrian

Whether or not plate tectonic models apply to the earliest history of the earth is a matter of strong debate among Precambrian geologists. One school of thought holds that Archean greenstone belts (metavolcanic and associated sedimentary rocks) were formed from, and deposited on, a primary simatic crust, possibly analagous to modern island arc or oceanic environments (Green and Baadsgaard, 1971; Ermanovics, 1973; Goodwin, 1974; Wilson et al., 1974; Hubregtse, 1976). Another group holds that Archean volcanism and sedimentation occurred on a silicic igneous/metamorphic basement (McGlynn and Henderson, 1970; Bell, 1971; Ayres, 1974; Frith and Doig, 1975; Henderson, 1975; Baragar and McGlynn, 1978). A succinct review of these schools of thought and a brief discussion of the evidence is given by Glikson (1978) and Baragar and McGlynn (1976).

Goodwin (1980) proposes that greenstone assemblages such as occur in the southern Canadian Shield, India, southern Africa, and Western Australia begin development by attenuation and fissuring of sialic crust over a time span of at least 1200 m.y. He envisages numerous small lithospheric plates that move under a regime of high geothermal gradient and are lacking in the conventional subduction zones observed in modern plate tectonic settings. Grachev and Fedorovsky (1981) also hold that greenstone belts represent zones of rifting.

Archean volcanic (greenstone) belts that are widely distributed in the Canadian Shield range from isolated ribbons a few kilometers long to large irregular belts up to 700 km long separated by later plutonic intrusives. One of the largest continuous greenstone belt is the Abitibi Volcanic Belt (Superior Province, Canada) that measures up to 650 km long and 200 km wide. The rocks within the volcanic belts are a diverse assemblage of plutonic intrusions, volcanic flows and volcaniclastic rocks commonly occurring in mafic to felsic cycles. In many sequences, volcanic flows are interbedded with volcaniclastic rocks such as graywackes, mudstone, conglomerate, tuff, and iron formation. Intrusive rocks include granitic plutons (common), and less voluminous but widespread mafic to ultramafic sills, dikes, and small irregular intrusions.

A fundamental feature of Archean volcanism in the Canadian Shield is a general upward compositional change from mafic to felsic rocks. Many of the Archean assemblages have a single mafic to felsic cycle whereas others have two or more cycles in sequences of up to 20,000 m thick. Rocks that compose these cycles are flows and pyroclastics that are most commonly of tholeiitic to calcalkalic affinities. Less commonly, komatiitic and compositionally primitive volcanic rocks occur locally in the lower parts of sequences, whereas highly differentiated volcanic rocks (e.g. trachyte) occur in the upper parts. In a typical cycle, basalt flows

Table 14-6. Archean volcanics of Canada. (After Jakeš and White, 1971)

| Volcanic belt | Basalt | Andesite | Dacite | Rhyolite |
|---|---|---|---|---|
| Birch-Uchi | 57.8 | 29.2 | 13.0 | |
| Lake of the Woods-Wabigoon | 55.7 | 26.4 | 13.3 | 4.6 |
| Timmins-Kirkland Lake-Noranda | 58.9 | 32.4 | 7.2 | 1.7 |
| Average Superior Province | 57.3 | 29.4 | 10.2 | 3.1 |

predominate in the lower stratigraphic parts, andesite flows and volcaniclastic rocks interfinger with basalt in increasing amounts upward, and predominantly felsic pyroclastic rocks occur in the upper stratigraphic levels.

According to Jakeš and White (1971), the relative proportion of rocks of tholeiitic association in modern island arcs are: basalt to andesite to dacite = 50:35:15. Archean assemblages of tholeiitic association are remarkably similar with basalt to andesite to felsics = 57:30:13 (Table 14-6), suggesting a similar magmatic development. Moreover, in the Abitibi volcanic belt (Superior Province, Canada) the rocks include all of the distinctive igneous suites of modern island arcs and show a chemical polarity of increasing alkalinity from north to south (i.e. from "oceanic" to "continental" crusts). Goodwin (1974, 1977) gives details of these associations within the Abitibi volcanic belt.

The volcanic belts have similarities to modern island arc regions, but whether this warrants the interpretation that accretion by plate margin tectonism and volcanism occurred in the same manner as at present is still open to question. Analogous features include (1) mafic to felsic volcanic cycles, (2) similar chemical compositions and abundances of volcanic rock classes, (3) widespread calc-alkalic volcanism including voluminous explosive felsic rocks, (4) predominantly subaqueous accumulations of volcanic and associated sedimentary rocks and (5) parallel arc-like arrangement of linear volcanic belts.

# References

Number following reference denotes chapter in which citation appears

Abich, H., 1882. Geologische Forschungen in den Kaukasischen Ländern (Geologie des Armenischen Hochlandes). Wien, Alfred Holder 2, 1–478. 8
Ailin-Pyzik, I.B. and Sommer, S.E., 1981. Microscale chemical effects of low temperature alteration of DSDP basaltic glasses. J. Geophys. Res. 86, 9503–9510. 12
Aki, K., Fehler, M. and Das, L., 1977. Source mechanisms of volcanic tremor: fluid-driven crack models and their application to the 1963 Kilauea eruption. J. Volcanol. Geotherm. Res. 2, 259–287. 4
Allen, C.C., 1980. Icelandic subglacial volcanism: thermal and physical studies. J. Geol. 88, 108–117. 10
Allen, E.T. and Zies, E.J., 1923. A chemical study of the fumaroles of the Katmai region. Natl. Geog. Soc. Contributed Tech. Papers, Katmai Ser. 1, 75–155. 8
Allen, J.R.L., 1963. The classification of cross-stratified units with notes on their origin. Sedimentology 2, 93–114. 5
Allen, J.R.L., 1982. Sedimentary structures. Their character and physical basis. Vol. II. Elsevier, Amsterdam, 1–663. 1,5,9
Almond, D.C., 1971. Ignimbrite vents in the Sabaloka cauldron, Sudan. Geol. Mag. 108, 159–176. 8
Anderson, A.T., 1974a. Before eruption $H_2O$ content of some high alumina magmas. Bull. Volcanol. 37, 530–552. 2,3
Anderson, A.T., 1974b. Chlorine, sulfur, and water in magmas and oceans. Geol. Soc. Amer. Bull. 85, 1485–1492. 2,3
Anderson, A.T., 1975. Some basaltic and andesitic gases. Rev. Geophys. Space Physics 13, 37–55. 3
Anderson, C.A., 1933. The Tuscan Formation of northern California. Univ. Calif. Publ. Geol. Sci. 23, 215–276. 11
Anderson, T. and Flett, J.S., 1903. Report on the eruption of the Soufrière in St. Vincent in 1902 and on a visit to Montagne Pelée in Martinique, Part I. Philos. Trans. Roy. Soc. London A. 200, 353–553. 8,10
Andrews, A.J., 1977. Low temperature fluid alteration of oceanic Layer 2 basalts, DSDP Leg 37. Can. J. Earth Sci. 14, 911–926. 12
Antweiler, R.C. and Drever, J.I., 1983. The weathering of a late Tertiary volcanic ash: importance of organic solutes. Geochim. Cosmochim. Acta 47, 623–629. 12
Aoki, K., Ishiwaka, K. and Kanisawa, S., 1981. Fluorine geochemistry of basaltic rocks from continental and oceanic regions and petrogenetic application. Contr. Mineral. Petrol. 76, 53–59. 3
Aramaki, S., 1956. The 1783 activity of Asama Volcano. Part I: Jap. J. Geol. Geog. 27, 189–229. 8
Aramaki, S., 1957. The 1783 activity of Asama Volcano. Part II: Jap. J. Geol. Geog. 28, 11–33. 8
Aramaki, S., 1963. Geology of Asama Volcano. Tokyo Univ. Fac. Sci. Jour., sec. II, 14, 233–433. 6,11,13
Aramaki, S. and Akimoto, S., 1957. Temperature estimation of pyroclastic deposits by natural remanent magnetism. Amer. J. Sci. 255, 619–627. 5,8,11
Aramaki, S. and Ui, T., 1966. The Aira and Ata pyroclastic flows and related caldera depressions in southern Kyushu, Japan. Bull. Volcanol. 29, 29–47. 2,7,8
Aramaki, S. and Yamasaki, M., 1963. Pyroclastic flows in Japan. Bull. Volcanol. 26, 89–99. 8
Arrhenius, G., 1963. Pelagic sediments. In Hill, M.N., ed., The Sea 3. Wiley-Interscience, New York, 655–727. 12
Atwater, T., 1970. Implications of plate tectonics for the tectonic evolution of western North America. Geol. Soc. Amer. Bull. 81, 3513–3536. 14
Ayres, L.D., 1974. Geology of the Trouts Lake area. Ontario Dept. Mines Geol. Rpt. 113, 1–199. 14

Ayres, L.D., 1977. Importance of stratigraphy in early Precambrian volcanic terranes: cyclic volcanism at Setting Net Lake, northwestern Ontario. Geol. Assoc. Can. Sp. Paper 16, 243–264.  13

Ayres, L.D., 1983. The physical form, environment, and genesis of Precambrian greenstone-belt volcanoes, with particular reference to the Archean of Canada. (personal communication).  13

Baak, J.A., 1949. A comparative study on Recent ashes of the Java volcanoes Smeru, Kelut and Merapi. Meded. Alg. Proefst. Landb. Buitenzorg. 83, 1–60.  1

Bagnold, R.A., 1954a. The physics of blown sand and desert dunes. Methuen, London, 1–265.  6

Bagnold, R.A., 1954b. Experiments on a gravity-free dispersion of large solid spheres in a Newtonian fluid under shear. Proc. Roy. Soc. London A. 225, 49–63.  6,11

Bagnold, R.A., 1955. Some flume experiments on large grains but little denser than the transporting fluid, and their implications. Proc. Instn. Civ. Engin., pt. III, no. 1, 4, 174–205.  11

Bailey, R.A., Dalrymple, G.B. and Lanphere, M.A., 1976. Volcanism, structure, and geochronology of Long Valley Caldera, Mono County, California. J. Geophys. Res. 81, 725–744.  2

Ballard, R.D. and Moore, J.G., 1977. Photographic atlas of the Mid-Atlantic Ridge. Springer-Verlag, Berlin, Heidelberg, New York, 1–114.  2

Ballard, R.D. and van Andel, Tj.H., 1977. Morphology and tectonics of the inner rift valley at lat. 36°50' N on the Mid-Atlantic Ridge. Geol. Soc. Amer. Bull. 88, 507–530.  2

Ballard, R.D., Holcomb, R.T. and van Andel, T.H., 1979. The Galapagos Rift at 86° W: 3. Sheet flows, collapse pits, and lava lakes of the rift valley. J. Geophys. Res. 84, 5407–5422.  2

Banks, N.G. and Hoblitt, R.P., 1981. Summary of temperature studies of 1980 deposits. In Lipman, P.W. and Mullineaux, D.R., eds., The 1980 eruptions of Mount St. Helens, Washington. U.S. Geol. Survey Prof. Paper 1250, 295–313.  8

Baragar, W.R.A. and McGlynn, J.C., 1976. Early Archean basement in the Canadian Shield: A review of the evidence. Geol. Survey Can. Sp. Paper 76-14.  14

Baragar, W.R.A. and McGlynn, J.C., 1978. On the basement of Canadian greenstone belts: discussion. Geoscience Can. 5, 13–15.  14

Baragar, W.R.A., Plant, A.G., Pringle, C.J. and Schau, M., 1977. Petrology and alteration of selected units of Mid-Atlantic Ridge basalts sampled from sites 332 and 335, DSDP Leg 37. Can. J. Earth Sci. 14, 837–874.  12

Barnes, I. and McCoy, G.A., 1979. Possible role of mantle-derived $CO_2$ in causing two "phreatic" explosions in Alaska. Geology 7, 434–435.  9

Basaltic Volcanism Study Project, 1981. Basaltic volcanism on the terrestrial planets. Pergamon Press, Inc., New York, 1–1286.  3

Bateson, J.H., 1965. Accretionary lapilli in a geosynclinal environment. Geol. Mag. 102, 1–7.  7

Batiza, R., 1977. Age, volume, compositional and spatial relations of small isolated oceanic central volcanoes. Mar. Geol. 24, 169–183.  10

Batiza, R., 1982. Abundances, distribution and sizes of volcanoes in the Pacific Ocean and implications for the origin of non-hotspot volcanoes. Earth Planet. Sci. Lett. 60, 195–206. 2

Bell, C.K., 1971. Boundary geology, upper Nelson River area, Manitoba and northeastern Ontario. In Turnock, A.C., ed., Geoscience Studies in Manitoba, Geol. Assoc. Can. Sp. Paper 9, 11–39.  14

Berger, W.H. and von Rad, U., 1972. Cretaceous and Cenozoic sediments from the Atlantic Ocean. In Hayes, D.E., Pimm, A.C. et al., eds., Init. Repts. Deep Sea Drilling Project 14, 787–954.  12

Best, M.G., 1982. Igneous and metamorphic petrology. W.H. Freeman and Co., San Francisco, 1–630.  2

Beverage, J.P. and Culbertson, J.K., 1964. Hyperconcentrations of suspended sediment. J. Hydraulics Div., Amer. Soc. Civ. Engin. Proc. 90, no. HY6, 117–128.  11

Bevins, R.E. and Roach, R.A., 1979. Pillow lava and isolated pillow breccia of rhyodacitic composition from the Fishguard Volcanic Group, Lower Ordovician, S.W. Wales, United Kingdom. J. Geol. 87, 193–201.  10

Binns, R.E., 1967. Drift pumice on postglacial raised shorelines of northern Europe. Acta Borealia, Tromsø Museum, A. Scientia 24, 1–63.  7

Binns, R.E., 1972. Composition and derivation of pumice on postglacial strandlines in northern Europe and the western Arctic. Geol. Soc. Amer. Bull. 83, 2303–2324.  7

Bischof, G., 1847–55. Lehrbuch der chemischen und physikalischen Geologie. Bonn. 1st ed. 2 vol., 1–3615.  12

Blackburn, E.A., Wilson, L. and Sparks, R.S.J., 1976. Mechanisms and dynamics of strombolian activity. J. Geol. Soc. London 132, 429–440.  3,4

Blackwelder, E., 1928. Mudflows as a geologic agent in semi-arid mountains. Geol. Soc. Amer. Bull. 39, 465–480.  11

Blatt, H., 1970. Determination of the mean sediment thickness in the crust: a sedimentologic model. Geol. Soc. Amer. Bull. 81, 255–262. 1

Blatt, H., Middleton, G. and Murray, R., 1972. Origin of sedimentary rocks. Prentice-Hall, Inc., Englewood Cliffs, N.J. 1–634. 5,12

Bloch, S. and Bischoff, J.L., 1979. The effect of low-temperature alteration of basalt on the oceanic budget of potassium. Geology 7, 193–196. 12

Blong, R.J., 1982. The time of darkness. Univ. of Wash. Press, Seattle and London, 1–257. 13

Bloomfield, K. and Valastro, S., Jr., 1977. Late Quaternary tephrochronology of Nevado de Toluca volcano, central Mexico. Overseas Geol. Min. Resour. 46, 1–15. 11

Bloomfield, K., Rubio, G.S. and Wilson, L., 1977. Plinian eruptions of Nevado de Toluca Volcano, Mexico. Geol. Rundsch. 66, 120–146. 6

Bogaard, P.v.d., 1978. Aufbau und Entstehung der Oberen Laacher Pyroklastika bei Mendig. Ruhr Univ. Bochum, Unpubl. Dipl. thesis, 1–64. 8

Bogaard, P.v.d., 1983. Die Eruption des Laacher See Vulkans. Ruhr Univ. Bochum Ph. D. diss. 1–348. 6,7

Bøggild, O.B., 1918. Den vulkanske aske i Moleret samt en oversigt over Danmarks aeldre Tertiaerbjergarter. Dan. geol. Unders. roekkez, 2, 33, 1–159. 7

Bohor, B.F., Phillips, R.E. and Pollastro, R.M., 1979. Altered volcanic ash partings in Wasatch Formation coal beds of the northern Powder River basin: Composition and geologic applications. U.S. Geol. Survey, Open File Rpt. 79-1203, 1–11. 12

Boles, J.R., 1974. Structure, stratigraphy and petrology of mainly Triassic rocks, Hokonui Hills, Southland, New Zealand. N. Z. J. Geol. Geophys. 17, 337–374. 12

Boles, J.R. and Coombs, D.S., 1975. Mineral reactions in zeolitic Triassic tuff, Hokonui Hills, New Zealand. Geol. Soc. Amer. Bull. 86, 163–173. 12

Boles, J.R. and Coombs, D.S., 1977. Zeolite facies alteration of sandstones in the Southland Syncline, New Zealand. Amer. J. Sci. 277, 982–1012. 12

Boles, J.R. and Surdam, R.C., 1979. Diagenesis of volcanogenic sediments in a Tertiary saline lake: Wagon Bed Formation, Wyoming. Amer. J. Sci. 279, 832–853. 12

Boles, J.R. and Wise, W.S., 1978. Nature and origin of deep-sea clinoptilolite. In Sand, L.B. and Mumpton, F.A., eds., Natural Zeolites: occurrence, properties, use. Pergamon Press, Oxford, 235–243. 12

Bonatti, E., 1963. Zeolites in Pacific pelagic sediments. Trans. N. Y. Acad. Sci. Ser.11, 25, 938–948. 12

Bonatti, E., 1965. Palagonite, hyaloclastites and alteration of volcanic glass in the ocean. Bull. Volcanol. 28, 257–269. 12

Bonatti, E., 1967. Mechanism of deep sea volcanism in the south Pacific. In Abelson, P., ed., Researches in Geochemistry, J. Wiley, New York, 2, 453–491. 12

Bond, A. and Sparks, R.S.J., 1976. The Minoan eruption of Santorini, Greece. J. Geol. Soc. London 132, 1–16. 11

Bond, G.C., 1973. A late Paleozoic volcanic arc in the eastern Alaska Range, Alaska. J. Geol. 81, 557–575. 10

Booth, B., 1973. The Granadilla pumice deposit of southern Tenerife, Canary Islands. Proc. Geol. Assoc. London 84, 353–370. 6

Booth, B., Croasdale, R. and Walker, G.P.L., 1978. A quantitative study of five thousand years of volcanism on São Miguel, Azores. Philos. Trans. Roy. Soc. London A 288, 271–319. 4,6

Born, A., 1923. Über die Erscheinungsformen eines submarinen Ergusses. Z. deutsch. Geol. Ges. 74, 101–117. 4

Bottinga, Y. and Weill, D.F., 1972. The viscosity of magmatic silicate liquids: A model for calculation. Amer. J. Sci. 272, 438–475. 3

Bottinga, Y., Weill, D.F. and Richet, P., 1981. Thermodynamic modeling of silicate melts. In Newton, R.C., Navrotsky, A. and Wood, B.J., eds., Thermodynamics of Minerals and Melts. Springer-Verlag, New York, Heidelberg, Berlin, 207–245. 3

Bouma, A.H., 1962. Sedimentology of some flysch deposits; A graphic approach to facies interpretation. Elsevier, Amsterdam, 1–168. 10

Bowles, F.A., Jack, R.N. and Carmichael, I.S.E., 1973. Investigation of deep-sea volcanic ash layers from equatorial Pacific cores. Geol. Soc. Amer. Bull. 84, 2371–2388. 7

Boyd, F.R., 1961. Welded tuff and flows in the rhyolite plateau of Yellowstone Park, Wyoming. Geol. Soc. Amer. Bull. 72, 387–426. 4,8

Boyd, F.R. and Kennedy, G.C., 1951. Some experiments and calculations relating to the origin of welded tuffs (Abst.). Amer. Geophys. Un. Trans. 32, 327–328. 8

Bramlette, M.N. and Bradley, W.H., 1942. Geology and biology of North Atlantic deep-sea cores between Newfoundland and Ireland, pt. 1, Lithology and geologic interpretations. U.S. Geol. Survey Prof. Paper 196A, 1–55. 7

Bramlette, M.N. and Posnjak, E., 1933. Zeolite alteration of pyroclasts. Amer. Mineral. 18, 167–171. 12

Brazier, S., Davis, A.N., Sigurdsson, H. and Sparks, R.S.J., 1982. Fall-out and deposition of volcanic ash during the 1979 explosive eruption of the Soufrière of St. Vincent. J. Volcanol. Geotherm. Res. 14, 335–359. 6

Brazier, S., Sparks, R.S.J., Carey, S.N., Sigurdsson, H. and Westgate, J.A., 1983. Bimodal grain size distribution and secondary thickening in air-fall ash layers. Nature 301, 115–119. 6

Brenchley, P.J., 1972. The Cwm Clwyd Tuff, North Wales: a paleogeographical interpretation of some Ordovician ash-shower deposits. Yorkshire Geol. Soc. Proc. 39, 199–224. 7

Brey, G., 1976. $CO_2$ solubility mechanisms in silicate melts at high pressures. Contr. Mineral. Petrol. 57, 215–221. 3

Brey, G. and Green, D.H., 1975. The role of $CO_2$ in the genesis of olivine melilitite. Contr. Mineral. Petrol. 49, 93–103. 3

Brey, G. and Green, D.H., 1976. Solubility of $CO_2$ in olivine melilitite at high pressures and role of $CO_2$ in the earth's upper mantle. Contr. Mineral. Petrol. 55, 217–230. 3

Brey, G. and Green, D.H., 1977. Systematic study of liquidus phase relations in olivine melilitite $+H_2O+CO_2$ at high pressures and petrogenesis of an olivine melilitite magma. Contr. Mineral. Petrol. 61, 141–162. 3

Brey, G. and Schmincke, H.-U., 1980. Origin and diagenesis of the Roque Nublo Breccia, Gran Canaria (Canary Islands) – Petrology of Roque Nublo volcanics, II. Bull. Volcanol. 43–1, 15–33. 3,11,12

Brinkley, S.R., Jr., Kirkwood, J.G., Lampson, C.W., Revelle, R. and Smith, S.B., 1950. Shock from underwater and underground blasts. In The Effects of Atomic Weapons, Los Alamos Scientific Lab., Los Alamos, New Mexico. U.S. Government Printing Office, 83–113. 9

Broscoe, A.J. and Thomson, S., 1969. Observations on an alpine mudflow, Steele Creek, Yukon. Can. J. Earth Sci. 6, 219–229. 11

Bryan, W.B., 1968. Low-potash dacite drift pumice from the Coral Sea. Geol. Mag. 105, 431–439. 7

Bryan, W.B., Stice, G.D. and Ewart, A., 1972. Geology, petrography and geochemistry of the volcanic islands of Tonga. J. Geophys. Res. 77, 1566–1585. 14

Buchanan, D.J., 1974. A model for fuel-coolant interactions. J. Phys. Dev. Appl. Phys. 7, 1441–1457. 4

Buchanan, D.J. and Dullforce, T.A., 1973. Mechanism for vapour explosions. Nature 245, 32–34. 4

Buck, P.S., 1976. An early Precambrian caldera in the Favourable Lake metavolcanic-metasedimentary belt, northwestern Ontario. Centre for Precambrian Studies, University of Manitoba, 1975, Ann. Rpt. 108–115. 13

Bull, W.B., 1964. Alluvial fans and near-surface subsidence in western Fresno County, California. U.S. Geol. Survey Prof. Paper 437-4, 1–70. 9,11

Buller, A.T. and McManus, J., 1973. Distinction among pyroclastic deposits from their grain–size frequency distribution. J. Geol. 81, 97–106. 5

Bunsen, R., 1847. Beitrag zur Kenntnis des isländischen Tuffgebirges. Annal. Chem. Pharm. 61,3, 265–279. 12

Bunsen, R., 1851. Über die Prozesse der vulkanischen Gesteinsbildungen Islands. Annal. Phys. Chem. 83, 197–272. 2,12

Burger, K., 1980. Kaolin-Kohlentonsteine im flözführenden Oberkarbon des Niederrheinisch-Westfälischen Steinkohlenreviers. Geol. Rundsch. 69, 488–531. 12

Burk, C.A. and Drake, C.L., eds. 1974. The geology of continental margins. Springer-Verlag, New York, Heidelberg, Berlin, 1–1009. 14

Burnham, C.W., 1975a. Water and magmas: A mixing model. Geochim. Cosmochim. Acta 39, 1077–1084. 3

Burnham, C.W., 1975b. Thermodynamics of melting in experimental silicate-volatile systems. Fortschr. Mineral. 52, 101–118. 3

Burnham, C.W., 1979. The importance of volatile constituents. In Yoder, H.S., Jr., ed., The evolution of igneous rocks. Princeton Univ. Press, Princeton, N.J., 439–482. 3

Burnham, C.W. and Jahns, R.H., 1962. A method for determining the solubility of water in silicate melts. Amer. J. Sci. 260, 721–745. 3

Burns, R.E. and ten co-authors, 1973. Site 203. In Burns, R.E., Andrews, J.E. et al., eds., Init. Rpts. Deep Sea Drilling Proj. 21, 17–32.   14

Busby-Spera, C., 1981a. Silicic ash-flow tuffs interbedded with submarine andesitic and sedimentary rocks in lower Mesozoic roof pendants, Sierra Nevada, California (Abst.). Geol. Soc. Amer. Abst. with Programs 13, 47.   10

Busby-Spera, C., 1981b. Early Mesozoic submarine epicontinental calderas in the southern Sierra Nevada, California (Abst.). EOS 62, 1061.   10

Byerly, G.R. and Sinton, J.M., 1979. Compositional trends in natural basaltic glasses from DSDP holes 417D and 418A. In Donnelly, T., Francheteau, J., Bryan, W., Robinson, P., Flower, M., Salisbury, M. et al., eds., Init. Rpts. Deep Sea Drilling Proj. 51, 52, 53, Pt. 2, 957–971.   12

Cady, W.M., 1975. Tectonic setting of the Tertiary volcanic rocks of the Olympic Peninsula, Washington. U.S. Geol. Survey J. Res. 3, 573–582.   14

Campbell, A.S. and Fyfe, W.S., 1965. Analcime-albite equilibria. Amer. J. Sci. 263, 807–816.   12

Car, D., 1979. An investigation of clastic rocks derived from an early Precambrian stratovolcano in the Lake of the Woods area, northwestern Ontario. Univ. of Manitoba, M.Sc. diss. 13, 1–111

Carey, S.N. and Sigurdsson, H., 1978. Deep–sea evidence for distribution of tephra from the mixed magma eruption of the Soufrière on St. Vincent, 1902: ash turbidites and air fall. Geology 6, 271–274.   7

Carey, S.N. and Sigurdsson, H., 1980. The Roseau Ash: deep-sea tephra deposits from a major eruption on Dominica, Lesser Antilles Arc. J. Volcanol. Geotherm. Res. 7, 67–86.   7,10,14

Carey, S.N. and Sigurdsson, H., 1982. Influence of particle aggregation on deposition of distal tephra from the May 18, 1980, eruption of Mount St. Helens Volcano. J. Geophys. Res. 87, 7061–7072.   6

Carlisle, D., 1963. Pillow breccias and their aquagene tuffs, Quadra Island, British Columbia. J. Geol. 71, 48–71.   9,10

Carlisle, D. and Susuki, T., 1974. Emergent and submergent carbonate-clastic sequences including the Upper Triassic Dilleri and Welleri Zones on Vancouver Island. Can. J. Earth Sci. 11, 254–279.   10

Carmichael, I.S.E., 1979. Glass and the glassy rocks. In Yoder, H.S., Jr., ed., The evolution of the igneous rocks. Princeton Univ. Press, Princeton, N.J., 233–244.   4

Carmichael, I.S.E., Turner, F.J., Verhoogen, J., 1974. Igneous Petrology. McGraw-Hill Book Co., New York, 1–739.   2,3

Carozzi, A.V., 1972. Microscopic sedimentary petrology. Krieger, R.E., Publ. Co. Inc., New York, Reprint with corrections 1–485.   1

Carozzi, A.V., ed., 1976. Sedimentary rocks: concepts and history: Benchmark papers in geology. Dowden, Hutchinson and Ross, Inc., Stroudsburg, Pa., 1–468.   1

Carter, R.M., Hicks, M.D., Norris, R.J. and Turnbull, I.M., 1978. Sedimentation patterns in an ancient arc-trench ocean basin complex: Carboniferous to Jurassic Rangitata orogen, New Zealand. In Stanley, D.J. and Kelling, G., eds., Sedimentation in submarine canyons, fans and trenches. Dowden, Hutchinson and Ross, Stroudsburg, Pa., 340–361.   14

Cas, R., 1978. Silicic lavas in Paleozoic flyschlike deposits in New South Wales, Australia: Behavior of deep subaqueous silicic flows. Geol. Soc. Amer. Bull. 89, 1708–1714.   10

Cas, R.A.F. and Jones, J.G., 1979. Paleozoic interarc basin in eastern Australia and a modern New Zealand analogue. N.Z. J. Geol. Geophys. 22, 71–85.   14

Casadevall, T.J. and Greenland, L.P., 1981. The chemistry of gases emanating from Mount St. Helens, May–September, 1980. In Lipman, P.W. and Mullineaux, D.R., eds., The 1980 eruptions of Mount St. Helens, Washington, U.S. Geol. Survey Prof. Paper 1250, 221–231.   8

Cerling, T.E., Biggs, D.L. and Vondra, C.F., 1975. Use of oxygen isotope ratios in correlation of tuffs, east Rudolf Basin, northern Kenya. Earth Planet. Sci. Lett. 25, 291–296.   13

Chadwick, R.A., 1971. Paleomagnetic criteria for volcanic breccia emplacement. Geol. Soc. Amer. Bull. 82, 2285–2294.   5

Chapin, C.E. and Lowell, G.R., 1979. Primary and secondary flow structures in ash-flow tuffs of the Gribbles Run paleovalley, central Colorado. Geol. Soc. Amer. Sp. Paper 180, 137–154.   8

Chesterman, C.W., 1956. Pumice, pumicite and volcanic cinders in California. Calif. Div. Mines Bull. 174, 3–97.   7

Chouet, B., Hamisevicz, N. and McGetchin, T.R., 1974. Photoballistics of volcanic jet activity at Stromboli, Italy. J. Geophys. Res. 79, 4961–4976.   4,6

Christiansen, R.L., 1980. Eruption of Mount St. Helens. Nature 285, 531–533.   11

Christiansen, R.L. and Lipman, P.W., 1972. Cenozoic volcanism and plate-tectonic evolution of the western United States. II. Late Cenozoic. Philos. Trans. Roy. Soc. London A 271, 249–284.   13,14

Christiansen, R.L. and Peterson, D.W., 1981. Chronology of the 1980 eruptive activity. In Lipman, P.W. and Mullineaux, D.R., eds., The 1980 eruptions of Mount St. Helens, Washington. U.S. Geol. Survey Prof. Paper 1250, 17–33.   4

Church, B.N. and Johnson, W.M., 1980. Calculation of the refractive index of silicate glasses from chemical composition. Geol. Soc. Amer. Bull. 91, 619–625.   7

Churkin, M., Jr., 1974. Paleozoic marginal ocean basin-volcanic arc systems in the Cordilleran foldbelt. In Dott, R.H., Jr. and Shaver, R.H., eds., Modern and ancient geosynclinal sedimentation. Soc. Econ. Paleont. Mineral. Sp. Publ. 19, 174–192.   14

Clarke, F.W., 1924. The data of geochemistry. U.S. Geol. Survey Bull. 770, 1–841.   1

Colgate, S.A. and Sigurgeirsson, T., 1973. Dynamic mixing of water and lava. Nature 244, 552–555.   9

Connelly, W., 1978. Uyak Complex, Kodiak Islands, Alaska: A Cretaceous subduction complex. Geol. Soc. Amer. Bull. 89, 755–769.   14

Cook, E.F., 1966, ed. Tufflavas and ignimbrites. Amer. Elsevier Pub. Co., Inc., New York, 1–212.   8

Cook, H.E., 1968. Ignimbrite flows, plugs and dikes in the southern part of the Hot Creek Range, Nye County, Nevada. Geol. Soc. Amer. Mem. 116, 107–152.   8

Cook, H.E., 1975. North American stratigraphic principles as applied to deep-sea sediments. Amer. Assoc. Petrol. Geol. Bull. 59, 817–837.   13

Coombs, D.S., 1954. The nature and alteration of some Triassic sediments from Southland, New Zealand. Roy. Soc. New Zealand Trans. 82, 65–109.   12

Coombs, D.S. and Landis, C.A., 1966. Pumice from the South Sandwich eruption of March, 1962, reaches New Zealand. Nature 209, 289–290.   7

Coombs, D.S., Ellis, A.J., Fyfe, W.S. and Taylor, A.M., 1959. The zeolite facies, with comments on the interpretation of hydrothermal synthesis. Geochim. Cosmochim. Acta 17, 53–107.   12

Correns, C.W., 1930. Über einen Basalt vom Boden des atlantischen Ozeans und seine Zersetzungsrinde. Chemie der Erde 5, 76–86.   12

Corwin, G. and Foster, H.L., 1959. The 1957 explosive eruption on Iwo Jima volcanic islands. Amer. J. Sci. 257, 161–171.   9

Cotton, C.A., 1944. Volcanoes as landscape forms. Whitcombe and Tombs Ltd., Auckland, 1–416.   2,11

Cousineau, P. and Dimroth, E., 1982. Interpretation of the relations between massive, pillowed and brecciated facies in an Archean submarine andesite volcano – Amulet Andesite, Rouyn-Noranda, Canada. J. Volcanol. Geotherm. Res. 13, 83–102.   10

Cox, K.G., Bell, J.D. and Pankhust, 1979. The interpretation of igneous rocks. George Allen and Unwin, London, 1–450.   2

Crandell, D.R., 1957. Some features of mudflow deposits (Abst.). Geol. Soc. Amer. Bull. 68, pt. 2, 1821.   11

Crandell, D.R., 1963. Surficial geology and geomorphology of the Lake Tapps quadrangle, Washington. U. S. Geol. Survey Prof. Paper 388-A, A1–A84.   11

Crandell, D.R., 1971. Postglacial lahars from Mount Rainier volcano, Washington. U. S. Geol. Survey Prof. Paper 677, 1–75.   5,11

Crandell, D.R. and Mullineaux, D.R., 1973. Pine Creek volcanic assemblage at Mount St. Helens, Washington. U.S. Geol. Survey Bull. 1383-A, A1–A23.   5,8,11

Crandell, D.R. and Waldron, H.H., 1956. A recent volcanic mudflow of exceptional dimensions from Mount Rainier, Washington. Amer. J. Sci. 254, 349–362.   11

Crisp, J.A., 1984. Rates of magmatism. J. Volcanol. Geotherm. Res. 20 (in press).   2

Crowe, B.M. and Fisher, R.V., 1973. Sedimentary structures in base-surge deposits with special reference to cross-bedding, Ubehebe Craters, Death Valley, California. Geol. Soc. Amer. Bull. 84, 663–682.   5,8,9

Crowe, B.M., Linn, G.W., Heiken, G. and Bevier, M.L., 1978. Stratigraphy of the Bandelier Tuff in the Pajarito Plateau; Applications to waste management. Los Alamo Sci. Lab., New Mex., Informal Rpt., LA-7225-MS, 1–57.   8

Cunningham, C.G. and Steven, T.A., 1979. Mount Belknap and Red Hills Calderas and associated rocks, Marysvale Volcanic Field, West-Central Utah. U. S. Geol. Survey Bull. 1468, 1–34.   13

Curry, R.R., 1966. Observation of alpine mudflows in the Ten-mile Range, central Colorado. Geol. Soc. Amer. Bull. 77, 771–776.   11

Curtis, G.H., 1954. Mode of origin of pyroclastic debris in the Mehrten Formation of the Sierra Nevada. Univ. Calif. Publ. Geol. Sci. 29, 453–502.   11

Curtis, G.H., 1968. The stratigraphy of the ejectamenta of the 1912 eruption of Mt. Katmai and Novarupta, Alaska. Geol. Soc. Amer. Mem. 116, 153–210.   8

Dakyn, J.R. and Greenly, E., 1905. On the probable Peléan origin of the felsitic slates of Snowdon. Geol. Mag. 2, 541–549. 8

Dalrymple, G.S., Cox, A. and Doell, R.R., 1965. Potassium-argon age and paleomagnetism of the Bishop Tuff, California. Geol. Soc. Amer. Bull. 76, 665–674. 13

Daly, R.A., 1933. Igneous rocks and the depths of the earth. McGraw-Hill Book Co., Inc., New York and London, 1–598. 9

Dana, J.D., 1890. Characteristics of volcanoes. Dodd, Mead and Co., New York, 1–399. 5

David, P.P., 1970. Discovery of Mazama ash in Saskatchewan, Canada. Can. J. Earth Sci. 7, 1579–1583. 13

Davies, D.K., Quearry, M.W. and Bonis, S.B., 1978. Glowing avalanches from the 1974 eruption of the volcano Fuego, Guatemala. Geol. Soc. Amer. Bull. 89, 369–384. 8

Davies, I.C. and Walker, R.G., 1974. Transport and deposition of resedimented conglomerates: The Cap Enrage Formation, Gaspé, Quebec. J. Sed. Petrol. 44, 1200–1216. 11

Dawson, J.B., 1962. The geology of Oldoinyo Lengai. Bull. Volcanol. 24, 349–387. 6

Dawson, J.B., 1964a. Carbonatitic volcanic ashes in northern Tanganyika. Bull. Volcanol. 27, 81–91. 9

Dawson, J.B., 1964b. Carbonatite tuff cones in northern Tanganyika. Geol. Mag. 101, 129–137. 9

Dawson, J.B., 1980. Kimberlites and their xenoliths. Springer-Verlag, Berlin, Heidelberg, New York, 1–252. 9

Decker, R. and Decker, B., 1981. Volcanoes. W.H. Freeman and Company, San Francisco, 1–244. 1

Deer, W.H., Howie, R.A. and Zussman, J., 1963. Rock-forming minerals. Vol. 4, Framework Silicates. John Wiley and Sons, Inc., New York, 1–435. 12

Deffeyes, K.S., 1959. Zeolites in sedimentary rocks. J. Sed. Petrol. 29, 602–609. 12

Delaney, P.T., 1982. Rapid intrusion of magma into wet rock: groundwater flow due to pore pressure increases. J. Geophys. Res. 87, 7739–7756. 4,9

Delaney, J.R., Muenow, D.W. and Graham, D.G., 1978. Abundance and distribution of water, carbon and sulfur in the glassy rims of submarine pillow basalts. Geochim. Cosmochim. Acta 30, 963–982. 3

DeRosen-Spence, A.F., Provost, G., Dimroth, E., Gochnauer, K. and Owen, V., 1980. Archean subaqueous felsic flows, Rouyn-Noranda, Quebec, Canada, and their Quaternary equivalents. Precambrian Res. 12, 43–77. 10

Dewey, J.F., 1963. The Lower Paleozoic Stratigraphy of central Murrisk County, Mayo, Ireland, and the evolution of the South Mayo Trough. Q. J. Geol. Soc. London 119, 313–343. 10

Dick, H.J.B., Honnorez, J. and Kirst, P.W., 1978. Origin of the abyssal basaltic sand, sandstone, and gravel from DSDP Hole 396B, Leg 46. In Dmitriev, L., Heirtzler, J., et al., eds., Init. Rpts. Deep Sea Drilling Proj. 46, 331–339. 10

Dickinson, W.R., 1962. Marine sedimentation of clastic volcanic strata (Abst.). Amer. Assoc. Petrol. Geol. Bull. 46, 263. 12

Dickinson, W.R., 1968. Sedimentation of volcaniclastic strata of the Pliocene Koroimavua Group in northwest Viti Levu, Fiji. Amer. J. Sci. 266, 440–453. 14

Dickinson, W.R., 1974a. Plate tectonics and sedimentation. Soc. Econ. Paleont. Mineral. Sp. Publ. 22, 1–27. 14

Dickinson, W.R., 1974b. Sedimentation within and beside ancient and modern magmatic arcs. In Dott, Jr., R.H. and Shaver, R.H., eds., Modern and ancient geosynclinal sedimentation, Soc. Econ. Paleont. Mineral. Sp. Publ. 19, 230–239. 13,14

Dickinson, W.R., ed., 1974c. Tectonics and Sedimentation. Soc. Econ. Paleont. Mineral Sp. Publ. 22, 1–204. 14

Dickinson, W.R., 1976. Sedimentary basins developed during evolution of Mesozoic-Cenozoic arc-trench systems in western North America. Can. J. Earth Sci. 13, 1268–1287. 14

Dickinson, W.R., 1977. Plate tectonic evolution of sedimentary basins. In Dickinson, W.R. and Yarborough, H., eds., Plate tectonics and hydrocarbon accumulation. Amer. Assoc. Petrol. Geol., Continuing Education Course Note Series 1, 1–62. 14

Dickinson, W.R. and Seely, D.R., 1979. Structure and stratigraphy of fore-arc regions. Amer. Assoc. Petrol. Geol. Bull. 63, 2–31. 14

Dickinson, W.R. and Rich, E.L., 1972. Petrologic intervals and petrofacies in the Great Valley sequence, Sacramento Valley, California. Geol. Soc. Amer. Bull. 83, 3007–3024. 13

Dickinson, W.R. and Vigrass, L.W., 1964. Pre-Cenozoic history of Suplee-Izee district, Oregon: Implications for geosynclinal theory. Geol. Soc. Amer. Bull. 75, 1037–1044. 13

Dimroth, E. and Demarcke, J., 1978. Petrography and mechanism of eruption of the Archean Dalembert tuff, Rouyn-Noranda, Quebec, Canada. Can. J. Earth Sci. 15, 1712–1723. 10,13

Dimroth, E. and Lichtblau, A.P., 1979. Metamorphic evolution of Archean hyaloclastites, Noranda area, Quebec, Canada. Part I: Comparison of Archean and Cenozoic sea-floor metamorphism. Can. J. Earth Sci. 16, 1315–1340.   12

Dimroth, E., Cousineau, P., Leduc, M. and Sanschagrin, Y., 1978. Structure and organization of Archean subaqueous basalt flows, Rouyn-Noranda area, Quebec, Canada. Can. J. Earth Sci. 15, 902–918.   10

Dimroth, E., Cousineau, P., Leduc, M., Sanschagrin, Y. and Provost, G., 1979. Flow mechanisms of Archean subaqueous basalt and rhyolite flows. Geol. Surv. Can. Current Res. Paper 79-1A, 207–211.   10

Donn, W.L. and Ninokovich, D., 1980. Rate of Cenozoic explosive volcanism in the North Atlantic Ocean inferred from deep sea cores. J. Geophys. Res. 85, 5455–5460.   7

Doremus, R.H.J., 1975. Interdiffusion of hydrogen and alkali ions in a glass surface. J. Noncrystal Solids 19, 137–144.   12

Dott, Jr., R.H. and Shaver, R.H., eds., 1974. Modern and ancient geosynclinal sedimentation. Soc. Econ. Paleont. Mineral. Sp. Publ. 19, 1–380.   13,14

Drever, J.I., 1972. Relations among pH, carbon dioxide pressure, alkalinity, and calcium concentration in waters saturated with respect to calcite at 25 °C and one atmosphere total pressure. Contrib. Geol. 11, 41–42.   12

Drexler, J.W., Rose, W.I., Jr., Sparks, R.S.J. and Ledbetter, M.T., 1978. Geochemical correlation of Pleistocene rhyolitic ashes in Guatemala with deep-sea ash layers of the Gulf of Mexico, equatorial Pacific and Caribbean Sea. Amer. Geophys. Un. Trans. 59, 1105.   7

Duda, A. and Schmincke, H.-U., 1978. Quaternary basanites, melilite nephelinites and tephrites from the Laacher See area. N. Jb. Min. Mh. 132, 1–33.   2

Duffield, W.A., Bacon, C.R. and Roquemore, G.R., 1979. Origin of reverse-graded bedding in air fall pumice, Coso Range, California. J. Volcanol. Geotherm. Res. 5, 35–48.   6

Duffield, W.A., Gibson, E.K., Jr. and Heiken, G.H., 1977. Some characteristics of Pele's Hair. U.S. Geol. Survey J. Res. 5, 93–101.   4,5

Dullforce, T.A., Buchanan, D.J. and Peckover, R.S., 1976. Self-triggering of small-scale fuel-coolant interactions. I. Experiments. J. Phys. Dev.: Appl. Phys. 9, 1295–1303.   4

Dunbar, C.O. and Rodgers, J., 1957. Principles of Stratigraphy. John Wiley and Sons, Inc., New York, 1–356.   13

Dymek, R.F., Boak, J.L. and Gromet, L.P., 1983. Average sedimentary rock rare earth element patterns and crustal evolution: some observations and implications from the 3800 M.y. old Isna Supracrustal Belt, West Greenland. In 1983 Archean Geochemistry – Early Crustal Genesis field workshop. The Lunar and Planetary Science Institute, Houston, 1–4.   1

Eaton, G.P., 1964. Windborne volcanic ash: A possible index to polar wandering. J. Geol. 72, 1–35.   6

Eden, H.F., McConnell, R.K., Jr. and Allen, R.V., 1967. Mechanics of ash flows (Abst.). Amer. Geophys. Un. Trans. 48, 228.   8

Eggler, D.H., 1972. Water-saturated and under-saturated melting relations in a Paricutin andesite and an estimate of water content in the natural magma. Contr. Mineral. Petrol. 34, 261–271.   2,3

Eggler, D.H., 1973. Role of $CO_2$ in melting processes in the mantle. Carnegie Inst. Wash. Yearb. 72, 457–467.   3

Eggler, D.H., 1974. Effect of $CO_2$ on the melting of peridotite. Carnegie Inst. Wash. Yearb. 73, 215–224.   3

Eggler, D.H. and Rosenhauer, M., 1978. Carbon dioxide in silicate melts. II. Solubilities of $CO_2$ and $H_2O$ in $CaMgSi_2O_6$ (diopside) liquids and vapors at pressures to 40 kb. Amer. J. Sci. 278, 64–94.   3

Eggleton, R.A. and Keller, J., 1982. The palagonitization of limburgite glass – a TEM study. N. Jb. Miner. Mh. 1982, 289–311.   12

Eichelberger, J.C. and Hayes, D.B., 1982. Magmatic model for the Mount St. Helens blast of May 18, 1980. J. Geophys. Res. 87, 7727–7738.   4

Eichelberger, J.C. and Koch, F.G., 1979. Lithic fragments in the Bandelier Tuff, Jemez Mountains, New Mexico. J. Volcanol. Geotherm. Res. 5, 115–134.   8

Einstein, A., 1906. Eine neue Bestimmung der Moleküldimensionen. Ann. Phys. (Leipzig) 19, 289–306.   3

Ekren, E.B. and Byers, F.M., Jr., 1976. Ash-flow fissure vent in west-central Nevada. Geology 4, 247–251.   8

Enos, P., 1977. Flow regimes in debris flows. Sedimentology 24, 133–142.   11

Epstein, P.S. and Plesset, M.S., 1950. On the stability of gas bubbles in liquid-gas solutions. J. Chem. Phys. 18, 1505–1509.   3

Ermanovics, I., 1973. Evidence for early Precambrian Mg-rich crust in northwestern Superior Province of the Canadian Shields. In Volcanic Rocks, Geol. Survey Can. Open File Rpt. 164, 163–176.   14
Escher, B.G., 1920. L'éruption de Gounoung Galounggoung en juillet 1918. Natuurk. Tijds. Ned.-Ind. 80, 260–264.   11
Escher, B.G., 1933. On a classification of central eruptions according to gas pressure of the magma and viscosity of the lavas. Leids. Geol. Meded., Decl. VI, Afl. I., 50–58.   8
Eugster, H.P. and Surdam, R.C., 1973. Depositional environment of the Green River Formation of Wyoming: a preliminary report. Geol. Soc. Amer. Bull. 84, 1115–1120.   12
Ewart, A., 1963. Petrology and petrogenesis of the Quaternary pumice ash in the Taupo area, New Zealand. J. Petrol. 4, 392–431.   5,13
Ewart, A. and Bryan, W.B., 1973. The petrology and geochemistry of the Tongan Islands. In Coleman, P.J., ed., The Western Pacific: island arcs, marginal seas, geochemistry. Univ. Western Australia Press, Nedlands, 503–522.   14
Ewart, A., Hildreth, W. and Carmichael, I.S.E., 1975. Quaternary acid magma in New Zealand. Contr. Mineral. Petrol. 51, 1–27.   3
Ewing, M., Heezen, B.C. and Ericson, D.B., 1959. Significance of the Worzel deep-sea ash. Nat. Acad. Sci. Proc. 45, 355–361.   7
Fairbridge, R.W., 1973. Glaciation and plate migration. In Tarling, D.H. and Runcorn, S.K., eds., Implications of continental drift. Academic Press, New York, 503–515.   4,5
Fairbridge, R.W. and Bourgeois, J., eds., 1978. The Encyclopedia of Sedimentology: Encyclopedia of Earth Sciences, vol. VI. Dowden, Hutchinson and Ross, Inc., Stroudsburg, Pa., 1–901.   1
Fedotov, S.A., 1978. Ascent of basic magmas in the crust and the mechanism of basaltic fissure eruptions. Int. Geol. Rev. 20, 33–48.   4
Fenner, C.N., 1923. The origin and mode of emplacement of the great tuff deposits in the Valley of Ten Thousand Smokes. Natl. Geog. Soc. Contributed Tech. Papers, Katmai Ser. 1, 1–74.   8
Fenner, C.N., 1937. Tuffs and other volcanic deposits of Katmai and Yellowstone Park. Amer. Geophys. Un. Trans., 18th Ann. Mtg., 236–289.   8
Fenner, C.N., 1948. Incandescent tuff flows in southern Peru. Geol. Soc. Amer. Bull. 59, 879–893.   8
Fernandez, H.E., 1969. Notes on the submarine ash flow tuff in Siargao Island, Surigao del Norte (Philippines). The Philippine Geologist 23, 29–36.   10
Fisher, R.V., 1958. Definition of volcanic breccia. Geol. Soc. Amer. Bull. 69, 1071–1073.   5
Fisher, R.V., 1960a. Criteria for recognition of laharic breccias, southern Cascade Mountains, Washington. Geol. Soc. Amer. Bull. 71, 127–132.   11
Fisher, R.V., 1960b. Classification of volcanic breccia. Geol. Soc. Amer. Bull. 71, 973–982.   5
Fisher, R.V., 1961a. Proposed classification of volcaniclastic sediments and rocks. Geol. Soc. Amer. Bull. 72, 1409–1414.   5
Fisher, R.V., 1961b. Stratigraphy of the Ashford area, southern Cascades, Washington. Geol. Soc. Amer. Bull. 72, 1395–1408.   14
Fisher, R.V., 1963. Bubble wall texture and its significance. J. Sed. Petrol. 33, 224–227.   5
Fisher, R.V., 1964a. Resurrected Oligocene hills, eastern Oregon. Amer. J. Sci. 262, 713–725.   6
Fisher, R.V., 1964b. Maximum size, median diameter, and sorting of tephra. J. Geophys. Res. 69, 341–355.   6
Fisher, R.V., 1965. Settling velocity of glass shards. Deep-Sea Res. 12, 345–353.   5,6,7
Fisher, R.V., 1966a. Geology of a Miocene ignimbrite layer, John Day Formation, eastern Oregon. Univ. Calif. Publ. Sci. Geol. 67, 1–58.   8,13
Fisher, R.V., 1966b. Mechanism of deposition from pyroclastic flows. Amer. J. Sci. 264, 350–363.   8
Fisher, R.V., 1966c. Rocks composed of volcanic fragments. Earth-Sci. Rev. 1, 287–298.   5,8
Fisher, R.V., 1966d. Textural comparison of John Day volcanic siltstone with loess and volcanic ash. J. Sed. Petrol. 36, 706–718.   5,6
Fisher, R.V., 1967. Early Tertiary deformation in north-central Oregon. Amer. Assoc. Petrol. Geol. Bull. 51, 111–123.   6
Fisher, R.V., 1968a. Puu Hou littoral cones, Hawaii. Geol. Rundsch. 57, 837–864.   9
Fisher, R.V., 1968b. Pyrogenic mineral stability, lower member of the John Day Formation, eastern Oregon. Univ. Calif. Publ. Geol. Sci. 75, 1–39.   13
Fisher, R.V., 1971. Features of coarse-grained, high-concentration fluids and their deposits. J. Sed. Petrol. 41, 916–927.   8,10,11
Fisher, R.V., 1975. Autocatalytic explosions and littoral cones (Abst.). Geol. Soc. Amer. Abst. with Programs 8, 317–318.   9

Fisher, R.V., 1977. Erosion by volcanic base-surge density currents: U-shaped channels. Geol. Soc. Amer. Bull. 88, 1287–1297.   9

Fisher, R.V., 1979. Models for pyroclastic surges and pyroclastic flows. J. Volcanol. Geotherm. Res. 6, 305–318.   5,8

Fisher, R.V., 1983. Flow transformations in sediment gravity flows. Geology 11, 273–274.   8

Fisher, R.V. and Charleton, D.W., 1976. Mid-Miocene Blanca Formation, Santa Cruz Island, California. In Howell, D.G., ed., Aspects of the geologic history of the California continental borderland. Pacific Section Amer. Assoc. Petrol. Geol., Misc. Publ. 24, 228–240.   10

Fisher, R.V. and Dimroth, E., 1978. Subaqueous volcanic rocks are examined. Geotimes 23, 16–18.   10

Fisher, R.V. and Heiken, G., 1982. Mt. Pelée, Martinique: May 8 and 20, 1902 pyroclastic flows and surges. J. Volcanol. Geotherm. Res. 13, 339–371.   5,8,9

Fisher, R.V. and Mattinson, J.M., 1968. Wheeler Gorge turbidite-conglomerate series: inverse grading. J. Sed. Petrol. 38, 1013–1023.   11

Fisher, R.V. and Rensberger, J.M., 1973. Physical stratigraphy of the John Day Formation. Univ. Calif. Publ. Geol. Sci. 101, 1–45.   6,13

Fisher, R.V. and Schmincke, H.-U., 1978. Les ignimbrites. In Girod, M., ed., Les Roches Volcaniques. Doin Ed., Paris, 193–209.   8

Fisher, R.V. and Waters, A.C., 1970. Base surge bed forms in maar volcanoes. Amer. J. Sci. 268, 157–180.   9

Fisher, R.V., Smith, A.L. and Roobol, M.J., 1980. Destruction of St. Pierre, Martinique by ash cloud surges, May 8 and 20, 1902. Geology 8, 472–476.   8,9,11

Fisher, R.V., Schmincke, H.-U. and Bogaard, P. v.d., 1983. Origin and emplacement of a pyroclastic flow and surge unit at Laacher See, Germany. J. Volcanol. Geotherm. Res. 17, 375–392.   8

Fiske, R.S., 1963. Subaqueous pyroclastic flows in the Ohanapecosh Formation, Washington. Geol. Soc. Amer. Bull. 74, 391–406.   4,10

Fiske, R.S., 1969. Recognition and significance of pumice in marine pyroclastic rocks. Geol. Soc. Amer. Bull. 80, 1–8.   5,7

Fiske, R.S. and Matsuda, T., 1964. Submarine equivalents of ash flows in the Tokiwa Formation, Japan. Amer. J. Sci. 262, 76–106.   4,10

Fiske, R.S. and Sigurdsson, H., 1979. Soufrière Volcano, St. Vincent: Its 1979 eruption observed from the ground, aircraft and satellites. Science 216, 1105–1106.   4

Fiske, R.S., Hopson, C.A. and Waters, A.C., 1963. Geology of Mount Rainier National Park, Washington. U. S. Geol. Survey Prof. Paper 444, 1–93.   10

Fleet, A.J. and McKelvey, B.C., 1978. Eocene explosive submarine volcanism, Ninetyeast Ridge, Indian Ocean. Mar. Geol. 26, 73–97.   10

Folk, R.L., 1966. A review of grain-size parameters. Sedimentology 6, 73–93.   5

Fornari, D.J., Malahoff, A. and Heezen, B.C., 1979. Visual observations of the volcanic micromorphology of Tortuga, Lorraine and Tutu seamounts; and petrology and chemistry of ridge and seamount features in and around the Panama Basin. Mar. Geol. 31, 1–30.   10

Fox, P.S. and Heezen, B.C., 1965. Sands of the Mid-Atlantic Ridge. Science 159, 1367–1370.   10

Francis, E.H., 1967. Review of Carboniferous-Permian volcanicity in Scotland. Geol. Rundsch. 57, 219–246.   13

Francis, E.H., 1970. Review of Carboniferous volcanism in England and Wales. J. Earth Sci., Leeds 8, 41–56.   13

Francis, E.H. and Howells, M.F., 1973. Transgressive welded ash flow tuffs among the Ordovician sediments on N.E. Snowdonia, N. Wales. J. Geol. Soc. London 129, 621–641.   10

Francis, E.H., Smart, J.G.O. and Raisbeck, D.E., 1968. Westphalian volcanism at the horizon of the Black Rake in Derbyshire and Nottinghamshire. Yorkshire Geol. Soc. Proc. 36, 395–416.   7

Francis, P.W., 1976. Volcanoes. Pelican Books, London, 1–368.   1

Francis, P.W., Roobol, M.J., Walker, G.P.L., Cobbold, P.R. and Coward, M., 1974. The San Pedro and San Pablo volcanoes and their hot avalanche deposits. Geol. Rundsch. 63, 357–388.   8

Frechen, J., 1971. Siebengebirge am Rhein, Laacher Vulkangebiet, Maargebiet der Westeifel, Vulkanologisch-petrographische Exkursionen: 2nd edition, Gebr. Borntraeger, Berlin, Stuttgart, 1–195.   9

Freundt, A., 1982. Stratigraphie des Brohltaltrass und seine Entstehung aus pyroklastischen Strömen des Laacher See Vulkans. Ruhr Univ., Bochum, Unpubl. Dipl. thesis, 1–319.   5

Friedman, G.M. and Sanders, J.E., 1978. Principles of Sedimentology. John Wiley and Sons, New York, 1–792.   1

Friedman, I., 1967. Water and deuterium in pumice from the 1959–1960 eruption of Kilauea Volcano, Hawaii. U. S. Geol. Survey Prof. Paper 575-B, 120–127.   3

Friedman, I. and Long, W., 1976. Hydration rate of obsidian. Science 191, 347–352. 12

Friedman, I. and Smith, R.L., 1958. The deuterium content of water in some volcanic glasses. Geochim. Cosmochim. Acta 15, 218–228. 12

Friedman, I. and Smith, R.L., 1960. A new dating method using obsidian: Part I, the development of the method. Amer. Antiquity 25, 476–522. 12

Friedman, I. and Trembour, F.W., 1978. Obsidian: The dating stone. Amer. Scientist 66, 44–51. 12

Friedman, I., Long, W., Smith, R.L., 1963. Viscosity and water content of rhyolite glass. J. Geophys. Res. 68, 6523–6535. 3,8

Friedman, I., Smith, R.L. and Long, W.D., 1966. Hydration of natural glass and formation of perlite. Geol. Soc. Amer. Bull. 77, 323–327. 12

Friedman, I., Lipman, P.W., Obradovich, J.D., Christiansen, R.L., 1974. Meteoric water in magmas. Science 184, 1069–1072. 2

Frith, R.A. and Doig, R., 1975. Pre-Kenoran tonalitic gneisses in the Grenville Province. Can. J. Earth Sci. 12, 844–849. 14

Froggatt, P.C., 1982. Review of methods of estimating rhyolitic tephra volumes; applications to the Taupo volcanic zone, New Zealand. J. Volcanol. Geotherm. Res. 14, 301–318. 6

Froggatt, P.C., Wilson, C.J.N. and Walker, G.P.L., 1981. Orientation of logs in the Taupo ignimbrite as an indicator of flow direction and vent position. Geology 9, 109–111. 8

Fryxell, R., 1965. Mazama and Glacier Peak volcanic ash layers: Relative ages. Science 147, 1288–1290. 6

Füchtbauer, H., 1974. Sediments and sedimentary rocks, 1. Schweizerbart'sche Verlagsbuchhandlung, Stuttgart, 1–464. 5

Fudali, R.F. and Melson, W.G., 1972. Ejecta velocities, magma chamber pressure, and kinetic energy associated with the 1968 eruption of Arenal volcano. Bull. Volcanol. 35, 383–401. 6

Fujii, N., 1975. Material and energy production from volcanoes. Bull. Volcanol. Soc. Jap. 20, 197–204 (in Japanese). 2,4

Fuller, R.E., 1931. The aqueous chilling of basaltic lava on the Columbia River Plateau. Amer. J. Sci. 21, 281–300. 4,10

Fuller, R.E., 1932. Concerning basalt glass. Amer. Mineral. 17, 104–107. 12

Furnes, H., 1972. Meta-hyaloclastite breccias associated with Ordovician pillow lavas in the Solund area, west Norway. Norsk Geol. Tidsskr. 52, 385–407. 10

Furnes, H., 1974. Volume relations between palagonite and authigenic minerals in hyaloclastites and its bearing on the rate of palagonitization. Bull. Volcanol. 38, 173–186. 12

Furnes, H., 1975. Experimental palagonitization of basaltic glasses of varied composition. Contr. Mineral. Petrol. 50, 105–113. 12

Furnes, H., 1978. Element mobility during palagonitization of a subglacial hyaloclastite in Iceland. Chem. Geol. 22, 249–264. 12

Furnes, H., 1980. Chemical changes during palagonitization of an alkali olivine basaltic hyaloclastite, Santa Maria, Azores. N. Jb. Min. Abh. 138, 14–30. 12

Furnes, H. and Friedleifsson, I.B., 1979. Pillow block breccia – occurences and mode of formation. N. Jb. Geol. Pal. Mh. 3, 147–154. 10

Furnes, H. and El-Anbaawy, I.H., 1980. Chemical changes and authigenic mineral formation during palagonitization of a basanite hyaloclastite, Gran Canaria, Canary Islands. N. Jb. Min. Abh. 139, 279–302. 12

Furnes, H., Friedleifsson, I.B. and Atkins, F.B., 1980. Subglacial volcanics – on the formation of acid hyaloclastites. J. Volcanol. Geotherm. Res. 8, 95–110. 10

Garcia, M., Liu, N.W.K. and Muenow, D.W., 1979. Volatiles in submarine volcanic rocks from the Mariana Island arc and trough. Geochim. Cosmochim. Acta 43, 305–312. 3

Garrels, R.M. and Mackenzie, F.T., 1971. Evolution of sedimentary rocks. W.W. Norton and Co., Inc., New York, 1–397. 1

Garrison, R.E., Espiritu, E., Horan, J.J. and Mack, L.E., 1979. Petrology, sedimentology and diagenesis of hemipelagic limestone and tuffaceous turbidites in the Aksitero Formation, central Luzon, Philippines. U.S. Geol. Survey Prof. Paper 1112, 1–16. 10

Gary, M., McAfee, R., Jr. and Wolf, C.L., eds., 1974. Glossary of geology. Amer. Geol. Inst., Washington, D.C., 1–805. 9,12

Gass, I.G., Harris, P.G. and Holgate, M.W., 1963. Pumice eruptions in the area of South Sandwich Islands. Geol. Mag. 100, 321–330. 7

Gerlach, T.M., 1981. Restoration of new volcanic gas analyses from basalts of the Afar region: further evidence of $CO_2$-degassing trends. J. Volcanol. Geotherm. Res. 10, 83–91. 3

Gerlach, T.M., 1983. Intrinsic chemical variations in high temperature volcanic gases from basic lavas. Bull. Volcanol. 45, 235–244. 3

Gibson, I.L., 1970. A pantelleritic welded ash-flow from the Ethiopian Rift Valley. Contr. Mineral. Petrol. 28, 89–111. 8

Gibson, I.L., 1972. The chemistry and petrogenesis of a suite of pantellerites from the Ethiopian Rift. J. Petrol. 13, 31–44. 8

Gieskes, J.M. and Lawrence, J.R., 1981. Alteration of volcanic matter in deep sea sediments: evidence from chemical composition of interstitial waters from deep sea drilling cores. Geochim. Cosmochim. Acta 45, 1687–1704. 12

Gilbert, C.M., 1938. Welded tuff in eastern California. Geol. Soc. Amer. Bull. 49, 1829–1862. 8

Gilbert, G.K., 1914. The transportation of debris by running water. U.S. Geol. Survey Prof. Paper 86, 1–263. 5

Gill, J.B., 1981. Orogenic andesites and plate tectonics. Springer-Verlag, Berlin, Heidelberg, New York, 1–390. 2

Glassley, W., 1974. Geochemistry and tectonics of the Crescent volcanic rocks, Olympic Peninsula, Washington. Geol. Soc. Amer. Bull. 85, 785–794. 14

Glikson, A.Y., 1978. On the basement of Canadian greenstone belts. Geoscience Can. 5, 3–12. 14

Goldschmidt, V.M., 1933. Grundlagen der quantitativen Geochemie. Fortschr. Min. Krist. Petrogr. 17, 112–156. 1

Goodell, P.C. and Waters, A.C., eds., 1981. Uranium in volcanic and volcaniclastic rocks. Amer. Assoc. Petrol. Geol., Studies in Geology 13, 1–331. 12

Goodwin, A.M., 1974. The most ancient continental margins. In Burk, C.A. and Drake, C.L., eds., The Geology of Continental Margins. Springer-Verlag, Berlin, Heidelberg, New York, 767–780. 14

Goodwin, A.M., 1977. Archean volcanism in Superior Province, Canadian Shield. Geol. Assoc. Can. Sp. Paper 16, 205–241. 14

Goodwin, A.M., 1980. Archean plates and greenstone belts (Abst.). EOS 67, 383. 14

Gorshkov, G.S., 1959. Gigantic eruption of the volcano Bezymianny. Bull. Volcanol. 20, 77–109. 8,11

Gorshkov, G.S. and Dubik, Y.M., 1970. Gigantic directed blast at Sheveluch Volcano (Kamchatka). Bull. Volcanol. 34, 261–268. 11

Grachev, A.F. and Fedorovsky, V.S., 1981. On the nature of greenstone belts in the Precambrian. Tectonophysics 73, 195–212. 14

Grange, L.I., 1931. Conical mounds on Egmont and Ruapehu volcanoes. N.Z. Bull. Sci. and Tech. 12, 376–384. 11

Grange, L.I., 1937. The geology of the Rotorua-Taupo Subdivision, Rotorua and Kaimanawa Divisions. N.Z. Geol. Survey Bull. 37, 1–138. 7

Green, D.C. and Baadsgaard, H., 1971. Temporal evolution and petrogenesis of an Archean crustal segment at Yellowknife, NWT, Canada. J. Petrol. 12, 177–217. 14

Griffiths, J.C., 1967. Scientific method in analyses of sediments. McGraw-Hill Book Co., New York, 1–508. 5

Griggs, R.F., 1922. The Valley of Ten Thousand Smokes (Alaska). Washington, Nat. Geogr. Soc. 1–340. 8

Grim, R.E. and Güven, N., 1978. Bentonites: Geology, Mineralogy, Properties and Use. Develop. in Sedim. 24, Elsevier, Amsterdam, 1–256. 12

Grippi, J. and Burke, K., 1980. Submarine-canyon complex among Cretaceous island-arc sediments, western Jamaica. Geol. Soc. Amer. Bull. Part I, 91, 179–184. 14

Grönvold, K., 1972. Structural and petrochemical studies in the Kerlingarfjöll region, southwest Iceland. Oxford Univ., Ph.D. diss., 1–208. 10

Gwinn, V.E. and Mutch, T.A., 1965. Intertongued Upper Cretaceous volcanic and non-volcanic rocks, central-western Montana. Geol. Soc. Amer. Bull. 76, 1125–1144. 13

Hahn, G.A., Rose, W.I., Jr. and Meyers, T., 1979. Geochemical correlation of genetically related rhyolitic ash-flow and air-fall ashes, central and western Guatemala and the equatorial Pacific. Geol. Soc. Amer. Sp. Paper 180, 101–112. 7

Hall, J.M. and Robinson, P.T., 1979. Deep crustal drilling in the north Atlantic ocean. Science 204, 573–586. 10

Hamilton, D.C., Burnham, C.W., Osborn, E.F., 1964. The solubility of water and effects of oxygen fugacity and water content on crystallization in mafic magmas. J. Petrol. 5, 21–39. 3

Hampton, M.A., 1972. The role of subaqueous debris flows in generating turbidity currents. J. Sed. Petrol. 42, 775–793. 11

Hand, B.M., 1969. Antidunes as trochoidal waves. J. Sed. Petrol. 39, 1302–1309. 5

Hand, B.M., 1974. Supercritical flow in density currents. J. Sed. Petrol. 44, 637–648.   5

Hansen, R., Lemke, R.W., Cattermole, J.M. and Gibbons, A.B., 1963. Stratigraphy and structure of the Rainier and USGS Tunnel areas Nevada Test Site. U.S. Geol. Survey Prof. Paper 382-A, 1–49.   7

Harris, D.M., 1981. The concentration of $CO_2$ in submarine tholeiitic basalts. J. Geol. 89, 689–701.   3

Harris, D.M., Rose, W.I., Jr., Roe, R. and Thompson, M.R., 1981. Radar observations of ash eruptions. In Lipman, P.W. and Mullineaux, D.R., eds., The 1980 eruptions of Mount St. Helens, Washington. U.S. Geol. Survey Prof. Paper 1250, 323–333.   4

Harris, D.M., Sato, M., Casadevall, T.J., Rose, W.I., Jr. and Bornhorst, T.J., 1981. Emission rates of $CO_2$ from plume measurements. In Lipman, P.W. and Mullineaux, D.R., The 1980 eruption of Mount St. Helens. U.S. Geol. Survey Prof. Paper 1250, 201–207.   3

Harris, P.G., Kennedy, W.Q. and Scarfe, C.M., 1970. Volcanism versus plutonism – the effect of chemical composition. In Newall, G. and Rast, N., eds., Mechanics of igneous intrusion. Geol. Jour. Spec. Issue 3, Liverpool, 187–200.   3

Harrison, S. and Fritz, W.J., 1982. Depositional features of March, 1982, Mount St. Helens sediment flows. Nature 299, 720–722.   11

Hartmann, W.K., 1967. Secondary volcanic impact craters at Kapoho, Hawaii and comparisons with lunar surfaces. Icarus 7, 66–75.   9

Haughton, D.R., Roeder, P.L. and Skinner, B.J., 1974. Solubility of sulfur in mafic magmas. Econ. Geol. 69, 451–467.   3

Hawkins, D.B. and Rustum, R., 1963. Experimental hydrothermal studies on rock alteration and clay mineral formation. Geochim. Cosmochim. Acta 27, 1047–1054.   12

Hawkins, J.W., 1974. Geology of the Lau Basin, a marginal sea behind the Tonga Arc. In Burk, C.A. and Drake, C.L., eds., The geology of continental margins. Springer-Verlag, Berlin, Heidelberg, New York, 505–520.   14

Hay, R.L., 1959a. Formation of the crystal-rich glowing avalanche deposits of St. Vincent, B.W.I. J. Geol. 67, 540–562.   6,8

Hay, R.L., 1959b. Origin and weathering of late Pleistocene ash deposits on St. Vincent, B.W.I. J. Geol. 67, 65–87.   12

Hay, R.L., 1963. Stratigraphy and zeolitic diagenesis of the John Day Formation of Oregon. Univ. Calif. Publ. Geol. Sci. 42, 199–262.   12,13

Hay, R.L., 1966. Zeolites and zeolitic reactions in sedimentary rocks. Geol. Soc. Amer. Sp. Paper 85, 1–130.   12

Hay, R.L., 1973. Lithofacies and environments of Bed 1, Olduvai Gorge, Tanzania. J. Quat. Res. 3, 541–560.   13

Hay, R.L., 1976. Geology of the Olduvai Gorge: A study of sedimentation in a semiarid basin. Univ. Calif. Press, Berkeley, 1–203.   13

Hay, R.L., 1978. Melilitite-carbonatite tuffs in the Laetolil beds of Tanzania. Contr. Mineral. Petrol. 67, 357–367.   6

Hay, R.L. and Iijima, A., 1968a. Nature and origin of palagonite tuffs of the Honolulu Group on Oahu, Hawaii. Geol. Soc. Amer. Mem. 116, 331–376.   12

Hay, R.L. and Iijima, A., 1968b. Petrology of palagonite tuffs of Koko Crater, Oahu, Hawaii. Contr. Mineral. Petrol. 17, 141–154.   12

Hay, R.L. and Jones, B.F., 1972. Weathering of basaltic tephra on the Island of Hawaii. Geol. Soc. Amer. Bull. 83, 317–332.   12

Hay, R.L., Hildreth, W. and Lambe, R.N., 1979. Globule ignimbrite of Mount Suswa, Kenya. Geol. Soc. Amer. Sp. Paper 180, 167–175.   5,8

Heald, E.F., Naughton, J.H., Barnes, I.L., 1963. The chemistry of volcanic gases. 2. Use of equilibrium calculations in the interpretation of volcanic gas samples. J. Geophys. Res. 68, 545–557.   3

Healy, J., Vucetich, C.G. and Pullar, W.A., 1964. Stratigraphy and chronology of Late Quaternary volcanic ash in Taupo, Rotorua, and Gisborne Districts. N.Z. Geol. Survey Bull. 73, 7–88.   13

Hedervari, P., 1963. On the energy and magnitude of volcanic eruptions. Bull. Volcanol. 25, 373–385.   4,6

Hedervari, P., 1982. A possible submarine volcano near the central part of Ninety-East Ridge, Indian Ocean. J. Volcanol. Geotherm. Res. 13, 199–212.   10

Heiken, G.H., 1971. Tuff rings: examples from Fort Rock Christmas Lake Valley, south-central Oregon. J. Geophys. Res. 76, 5615–5626.   9

Heiken, G.H., 1972. Morphology and petrography of volcanic ashes. Geol. Soc. Amer. Bull. 83, 1961–1988.   5,9,12

Heiken, G.H., 1974. An atlas of volcanic ash. Smithsonian Contr. Earth Sciences 12, 1–101. 5,9
Heiken, G., 1978. Characteristics of tephra from Cinder Cone, Lassen Volcanic National Park, California. Bull. Volcanol. 41-2, 1–12. 5
Heiken, G., 1979. Pyroclastic flow deposits. Amer. Scientist 67, 564–571. 8
Heiken, G. and Wohletz, K.H., 1984. Volcanic ash. Univ. Calif. Press, Berkeley (in press). 5,8
Heiken, G., McKay, D.S. and Brown, R.W., 1974. Lunar deposits of possible pyroclastic origin. Geochim. Cosmochim. Acta 38, 1703–1718. 5
Hein J.R. and Scholl, D.W., 1978. Diagenesis and distribution of Late Cenozoic volcanic sediment in the southern Bering Sea. Geol. Soc. Amer. Bull. 89, 197–210. 12
Hein, J.R., Scholl, D.W. and Miller, J., 1978. Episodes of Aleutian Ridge explosive volcanism. Science 199, 137–141. 7
Hekinian, R. and Hoffert, M., 1975. Rate of palagonitization and manganese coating on basaltic rocks from the Rift Valley in the Atlantic Ocean near 36°50′ N. Mar. Geol. 19, 91–109. 12
Henderson, J.B., 1975. Sedimentological studies in the Yellowknife Supergroup, District of Mackenzie. Geol. Survey Can. Paper 70–26. 14
Hendry, H.E., 1976. The orientation of discoidal clasts in resedimented conglomerates, Cambro-Ordovician, Gaspé, eastern Quebec. J. Sed. Petrol. 46, 48–55. 11
Hentschel, H., 1963. In-situ Brekzien der Unter-Karbonischen Pillowdiabase des Dillgebietes im Rheinischen Schiefergebirge. Bull. Volcanol. 25, 97–107. 4,10
Hess, H.H., 1946. Drowned ancient islands of the Pacific basin. Amer. J. Sci. 244, 772–791. 10
Hess, H.H., 1948. Major structural features of the western and North Pacific. Geol. Soc. Amer. Bull. 59, 417–446. 14
Hess, P.C., 1971. Polymer model of silicate melt. Geochim. Cosmochim. Acta 35, 289–306. 3
Hewett, D.F., 1917. The origin of bentonite and the geologic range of related materials in Big Horn basin. Wyoming. Washington Acad. Sci. Proc. 7, 196–198. 12
Hildreth, W., 1979. The Bishop Tuff: evidence for the origin of compositional zonation in magma chambers. Geol. Soc. Amer. Sp. Paper 180, 43–75. 2,8
Hildreth, W., 1981. Gradients in silicic magma chambers: implications for lithospheric magmatism. J. Geophys. Res. 86, 10153–10192. 2,3
Hildreth, W., 1983. The compositionally zoned eruption of 1912 in the Valley of Ten Thousand Smokes, Katmai National Park, Alaska. J. Volcanol. Geotherm. Res. 18, 1–56. 2
Hoblitt, R.P. and Kellogg, K.S., 1979. Emplacement temperatures of unsorted and unstratified deposits of volcanic rock debris as determined by paleomagnetic techniques. Geol. Soc. Amer. Bull. Part I, 90, 633–642. 8,11
Hoblitt, R.P., Miller, C.D. and Vallance, J.W., 1981. Origin and stratigraphy of the deposit produced by the May 18 directed blast. In Lipman, P.W. and Mullineaux, D.R., eds., The 1980 eruptions of Mount St. Helens, Washington. U.S. Geol. Survey Prof. Paper 1250, 401–419. 8
Hoffmeister, J.E., Ladd, H.S. and Alling, H.L., 1929. Falcon Island. Amer. J. Sci. 18, 461. 9
Hogan, L.G., Scheiddegger, K.F., Kulm, L.D., Dymond, J. and Mikkelsen, N., 1978. Biostratigraphic and tectonic implications of $^{40}Ar-^{39}Ar$ dates of ash layers from the northeast Gulf of Alaska. Geol. Soc. Amer. Bull. 89, 1259–1264. 7
Holloway, J.R., 1976. Fluids in the evolution of granitic magmas: consequences of finite $CO_2$ solubility. Geol. Soc. Amer. Bull. 87, 1513–1518. 3
Holloway, J.R., 1981. Volatile interactions in magma. In Newton, R.C., Navrotsky, A. and Wood, B.J., eds., Thermodynamics of minerals and melts. Springer-Verlag, Berlin, Heidelberg, New York, 273–293. 3
Holmes, A., 1913. The age of the earth. Harper and Row, London and New York, 1–195. 1
Honnorez, J., 1961. Sur l'origine sur les hyaloclastites. Bull. Soc. Belge Geol. 70, 407–412. 4
Honnorez, J., 1972. La Palagonitisation: l'alteration sous-marine du verre volcanique basique de Palagonia (Sicile). Vulkaninstitut I. Friedländer No. 9, Birkhäuser Verlag, Basel, Stuttgart, 1–132. 4,10,12
Honnorez, J., 1978. Generation of phillipsites by palagonitization of basaltic glass in sea water and the origin of K-rich deep sea sediments. In Sand, L.B. and Mumpton, F.A., eds., Natural zeolites: occurrence, properties and use. Pergamon Press, New York, 245–258. 12
Honnorez, J. and Kirst, P., 1975. Submarine basaltic volcanism: morphometric parameters for discriminating hyaloclastites from hyalotuffs. Bull. Volcanol. 39, 1–25. 9
Hoppe, H.-G., 1940. Untersuchungen an Palagonittuffen und über ihre Bildungsbedingungen. Chemie der Erde 13, 484–514. 12

Horn, D.R., Delach, M.N. and Horn, B.M., 1969. Distribution of volcanic ash layers and turbidites in the north Pacific. Geol. Soc. Amer. Bull. 80, 1715–1724.   7

Horn, D.R., Horn, B.M. and Delach, M.N., 1970. Sedimentary provinces of the north Pacific. Geol. Soc. Amer. Mem. 126, 1–21.   7

Horn, M.K. and Adams, J.A.S., 1966. Computer-derived geochemical balances and element abundances. Geochim. Cosmochim. Acta 30, 279–290.   1

Howells, M.F. and Leveridge, B.E., 1980. The Capel Curig Volcanic Formation. Inst. Geol. Sci., London, Rpt. 80/6, 1–23.   10

Howells, M.F., Leveridge, B.E. and Evans, C.D.R., 1973. Ordovician ash-flow tuffs in eastern Snowdonia. Inst. Geol. Sci., London, Rpt. 73/3, 1–33.   10

Howells, M.F., Leveridge, B.E., Addison, R., Evans, C.D.R. and Nutt, M.J.C., 1979. The Capel Curig volcanic formation, Snowdonia, North Wales; variations in ash-flow tuffs related to emplacement environment. In The Caledonides of the British Isles. Geol. Soc. London 611–618.   10

Howorth, R., 1975. New formations of Late Pleistocene tephras from the Okataina volcanic centre, New Zealand. N.Z. J. Geol. Geophys. 18, 683–712.   6

Hsü, K.J., 1975. Catastrophic debris streams (Sturzstroms) generated by rockfalls. Geol. Soc. Amer. Bull. 86, 129–140.   8

Huang, T.C. and Watkins, M.D., 1976. Volcanic dust in deep-sea sediments: relationship of microfeatures to explosivity estimates. Science 193, 576–579.   5

Huang, T.C., Varner, J.R. and Wilson, L., 1980. Micropits on volcanic glass shards: Laboratory simulation and possible origin. J. Volcanol. Geotherm. Res. 8, 59–68.   5

Huang, T.C., Watkins, N.D. and Shaw, D.M., 1974. Atmospherically transported volcanic glass in deep-sea sediments: Development of a separation and counting technique. Deep-Sea Res. 22, 185–196.   7

Huang, T.C., Watkins, N.D. and Shaw, D.M., 1975. Atmospherically transported volcanic glass in deep-sea sediments: volcanism in sub-antarctic latitudes of the south Pacific during late Pleistocene time. Geol. Soc. Amer. Bull. 86, 1305–1315.   7

Huang, T.C., Watkins, N.D. and Wilson, L., 1979. Deep-sea tephra from the Azores during the past 300,000 years: eruptive cloud height and ash volume estimates. Geol. Soc. Amer. Bull., Pt. II, 90, 235–288.   7

Huang, T.C., Carey, S., Sigurdsson, H. and Davis, A., 1979. Correlations and contrasts of deep-sea ash deposits from the Lesser Antilles in the western equatorial Atlantic and eastern Caribbean at latitude 14° N. EOS 59, 1119.   7

Huang, T.C., Watkins, N.D., Shaw, D.M. and Kennett, J.P., 1973. Atmospherically transported volcanic dust in South Pacific deep sea sedimentary cores at distances over 3000 km from the eruptive source. Earth Planet. Sci. Lett. 20, 119–124.   7

Huber, N.K. and Rinehart, D.D., 1966. Some relationships between the refractive index of fused glass beads and the petrologic affinity of volcanic rock suites. Geol. Soc. Amer. Bull. 77, 101–110.   7

Hubert, J.F., Suchecki, R.K. and Callahan, R.K., 1975. Mass-flow megabreccia limestone, Cambrian-Ordovician continental slope, Newfoundland (Abst.). Ann. Mtg. Abs. Amer. Assoc. Petrol. Geol. 2, 38.   11

Hubregtse, J.J.M.W., 1976. Volcanism in the western Superior Province in Canada. In Windley, B.F., ed., Early History of the Earth. John Wiley and Sons, London, 278–287.   14

Hulme, G., 1974. The interpretation of lava flow morphology. Geophys. J. Roy. Astron. Soc. 39, 361–383.   3

Iijima, A., 1978. Geological occurrences of zeolites in marine environments: In Sand, L.B. and Mumpton, F.A., eds., Natural Zeolites: occurrence, properties, use. Pergamon Press, Oxford, 175–198.   12

Iijima, A. and Harada, K., 1968. Authigenic zeolites in palagonite tuffs on Oahu, Hawaii. Amer. Mineral. 54, 182–197.   12

Iijima, A. and Utada, M., 1972. A critical review of the occurrence of zeolites in sedimentary rocks in Japan. Jap. J. Geol. Geogr. 42, 61–84.   12

Illies, H., 1959. Die Entstehungsgeschichte eines Maares in Südchile. Geol. Rundsch. 48, 232–247.   9

Ingersoll, R.V., 1978. Petrofacies and petrologic evolution of the late Cretaceous fore-arc basin, northern and central California. J. Geol. 86, 335–352.   13

Ingram, R.L., 1954. Terminology for the thickness of stratification and parting units in sedimentary rocks. Geol. Soc. Amer. Bull. 65, 937–938.   5

Inman, D.L., 1952. Measures of describing the size distribution of sediments. J. Sed. Petrol. 22, 125–145.   5,6

Izett, G.A., 1981. Volcanic ash beds: recorders of Upper Cenozoic silicic pyroclastic volcanism in the western United States. J. Geophys. Res. 86, 10200–10222.  1,5,6,13

Izett, G.A. and Naeser, C.W., 1976. Age of the Bishop Tuff of eastern California as determined by the fission-track method. Geology 4, 587–590.  6

Izett, G.A., Wilcox, R.E. and Borchardt, G.A., 1972. Correlation of a volcanic ash bed in Pleistocene deposits near Mount Blanco, Texas, with the Guaje Pumice Bed of the Jemez Mountains, New Mexico. Quat. Res. 2, 554–578.  13

Jaeger, J.C., 1968. Cooling and solidification of igneous rocks. In Hess, H.H., ed., Basalts. Wiley-Interscience, New York, 503–536.  8

Jaggar, T.A., 1949. Steam blast volcanic eruptions. Hawaiian Volcano Observatory, 4th Sp. Rpt., 1–137.  4,9

Jakeš, P. and White, A.J.R., 1971. Composition of island arcs and continental growth. Earth Planet. Sci. Lett. 12, 224–230.  14

Jakobsson, S.P., 1972. On the consolidation and palagonitization of the tephra of the Surtsey volcanic island, Iceland. Surtsey Res. Progr. Paper 6, 121–128.  12

Jakobsson, S.P., 1978. Environmental factors controlling the palagonitization of the Surtsey tephra, Iceland. Bull. Geol. Soc. Denmark Sp. Issue 27, 91–105.  12

Janda, R.J., Scott, K.M., Nolan, K.M. and Martinson, H.A., 1981. Lahar movement, effects, and deposits. In Lipman, P.W. and Mullineaux, D.R., eds., The 1980 eruptions of Mount St. Helens. U.S. Geol. Survey Prof. Paper 1250, 461–478.  11

Janda, R.J., Nolan, K.M., Glicken, H., Abbott, L.S. and Voight, B., 1980. Processes responsible for massive river sedimentation following the May 18, 1980 eruption of Mount St. Helens (Abst.) EOS 61, 955.  11

Jeans, C.V., Merriman, R.J., Mitchell, J.G. and Bland, D.J., 1982. Volcanic clays in the Cretaceous of southern England and northern Ireland, Clay Minerals 17, 1205–156.  12

Jenner, G.A., Fryer, B.J. and McLennan, S.M., 1981. Geochemistry of the Archean Yellowknife Supergroup. Geochim. Cosmochim. Acta 45, 1111–1129.  1

Jezek, P.A., 1976. Compositional variation within and among volcanic ash layers in the Fiji Plateau area. J. Geol. 84, 595–616.  7

Jezek, P.A. and Noble, D.C., 1978. Natural hydration and ion exchange of obsidian: an electron microprobe study. Amer. Mineral. 63, 266–273.  12

Johansson, C.E., 1965. Structural studies of sedimentary deposits. Geol. Foren. Stockholm Forhandlingar 87, 3–61.  5

Johnson, A.M., 1965. A model for debris flow. Penns. State Univ., State College, Pa., Ph.D. diss., 1–232.  11

Johnson, A.M., 1970. Physical processes in geology. Freeman, Cooper and Co., San Francisco, 1–577.  8,11

Johnson, R.W., 1968. Volcanic globule rock from Mount Suswa, Kenya. Geol. Soc. Amer. Bull. 79, 647–652.  8

Johnson, R.W., Davies, R.A. and White, A.J.R., 1972. Ulawun Volcano, New Britain. Austr. Bur. Min. Resour. Geol. Geophys. Bull. 142, 1–42.  8

Johnston, D.A., 1980. Volatile distribution of chlorine to the stratosphere – more significant to ozone than previously estimated? Science 209, 491–493.

Johnston, D.A. and Schmincke, H.-U., 1977. Triggering of explosive volcanic eruptions by mixing of basaltic and silicic magmas (Abst.). Geol. Soc. Amer. Abst. with Programs 9, 1041.  8

Jones, E.J.W., 1973. Volcanic glass in abyssal clays at DSDP Leg 20 drilling sites, northwest Pacific. In Heezen, B.C. and MacGregor, I.O., et al., eds., Init. Rpts. Deep Sea Drilling Proj. 20, 389–416.  12

Jones, J.G., 1966. Intraglacial volcanoes of southwest Iceland and their significance in the interpretation of the form of marine basaltic volcanoes. Nature 212, 586–588.  10

Jones, J.G., 1969a. Intraglacial volcanoes of the Laugarvatn region, south-west Iceland, I. Q. J. Geol. Soc. London 124, 197–211.  10

Jones, J.G., 1969b. Pillow lavas as depth indicators. Amer. J. Sci. 267, 181–195.  9

Jones, J.G., 1970. Intraglacial volcanoes of the Laugarvatn region, southwest Iceland, II. J. Geol. 78, 127–140.  10

Jonsson, G., 1961. Some observations on the occurrence of sideromelane and palagonite. Bull. Geol. Inst. Univ. Uppsala 40, 81–86.  12

Judd, J.W., 1888. On the volcanic phenomena of the eruption (of Krakotoa), and on the nature and distribution of the ejected materials, pp. 1–56, Part I. In Symons, G.J., ed., The eruption of Krakatoa and subsequent phenomena. Rpt. Krakatoa Committee Roy. Soc. London, 1–494.  6

Juteau, T., Noack, Y., Whitechurch, H. and Courtois, C., 1979. Mineralogy and geochemistry of alteration products in Holes 417A and 417D basement samples. In Donnelly, T., Francheteau, J., Bryan, W., Robinson, P., Flower, M., Salisbury, M., et al., eds., Init. Rpts. Deep Sea Drilling Proj. 51, 52, 53, Pt. 2, 1273–1297.  12

Kadik, A.A., Lukanin, O.A., Lebedev, Y.B. and Korovushkina, E.Y., 1972. Solubility of $H_2O$ and $CO_2$ in granite and basalt melts at high pressures. Geochem. Internat. 6, 1041–1050.  3

Karakuzu, F., Schmincke, H.-U. and Wörner, G., 1982. Aufbau und chemische Entwicklung des quartären basanitisch-tephritischen Rothenberg Vulkans (Osteifel) (Abst.). Fortschr. Min. 60, Beih. I, 109.  2

Karig, D.E., 1970. Ridges and basins of the Tonga-Kermadec island arc system. J. Geophys. Res. 75, 239–254.  14

Karig, D.E. and Sharman, G.F., III, 1975. Subduction and accretion in trenches. Geol. Soc. Amer. Bull. 86, 377–389.  14

Kastner, M., 1976. Diagenesis of basal sediments and basalts of sites 322 and 323, Leg 35, Bellinghausen Abyssal Plain. In Hollister, C. and Cradock, C., eds., Init. Rpts. Deep Sea Drilling Proj. 35, 513–528.  12

Kastner, M., 1979. Zeolites. In Marine Minerals. Min. Soc. Amer. Short Course, 111–122.  12

Kastner, M. and Gieskes, J., 1976. Interstitial water profiles and sites of diagenetic reactions. Leg 35, DSDP, Bellinghausen Abyssal Plain. Earth Planet. Sci. Lett. 33, 11–20.  12

Kastner, M. and Siever, R., 1979. Low temperature feldspars in sedimentary rocks. Amer. J. Sci. 279, 435–479.  12

Kastner, M. and Stonecipher, S.A., 1978. Zeolites in pelagic sediments of the Atlantic, Pacific and Indian Oceans: In Sand, L.B. and Mumpton, F.A., eds., Natural zeolites: occurrence, properties, use. Pergamon Press, Oxford, 199–220.  12

Kato, I., Murai, I., Yamazaki, T. and Abe, M., 1971. Subaqueous pyroclastic flow deposits in the upper Donzurubo Formation, Nijo-san district, Osaka, Japan. J. Geol. Soc. Jap. 77, 193–206.  10

Katsui, Y., 1959. On the Shikotsu pumice fall deposit. Bull. Volcanol. Soc. Jap., 4, 33–48.  6

Katsui, Y., 1963. Evolution and magmatic history of some Krakatoan calderas in Hokkaido, Japan. J. Fac. Sci., Hokkaido Univ., ser. 4, 11, 631–650.  8

Katsura, T. and Nagashima, S., 1974. Solubility of sulfur in magmas. Geochim. Cosmochim. Acta 38, 517–531.  3

Kay, M., 1951. North American geosynclines. Geol. Soc. Amer. Mem. 48, 1–143.  14

Keller, J., 1981. Carbonatitic volcanism in the Kaiserstuhl alkaline complex: evidence for highly fluid carbonatitic melts at the earth's surface. J. Volcanol. Geotherm. Res. 9, 423–432.  6

Keller, J., Ryan, W.B.F., Ninkovich, D. and Altherr, R., 1978. Explosive volcanic activity in the Mediterranean over the past 200,000 years as recorded in deep-sea sediments. Geol. Soc. Amer. Bull. 89, 591–604.  7

Kennedy, G.C., 1955. Some aspects of the role of water in rock melts. In Poldervaart, A., ed., Crust of the Earth. Geol. Soc. Amer. Sp. Paper 62, 489–504.  3,8

Kennedy, J.F., 1961. Stationary waves and antidunes in alluvial channels. Calif. Inst. Tech. W.M. Keck Lab. Hydraulics and Water Resources, Rpt. KH-R-2, 1–146.  5

Kennett, J.P., 1981. Marine tephrochronology. In Emiliani, C., ed., The Oceanic Lithosphere. The Sea. Wiley-Interscience Publ., New York, 1373–1436.  7

Kennett, J.P. and Thunell, R.C., 1975. Global increase in Quaternary volcanism. Science 187, 497–503.  7

Kennett, J.P. and Thunell, R.C., 1977. On explosive volcanism and climatic implications. Science 196, 1231–1234.  7

Kennett, J.P., McBirney, A.R. and Thunell, R.C., 1977. Episodes of volcanism in the circum-Pacific region. J. Volcanol. Geotherm. Res. 2, 145–163.  7

Khitarov, N.J., Khundaze, A.G., Senderov, E.E. and Shibayeva, N.P., 1970. The effects of volcanic rocks on the compositions of hydrothermal solutions. Geochem. Int. 6, 469–482.  12

Kieffer, S.W., 1981. Blast dynamics at Mount St. Helens on 18 May 1980. Nature 291, 568–570.  4

Kieffer, S.W., 1982. Dynamics and thermodynamics of volcanic eruptions: implications for the plumes on Io. In Morrison, D., ed., Satellites of Jupiter. Univ. Arizona Press, Tucson, Ariz. 647–723.  4

Kienle, J. and Swanson, S.E., 1980. Volcanic hazards from future eruptions of Augustine Volcano, Alaska. Univ. Alaska Geophys. Inst. UAG R-275, 1–122.  8,10

Kienle, J., Kyle, P.R., Self, S., Motyka, R.J. and Lorenz, V., 1980. Ukinrek Maars, Alaska, I. April 1977 eruption sequence, petrology and tectonic setting. J. Volcanol. Geotherm. Res. 7, 11–37.   9

Killingley, J.S. and Muenow, D.W., 1975. Volatiles from Hawaiian submarine basalts determined by dynamic high temperature mass spectrometry. Geochim. Cosmochim. Acta 39, 1467–1473.   3

Kirkman, J.H., 1976. Clay mineralogy of thirteen paleosols developed in Holocene and late Pleistocene tephras of central North Island, New Zealand. N.Z. J. Geol. Geophys. 19, 179–187.   12

Kirkman, J.H., 1980. Mineralogy of the Kauroa Ash Formation of south-west and west Waikato, North Island, New Zealand, N.Z. J. Geol. Geophys. 23, 113–120.   12

Kittleman, L.R., 1964. Application of Rosin's distribution in size-frequency analysis of clastic rocks. J. Sed. Petrol. 34, 483–502.   5

Kittleman, L.R., 1973. Mineralogy, correlation, and grain-size distributions of Mazama tephra and other post-glacial pyroclastic layers; Pacific Northwest. Geol. Soc. Amer. Bull. 84, 2957–2980.   6

Kjartansson, G., 1951. The eruption of Hekla 1947–1948, II. 4, Water flood and mudflows. Soc. Scientiarum Islandica, Reykjavik, 1–51.   11

Knight, W.C., 1898. Mineral soap. Eng. Mining J., 66, 481.   12

Knox, J.B. and Short, N.M., 1963. A diagnostic model using ash fall data to determine eruption characteristics and atmospheric conditions during a major volcanic event. Univ. Calif. Lawrence Radiation Lab., Contract W-7405-eng-4B, UCRL 7197, 1–29.   6

Kobayashi, K., Shimizu, H., Kitazawa, K. and Kobashi, T., 1967. The pumice-fall deposit "PM-1" supplied from Ontake Volcano. J. Geol. Soc. Jap., 73, 291–308.   6

Koch, A.J. and McLean, H., 1975. Pleistocene tephra and ash-flow deposits in the volcanic highlands of Guatemala. Geol. Soc. Amer. Bull. 86, 529–541.   8

Kokelaar, B.P., 1982. Fluidization of wet sediments during the emplacement and cooling of various igneous bodies. J. Geol. Soc. London 139, 21–33.   9

Korringa, M.K. and Noble, D.C., 1970. Ash-flow eruptions from a linear vent area without caldera collapse. Geol. Soc. Amer. Abst. with Programs 2, 108–109.   8

Kozu, S., 1934. The great activity of Komagatake in 1929. Tschermak's Mineral. Petr. Mitt. 45, 133–174.   8

Krinsley, D. and Margolis, S.V., 1969. A study of quartz sand grain surface textures with the scanning electron microscope. Trans. New York Acad. Sci., Series II 31, 457–477.   5

Krumbein, W.C., 1941. Measurement and geological significance of shape and roundness of sedimentary particles. J. Sed. Petrol. 11, 64–72.   5

Krumbein, W.C. and Graybill, F.A., 1965. An introduction to statistical models in geology. McGraw-Hill Book Co., New York. 1–475.   5

Krumbein, W.C. and Tisdel, F.W., 1940. Size distributions of source rocks of sediments. Amer. J. Sci. 238, 296–305.   5

Krynine, P.D., 1948. The megascopic study and field classification of the sedimentary rocks. J. Geol. 56, 130–165.   1

Kuenen, Ph.H., 1941. Geochemical calculations concerning the total mass of sediments in the earth. Amer. J. Sci. 239, 161–190.   1

Kuenzi, W.D., Horst, O.H. and McGehee, R.V., 1979. Effect of volcanic activity on fluvial-deltaic sedimentation on a modern arc-trench gap, southwestern Guatemala. Geol. Soc. Amer. Bull. Pt. I 90, 827–838.   13,14

Kuniyoshi, S. and Liou, J.G., 1976. Burial metamorphism of the Karmutsen volcanic rocks, northeastern Vancouver Island, British Columbia. Amer. J. Sci. 276, 1096–1119.   12

Kuno, H., 1941. Characteristics of deposits formed by pumice flows and those by ejected pumice. Tokyo Univ. Earthq. Res. Inst. Bull. 19, 144–149.   8

Kuno, H., Ishikawa, T., Katsui, Y., Yagi, K., Yamasaki, M. and Taneda, S., 1964. Sorting of pumice and lithic fragments as a key to eruptive and emplacement mechanism. Jap. J. Geol. Geog. 35, 223–238.   6,8

Kushiro, I., 1974. Melting of hydrous upper mantle and a possible generation of andesitic magma: an approach from synthetic systems. Earth Planet. Sci. Lett. 22, 294–299.   2

Kushiro, I., Yoder, H.S., Jr. and Mysen, B.O., 1976. Viscosities of basalt and andesite melts at high pressures. J. Geophys. Res. 81, 6351–6356.   3

Lacroix, A., 1904. La Montagne Pelée et ses eruptions. Masson et Cie, Paris, 1–662.   4,8,10

Lacroix, A., 1930. Remarques sur les matériaux de projection des volcans et sur la genese des roches pyroclastiques qu'ils constituent. Livre Jubilaire de Centenaire Soc. Geol. France 2, 431–472.   8

Lajoie, J., 1979. Facies models 15. Volcaniclastic rocks. Geoscience Can. 6, no. 3, 129–139.   13

Lamb, H.H., 1970. Volcanic dust in the atmosphere, with a chronology and assessment of its meteorological significance. Phil. Trans. Roy. Soc. London A 266, 425–533.   6

Larsen, E.S., Jr. and Cross, W., 1956. Geology and petrology of the San Juan region, southwestern Colorado. U.S. Geol. Survey Prof. Paper 258, 1–303.   13

Larsson, W., 1937. Vulkanische Asche vom Ausbruch des chilenischen Vulkans Quizapu (1932) in Argentinien gesammelt. Eine Studie über äolische Differentiation. Geol. Inst. Upsala Bull. 26, 27–52.   6

Laurent, M., 1956. Le laitier des hauts fourneaux, Ciments de laitier. In Techniques de l'Ingénieur. Dunod, Paris, M 1850, 1–8.   4

Laursen, T. and Lanford, A., 1978. Hydration of obsidian. Nature 276, 153–156.   12

Ledbetter, M.T. and Sparks, R.S.J., 1979. Duration of large-magnitude explosive eruptions deduced from graded bedding in deep-sea ash layers. Geology 7, 240–244.   7

Leith, C.K. and Mead, W.J., 1915. Metamorphic geology. Holt, Rinehart and Winston, New York, 1–337.   1

Le Maitre, R.W., 1976. The chemical variability of some common igneous rocks. J. Petrol. 17, 589–637.   1,2

Lemke, R.W., Mudge, M.R., Wilcox, R.E. and Powers, H.A., 1975. Geologic setting of the Glacier Peak and Mazama ash-bed markers in west-central Montana. U.S. Geol. Survey Bull. 1395-H, H1–H31.   6

Lerbekmo, J.F. and Campbell, F.A., 1969. Distribution, composition, and source of the White River Ash, Yukon Territory. Can. J. Earth Sci. 6, 109–116.   6

Lerbekmo, J.F., Westgate, J.A., Smith, D.G.W. and Denton, G.H., 1975. New data on the character and history of the White River volcanic eruption, Alaska. In Cresswell, M.M. and Suggate, R.P., eds., Quaternary studies. Roy. Soc. N.Z., Wellington, 203–209.   1,6

Lewis, K.B. and Kohn, B.P., 1973. Ashes, turbidites, and rates of sedimentation on the continental slope of Hawkes Bay. N.Z.J. Geol. Geophys. 16, 439–454.   7

Levi, B., 1970. Burial metamorphic episodes in the Andean geosyncline, Central Chile. Geol. Rundsch. 59, 994–1013.   12

Lichtblau, A.P. and Dimroth, E., 1980. Stratigraphy and facies at the south margin of the Archean Noranda Caldera, Noranda, Quebec. In Current Research, Pt. A, Geol. Survey Can. Paper 80-1A, 69–79.   13

Lindsay, J.F., 1966. Carboniferous subaqueous mass-movement in the Manning-Macleay Basin, Kempsey, New South Wales. J. Sed. Petrol. 36, 719–732.   11

Lindsay, J.F., 1968. The development of clast fabric in mud flows. J. Sed. Petrol. 38, 1242–1253.   11

Lindström, M., 1974. Volcanic contribution to Ordovician pelagic sediments. J. Sed. Petrol. 44, 287–291.   1

Lipman, P.W., 1965. Chemical comparison of glassy and crystalline volcanic rocks. U.S. Geol. Survey Bull. 1201-D, D1–D24.   12

Lipman, P.W., 1967. Mineral and chemical variations within an ash–flow sheet from Aso caldera, southwestern Japan. Contr. Mineral. Petrol. 16, 300–327.   6,8

Lipman, P.W., 1968. Geology of Summer Coon volcanic center, eastern San Juan Mountains, Colorado. In Epis, R.C., ed., Cenozoic volcanism in the southern Rocky Mountains. Colo. School of Mines Qt. 63, 211–236.   13

Lipman, P.W., 1975. Evolution of the Platoro Caldera complex and related volcanic rocks, southeastern San Juan Mountains, Colorado. U. S. Geol. Survey Prof. Paper 852, 1–128.   13

Lipman, P.W. and Christiansen, R.L., 1964. Zonal features of an ash-flow sheet in the Piapi Canyon Formation, southern Nevada. U. S. Geol. Survey Prof. Paper 501-B, 74–78.   8

Lipman, P.W. and Friedman, I., 1975. Interaction of meteoric water with magma: An oxygen-isotope study of ash-flow sheets from southern Nevada. Geol. Soc. Amer. Bull. 86, 695–702.   3

Lipman, P.W. and Mullineaux, eds., 1981. The 1980 eruptions of Mount St. Helens. U.S. Geol. Survey Prof. Paper 1250, 1–844.   1,3,4,6,8,11

Lipman, P.W. and Steven, T.A., 1970. Reconnaissance geology and economic significance of the Platoro Caldera, south-eastern San Juan Mountains, Colorado. In Geol. Survey Res. 1970. U.S. Geol. Survey Prof. Paper 700-C, C19–C29.   13

Lipman, P.W., Christiansen, R.L. and O'Connor, J.T., 1966. A compositionally zoned ash-flow sheet in southern Nevada. U.S. Geol. Survey Prof. Paper 524-F, 1–47.   2,8

Lipman, P.W., Prostka, H.J. and Christiansen, R.L., 1972. Cenozoic volcanism and plate-tectonic evolution of the Western United States. I. Early and Middle Cenozoic. Phil. Trans. Roy. Soc. London A 271, 217–248.   13,14

Lipman, P.W., Steven, T.A. and Mehnert, H.H., 1970. Volcanic history of the San Juan Mountains, Colorado, as indicated by potassium–argon dating. Geol. Soc. Amer. Bull. 81, 2329–2352.  13

Lipman, P.W., Doe, B.R., Hedge, C.E. and Steven, T.A., 1978. Petrologic evolution of the San Juan volcanic field, southwestern Colorado: Pb and Sr isotope evidence. Geol. Soc. Amer. Bull, 89, 59–82.  13

Lipman, P.W., Steven, T.A., Luedke, R.G. and Burbank, W.S., 1973. Revised volcanic history of the San Juan, Uncompahgre, Silverton, and Lake City calderas in the western San Juan Mountains, Colorado. U.S. Geol. Survey J. Res. 1, 627–642.  13

Lipple, S.L., 1972. Silica-rich pillow lavas near Soansville, Marble Bar, 1:250,000 Sheet. Western Austr. Geol. Survey Ann. Rpt. 52–57.  10

Lirer, L., Pescatore, T., Booth, B. and Walker, G.P.L., 1973. Two Plinian pumice-fall deposits from Somma-Vesuvius, Italy. Geol. Soc. Amer. Bull. 84, 759–772.  4,6

Lisitzin, A.P., 1972. Sedimentation in the world ocean. Soc. Econ. Paleont. Mineral. Sp. Publ. 17, 1–208.  7

Lloyd, E.F., 1972. Geology and hot springs of Orakeikorako. N.Z. Geol. Survey Bull. 85, 1–164.  4

Locardi, E. and Mittempergher, M., 1967. On the genesis of ignimbrites; how ignimbrites and other pyroclastic products originate from a flowing melt. Bull. Volcanol. 31, 131–152.  8

Lonsdale, P.F., 1975. Sedimentation and tectonic modification of the Samoan archipelagic apron. Amer. Assoc. Petrol. Geol. Bull. 59, 780–798.  10

Lonsdale, P., 1977. Abyssal pahoehoe with lava coils at the Galapagos Rift. Geology 5, 147–152.  2.

Lonsdale, P. and Batiza, R., 1980. Hyaloclastite and lava flows on young seamounts examined with a submersible. Geol. Soc. Amer. Bull., Part I, 91, 545–554.  10

Lonsdale, P. and Spiess, F.N., 1979. A pair of young cratered volcanoes on the east Pacific Rise. J. Geol. 87, 157–173.  10

Lorenz, V., 1970. Some aspects of the eruption mechanism of the Big Hole maar, central Oregon. Geol. Soc. Amer. Bull. 81, 1823–1830.  9

Lorenz, V., 1973. On the formation of maars. Bull. Volcanol. 37, 183–204.  9

Lorenz, V., 1974. Vesiculated tuffs and associated features. Sedimentology 21, 273–291.  9

Lorenz, V., 1975. Formation of phreatomagmatic maar-diatreme volcanoes and its relevance to kimberlite diatremes. Phys. Chem. Earth 9, 17–27.  9

Lorenz, V., 1980. Explosive volcanism of alkalibasaltic to kimberlitic melts. Lithos 13, 217–219.  9

Lorenz, V., McBirney, A.R. and Williams, H., 1971. An investigation of volcanic depressions, Part III, Maars, tuff-rings, tuff-cones, and diatremes. NASA Progress Rpt. (NGR-38-003 012), Houston, Texas, 1–198.  9

Losacco, U. and Parea, G.C., 1969. Saggio di un atlante di strutture sedimentarie e postsedimentarie osservate nelle piroclastiti del Lazio. Atti Soc. Nat. e Matem. di Modena 94, 1–30.  9

Lowe, D.R., 1975. Water escape structures in coarse-grained sediments. Sedimentology 22, 157–204.  7

Lowe, D.R. and LoPiccolo, R.D., 1974. The characteristics and origins of dish and pillar structures. J. Sed. Petrol. 44, 484–501.  7

Lowman, R.D.W. and Bloxam, T.W., 1981. The petrology of the Lower Paleozoic Fishguard Volcanic Group and associated rocks E of Fishguard, N. Pembrokeshire (Dyfed), South Wales. J. Geol. Soc. London 138, 47–68.  10

Luedke, R.G. and Burbank, W.S., 1963. Tertiary volcanic stratigraphy in the western San Juan Mountains. In Short papers in geology and hydrology. U.S. Geol. Survey Prof. Paper 475-C, C39–C44.  13

Luedke, R.G. and Burbank, W.S., 1968. Volcanism and cauldron development in the western San Juan Mountains, Colorado. In Epis, R.C., ed., Cenozoic volcanism in southern Rocky Mountains. Colo. School Mines Quart. 63, 175–208.  13

Lydon, P.A., 1969. Geology and lahars of the Tuscan Formation, northern California. Geol. Soc. Amer. Mem. 116, 441–475.  11

Macdonald, G.A., 1967. Forms and structures of extrusive basaltic rocks. In Hess, H.H. and Poldervaart, A., eds., The Poldervaart Treatise on Rocks of Basaltic Composition, Wiley-Interscience, New York, 1–61.  5

Macdonald, G.A., 1972. Volcanoes. Prentice-Hall, Inc., Englewood Cliffs, N.J., 1–510.  1,2,4,5,8,9,11

Macgregor, A.G., 1952. Eruptive mechanisms – Mt. Pelée, The Soufrière of St. Vincent (West Indies) and the Valley of Ten Thousand Smokes (Alaska). Bull. Volcanol. 12, 49–74.  8

Macgregor, A.G., 1955. Classification of nuée ardente eruptions. Bull. Volcanol. 16, 7–11.　8
Mackin, J.H., 1960. Structural significance of Tertiary volcanic rocks in southwestern Utah. Amer. J. Sci. 258, 81–131.　1
Malfait, B.T. and Dinkelman, M.G., 1972. Circum-Caribbean tectonic and igneous activity and the evolution of the Caribbean plate. Geol. Soc. Amer. Bull. 83, 251–272.　14
Mariner, R.H., 1971. Experimental evaluation of authigenic mineral reactions in the Pliocene Moonstone formation. Univ. Wyoming Ph.D. diss., 1–133.　12
Mariner, R.H. and Surdam, R.C., 1970. Alkalinity and formation of zeolites in saline alkaline lakes. Science 170, 977–980.　12
Markhinin, E.K. and 12 others, 1974. The eruption of Mt. Tyatya volcano in the Kurile Islands in July of 1973. Sov. Geol. Geophys. 15, 14–24.　9
Marshall, P., 1935. Acid rocks of the Taupo-Rotorua volcanic district. Trans. Roy. Soc. N.Z. 64, 323–366.　8
Martin, R.C. and Malahoff, A., 1965. Some recent Russian studies of ignimbritic rocks. N.Z. J. Geol. Geophys. 8, 706–735.　8
Mason, A.C. and Foster, H.L., 1956. Extruded mudflow hills of Nirasaki, Japan. J. Geol. 64, 74–83.　11
Mathews, D.H., 1962. Altered lavas from the floor of the eastern north Atlantic. Nature 194, 368–369.　12
Mathews, W.H., 1947. "Tuyas", flat-topped volcanoes in northern British Columbia. Amer. J. Sci. 245, 560–570.　10
Mathews, W.H., 1951. A useful method for determining approximate composition of fine-grained igneous rocks. Amer. Mineral. 36, 92–101.　7
Mathez, E.A., 1976. Sulfur solubility and magmatic sulfides in submarine basalts. J. Geophys. Res. 81, 4269–4276.　3
Mathisen, M.E. and Vondra, C.F., 1983. The fluvial and pyroclastic deposits of the Cagayan Basin, Northern Luzon, Philippines – an example of non-marine volcaniclastic sedimentation in an inter-arc basin. Sedimentology 30, 369–392.　14
Matsuo, S., 1961. On the chemical nature of fumarolic gases of volcano Showashinzan, Hokkaido, Japan. J. Sci. Nagoya Univ. 9, 80–100.　3
Matthews, R.K., 1974. Dynamic stratigraphy. Prentice-Hall, Inc., Englewood Cliffs, N.J., 1–370.　13
Mattinson, J.M. and Fisher, R.V., 1970. Impossibility of Bernoulli pressure forces on particles suspended in boundary layers: Reply. J. Sed. Petrol. 30, 520–521.　11
Mattson, P.H. and Alvarez, W., 1973. Base surge deposits in Pleistocene volcanic ash near Rome. Bull. Volcanol. 37, 553–572.　9
Maury, R., 1973. La matière organique des bois fossiles, indicatrice des conditions thermiques de mise en place des brèches volcaniques. C.R. Acad. Sci. Paris 276, Ser. D, 917–920.　8
McBirney, A.R., 1963. Factors governing the nature of submarine volcanism. Bull. Volcanol. 26, 455–469.　1,3,4,10
McBirney, A.R., 1968. Compositional variations of the climactic eruption of Mount Mazama. Andesite Conference Guidebook, Ore. State Geol. Min. Ind. Bull. 62, 53–56.　8
McBirney, A.R., 1973. Factors governing the intensity of explosive andesitic eruptions. Bull. Volcanol. 36, 443–453.　3
McBirney, A. and Murase, T., 1971. Factors governing the formation of pyroclastic rocks. Bull. Volcanol. 34, 372–384.　3
McBirney, A., Sutter, J.F., Naslund, H.R., Sutton, K.G. and White, C.M., 1974. Episodic volcanism in the central Oregon Cascade range. Geology 2, 585–589.　7
McCall, G.J.H., 1964. Froth flows in Kenya. Geol. Rundsch. 54, 1148–1195.　8
McCoy, Jr., F.W., 1974. Late Quaternary sedimentation in the eastern Mediterranean Sea. Harvard Univ., Ph.D. diss. 1–132.　7
McGetchin, T.R. and Ullrich, G.W., 1973. Xenoliths in maars and diatremes with inferences for Moon, Mars and Venus. J. Geophys. Res. 78, 1832–1853.　9
McGetchin, T.R., Settle, M. and Chouet, B.A., 1974. Cinder cone growth modeled after northeast crater, Mount Etna, Sicily. J. Geophys. Res. 79, 3257–3272.　6
McGlynn, J.C. and Henderson, J.B., 1970. Archean sedimentation and volcanism in the Slave Province. Geol. Survey Can. Paper 70-40, 31–44.　14
McKee, E.D. and Weir, G.W., 1953. Terminology for stratification and cross-stratification in sedimentary rocks. Geol. Soc. Amer. Bull. 64, 381–390.　5
McTaggart, K.C., 1960. The mobility of nuées ardentes. Amer. J. Sci. 258, 369–382.　8

McTaggart, K.C., 1962. Nuées ardentes and fluidization – a reply. Amer. J. Sci. 260, 470–476.   8
Mead, W.J., 1907. Redistribution of elements in the formation of sedimentary rocks. J. Geol. 15, 238–256.   1
Mellis, O., 1954. Volcanic ash-horizons in deep-sea sediments from the eastern Mediterranean. Deep-Sea Res. 2, 89–92.   7
Melson, W.G. and Thompson, G., 1973. Glassy abyssal basalts, Atlantic sea-floor near St. Paul's Rocks: Petrography and composition of secondary clay minerals. Geol. Soc. Amer. Bull. 84, 703–716.   12
Menard, H.W., 1956. Archipelagic aprons. Amer. Assoc. Petrol. Geol. Bull. 40, 2195–2210.   10
Mercalli, G., 1907. I volcani attivi della Terra. Ulrico Hoepli, Milan, 1–421.   4
Merriam, J.C., 1901. A contribution to the geology of the John Day Basin (Oregon). Univ. Calif. Publ. Geol. Sci. 2, 269–314.   13
Mertes, H., 1983. Aufbau und Genese des Westeifeler Vulkanfeldes. Bochumer Geol. und Geotechn. Arb. 9, 1–415.   9
Meyer, J.D., 1972. Glass crust on intratelluric phenocrysts in volcanic rocks as a measure of eruptive violence. Bull. Volcanol. 35, 358–368.   5,6
Michael, P.J., 1983. Chemical differentiation of the Bishop Tuff and other high-silica magmas through crystallization processes. Geology 11, 31–34.   2
Michel, R., 1953. Contribution à l'étude pétrographique des pépérites et du volcanisme tertiare de la Grande Limagne. Publ. Fac. Sci. Univ. Clermont 1, 1–140.   9
Michel-Levy, A., 1890. Situation stratigraphique de régions volcaniques de l'Auvergne. La Chaine des puys. Le Mont Doré et ses alentours. Bull. Soc. Geol. France 18, 688–814.   9
Middleton, G.V., 1965. Antidune cross-bedding in a large flume. J. Sed. Petrol. 35, 922–927.   5
Middleton, G.V., 1967. Experiments on turbidity currents, III. Can. J. Earth Sci. 4, 475–505.   8,11
Middleton, G.V., 1970. Experimental studies related to problems of flysch sedimentation. Geol. Assoc. Can. Sp. Paper 7, 253–272.   11
Miller, T.P. and Smith, R.L., 1977. Spectacular mobility of ash flows around Aniakchak and Fisher calderas, Alaska. Geology 5, 173–176.   8
Mimura, K. and MacLeod, N.S., 1978. Source directions of pumice and ash deposits near Bend, Oregon (Abst.). Geol. Soc. Amer. Abst. with Programs 10, 137.   8
Minakami, T., 1942. On the distribution of volcanic ejecta, II. The distribution of Mt. Asama pumice in 1783. Tokyo Univ. Earthq. Res. Inst. Bull. 20, 93–106.   6
Mitchell, A.H.G., 1970. Facies of an early Miocene volcanic arc, Malekula Island, New Hebrides. Sedimentology 14, 201–243.   14
Mohr, E.C.J. and Van Baren, F.A., 1954. Tropical soils. Interscience Publ. Inc., New York, 1–498.   1
Momose, K., Kobayashi, K., Minagawa, K. and Michida, M., 1968. Identification of tephra by means of ferro-magnetic minerals in pumice. Tokyo Univ Earthq. Res. Inst. Bull. 46, 1275–1292.   13
Moore, J.G., 1965. Petrology of deep-sea basalt near Hawaii. Amer. J. Sci. 263, 40–52.   3,10
Moore, J.G., 1966. Rate of palagonitization of submarine basalt adjacent to Hawaii. U.S. Geol. Survey Prof. Paper 550D, 163–171.   12
Moore, J.G., 1967. Base surge in recent volcanic eruptions. Bull. Volcanol. 30, 337–363.   4,9
Moore, J.G., 1970. Water content of basalt erupted on the ocean floor. Contr. Mineral. Petrol. 28, 272–279.   3,4
Moore, J.G., 1975. Mechanism of formation of pillows. Amer. Scientist 63, 269–277.   10
Moore, J.G., 1979. Vesicularity and $CO_2$ in mid-ocean ridge basalt. Nature 282, 250–253.   3,4
Moore, J.G. and Ault, W.V., 1965. Historic littoral cones in Hawaii. Pacific Sci. 19, 3–11.   9
Moore, J.G. and Fabbi, B.P., 1971. An estimate of the juvenile sulfur content of basalt. Contr. Mineral. Petrol. 23, 118–127.   3
Moore, J.G. and Fiske, R.S., 1969. Volcanic substructure inferred from dredge samples and ocean-bottom photographs, Hawaii. Geol. Soc. Amer. Bull. 80, 1191–1202.   10
Moore, J.G. and Melson, W.G., 1969. Nuées ardentes of the 1968 eruption of Mayon Volcano, Philippines. Bull. Volcanol. 33, 600–620.   8,11
Moore, J.G. and Peck, D.L., 1962. Accretionary lapilli in volcanic rocks of the western continental United States. J. Geol. 70, 182–194.   9
Moore, J.G. and Schilling, J.G., 1973. Vesicles, water, and sulfur in Reykjanes Ridge basalts. Contr. Mineral. Petrol. 41, 105–118.   3,4,10
Moore, J.G. and Sisson, T.W., 1981. Deposits and effects of the May 18 pyroclastic surge. In Lipman, P.W. and Mullineaux, D.R., eds., The 1980 eruptions of Mount St. Helens, Washington. U.S. Geol. Survey Prof. Paper 1250, 421–438.   6,8

Moore, J.G., Batchelder, J.N. and Cunningham, C.G., 1977. $CO_2$-filled vesicles in mid-ocean basalt. J. Volcanol. Geotherm. Res. 2, 309–327.  3,4

Moore, J.G., Nakamura, K. and Alcaraz, A., 1966. The 1965 eruption of Taal Volcano. Science 151, 955–960.  4,9

Moore, J.G., Phillips, R.L., Grigg, R.W., Peterson, D.W. and Swanson, D.A., 1973. Flow of lava into the sea 1969–1971, Kilauea Volcano, Hawaii. Geol. Soc. Amer. Bull. 84, 537–546.  10

Morgenstein, M. and Riley, T.J., 1975. Hydration-rind dating of basaltic glass: A new method for archeological chronologies. Asian Perspectives 17, 145–159.  12

Muenow, D.W., Lin, N.W.K., Garcia, M.O. and Saunders, A.D., 1980. Volatiles in submarine volcanic rocks from the spreading axis of the East Scotia Sea back-arc basin. Earth Planet. Sci. Lett. 47, 272–278.  3

Muffler, L.J.P., Short, J.M., Keith, T.E.C. and Smith, V.C., 1969. Chemistry of fresh and altered basaltic glass from the Upper Triassic Hound Island Volcanics, southeastern Alaska. Amer. J. Sci. 267, 196–209.  12

Muffler, L.J.P., White, D.E. and Truesdell, A.H., 1971. Hydrothermal explosion craters in Yellowstone National Park. Geol. Soc. Amer. Bull. 82, 723–740.  4,9

Müller, G., 1967. Methods in Sedimentary Petrology. Part I, Sedimentary Petrology by Engelhardt, W.V., Füchtbauer, H. and Müller, G. Hafner Pub. Co., New York and London, 1–283.  5

Müller, G. and Veyl, G., 1957. The birth of Nilahue, a new maar type volcano at Rininahue, Chile. 20th Int. Geol. Cong., Mexico, Sect. I., pt. 2, Cenozoic Volcanism, 375–396.  9

Mullineaux, D.R., 1974. Pumice and other pyroclastic deposits in Mount Rainier National Park, Washington. U.S. Geol. Survey Bull. 1326, 1–83.  6,13

Mullineaux, D.R. and Crandell, D.R., 1962. Recent lahars from Mount St. Helens, Washington. Geol. Soc. Amer. Bull. 73, 855–869.  8,11

Mumpton, F.A., 1978. Natural zeolites: a new industrial mineral commodity. In Sand, L.B. and Mumpton, F.A., eds., Natural zeolites: occurrence, properties, use. Pergamon Press, Oxford, 3–27.  12

Murai, I., 1960. On the mudflows of the 1926 eruption of Tokachidake, central Hokkaido, Japan. Tokyo Univ. Earthq. Res. Inst. Bull. 38, 55–70.  11

Murai, I., 1961. A study of the textural characteristics of pyroclastic flow deposits in Japan. Tokyo Univ. Earthq. Res. Inst. Bull. 39, 133–248.  5,6,8,10

Murase, T., 1962. Viscosity and related properties of volcanic rocks at 800° to 1400 °C. Jour. Fac. Sci., Hokkaido Univ., Ser. VII, 1, 487–584.  3

Murase, T. and McBirney, A.R., 1973. Properties of some common igneous rocks and their melts at high temperature. Geol. Soc. Amer. Bull. 84, 3563–3592.  3

Murata, K.J., Donoli, C. and Saenz, R., 1966. The 1963–1965 eruption of Irazu Volcano, Costa Rica. Bull. Volcanol. 29, 765–796.  6

Murck, B.W., Burns, R.C. and Hollister, I.S., 1978. Phase equilibria of fluid inclusions in ultramafic xenoliths. Amer. Mineral. 63, 40–46.  3

Murray, J. and Renard, A.J., 1884. On the microscopic characters of volcanic ashes and cosmic dust and their distribution in the deep-sea deposits. Roy. Soc. Edinburgh Proc. 12, 474–495.  6,7

Murray, J. and Renard, A.F., 1891. Deep sea deposits: Scientific Reports of the Voyage of H.M.S. Challenger, 1–525.  12

Mutti, E., 1965. Submarine flood tuffs (ignimbrites) associated with turbidites in Oligocene deposits of Rhodes Island (Greece). Sedimentology 5, 265–288.  10

Mutti, E. and Ricci-Lucchi, F., 1972. Le torbiditi dell'Appennino settentrionale: introduzione all'analisi de facies. Mem. Soc. Geol. Ital. 11, 161–199.  14

Mysen, B.O., 1977. The solubility of $H_2O$ and $CO_2$ under predicted magma genesis conditions and some petrological and geophysical considerations. Rev. Geophys. Space Phys. 15, 351–361.  3

Mysen, B.O. and Virgo, D., 1980. The solubility behavior of $CO_2$ in melts on the join $NaAl_3O_8$-$CaAl_2$-$Si_2O_8$-$CO_2$ at high pressures and temperatures: a Raman spectroscopic study. Amer. Mineral. 65, 1166–1175.  3

Mysen, B.O., Virgo, D. and Kushiro, I., 1981. The structural role of aluminum in silicate melts – a Raman spectroscopic study at 1 atmosphere. Amer. Mineral. 66, 678–701.  3

Mysen, B.O., Virgo, D. and Scarfe, C.M., 1980. Relations between the anionic structure and viscosity of silicate melts – a Raman spectroscopic study. Amer. Mineral. 65, 690–710.  3

Mysen, B.O., Virgo, D. and Seifert, F., 1982. The structure of silicate melts: Implications for chemical and physical properties of natural magma. Rev. Geophys. Space Phys. 20, 353–383.  3

Naeser, C.W., Briggs, N.D., Obradovich, J.D. and Izett, G.A., 1981. Geochronology of Quaternary tephra deposits. In Self, S. and Sparks, R.S.J., eds., Tephra studies. D. Reidel Publ. Co., Dordrecht, Holland, 13–47.   13

Nagasawa, K., 1978. Weathering of volcanic ash and other pyroclastic materials. In Sudo, T. et al., eds., Clays and Clay Minerals of Japan. Developments in Sedimentology. Elsevier, Amsterdam, 105–125.   12

Nairn, I.A., 1972. Rotoehu ash and the Rotoiti Breccia Formation, Taupo volcanic zone, New Zealand. N.Z. J. Geol. Geophys. 15, 251–261.   6

Nairn, I.A., 1979. Rotomahana-Waimangu eruption, 1886. Base-surge and basalt magma. N.Z. J. Geol. Geophys. 22, 363–378.   9

Nairn, I.A. and Self, S., 1978. Explosive eruptions and pyroclastic avalanches from Ngauruhoe in February, 1975. J. Volcanol. Geotherm. Res. 3, 39–60.   4,8

Nairn, I.A. and Wiradiradja, S., 1980. Late Quaternary hydrothermal explosion at Kawerau geothermal field, New Zealand. Bull. Volcanol. 43, 1–13.   4

Nairn, I.A., Wood, C.P. and Hewson, C.A.Y., 1979. Phreatic eruptions of Ruapehu: April 1975. N.Z. J. Geol. Geophys. 22, 155–173.   9

Nairn, I.A., Hewson, C.A.Y., Latter, J.H. and Wood, C.P., 1976. Pyroclastic eruptions of Ngauruhoe Volcano, Central North Island, New Zealand, 1974 January and March. In Johnson, R.W., ed., Volcanism in Australasia. Elsevier, Amsterdam, 385–405.   8

Nakamura, K., 1964. Volcano-stratigraphic study of Oshima Volcano Izu. Tokyo Univ. Earthq. Res. Inst. Bull. 42, 649–728.   4,13

Nakamura, K., 1965. Energies dissipated with volcanic activities – classification and evaluation (in Japanese). Tokyo Univ. Earthq. Res. Inst. Bull. 43, 81–90.   4

Nakamura, K., 1966. The magmatophreatic eruptions of Taal Volcano in 1965, Philippines. Geology Magazine (Japan) 75, no. 2 (751), 93–104 (in Japanese).   9

Nakamura, K., 1974. Preliminary estimate of global volcanic production rate. In Furumoto, S., Minikami, T. and Yuhara, K., eds., The utilization of volcano energy. Sandia Laboratories, Albuquerque, New Mexico, 273–284.   1,2,4

Nakamura, K. and Krämer, F., 1970. Basaltic ash flow deposits from a maar in West-Eifel, Germany. N. Jb. Geol. Pal. Mh. 8, 491–501.   9

Nayudu, Y.R., 1962. A new hypothesis of origin of guyots and seamount terraces. In The crust of the Pacific basin. Amer. Geophys. Un. Monog. 6, 171–180.   12

Nayudu, Y.R., 1964a. Palagonite tuffs (hyaloclastites) and the products of post eruptive processes. Bull. Volcanol. 27, 391–410.   12

Nayudu, Y.R., 1964b. Volcanic ash deposits in the Gulf of Alaska and problems of correlation of deep-sea ash deposits. Mar. Geol. 1, 194–212.   7

Neall, V.E., 1976. Lahars – Global occurrence and annotated bibliography. Victoria Univ., Wellington, N.Z. Geol. Dept. Publ. 5, 1–18.   11

Neeb, G.A., 1943. The composition and distribution of the samples, II. In Snellius Expedition, v. 5, Geol. Research, pt. 3, Bottom Samples, E. J. Brill, Leiden, 55–238.   7

Newell, R.E. and Walker, G.P.L., eds., 1981. Volcanism and climate. J. Volcanol. Geotherm. Res. 11, 1–92.   7

Newhall, C.G. and Self, S., 1982. The Volcanic Explosivity Index (VEI): an estimate of explosive magnitude for historical volcanism. J. Geophys. Res. 87, 1231–1238.   4

Newton, R.C., Navrotsky, A. and Wood, B.J., eds., 1981. Thermodynamics of minerals and melts. Springer-Verlag, Berlin, Heidelberg, New York, 1–304.   3

Nicholls, I.A., Ringwood, A.E., 1973. Effect of water on olivine stability in tholeiites and the production of silica-saturated magmas in the island-arc environment. J. Geol. 81, 285–300.   3

Niem, A.R., 1977. Mississippian pyroclastic flow and ash-fall deposits in the deep-marine Ouachita flysch basin, Oklahoma and Arkansas. Geol. Soc. Amer. Bull. 88, 49–61.   10

Ninkovich, D. and Donn, W.L., 1976. Explosive Cenozoic volcanism and climatic implications. Science 194, 899–906.   7

Ninkovich, D. and Heezen, B.C., 1965. Santorini tephra. In Submarine geology and geophysics (Colston Papers, no. 17) 27, 413–453.   7

Ninkovich, D. and Shackleton, N.J., 1975. Distribution, stratigraphic position and age of ash layer "L", in the Panama Basin region. Earth Planet. Sci. Lett. 27, 20–34.   7

Ninkovich, D., Sparks, R.S.J. and Ledbetter, M.T., 1978. The exceptional magnitude and intensity of the Toba eruption, Sumatra: An example of the use of deep-sea tephra layers as a geological tool. Bull. Volcanol. 41, 286–298.   7

Ninkovich, D., Heezen, B.C., Conolly, J.R. and Burckle, L.H., 1964. South Sandwich tephra in deep-sea sediments. Deep-Sea Res. 11, 605–619.   7

Ninkovich, D., Opdyke, N., Heezen, B.C. and Foster, H.J., 1966. Paleomagnetic stratigraphy, rates of deposition, and tephra chronology in North Pacific deep-sea sediments. Earth Planet. Sci. Lett. 1, 476–492.   7

Noack, Y., 1981. La palagonite. Bull. Mineral. 104, 36–46.   12

Noack, Y. and Crovisier, J.-L., 1980. Evolution de la densité et de la réfractivité spécifique lors de l'altération sousmarine des verres basaltiques. Bull. Mineral. 103, 523–527.   12

Noble, D.C., 1967. Sodium, potassium and ferrous iron contents of some secondarily hydrated natural silicic glasses. Amer. Mineral. 52, 280–285.   8,12

Noble, D.C., 1968. Stress-corrosion failure and the hydration of glassy silicic rocks. Amer. Mineral. 53, 1756–1759.   12

Noble, D.C. and Parker, D.F., 1974. Peralkaline silicic volcanic rocks of the western United States. Bull. Volcanol. 33, 803–827.   8

Noble, D.C., McKee, E.H., Smith, J.G. and Korringa, M.K., 1970. Stratigraphy and geochronology of Miocene volcanic rocks in northwestern Nevada. In Geol. Survey Res. 1970. U.S. Geol. Survey Prof. Paper 700–D, D23–D32.   8

Noe-Nygaard, A., 1940. Sub-glacial volcanic activity in ancient and recent times. Fol. Geogr. Dan. 1, no. 2, 5–67   12

Noll, H., 1967. Maare und maar-ähnliche Explosionskrater in Island. Köln Univ. Geol. Inst. Sonderveröffentlichungen, 1–117.   3,9

Nordlie, B.E., 1971. The composition of the magmatic gas of Kilauea and its behavior in the near-surface environment. Amer. J. Sci. 271, 417–463.   3

Norin, R., 1940. Problems concerning volcanic ash layers of the Lower Tertiary of Denmark. Geol. Foren. Forh. 62, 31–44.   7

Norin, E., 1958. The sediments of the central Tyrrhenian Sea. In Pettersson, H., ed., Reports of the Swedish Deep-Sea Expedition, 1947–1948, 8, Sediment cores from the Mediterranean Sea and the Red Sea. Elanders Boktryckeri Aktiebolag, Göteborg, 1–136.   7

North American Commission on Stratigraphic Nomenclature, 1983. North American Stratigraphic Code. Amer. Assoc. Petrol. Geol. Bull. 67, 841–875.   13

Ollier, C.D., 1967. Maars, their characteristics, varieties and definition. Bull. Volcanol. 31, 45–73.   3,9

Packham, G.H., 1968. The Lower and Middle Palaeozoic stratigraphy and sedimentary tectonics of the Sofala-Hill End-Euchareena region, N.S.W. Proc. Linnean Soc. New South Wales 93, Part 1, 111–163.   13

Pantó, G., 1962. The role of ignimbrites in the volcanism of Hungary. Acta Geol. Budapest 6, 307–331.   8

Parsons, W.H., 1969. Criteria for the recognition of volcanic breccias: Review. Geol. Soc. Amer. Mem. 115, 263–304.   11,13

Peacock, M.A., 1926. The petrology of Iceland. Part I. The basic tuffs. Trans. Roy. Soc. Edinb. 55, 51–76.   12

Peacock, M.A. and Fuller, R.R., 1928. Chlorophaeite, sideromelane and palagonite from the Columbia River Plateau. Amer. Mineral. 13, 360–383.   12

Peck, D.L., Griggs, A.B., Schlicker, H.G., Wells, F.G. and Dole, H.M., 1964. Geology of the central and northern parts of the western Cascade Range in Oregon. U.S. Geol. Survey Prof. Paper 449, 1–56.   13,14

Peckover, R.S., Buchanan, D.J. and Ashby, D.E.T.F., 1973. Fuel-coolant interactions in submarine volcanism. Nature 245, 307–308.   4

Pedersen, A.K., Engell, J. and Rønsbo, J.G., 1975. Early Tertiary volcanism in the Skagerrak: New chemical evidence from ash-layers of the mo-clay of northern Denmark. Lithos 8, 255–268.   7

Pedersen, G.K. and Surlyk, F., 1977. Dish structures in Eocene volcanic ash layers, Denmark. Sedimentology 24, 581–590.   7

Penck, A., 1879. Über Palagonit- und Basalttuffe. Z. deutsch. Geol. Ges. 31, 504–577.   12

Perlaki, E., 1966. Pumice and scoria: their nature, criteria, structure and genesis. Acta Geol. Hung. 10, 13–29.   5

Perret, F.A., 1937. The eruption of Mt. Pelée 1929–1932. Carnegie Inst. Wash. Publ. 458, 1–126.   8

Perret, F.A., 1950. Volcanological observations. Carnegie Inst. Wash. Publ. 549, 1–162.   5

Peterson, D.W., 1970. Ash-flow deposits – their character, origin, and significance. J. Geol. Education 18, 66–76.   8

Peterson, D.W., 1976. Processes of volcanic island growth, Kilauea Volcano, Hawaii, 1969–1973. In Ferran, O.G., ed., Proc., Andean and Anarctic Volcanology Problems. Internat. Assoc. Volcanol. Chem. Earth's Inter. Sp. Series, 172–189.   9

Peterson, D.W., 1979. Significance of the flattening of pumice fragments in ash-flow tuffs. Geol. Soc. Amer. Sp. Paper 180, 195–204.   8

Pettijohn, F.J., 1975. Sedimentary rocks. 3rd Edition, Harper and Row, Publishers, New York, 1–628.   1,5,13

Pettijohn, F.J., Potter, P.E. and Siever, R., 1972. Sand and Sandstone. Springer-Verlag Berlin, Heidelberg, New York, 1–618.   1,5,13

Pike, R.J., 1974. Craters on Earth, Moon, and Maars: Multivariate classification and mode of origin. Earth Planet. Sci. Lett. 22, 245–255.   9

Pimm, A.C., 1974. Sedimentology and history of the northeastern Indian Ocean from Late Cretaceous to Recent. In Borch et al., eds., Init. Repts. Deep Sea Drilling Proj. 22, 717–803.   10

Piper, D.Z., 1974. Rare earth elements in the sedimentary cycle: a summary. Chem. Geol. 14, 285–304.   12

Pirsson, L.V., 1915. The microscopical characters of volcanic tuffs – a study for students. Amer. J. Sci. 40, 181–211.   5

Polak, B.G., 1967. An energy appraisal of volcanic and hydrothermal phenomena (on the example of Kamchatka). Bull. Volcanol. 30, 129–138.   4

Poldervaart, A., ed., 1955. Chemistry of the earth's crust – a symposium. Geol. Soc. Amer. Sp. Paper 62, 119–144.   1

Porter, S.C., 1973. Stratigraphy and chronology of late Quaternary tephra along the South Rift Zone of Mauna Kea Volcano, Hawaii. Geol. Soc. Amer. Bull. 84, 1923–1940.   6

Porter, S.C., 1981. Recent glacier variations and volcanic eruptions. Nature 291, 139–142.   7

Press, F. and Siever, R., 1978. Earth. W.H. Freeman and Co., San Francisco, 1–613.   2

Price, N.R. and Duff, P.M.D., 1969. Mineralogy and chemistry of tonsteins from Carboniferous sequences in Great Britain. Sedimentology 13, 45–69.   12

Provost, G., 1978. Les rhyolites du Complexe "Don", Région de Rouyn-Noranda, Abitibi-Ouest. Polytechn. School, Montreal, M.Sc. thesis, 1–87.   10

Quinlivan, W.D. and Rogers, C.L., 1974. Geologic map of the Tybo quadrangle, Nye County, Nevada. U.S. Geol. Survey Misc. Geol. Inv. Map I–821.   8

Reading, H.G., ed., 1978. Sedimentary environments and facies. Elsevier, Amsterdam, 1–557.   13

Reynolds, R.L., 1975. Paleomagnetism of the Yellowstone tuffs and their associated airfall ashes. Univ. Colorado, Boulder, Ph.D. diss., 1–268.   13

Ricci-Lucchi, F., 1975. Depositional cycles in two turbidite formations of the northern Apennines (Italy). J. Sedim. Petrol. 45, 3–43.   14

Richards, A.F., 1958. Trans-Pacific distribution of floating pumice from Isla San Benedicto, Mexico. Deep-Sea Res. 5, 29–35.   7

Richards, A.F., 1959. Geology of the Islas Revillagigedo, Mexico, 1. Birth and development of Volcan Barcena, Isla San Benedicto (1). Bull. Volcanol. 22, 73–123.   9

Richardson, D. and Ninkovich, D., 1976. Use of $K_2O$, Rb, Zr, and Y versus $SiO_2$ in volcanic ash layers of the eastern Mediterranean to trace their sources. Geol. Soc. Amer. Bull. 87, 110–116.   7

Riehle, J.R., 1973. Calculated compaction profiles of rhyolitic ash-flow tuffs. Geol. Soc. Amer. Bull. 84, 2193–2216.   8,10

Ringwood, A.E., 1975. Composition and petrology of the earth's mantle. McGraw-Hill, Inc., New York, 1–618.   2,3,9

Ritchey, J.L. and Eggler, D.H., 1978. Amphibole stability in a differentiated calc-alkaline magma chamber: An experimental investigation. Carnegie Inst. Wash. Yearb. 77, 790–793.   3

Rittmann, A., 1958. Il meccanismo di formazione delle lave a pillows e dei cosidetti tufi palagonitici. Atti Acc. Gioenia 4, 310–317.   9,10

Rittmann, A., 1962. Volcanoes and their activity. John Wiley and Sons, New York, 1–305.   1,4,9

Roedder, E., 1972. Composition of fluid inclusions. U.S. Geol. Survey Prof. Paper 440-JJ, 1–64.   3

Roedder, E. and Coombs, D.S., 1967. Immiscibility in granitic melts, indicated by fluid inclusions in ejected granitic blocks from Ascension Island: J. Petrol. 8, 417–451.   3

Roedder, E. and Smith, R.L., 1965. Liquid water in pumice vesicles, a crude but useful dating method (Abst.). Geol. Soc. Amer. Sp. Paper 82, Abst. for 1964, 164.   13

Roen, J.B. and Hosterman, J.W., 1982. Misuse of the term "bentonite" for ash beds of Devonian age in the Appalachian basin. Geol. Soc. America Bull. 93, 921–925.   12

Rohlof, K.J., 1969. Analysis of the exterior ballistics of block ejecta at Nanwaksjiak Crater, Nunivak Island, Alaska. Air Force Institute of Technology, Wright Patterson AFB, Ohio, M. S. thesis, 1–118.  9

Ronov, A.B., 1964. Common tendencies in the chemical evolution of the earth's crust, ocean and atmosphere. Geokhimiya 8, 715–743.  1

Ronov, A.B., 1968. Probable changes in the composition of sea water during the course of geologic time. Sedimentology 10, 25–43.  1

Roobol, M.J. and Smith, A.L., 1975. A comparison of the recent eruptions of Mt. Pelée, Martinique and Soufrière, St. Vincent. Bull. Volcanol. 39, 1–27.  11

Rose, W.I., Jr., 1972. Notes on the 1902 eruption of Santa Maria volcano, Guatemala. Bull. Volcanol. 36, 29–45.  6

Rose, W.I., Jr., 1977. Scavenging of volcanic aerosol by ash – atmospheric and volcanologic implications. Geology 5, 621–624.  3

Rose, W.I., Jr., Grant, N.K. and Easter, J., 1979. Geochemistry of the Los Chocoyos Ash, Guatemala. Geol. Soc. Amer. Sp. Paper 180, 87–99.  8

Rose, W.I., Jr., Pearson, T. and Bonis, S., 1977. Nuée ardente eruption from the foot of a dacite lava flow, Santiaguito Volcano, Guatemala. Bull. Volcanol. 40, 1–16.  8

Rose, W.I., Jr., Penfield, G.T., Drexler, J.W. and Larson, P.B., 1980. Geochemistry of the andesite flank lavas of three composite cones within the Atitlán Cauldron, Guatemala. Bull. Volcanol. 43, 131–153.  2

Rosholt, J.N., Ptijana, and Noble, D.C., 1971. Mobility of uranium and thorium in glassy and crystallized silicic volcanic rocks. Econ. Geol. 66, 1061–1069.  12

Ross, C.S., 1955. Provenance of pyroclastic materials. Geol. Soc. Amer. Bull. 66, 427–434.  1,12

Ross, C.S., 1964. Volatiles in volcanic glasses and their stability relations. Amer. Mineral. 49, 258–269.  3

Ross, C.S. and Smith, R.L., 1955. Water and other volatiles in volcanic glass. Amer. Mineral. 40, 1071–1089.  3,7,8,12

Ross, C.S. and Smith, R.L., 1961. Ash-flow tuffs: their origin, geologic relations and identification. U.S. Geol. Survey Prof. Paper 366, 1–77.  5,8,10

Ross, C.S., Miser, H.D., Stephenson, L.W., 1928. Water-laid volcanic rocks of early upper Cretaceous age in southwestern Arkansas, southeastern Oklahoma, and northeastern Texas. U.S. Geol. Survey Prof. Paper 154-F, 175–202.  5

Rothe, P. and Koch, R., 1978. Miocene volcanic glass from DSDP Sites 368, 369 and 370. In Lancelot, Y., Seibold, E., et al., Init. Rpts. Deep Sea Drilling Proj. 41, 1061–1064.  7

Rowley, P.D., Kuntz, M.A. and MacLeod, N.S., 1981. Pyroclastic-flow deposits. In Lipman, P.W. and Mullineaux, D.R., eds., The 1980 eruptions of Mount St. Helens, Washington. U.S. Geol. Survey Prof. Paper 1250, 489–512.  8,9

Rubel, D.H., 1971. Independence Volcano: A major Eocene eruptive center, northern Absaraka volcanic province. Geol. Soc. Amer. Bull. 82, 2473–2494.  13

Ruddiman, W.F. and Glover, L.K., 1972. Vertical mixing of ice-rafted volcanic ash in North Atlantic sediments. Geol. Soc. Amer. Bull. 83, 2817–2836.  7

Ryan, M.P. and Sammis, Ch.A., 1981. The glass transition in basalt. J. Geophys. Res. 86, 9519–9535.  4

Saemundsson, K., 1972. Notes on the geology of the Torfajokull central volcano. Natturufraedingnum 42, 81–99.  10

Sakai, H., Casadevall, T.J. and Moore, J.G., 1982. Chemistry and isotope values of sulfur in basalts and volcanic gases at Kilauea Volcano, Hawaii. Geochim. Cosmochim. Acta 46, 729–738.  3

Salmi, M., 1948. The Hekla ashfalls in Finland A.D. 1947. Geol. Forskning Sanstaten Bull. 142, 87–96.  6

Sanders, J.E., 1965. Primary sedimentary structures formed by turbidity currents and related resedimentation mechanisms. Soc. Econ. Paleontol. Mineral. Sp. Publ. 12, 192–219.  11

Sapper, K., 1904. Die vulkanischen Ereignisse in Mittelamerika im Jahre 1902. N. Jb. Min. Geol. Pal., 39–90.  6

Sapper, K., 1927. Vulkankunde. J. Engelhorns Nachf., Stuttgart, 1–424.  1,2,3

Sarna-Wojcicki, A.M., Meyer, C.E., Woodward, M.J. and Lamothe, P.J., 1981. Composition of air-fall ash erupted on May 18, May 25, June 12, July 22 and August 7. In Lipman, P.W. and Mullineaux, D.R., eds., The 1980 eruptions of Mount St. Helens, Washington. U.S. Geol. Survey Prof. Paper 1250, 667–681.  6

Sarna-Wojcicki, A.M., Shipley, S., Waitt, R.B., Jr., Dzurisin, D. and Wood, S.H., 1981. Areal distribution, thickness, mass, volume, and grain size of air-fall ash from six major eruptions of 1980. In Lipman, P.W. and Mullineaux, D.R., eds., The 1980 eruptions of Mount St. Helens, Washington. U.S. Geol. Survey Prof. Paper 1250, 577–600.  6,13

Scheidegger, A.E. and Potter, P.E., 1968. Textural studies of grading: volcanic ashfalls. Sedimentology 11, 163–170.  6

Scheidegger, K.F. and Kulm, L.D., 1975. Late Cenozoic volcanism in the Aleutian arc; information from ash layers in the northeastern Gulf of Alaska. Geol. Soc. Amer. Bull. 86, 1407–1412.  10

Scheidegger, K.F., Jezek, P.A. and Ninkovich, D., 1978. Chemical and optical studies of glass shards in Pleistocene and Pliocene ash layers from DSDP Site 192, northwest Pacific Ocean. J. Volcanol. Geotherm. Res. 4, 99–116.  7

Scheidegger, K.F., Corliss, J.B., Jezek, P.A. and Ninkovich, D., 1980. Compositions of deep-sea ash layers derived from north Pacific island arcs: Variations in time and space. J. Volcanol. Geotherm. Res. 7, 107–137.  7

Scherp, A. and Grabert, H., 1983. Unterdevonische Schmelztuffe im rechtsrheinischen Schiefergebirge. N. Jb. Geol. Pal. Mh., 47–58.  10

Schiener, E.J., 1970. Sedimentology and petrography of three tuff horizons in the Caradocian sequence of the Bala area (North Wales). Geol. J. 7, 25–46.  7

Schmid, R., 1981. Descriptive nomenclature and classification of pyroclastic deposits and fragments: Recommendations of the IUGS Subcommission on the Systematics of Igneous Rocks. Geology 9, 41–43.  5

Schmincke, H.-U., 1967a. Fused tuff and pépérites in south central Washington. Geol. Soc. Amer. Bull. 78, 319–330.  8,9

Schmincke, H.-U., 1967b. Graded lahars in the type section of the Ellensburg Formation, south-central Washington. J. Sed. Petrol. 37, 438–448.  10,11

Schmincke, H.-U., 1969a. Ignimbrite sequence on Gran Canaria. Bull. Volcanol. 33, 1199–1219.  3,8

Schmincke, H.-U., 1969b. Petrologie der phonolithischen bis rhyolithischen Vulkanite auf Gran Canaria, Kanarische Inseln, Habilitationsschrift, Universität Heidelberg, 1–151.  2,8

Schmincke, H.-U., 1970. Base Surge – Ablagerungen des Laacher See-Vulkans. Aufschluss 21, 350–364.  8,9

Schmincke, H.-U., 1972. Froth flows and globule flows in Kenya. Naturwissenschaften 11, 1–2.  8

Schmincke, H.-U., 1973. Magmatic evolution and tectonic regime in the Canary, Madeira and Azores Island Groups. Geol. Soc. Amer. Bull. 84, 633–648.  8,10

Schmincke, H.-U., 1974a. Pyroclastic rocks. In Füchtbauer, H., Sediments and Sedimentary rocks. Schweizerbart'sche Verlagsbuchhandlung, Stuttgart, 160–189.  1,8,9,11

Schmincke, H.-U., 1974b. Volcanological aspects of peralkaline silicic welded ash-flow tuffs. Bull. Volcanol. 38, 594–636.  5,6,8

Schmincke, H.-U., 1976. Geology of the Canary Islands. In Kunkel, G., ed., Biogeography and Ecology in the Canary Islands. W. Junk, The Hague, 67–184.  3,6,8,11

Schmincke, H.-U., 1977a. Eifel-Vulkanismus östlich des Gebietes Rieden-Mayen. Fortschr. Miner. 55, 1–31.  6

Schmincke, H.-U., 1977b. Phreatomagmatische Phasen in quartären Vulkanen der Osteifel. Geol. Jahrb. 39, 3–45.  3,4,6,9,13

Schmincke, H.-U., 1981. Ash from vitric muds in deep sea cores from the Mariana Trough and fore-arc regions (South Philippine Sea) (sites 453, 454, 455, 458, 459). In Hussong, D.M., Uyeda, S., et al., eds., Init. Rpts. Deep Sea Drilling Proj. 60, 473–481.  7

Schmincke, H.-U., 1982a. Volcanic and chemical evolution of the Canary Islands. In von Rad, U., Hinz, K., Sarnthein, M. and Seibold, E., eds., Geology of the Northwest African Continental Margin. Springer-Verlag, Berlin, Heidelberg, New York, 273–308.  2,7,10

Schmincke, H.-U., 1982b. Vulkane und ihre Wurzeln. Rhein. Westf. Akad. Wiss. Westd. Verlag Opladen, Vorträge N 315, 35–78.  2

Schmincke, H.-U., 1983. Rhyolitic and basaltic ashes from the Galapagos Mounds Area, Leg 70. In Cann et al., eds., Init. Rpts. Deep Sea Drilling Proj. 69, 451–457.  2,7

Schmincke, H.-U. and Johnston, D.A., 1977. Contrasting pyroclastic flow deposits of the 1976 eruption of Augustine Volcano, Alaska (Abst.). Geol. Soc. Amer. Abst. with Programs 9, 1161.  8

Schmincke, H.-U. and Pritchard, G., 1981. Carboniferous volcanic glass in submarine hyaloclastites from the Lahn-Dill area, Germany. Naturwissenschaften 68, 615–616.  12

Schmincke, H.-U. and von Rad, U., 1979. Neogene evolution of Canary Island volcanism inferred from ash layers and volcaniclastic sandstones of DSDP site 397 (Leg 47A). In von Rad, U., Ryan, W.B.F., et al., eds., Init. Rpts. Deep Sea Drilling Proj. 47, pt. I, 703–725.  7,10,13

Schmincke, H.-U. and Staudigel, H., 1976. Pillow lavas on central and eastern Atlantic Islands (La Palma, Gran Canaria, Porto Santo, Santa Maria). Bull. Soc. Geol. France 7, 871–883.  10

Schmincke, H.-U. and Swanson, D.L., 1967. Laminar viscous flowage structures in ash-flow tuffs from Gran Canaria, Canary Islands. J. Geol. 75, 641–664.  8

Schmincke, H.-U., Brey, G. and Staudigel, H., 1974. Craters of phreatomagmatic origin on Gran Canaria, Canary Islands. Naturwissenschaften 61, 125.  9

Schmincke, H.-U., Fisher, R.V. and Waters, A.C., 1973. Antidune and chute and pool structures in the base surge deposits of the Laacher See area, Germany. Sedimentology 20, 553–574.  5,8,9,13

Schmincke, H.-U., Robinson, P.T., Ohnmacht, W. and Flower, M.F.J., 1978. Basaltic hyaloclastites from Hole 396B, DSDP Leg 46. In Dmitriev, L., Heirtzler, J., et al., eds., Init. Rpts. Deep Sea Drilling Proj. 46, 341–355.  4,9,10,12

Schmincke, H.-U., Viereck, L.G., Griffin B.J. and Pritchard, R.G., 1982. Volcaniclastic rocks of the Reydarfjordur drill hole, Eastern Iceland I. Primary features. J. Geophys. Res. 87, 6437–6458.  10,12

Schmincke, H.-U., Rautenschlein, M., Robinson, P.T. and Mehegan, J.M., 1983. The Troodos Extrusive Series of Cyprus: a comparison with oceanic crust. Geology 11, 410–412.  10

Schuchert, C., 1931. Geochronology or the age of the earth on the basis of sediments and life. In The age of the earth. U.S. Nat. Res. Counc. Bull. 80, 10–64.  1

Schultz, C.H., 1972. Eruption at Deception Island, Anarctica, August 1970. Geol. Soc. Amer. Bull. 83, 2837–2842.  9

Scrope, J.P., 1862. Volcanoes. The character of their phenomena, their share in the structure and composition of the surface of the globe, and their relation to its internal forces. 2nd ed., Longman, Green, Longmans and Roberts, London, 1–490.  3,9

Seely, D.R., Vail, P.R. and Walton, G.G., 1974. Trench slope model. In Burk, C.A. and Drake, C.L., eds., The geology of continental margins. Springer-Verlag, New York, Heidelberg, Berlin, 249–260.  14

Segerstrom, K., 1950. Erosion studies at Parícutin, State of Michoacán, Mexico. U.S. Geol. Survey Bull. 965-A, 1–164.  6

Sekine, T., Katsura, T., Aramaki, S., 1979. Water-saturated phase relations of some andesites with application to the estimation of the initial temperature and water pressure at the time of eruption. Geochim. Cosmochim. Acta 43, 1367–1376.  3

Self, S., 1972. The Lajes ignimbrite, Ilha Terceira, Azores. Comunicoes Servs. Geol. Port. LV, 165–180.  8

Self, S., 1976. The Recent volcanology of Terceira, Azores. J. Geol. Soc. London 132, 645–666.  6,8

Self, S. and Rampino, M.R., 1981. The 1883 eruption of Krakatau. Nature 294, 699–704.  10

Self, S. and Sparks, R.S.J., 1978. Characteristics of wide-spread pyroclastic deposits formed by the interaction of silicic magma and water. Bull. Volcanol. 41-3, 1–17.  4,6,7,9

Self, S. and Sparks, R.S.J., eds., 1981. Tephra studies. D. Reidel Publ. Co., Dordrecht, Holland, 1–481.  1,13

Self, S., Kienle, J. and Huot, J.P., 1980. Ukinrek Maars, Alaska, II. Deposits and formation of the 1977 craters. J. Volcanol. Geotherm. Res. 7, 39–65.  5,9

Self, S., Wilson, L. and Nairn, I.A., 1979. Vulcanian explosion mechanisms. Nature 277, 440–443.  4

Self, S., Sparks, R.S.J., Booth, B. and Walker, G.P.L., 1974. The 1973 Heimaey Strombolian scoria deposit, Iceland. Geol. Mag. 111, 534–548.

Selley, R.C., 1978. Ancient sedimentary environments. 2nd Ed., Cornell Univ. Press, New York, 1–287.  13

Settle, M., 1978. Volcanic eruption clouds and the thermal power output of explosive eruptions. J. Volcanol. Geotherm. Res. 3, 1727–1739.  7

Settle, M., 1979. The structure and emplacement of cinder cone fields. Amer. J. Sci. 279, 1089–1107.

Seyfried, W.E. and Bischoff, J.L., 1979. Basalt-sea water interaction: trace element and strontium isotopic variations in experimentally altered glassy basalt. Earth Planet. Sci. Lett. 44, 463–472.  12

Shackleton, N.J. and Opdyke, N.D., 1973. Oxygen isotope and paleomagnetic stratigraphy of equatorial Pacific core V28-238: Oxygen isotope temperatures and ice volumes on a $10^8$ and $10^6$ year scale. Quat. Res. 3, 39–55.  7

Shackleton, N.J. and Opdyke, N.D., 1976. Oxygen isotope and paleomagnetic stratigraphy of Pacific core V28-239: Late Pliocene to Latest Pleistocene. Geol. Soc. Amer. Mem. 145, 449–464.  7

Sharp, R.P. and Nobles, L.H., 1953. Mudflows of 1941 at Wrightwood, southern California. Geol. Soc. Amer. Bull. 64, 547–560. 9,11

Shaw, D.M., Watkins, N.D. and Huang, T.C., 1974. Atmospherically transported volcanic dust in deep sea sediments: theoretical considerations. J. Geophys. Res. 79, 3087–3097. 6,7

Shaw, H.R., 1963. Obsidian-$H_2O$ viscosities at 1000 and 2000 bars in the temperature range 700° to 900 °C. J. Geophys. Res. 68, 6337–6343. 3

Shaw, H.R., 1965. Comments on viscosity, crystal settling, and convection in granitic magmas. Amer. J. Sci. 263, 120–152. 3

Shaw, H.R., 1969. Rheology of basalt in the melting range. J. Petrol. 10, 510–535. 3

Shaw, H.R., 1972. Viscosities of magmatic silicate liquids: An empirical method of prediction. Amer. J. Sci. 272, 870–893. 3

Shaw, H.R., 1974. Diffusion of $H_2O$ in granitic liquids: Part I, Experimental data; Part II, Mass transfer in magma chambers. In Hoffmann, A.W., Giletti, B.J., Yoder, H.S., Jr., Yund, R.A., eds., Geochemical transport and kinetics. Carnegie Inst. Wash. Publ. 634, 139–170. 3

Shaw, H.R., 1980. The fracture mechanism of magma transport from the mantle to the surface. In Hargraves, R.B., ed., Physics of magmatic processes. Princeton Univ. Press, Princeton, N.J., 201–264. 4

Shaw, H.R., Peck, D.L., Wright, T.L. and Okamura, R., 1968. The viscosity of basaltic magma: An analysis of field measurements in Makaopuhi lava lake, Hawaii. Amer. J. Sci. 266, 225–264. 3

Sheimovich, V.S., 1979. Problems of ignimbrite petrology. Bull. Volcanol. 42, 218–232. 8

Shepherd, E.S., 1938. The gases in rocks and some related problems. Amer. J. Sci. 35A, 311–351. 3

Shepherd, J.B. and Sigurdsson, H., 1982. Mechanism of the 1979 explosive eruption of Soufrière Volcano, St. Vincent. J. Volcanol. Geotherm. Res. 13, 119–130. 4

Sheppard, R.A., 1969. Diagenesis of tuffs in the Barstow formation, Mud Hills, San Bernardino County, California. U.S. Geol. Survey Prof. Paper 634, 1–33. 12

Sheppard, R.A., 1971. Zeolites in sedimentary deposits of the United States – a review. In Gould, R.F., ed., Molecular Sieve Zeolites I. Adv. Chem. Ser. 101, Amer. Chem. Soc. Washington, 279–310. 12

Sheppard, R.A., 1973. Zeolites in sedimentary rocks. U.S. Geol. Survey Prof. Paper 820, 689–695. 12

Sheppard, R.A. and Gude, A.J., 3rd, 1968. Distribution and genesis of authigenic silicate minerals in tuffs of Pleistocene Lake Tecopa, Inyo County, Calif. U.S. Geol. Survey Prof. Paper 597, 1–38. 12

Sheppard, R.A. and Gude, A.J., 3rd, 1973. Zeolites and associated authigenic silicate minerals in tuffaceous rocks of the Big Sandy Formation, Mohave County, Arizona. U.S. Geol. Survey Prof. Paper 830, 1–36. 12

Sheridan, M.F., 1970. Fumarolic mounds and ridges of the Bishop Tuff, California. Geol. Soc. Amer. Bull. 81, 851–868. 8

Sheridan, M.F., 1971. Particle-size characteristics of pyroclastic tuffs. J. Geophys. Res. 76, 5627–5634. 5,6,8,9

Sheridan, M.F., 1979. Emplacement of pyroclastic flows: A review. Geol. Soc. Amer. Sp. Paper 180, 125–136. 8

Sheridan, M.F. and Marshall, J.R., 1983. Interpretation of pyroclast surface features using SEM images. J. Volcanol. Geotherm. Res. 16, 153–159. 5

Sheridan, M.F. and Ragan, D.M., 1977. Compaction of ash-flow tuffs. In Chilingarian, G.V. and Wolf, K.H., eds., Compaction of coarse-grained sediments, II; Developments in sedimentology 18B. Elsevier, Amsterdam, 677–713. 8

Sheridan, M.F. and Updike, R.G., 1975. Sugarloaf Mountain tephra – a Pleistocene rhyolitic deposit of base-surge origin. Geol. Soc. Amer. Bull. 86, 571–581. 9

Sheridan, M.F. and Wohletz, K.H., 1981. Hydrovolcanic explosions I. The systematics of water-pyroclast equilibration. Science 212, 1387–1389. 9

Shimozuru, D., 1968. Discussion on the energy partitioning of volcanic eruptions. Bull. Volcanol. 32, 383–394. 4

Sigurdsson, H. and Loebner, B., 1981. Deep sea record of Cenozoic explosive volcanism in the North Atlantic. In Self, S. and Sparks, R.S.J., eds., Tephra studies. D. Reidel Publ. Co., Dordrecht, Holland, 289–316. 7

Sigurdsson, H. and Sparks, R.S.J., 1978. Rifting episodes in north Iceland in 1874–75 and the eruption of Askja and Sveinagja. Bull. Volcanol. 41, 1–19. 2

Sigurdsson, H. and Sparks, R.S.J., 1981. Petrology of rhyolitic and mixed magma ejecta from the 1875 eruption of Askja, Iceland. J. Petrol. 22, 41–84. 2

Sigurdsson, H., Sparks, R.S.J., Carey, S.N. and Huang, T.C., 1980. Volcanogenic sedimentation in the Lesser Antilles Arc. J. Geol. 88, 523–540. 14

Sigvaldasson, G.E., 1968. Structure and products of subaquatic volcanoes in Iceland. Contr. Mineral. and Petrol. 18, 1–16.   10
Simkin, T., Siebert, L., McClelland, L., Bridge, D., Newhall, C. and Latter, J.H., 1981. Volcanoes of the world. Smithsonian Institution, Hutchinson Ross Publ. Co., Stroudsberg, Pa., 1–232.   1,2,4,13
Simon, M. and Schmincke, H.-U., 1983. Late Cretaceous volcaniclastic rocks from the Walvis Ridge, Southeast Atlantic, Leg 74. In Moore, T.C., Jr., Rabinowitz, P.D., et al., eds., Init. Rpts. Deep Sea Drilling Proj. 74, 765–792.   10
Simons, D.B. and Richardson, E.V., 1961. Forms of bed roughness in alluvial channels. Amer. Soc. Civil Eng. Proc. 87, no. HY3, 87–105.   5
Sisson, T.W., 1982. Sedimentary characteristics of the airfall deposit produced by the major pyroclastic surge of May 18, 1980 at Mount St. Helens, Washington. Univ. Calif. Santa Barbara, M.A. thesis, 1–145.   7
Skipper, K., 1971. Antidune cross-stratification in a turbidite sequence, Cloridorme Formation, Gaspé, Quebec. Sedimentology 17, 51–68.   5
Slaughter, M. and Earley, J.W., 1965. Mineralogy and geological significance of the Mowry Bentonites, Wyoming. Geol. Soc. Amer. Sp. Paper 83, 1–116.   7,12
Slaughter, M. and Hamil, M., 1970. Model for deposition of volcanic ash and resulting bentonite. Geol. Soc. Amer. Bull. 81, 961–968.   6
Smedes, H.W. and Prostka, H.J., 1972. Stratigraphic framework of the Absaroka Volcanic Supergroup in the Yellowstone National Park region. U.S. Geol. Survey Prof. Paper 729-C, C1–C33.   13
Smith, A.L. and Roobol, M.J., 1982. Andesite pyroclastic flows. In Thorpe, R.S., ed., Andesites. John Wiley and Sons, New York, 415–433.   8
Smith, D.G.W., 1967. The petrology and mineralogy of some lower Devonian bentonites from Gaspé, Quebec. Can. Mineral. 9, 141–165.   12
Smith, R.L., 1960a. Ash flows. Geol. Soc. Amer. Bull. 71, 795–842.   8,10
Smith, R.L., 1960b. Zones and zonal variations in welded ash flows. U.S. Geol. Survey Prof. Paper 354-F, 149–159.   8
Smith, R.L., 1979. Ash-flow magmatism. Geol. Soc. Amer. Sp. Paper 180, 5–27.   2,6,8
Smith, R.L. and Bailey, R.A., 1966. The Bandelier Tuff: a study of ash-flow eruption cycles from zoned magma chambers. Bull. Volcanol. 29, 83–104.   2,8
Smith, R.L. and Bailey, R.A., 1968. Resurgent cauldrons. Geol. Soc. Amer. Mem. 116, 613–662.   13
Snavely, P.D., Jr., MacLeod, N.S. and Wagner, H.C., 1968. Tholeiitic and alkalic basalts of the Eocene Siletz River Volcanics, Oregon Coast Range. Amer. J. Sci. 266, 454–481.   14
Sommer, M.A., 1977. Volatiles $H_2O$, $CO_2$, and $CO$ in silicate melt inclusions in quartz phenocrysts from the rhyolitic Bandelier air-fall and ash-flow tuff, New Mexico. J. Geol. 85, 423–432.   3,4
Sonayada, T.J., 1971. Normal and accelerated sinking of phytoplankton in the sea. Mar. Geol. 11, 105–122.   7
Sorem, R.K., 1982. Volcanic ash clusters: tephra rafts and scavengers. J. Volcanol. Geotherm. Res. 13, 63–71.   6
Southard, J.B., 1970. Lift forces on suspended particles in laminar flow: experiments and sedimentological interpretation. J. Sed. Petrol. 40, 320–324.   11
Southard, J.B., 1971. Representation of bed configuration in depth-velocity-size diagrams. J. Sed. Petrol. 41, 903–915.   5
Sparks, R.S.J., 1975. Stratigraphy and geology of the ignimbrites of Vulsini Volcano, Central Italy. Geol. Rundsch. 64, 497–523.   8
Sparks, R.S.J., 1976. Grain size variations in ignimbrites and implications for the transport of pyroclastic flows. Sedimentology 23, 147–188.   5,8,11
Sparks, R.S.J., 1978. The dynamics of bubble formation and growth in magmas: A review and analysis. J. Volcanol. Geotherm. Res. 3, 1–37.   3,4
Sparks, R.S.J., 1979. Gas release rates from pyroclastic flows: An assessment of the role of fluidization in their emplacement. Bull. Volcanol. 41, 1–9.   8
Sparks, R.S.J. and Huang, T.C., 1980. The volcanological significance of deep-sea ash layers associated with ignimbrites. Geol. Mag. 117, 425–436.   7
Sparks, R.S.J. and Walker, G.P.L., 1973. The ground surge deposit: a third type of pyroclastic rock. Nature 241, 62–64.   8
Sparks, R.S.J. and Walker, J.P.L., 1977. The significance of vitric-enriched air-fall ashes associated with crystal-enriched ignimbrites. J. Volcanol. Geotherm. Res. 2, 329–341.   6,8
Sparks, R.S.J. and Wilson, L., 1976. A model for the formation of ignimbrite by gravitational column collapse. J. Geol. Soc. London 132, 441–451.   4,8

Sparks, R.S.J. and Wilson, L., 1982. Explosive volcanic eruptions. V. Observations of plume dynamics during the 1979 Soufrière eruption, St. Vincent. Geophys. J. Roy. Astron. Soc. 69, 551–570.   4

Sparks, R.S.J. and Wright, J.V., 1979. Welded air-fall tuffs. Geol. Soc. Amer. Sp. Paper 180, 155–166.   8

Sparks, R.S.J., Self, S. and Walker, G.P.L., 1973. Products of ignimbrite eruption. Geology 1, 115–118.   8

Sparks, R.S.J., Sigurdsson, H. and Carey, S.N., 1980a. The entrance of pyroclastic flows into the sea, I. Oceanographic and geologic evidence from Dominica, Lesser Antilles. J. Volcanol. Geotherm. Res. 7, 87–96.   10,14

Sparks, R.S.J., Sigurdsson, H. and Carey, S.N., 1980b. The entrance of pyroclastic flows into the sea, II. Theoretical considerations on subaqueous emplacement and welding. J. Volcanol. Geotherm. Res. 7, 97–105.   10

Sparks, R.S.J., Wilson, L. and Hulme, G., 1978. Theoretical modeling of the generation, movement and emplacement of pyroclastic flows by column collapse. J. Geophys. Res. 83, 1727–1739.   4,6,8

Spears, D.A. and Rice, C.M., 1973. An upper Carboniferous limestone of volcanic origin. Sedimentology 20, 281–294.   12

Spears, D.A. and Kanaris-Sotiriou, 1979. A geochemical and mineralogical investigation of some British and other European tonsteins. Sedimentology 26, 407–425.   12

Spera, F.J., 1980. Aspects of magma transport. In Hargraves, R.B., ed., Physics of magmatic processes. Princeton Univ. Press. Princeton, N.J., 265–323.   3,4

Spera, F.J., 1984. Ascent of basaltic magma. Contr. Mineral. Petrol. (in press).   3

Spera, F. and Bergman, S.C., 1980. Carbon dioxide in igneous petrogenesis. I, Aspects of the dissolution of $CO_2$ in silicate liquids. Contr. Mineral. Petrol. 74, 55–66.   3

Spooner, E.T.C., 1976. The strontium isotopic composition of seawater, and seawater-oceanic crust interaction. Earth Planet. Sci. Lett. 31, 167–174.   12

Stanley, D.J. and Taylor, P.T., 1977. Sediment transport down a seamount flank by a combined current and gravity process. Mar. Geol. 23, 77–88.   10

Stanley, K.O., 1976. Sandstone petrofacies in the Cenozoic High Plains sequence, eastern Wyoming and Nebraska. Geol. Soc. Amer. Bull. 87, 297–309.   13

Stanton, W.I., 1960. The lower Paleozoic rocks of south-west Murrisk, Ireland. Q. J. Geol. Soc. London 116, 269–296.   10

Staudigel, H. and Hart, S.R., 1983. Alteration of basaltic glass. Mechanisms and significance for the oceanic crust-seawater budget. Geochim. Cosmochim. Acta 47, 337–350.   12

Staudigel, H. and Schmincke, H.-U., 1984. The Pliocene seamount series of La Palma (Canary Islands). J. Geophys. Res. 89 11195–11215.   10

Staudigel, H., Frey, F. and Hart, S.R., 1979. Incompatible trace element geochemistry and $^{87}Sr/^{86}Sr$ in basalts and corresponding glasses and palagonites. In Donelly, T., Francheteau, H., et al., eds., Init. Rpts. Deep Sea Drilling Proj. 51–53, 1137–1143.   10

Staudigel, H., Hart, S.R. and Richardson, S.H., 1981. Alteration of the oceanic crust: processes and timing. Earth Planet. Sci. Lett. 52, 311–327   10

Stauffer, P.H., 1967. Grain-flow deposits and their implications, Santa Ynez Mountains, California. J. Sed. Petrol. 37, 487–508.   7

Stearns, H.T., 1925. The explosive phase of Kilauea Volcano, Hawaii in 1924. Bull. Volcanol. 5, 1–16.   13

Stearns, H.T. and Clark, W.D., 1930. Geology and groundwater resources of the Kau District, Hawaii. U.S. Geol. Survey Water Supp. Paper 616, 1–194.   9

Stearns, H.T. and Macdonald, G.A., 1946. Geology and ground-water resources of the Island of Hawaii. Hawaii Div. Hydrogr. Bull. 9, 1–363.   9

Stearns, H.T. and Vaksvik, K.N., 1935. Geology and ground-water resources of the Island of Oahu, Hawaii. Hawaii Terr. Dept. Public Lands Div., Hydrog. Bull. 1, 1–479.   9

Steen, V.C. and Fryxell, R., 1965. Mazama and Glacier Peak pumice: Uniformity of refractive index after weathering. Science 150, 878–880.   13

Steen-McIntyre, V.C., 1975. Hydration and superhydration of tephra glass – a potential tool for estimating age of Holocene and Pleistocene ash beds. In Suggate, R.P. and Cresswell, M.M., eds., Quaternary studies, Wellington, Roy. Soc. N.Z., 271–278.   7,12

Steen-McIntyre, V.C., 1977. A manual for tephrochronology. Idaho Springs, Colorado, 1–167.   13

Stehn, C.E., 1934. Die semivulkanische Explosion des Pematang Bata, 1933. Natuurk. Tijdschr. Nederl.-Indie 94, 46–69.   9

Stehn, C.E., 1936. Beobachtungen an Glutwolken während der erhöhten Tätigkeit des Vulkans Merapi in Mittel-Java in den Jahren 1933–1935. Ned.-Indie. Natuurwet Congres, Batavia, 647–656. 8

Steininger, J., 1819. Geognostische Studien am Mittelrhein: Mainz, 1–223. 9

Steven, T.A., 1975. Middle Tertiary volcanic field in the southern Rocky Mountains. In Curtis, B.F., ed., Cenozoic history of the southern Rocky Mountains. Geol. Soc. Amer. Mem. 144, 75–94. 13

Steven, T.A. and Lipman, P.W., 1968. Central San Juan cauldron complex, Colorado. In Epis, R.C., ed., Cenozoic volcanism in the southern Rocky Mountains. Colo. School Mines Qt. 63, 241–258. 13

Steven, T.A. and Lipman, P.W., 1976. Calderas of the San Juan volcanic field, southwestern Colorado. U.S. Geol. Survey Prof. Paper 958, 1–35. 8,13

Steven, T.A. and Ratté, J.C., 1964. Revised Tertiary volcanic sequence in the central San Juan Mountains, Colorado. In Short papers in geology and hydrology. U.S. Geol. Survey Prof. Paper 475-D, D54–D63. 13

Steven, T.A. and Ratté, J.C., 1965. Geology and structural control of ore deposition in the Creede district, San Juan Mountains, Colorado. U.S. Geol. Survey Prof. Paper 487, 1–90. 13

Steven, T.A., Mehnert, H.H. and Obradovich, J.D., 1967. Age of volcanic activity in the San Juan Mountains, Colorado. In Geological Survey Research 1967. U.S. Geol. Survey Prof. Paper 575-D, D47–D55. 13

Steven, T.A., Cunningham, C.G., Naeser, C.W. and Mehnert, H.H., 1979. Revised stratigraphy and radiometric ages of volcanic rocks and mineral deposits in the Marysvale area, West Central Utah. U. S. Geol. Survey Bull. 1469, 1–40. 13

Steven, T.A., Lipman, P.W., Hail, W.J., Jr., Barker, F. and Luedke, R.G., 1974. Geologic map of the Durango quadrangle, southwestern Colorado. U. S. Geol. Survey Misc. Geol. Inv. Map I–764. 13

Stewart, R.J., 1975. Late Cenozoic explosive eruptions in the Aleutian and Kuril Island arcs. Nature 258, 505–507. 7

Stewart, R.J., 1978. Neogene volcaniclastic sediments from Atka Basin, Aleutian Ridge. Amer. Assoc. Petrol. Geol. Bull. 62, 87–97. 14

Stith, J.L., Hobbs, P.V. and Radke, L.F., 1977. Observations of a nuée ardente from St. Augustine volcano. Geophys. Res. Lett. 4, 259–262. 8

Stoiber, R.E., Williams, S.N., Malinconico, L.L., Johnston, D.A. and Casadevall, T.S., 1981. Mt. St. Helens: evidence of increased magmatic gas component. J. Volcanol. Geotherm. Res. 11, 203–212. 3

Stokes, K.R., 1971. Further investigations into the nature of the materials chlorophaeite and palagonite. Min. Mag. 38, 205–214. 12

Stolper, E., 1982. The speciation of water in silicate melts. Geochim. Cosmochim. Acta 46, 2609–2620. 3

Stonecipher, S.A., 1978. Chemistry of deep-sea phillipsite, clinoptilolite and host sediments. In Sand, L.B. and Mumpton, F.A., eds., Natural Zeolites: occurrence, properties, use. Pergamon Press, Oxford, 221–234. 12

Stuart, C.J. and Brenner, M.G., 1979. "Low-regime" base surge dunes – an example from Kilbourne and Hunt's Holes, south-central New Mexico (Abst.): Geol. Soc. America Abst. with Programs 11, 525. 9

Stumm, W. and Morgan, J.J., 1981. Aquatic chemistry. John Wiley and Sons, New York, 1–780. 12

Surdam, R.C., 1972. Economic potential of zeolite-rich sedimentary rocks in Wyoming. Wyo. Geol. Assoc. Sci. Bull. 1972, 5–8. 12

Surdam, R.C., 1973. Low-grade metamorphism of tuffaceous rocks in the Karmutsen Group, Vancouver Island, British Columbia. Geol. Soc. Amer. Bull. 84, 1911–1922. 12

Surdam, R.C. and Eugster, H.P., 1976. Mineral reactions in the sedimentary deposits of the Lake Magadi region, Kenya. Geol. Soc. Amer. Bull. 87, 1739–1752. 12

Surdam, R.C. and Sheppard, R.A., 1978. Zeolites in saline, alkaline lake deposits. In Sand, L.B. and Mumpton, F.A., eds., Natural Zeolites: occurrence, properties, use. Pergamon Press, Oxford, 145–174. 12

Sutherland, F.L., 1965. Dispersal of pumice, supposedly from the March, 1962, South Sandwich Islands eruption, on southern Australian shores. Nature 207, 1332–1335. 7

Suzuki, T., Katsui, Y. and Nakamura, T., 1973. Size distribution of the Tarumai Ta-b pumice-fall deposit. Bull. Volcanol. Soc. Japan 18, 47–64. 6,7

Swanson, D.A. and Christiansen, R.L., 1973. Tragic base surge in 1790 at Kilauea Volcano. Geology 1, 83–86. 9

Swanson, D.A. and Fabbi, B.P., 1973. Loss of volatiles during fountaining and flowage of basaltic lava at Kilauea volcano, Hawaii. U.S. Geol. Survey J. Res. 1, 649–658.   3

Swineford, A. and Frye, J.C., 1946. Petrographic comparison of Pliocene and Pleistocene volcanic ash from western Kansas. Kans. Geol. Survey Bull. 64, Pt. I, 1–32.   5

Szadeczky-Kardoss, E.V., 1933. Die Bestimmung des Abrollungsgrades. Centralbl. Min. Geol. Paläont. Abt. B, 389–401.   9

Tanakadate, H., 1935. Evolution of a new volcanic islet near Iwo Jima. Proc. Imper. Acad. 9, 152–154.   4

Tassé, N., Lajoie, J. and Dimroth, E., 1978. The anatomy and interpretation of an Archean volcaniclastic sequence, Noranda region, Quebec. Can. J. Earth Sci. 15, 874–888.   10,13

Taylor, G.A., 1956. Review of the volcanic activity in the Territory of Papua-New Guinea, the Solomon and New Hebrides Islands, 1951–1953. Bull. Volcanol. 18, 25–37.   8

Taylor, G.A., 1958. The 1951 eruption of Mount Lamington, Papua. Austr. Bur. Min. Resour. Geol. Geophys. Bull. 38, 1–117.   8

Taylor, H.P., Jr., 1974. The application of oxygen and hydrogen isotope studies to problems of hydrothermal alteration and ore deposition. Econ. Geol. 69, 843–883.   2

Taylor, M. and Brown, G.E., 1979. Structure of mineral glasses. I. The feldspar glasses $NaAlSi_3O_8$, $KAlSi_3O_8$, $CaAl_2Si_2O_8$. Geochim. Cosmochim. Acta 43, 61–77.   3

Taylor, N.H., 1933. Soil processes in volcanic ash beds, Part I. N.Z. J. Sci. Tech. 14, 193–202.   6

Taylor, S.R. and McLennan, S.M., 1981. The composition and evolution of the continental crust: rare earth element evidence from sedimentary rocks. Phil. Trans. Roy. Soc. London A 301, 381–399.   1

Tazieff, H., 1970. New investigations on eruptive gases. Bull. Volcanol. 34, 421–438.   3

Tazieff, H., 1972. About deep-sea volcanism. Geol. Rundsch. 61, 470–480.   9

Thompson, A.B., 1971. Analcite-albite equilibria at low temperatures. Amer. J. Sci. 271, 79–92.   12

Thorarinsson, S., 1944. Tefrokronologiska studier pa Island. Geogr. Analer 1–203 (English summary, p. 204–215).   13

Thorarinsson, S., 1954. The eruption of Hekla 1947–48, part 2, Ch. 3, The tephra-fall from Hekla on March 29, 1947. Soc. Sci. Islandica, Reykjavik, 1–68.   5,6

Thorarinsson, S., 1967a. Surtsey – The new island in the North Atlantic. Viking Press, New York, 1–47.   9

Thorarinsson, S., 1967b. The eruption of Hekla 1947–1948, I. The eruptions of Hekla in historical times. A tephrochronological study. Visindafelag Islendinga, Reykjavik, 1–183.   6

Thorarinsson, S., 1974. The terms tephra and tephrochronology. In Westgate, J.A. and Gold, C.M., eds., World bibliography and index of Quaternary tephrochronology. Printing Services Dept., Univ. of Alberta (Canada), 1–528.   13

Thorarinsson, S., 1981. Tephra studies and tephrochronology: A historical review with special reference to Iceland. In Self, S. and Sparks, R.S.J., eds., Tephra studies. D. Reidel Publ. Co., Dordrecht, Holland, 1–12.   13

Thorpe, R.S., ed., 1982. Andesites. Orogenic andesites and related rocks. John Wiley and Sons, New York, 1–724.   2

Thoulet, J., 1904. L'Océan, ses lois et ses problèmes. Hachette, Paris, 1–133.   4

Thoulet, J., 1922. Eruptions volcaniques sous-marines profondes. C.R. Acad. Sci., Paris 14, 1068–1070.   4

Thunell, R., Federman, A., Sparks, S. and Williams, D., 1979. The age, origin and volcanological significance of the Y-5 ash layer in the Mediterranean. Quat. Res. 12, 241–253.   7

Tokunaga, T. and Yokoyama, S., 1979. Mode of eruption and volcanic history of Mukaiyama Volcano, Nii-Jima. Geog. Rev. Jap. 52, 111–125 (in Japanese).   5

Truesdell, A.H., 1966. Ion-exchange constants of natural glasses by the electrode method. Amer. Mineral. 51, 110–122.   12

Tsong, T.S.T., Houser, C.A., Yusef, N.A., Messier, R.F., White, W.B. and Michels, J.W., 1978. Obsidian hydration profiles measured by sputter-induced optical emission. Science 201, 334–339.   12

Tsuya, H., 1930. The eruption of Komagatake, Hokkaido, in 1929. Tokyo Univ. Earthq. Res. Inst. Bull. 8, 238–270.   8

Tsuya, H., 1955. Geological and petrological studies of volcano Fuji; 5. On the 1707 eruption of Volcano Fuji. Tokyo Univ. Earthq. Res. Inst. Bull. 33, 341–383.   6

Tucker, M.E., 1981. Sedimentary petrology: an introduction. Blackwell Scientific Publ., Oxford, 1–252.   1

Turner, D.L., 1970. Potassium-argon dating of Pacific coast Miocene foraminiferal stages. Geol. Soc. Amer. Sp. Paper 124, 91–129.   13

Tuttle, O.F. and Bowen, N.L., 1958. Origin of granite in the light of experimental studies in the system NaAlSi$_3$O$_8$-KAlSi$_3$O$_8$-SiO$_2$-H$_2$O. Geol. Soc. Amer. Mem. 74, 1–153.   2

Urbain, G., Bottinga, Y. and Richet, P., 1982. Viscosity of liquid silica, silicates and alumino-silicates. Geochim. Cosmochim. Acta 46, 1061–1072.   3

Utada, M., 1970. Occurrence and distribution of authigenic zeolites in the Neogene pyroclastic rocks in Japan. Sci. Papers Coll. Gen. Educ. Univ. Tokyo 20, 191–262.   12

Utada, M., 1971. Zeolitic zoning of the Neogene pyroclastic rocks in Japan. Sci. Papers Gen. Educ. Univ. Tokyo 21, 189–221.   12

Vallier, T.L. and Kidd, R.B., 1977. Volcanogenic sediments in the Indian Ocean. In Heirtzler, J.R., et al., eds., Indian Ocean geology and biostratigraphy. Amer. Geophys. Un. 87–118.   10

Vallier, T.L., Bohrer, D., Moreland, G., McKee, E.E., 1977. Origin of basalt microlapilli in lower Miocene pelagic sediment, northeastern Pacific Ocean. Geol. Soc. Amer. Bull. 88, 787–796.   9,10

Van Andel, Tj.H. and Ballard, R.D., 1979. The Galapagos Rift at 86° W:2. Volcanism, structure and evolution of the rift valley. J. Geophys. Res. 84, 5390–5406.   2,10

Van Bemmelen, R.W., 1949. The Geology of Indonesia. General Geology. Govt. Printing Office, The Hague, 1A, 1–732.   7,8,11

Van Houten, F.B., 1976. Late Cenozoic volcaniclastic Andean foredeep, Columbia. Geol. Soc. Amer. Bull. 87, 481–495.   14

Verbeek, R.D.M., 1886. Krakatau. Batavia, Java, 1–495.   8

Verhoogen, J., 1946. Volcanic heat. Amer. J. Sci. 244, 745–771.   4

Verhoogen, J., 1951. Mechanics of ash formation. Amer. J. Sci. 249, 729–739.   3

Vessel, R.K. and Davies, D.K., 1981. Nonmarine sedimentation in an active fore arc basin. Soc. Econ. Paleont. Mineral. Publ. 31, 31–45.   13,14

Viereck, L.G., Griffin, B.J., Schmincke, H.-U. and Pritchard, R.G., 1982. Volcaniclastic rocks of the Reydarfjordur Drill Hole, eastern Iceland. 2. Alteration. J. Geophys. Res. 87, 6459–6476.   12

Vitaliano, D.B., ed., 1982. World bibliography and index of Quaternary tephrochronology. Supplement No. 1. Geobooks. Regency House, Norwich, England, 1–194.   13

Vlodavetz, V.I. and Piip, B.I., 1959. Kamchatka and continental areas of Asia. In Catalogue of the active volcanoes of the world, Part 7, I.A.V.C.E.I., Rome, 1–110.   9

Vogt, P.R., 1979. Global magmatic episodes: new evidence and implication for the steady-state mid-oceanic ridge. Geology 7, 93–98.   7

Voight, B., Glicken, H., Janda, R.J. and Douglass, P.M., 1981. Catastrophic rockslide avalanche of May 18. In Lipman, P.W. and Mullineaux, D.R., eds., The 1980 eruptions of Mount St. Helens. U.S. Geol. Survey Prof. Paper 1250, 347–377.   11

Völzing, K., 1907. Der Trass des Brohltales. Jb. Preuss. Geol. LA 28, 1–56.   8

Von Fritsch, R. and Reiss, W., 1868. Geologische Beschreibung der Insel Tenerife. Verlag von Wurster and Co., Winterthur, 1–494.   8

Von Waltershausen, W.S., 1845. Über die submarinen Ausbrüche in der tertiären Formation des Val di Noto im Vergleich mit verwandten Erscheinungen am Ätna. Gött. Stud. 1, 371–431.   12

Von Wolff, F., 1913, 1914. Der Vulkanismus. I. Band: Allgemeiner Teil; Ferdinand Enke Verlag, Stuttgart, 1–711.   4

Wadell, H., 1932. Volume, shape, and roundness of rock particles. J. Geol. 40, 443–451.   5

Waff, H.S., 1975. Pressure-induced coordination changes in magmatic liquids. Geophys. Res. Lett. 2, 193–196.   3

Waitt, R.B., Jr., 1981. Devastating pyroclastic density flow and attendant air fall of May 18 – stratigraphy and sedimentology of deposits. In Lipman, P.W. and Mullineaux, D.R., eds., The 1980 eruptions of Mount St. Helens, Washington. U.S. Geol. Survey Prof. Paper 1250, 439–458.   8

Waitt, R.B., Jr. and Dzurisin, D., 1981. Proximal air-fall deposits from the May 18 eruption – stratigraphy and field sedimentology. In Lipman, P.W. and Mullineaux, D.R., eds., The 1980 eruptions of Mount St. Helens, Washington. U.S. Geol. Survey Prof. Paper 1250, 601–616.   6,13

Waldron, H.H., 1967. Debris flow and erosion control problems caused by the ash eruptions of Irazu Volcano, Costa Rica. U. S. Geol. Survey Bull. 1241-I, 1–35.   6,11

Walker, G.P.L., 1969. The breaking of magma. Geol. Mag. 106, 166–173.   9

Walker, G.P.L., 1971. Grain-size characteristics of pyroclastic deposits. J. Geol. 79, 696–714.   5,6,8,10

Walker, G.P.L., 1972. Crystal concentration in ignimbrites. Contr. Mineral. Petrol. 36, 135–146.   6,8

Walker, G.P.L., 1973. Explosive volcanic eruptions – a new classification scheme. Geol. Rundsch. 62, 431–446.   4,6,9

Walker, G.P.L., 1979. A volcanic ash generated by explosions where ignimbrite entered the sea. Nature 281, 642–646. 10

Walker, G.P.L., 1980. The Taupo Pumice: product of the most powerful known (ultraplinian) eruption J. Volcanol. Geotherm. Res. 8, 69–94. 4,6

Walker, G.P.L. and Blake, D.H., 1966. Formation of a palagonite breccia mass beneath a valley glacier. Q. J. Geol. Soc. London 122, 45–61. 9

Walker, G.P.L. and Croasdale, R., 1971. Two plinian-type eruptions in the Azores. J. Geol. Soc. London 127, 17–55. 6,8

Walker, G.P.L. and Croasdale, R., 1972. Characteristics of some basaltic pyroclastics. Bull. Volcanol. 35, 303–317. 2,4,5,9,10

Walker, G.P.L., Heming, R.F. and Wilson, C.J.N., 1980. Low-aspect ratio ignimbrites. Nature 283, 286–287. 8

Walker, G.P.L., Self, S. and Froggatt, P.C., 1981. The ground layer of the Taupo ignimbrite: a striking example of sedimentation from a pyroclastic flow. J. Volcanol. Geotherm. Res. 10, 1–11. 8

Walker, G.P.L., Wilson, C.J.N. and Froggatt, P.C., 1980. Fines-depleted ignimbrite in New Zealand – the product of a turbulent pyroclastic flow. Geology 8, 245–249. 8

Walker, G.P.L., Wilson, L. and Bowell, E.L.G., 1971. Explosive volcanic eruptions – I. The rate and fall of pyroclasts. Geophys. J. Roy. Astron. Soc. 22, 377–383. 4,5,6

Walker, G.P.L., Heming, R.F., Sprod, T.J. and Walker, H.R., 1981. Latest major eruptions of Rabaul Volcano. In Johnson, R.W., ed., Cooke-Ravian Volume of Volcanological papers: Geol. Survey Papua New Guinea, Mem. 10, 181–193. 8

Walker, R.G., 1975. Generalized facies models for resedimented conglomerates of turbidite association. Geol. Soc. Amer. Bull. 86, 737–748. 10

Walker, R.G., ed., 1979. Facies models. Geoscience Can., Reprint Series 1, 1–211. 13

Waters, A.C. and Fisher, R.V., 1970. Maar volcanoes. In Proc. Second Columbia River Basalt Symposium, Cheney, Washington, Eastern Washington State College Press, 157–170. 5,9

Waters, A.C. and Fisher, R.V., 1971. Base surges and their deposits: Capelinhos and Taal volcanoes. J. Geophys. Res. 76, 5596–5614. 4,5,9

Watkins, N.D. and Huang, T.C., 1977. Tephras in abyssal sediment east of the North Island, New Zealand: Chronology, paleowind velocity, and paleoexplosivity. N.Z. J. Geol. Geophys. 20, 179–198. 7

Watkins, N.D., Sparks, R.S.J., Sigurdsson, H., Huang, T.C., Federman, A., Carey, S. and Ninkovich, D., 1978. Volume and extent of the Minoan tephra from Santorini: new evidence from deep-sea sediment cores. Nature 271, 122–126. 7

Watson, E.B., Sneeriger, M.A. and Ross, A., 1982. Diffusion of dissolved carbonate in magmas: experimental results and applications. Earth Planet. Sci. Lett. 61, 346–358. 3,4

Weaver, C.E., 1953. Mineralogy and petrology of some Ordovician K-bentonites and related limestones. Geol. Soc. Amer. Bull. 64, 921–943. 12

Weaver, C.F., 1963. Interpretative value of heavy minerals from bentonites. J. Sed. Petrol. 33, 343–349. 12

Wendtlandt, R.F. and Harrison, W.J., 1979. Rare earth partitioning between immiscible carbonate and silicate liquids and $CO_2$ vapor: results and implications for the formation of light rare earth-enriched rocks. Contr. Mineral. Petrol. 69, 409–419. 3

Wentworth, C.K., 1922. A scale of grade and class terms for clastic sediments. J. Geol. 30, 377–392. 5

Wentworth, C.K., 1926. Pyroclastic geology of Oahu. Bishop Museum Bull. 30, 1–121. 9

Wentworth, C.K., 1938. Ash formations of the island of Hawaii. 3rd Sp. Rpt., Hawaiian Volcano Observatory, Honolulu, Hawaii, 1–183. 4,5,6,9,12

Wentworth, C.K. and Macdonald, G.A., 1953. Structures and forms of basaltic rocks in Hawaii. U.S. Geol. Survey Bull. 994, 1–98. 4,5,6

Wentworth, C.M., Jr., 1967. Dish structure; a primary sedimentary structure in coarse turbidites. Amer. Assoc. Petrol. Geol. Bull. 51, 485. 7

Westgate, J.A. and Gold, C.M., 1974. World bibliography and index of Quaternary tephrochronology. Printing Services Dept., Univ. of Alberta (Canada), 1–528. 13

Westgate, J.A. and Gorton, M.P., 1981. Correlation techniques in tephra studies. In Self, S. and Sparks, R.S.J., eds., Tephra studies. D. Reidel Publ. Co., Dordrecht, Holland, 73–94. 13

Wheeler, H.E. and Mallory, V.S., 1953. Designation of stratigraphic units. Amer. Assoc. Petrol. Geol. Bull. 37, 2407–2421. 13

White, D.E., 1967. Some principles of geyser activity, mainly from Steamboat Springs, Nevada. Amer. J. Sci. 265, 641–684. 4

White, D.E. and Waring, G.A., 1963. Volcanic emanations. Data of Geochemistry. U. S. Geol. Survey Prof. Paper 440-K, 1–29. 3

Wickman, F.E., 1954. The "total" amount of sediment and the composition of the "average igneous rock." Geochim. Cosmochim. Acta 5, 97–110. 1

Wilcox, R.E., 1959. Some effects of recent volcanic ash falls with especial reference to Alaska. U. S. Geol. Survey Bull. 1028-N, 409–476. 6

Wilcox, R.E., 1965. Volcanic ash chronology. In Wright, H.E., Jr. and Frey, D.G., eds., Quaternary of the United States. Princeton Univ. Press, Princeton, N.J., 807–816. 13

Williams, C.E. and Curtis, R., 1964. The eruption of Lopevi Volcano, New Hebrides, July 1960. Bull. Volcanol. 27, 423–433. 8

Williams, H., 1941. Calderas and their origin. Univ. Calif. Publ. Geol. Sci. 25, 239–346. 7

Williams, H., 1942. The geology of Crater Lake National Park. Carnegie Inst. Wash. Publ. 540, 1–162. 8

Williams, H., 1952. The great eruption of Coseguina, Nicaragua in 1835. Univ. Calif. Publ. Geol. Sci. 29, 21–46. 6

Williams, H., 1957. Glowing avalanche deposits of the Sudbury Basin. Ontario Dept. Mines, 65th Ann. Rpt., 57–89. 8

Williams, H., 1960. Volcanic history of the Guatemalan Highlands. Univ. Calif. Publ. Geol. Sci. 38, 1–87. 8

Williams, H. and Goles, G., 1968. Volume of the Mazama ash-fall and the origin of Crater Lake Caldera. In Dole, H.M., ed., Andesite Conference Guidebook, Oregon State Dept. Geol. Miner. Indust. Bull. 62, 37–41. 1,6

Williams, H. and McBirney, A.R., 1969. An investigation of volcanic depressions Part II. Subaerial pyroclastic flows and their deposits. NASA Progress Rpt. (NGR-38-033-012), 41–92. 4

Williams, H. and McBirney, A., 1979. Volcanology. Freeman, Cooper and Co., San Francisco, 1–391. 1,3,4,8

Williams, H., Turner, F.J. and Gilbert, C.M., 1982. Petrography: An introduction to the study of rocks in thin section. 2nd ed. W.H. Freeman and Co., San Francisco, 1–626. 1

Williamson, I.A., 1970. Tonsteins – their nature, origins and use. Part I and II. Mining Mag. 122. I, 119–125, II, 203–211. 12

Wilson, C.J.N., 1980. The role of fluidization in the emplacement of pyroclastic flows: an experimental approach. J. Volcanol. Geotherm. Res., 8, 231–249. 8

Wilson, C.J.N. and Walker, G.P.L., 1981. Violence in pyroclastic flow eruptions. In Self, S. and Sparks, R.S.J., eds., Tephra studies. D. Reidel Publ. Co., Dordrecht, Holland, 441–448. 8

Wilson, C.J.N. and Walker, G.P.L., 1982. Ignimbrite depositional facies: the anatomy of a pyroclastic flow. J. Geol. Soc. London 139, 581–592. 8

Wilson, H.D.B., Morrice, M.G. and Ziehelke, H.V., 1974. Archean continents. Geoscience Can. 1, 12–20. 14

Wilson, L., 1972. Explosive volcanic eruptions – II. The atmospheric trajectories of pyroclasts. Geophys. J. Roy. Astron. Soc. 30, 381–392. 6

Wilson, L., 1976. Explosive volcanic eruptions – III. Plinian Eruption Columns. Geophys. J. Roy. Astron. Soc. 45, 543–556. 3,4,6,8

Wilson, L., 1980. Relationships between pressure, volatile content and ejecta velocity in three types of volcanic explosions. J. Volcanol. Geotherm. Res. 8, 297–313. 4

Wilson, L. and Head, J.W., III, 1981a. Ascent and eruption of basaltic magma on the earth and moon. J. Geophys. Res. 86, 2971–3001. 4

Wilson, L. and Head, J.W., III, 1981b. Morphology and rheology of pyroclastic flows and their deposits, and guidelines for future observations. In Lipman, P.W. and Mullineaux, D.R., eds., The 1980 eruptions of Mount St. Helens, Washington. U.S. Geol. Survey Prof. Paper 1250, 513–524. 3,8

Wilson, L. and Head, J.W., III, 1983. A comparison of volcanic eruption processes on Earth, Moon, Maars, Io and Venus. Nature 302, 663–669. 4

Wilson, L. and Huang, T.C., 1979. The influence of shape on the atmospheric settling velocity of volcanic ash particles. Earth Planet. Sci. Lett. 44, 311–324. 5,6

Wilson, L., Sparks, R.S.J. and Walker, G.P.L., 1980. Explosive volcanic eruptions, IV. The control of magma properties and conduit geometry on eruption column behavior. Geophys. J. Roy. Astron. Soc. 63, 117–148. 4

Wilson, L., Sparks, R.S.J., Huang, T.C. and Watkins, N.D., 1978. The control of volcanic column eruption heights by eruption energetics and dynamics. J. Geophys. Res. 83, 1829–1836. 4,7,8

Winkler, H.G.F., 1962. Viel Basalt und wenig Gabbro – wenig Rhyolith und viel Granit. Beitr. Min. Pet. 8, 222–231.  3

Winkler, H.G.F., 1979. Petrogenesis of metamorphic rocks. 5th ed. Springer-Verlag, Berlin, Heidelberg, New York, 1–348.  12

Winn, R.D., Jr., 1978. Upper Mesozoic flysch of Tierra del Fuego and South Georgia Island: A sedimentological approach to lithospheric plate restoration. Geol. Soc. Amer. Bull. 89, 533–547.  14

Winter, J., 1981. Exakte tephro-stratigraphische Korrelation mit morphologisch differenzierten Zirkonpopulationen (Grenzbereich Unter/Mitteldevon, Eifel-Ardennen). N. Jb. Geol. Pal. Abh. 162, 1–56.  12,13

Wohletz, K.H., 1980. Explosive hydromagmatic volcanism. Arizona State Univ., Ph.D. diss., 1–303.  9

Wohletz, K.H. and Sheridan, M.F., 1979. A model of pyroclastic surge. Geol. Soc. Amer. Sp. Paper 180, 177–194.  9

Wohletz, K.H. and Sheridan, M.F., 1983. Hydrovolcanic explosions II. Evolution of basaltic tuff rings and tuff cones. Amer. J. Sci. 283, 385–413.  9

Wolery, T.J. and Sleep, N.H., 1976. Hydrothermal circulation and geochemical flux at mid-ocean ridges. J. Geol. 84, 249–275.  12

Wolf, K.H. and Ellison, B., 1971. Sedimentary geology of the zeolitic volcanic lacustrine Pliocene Rome beds, Oregon, 1. Sed. Geol. 6, 271–302.  13

Wolf, T., 1878. Der Cotopaxi und seine letzte Eruption am 26. Juni 1877. N. Jb. Min. Geol. 113–167.  8,11

Wolff, J.A. and Wright, J.V., 1981. Rheomorphism of welded tuffs. J. Volcanol. Geotherm. Res. 10, 13–34.  8

Womer, M.B., Greely, R. and King, J.S., 1980. The geology of Split Butte – A maar of the south-central Snake River Plain, Idaho. Bull. Volcanol. 43–3, 453–471.  9

Wood, B.R. and Carmichael, I.S.E., 1973. $P_{total}$, $P_{H_2O}$ and the occurrence of cummingtonite in volcanic rocks. Contr. Mineral. Petrol. 40, 149–158.  3

Wörner, G. and Schmincke, H.-U., 1984. The Laacher See Volcano I: Mineralogical and chemical zonation of the Laacher See tephra sequence. J. Petrol. 25, 805–835.  2

Worzel, L.J., 1959. Extensive deep-sea sub-bottom reflections identified as white ash. U.S. Natl. Acad. Sci. Proc. 45, 349–355.  7

Wright, J.V., 1978. Remanent magnetism of poorly sorted deposits from the Minoan eruption of Santorini. Bull. Volcanol. 41, 1–5.  8,11

Wright, J.V., 1979. Formation, transport and deposition of ignimbrites and welded tuffs. Imperial College, London, Ph.D. diss., 1–451.  5,8

Wright, J.V., 1980. Stratigraphy and geology of the welded air fall tuffs of Pantelleria, Italy. Geol. Rundsch. 69, 263–291.  8

Wright, J.V. and Coward, M.P., 1977. Rootless vents in welded ash flow tuffs from northern Snowdonia, Northern Wales, indicating deposition in a shallow water environment. Geol. Mag. 114, 133–140.  10

Wright, J.V. and Mutti, E., 1981. The Dali Ash, Island of Rhodes, Greece: a problem in interpreting submarine volcanigenic sediments. Bull. Volcanol. 44-2, 153–167.  10

Wright, J.V. and Walker, G.P.L., 1977. The ignimbrite source problem: significance of a co-ignimbrite lag-fall deposit. Geology 5, 729–732.  8

Wright, J.V., Self, S. and Fisher, R.V., 1980a. Towards a facies model for ignimbrite-forming eruptions. In Self, S. and Sparks, R.S.J., eds., Tephra studies. D. Reidel Publ. Co., Dordrecht, Holland, 433–439.  8

Wright, J.V., Smith, A.L. and Self, S., 1980b. A working terminology of pyroclastic deposits. J. Volcanol. Geotherm. Res. 8, 315–336.  4,8

Wrucke, C.T., Churkin, M., Jr. and Heropoulos, C., 1978. Deep-sea origin of Ordovician pillow basalt and associated sedimentary rocks, northern Nevada. Geol. Soc. Amer. Bull. 89, 1272–1280.  14

Wyllie, P.J., 1971. The dynamic earth: Textbook in Geosciences. J. Wiley and Sons, New York, 1–416.  2

Wyllie, P.J., 1979. Magmas and volatile components. Amer. Mineral. 64, 469–500.  3,9

Yagi, K., 1966. Experimental study on pumice and obsidian. Bull. Volcanol. 29, 559–572.  8

Yamada, E., 1973. Subaqueous pumice flow deposits in the Onikobe Caldera, Miyagi Prefecture, Japan. J. Geol. Soc. Jap. 79, 585–597.  10

Yamasaki, N., 1911. The condition of the eruption of Mt. Asama in 1783. Rpt. Earthq. Inv. Comm. 73, 20–28 (in Japanese).  8

Yamazaki, T., Kato, I., Muroi, I. and Abe, M., 1973. Textural analysis and flow mechanism of the Donzurubo subaqueous pyroclastic flow deposits. Bull. Volcanol. 37, 231–244. 10

Yoder, H.S., Jr., 1976. Generation of basaltic magma. Natl. Acad. Sci. Washington, 1–264. 2

Yokoyama, I., 1956. Energetics in active volcanoes. 1st paper. Tokyo Univ. Bull. Earthq. Res. Inst. 34, 190–195. 4

Yokoyama, I., 1957a. Energetics in active volcanoes. 2nd paper. Tokyo Univ. Bull. Earthq. Res. Inst. 35, 75–97. 4,6

Yokoyama, I., 1957b. Energetics in active volcanoes. 3rd paper. Tokyo Univ. Bull. Earthq. Res. Inst. 35, 99–108. 4

Yokoyama, I., 1972. Heat and mass transfer through volcanoes. Revista Italiana di Geofisica 21, 165–169. 4

Yokoyama, S., 1974. Mode of movement and emplacement of Ito pyroclastic flow from Aira Caldera, Japan. Sci. Reports of the Tokyo Kyoiku Daigaku, Sec. C (Geography, Geology and Mineralogy) 12, 17–62. 8

Yokoyama, S. and Tokunaga, T., 1978. Base-surge deposits of Mukaiyama Volcano, Nii-jima, Izu Islands. Bull. Volcanol. Soc. Jap. 23, no. 4, 249–262 (in Japanese). 9

Zen, M.T. and Hadikusumo, D., 1965. The future danger of Mt. Kelut (eastern Java-Indonesia). Bull. Volcanol. 28, 275–282. 11

Zielinski, R.A., 1980. Stability of glass in the geologic environment: some evidence from studies of natural silicate glasses. Nucl. Technology 15, 197–200. 12

Zielinski, R.A., 1982. The mobility of uranium and other elements during alteration of rhyolitic ash to montmorillonite: a case study in the Troublesome Formation, Colorado. Chem. Geol. 35, 185–204. 12

Zielinski, R.A., Lindsey, D.A. and Rosholt, J.N., 1980. The distribution and mobility of uranium in glassy and zeolitized tuff, Keg Mountain area, Utah, U.S.A. Chem. Geol. 29, 139–162. 12

Zielinski, R.A., Lipman, P.W. and Millard, H.T. Jr., 1977. Minor element abundances in obsidian, perlite, and felsite in calc-alkalic rhyolites. Amer. Mineral. 62, 426–437. 12

Zies, E.G., 1924. The fumarolic incrustations in the Valley of Ten Thousand Smokes. Natl. Geog. Soc. Contributed Tech. Papers, Katmai Ser. 1, no. 3, 159–179. 8

Zies, E.G., 1929. The Valley of Ten Thousand Smokes; I, The fumarolic incrustations and their bearing on ore deposition; II, The acid gases contributed to the sea during volcanic activity. Natl. Geog. Soc. Contributed Tech. Papers, Katmai Ser. 1, no. 4, 1–79. 8

Zingg, Th., 1935. Beitrag zur Schotteranalyse. Schweiz. Min. Petr. Mitt. 15, 39–140. 5

Zuniga, C.C., 1956. Erupcion del Volcan Nilahue. Inst. Geol. Univ. Chile, Publ. 7, 73–96. 9

# Subject Index

Asterisks (★) refer to figures or tables
Numbers in *italics* refer to definitions or main discussions

Abrasion index  105, 106★
Accidental clasts  *239*
  ejection velocity  241
  maximum size  239, 241★
  shape  239
Accretionary lapilli  91, 93★, 94★, 124, 127, 143★, 154, 168, 238, 239, 242, 263, 370
Adularia (see feldspar)
Aenigmatite  22
Agglomerate  90–92, 400
Agglutinate  19, 91★, *96*, 124, 131★, 229
Albite (see feldspar)
Alignment bedding  107★, 114, 115, 116★
Alkali basalt  17–*19*, 22, 39–44, 262, 316★, 339
  volatile composition  39
Alkali zeolites  320, 341, 342
Alkali-silica variation diagrams  16★
Alkaline saline lake deposits  342, 343★
Allanite  29, 48, 338, 355★
Alloclastic  89
Amphibole  18, 19, 22, 32★, 43, 48, 105, 159, 197, 388★, 398★
Analcime  319★, *320*, 331, 333★, 335★, 341★, 342★
Andesite  17★, 20, 21, 42, 43
  crystallization temperature  43
  explosive behavior  20
  volcanoes  15, 21, 42
Andesitic arc magmatism  399
Anorthoclase (see feldspar)
Apatite  19, 22, 29, 157, 338, 355★
Aquagene tuff  233
Arc type basalt  340
Arc-trench gap  389
Archean  5, 7★, 267, 317, 357
Archean greenstone-belt  367, *378–382*, 408, 409
Armored lapilli  91, 94★, *239*
Ascent rate of magma  69, 71★, 74, 241, 242
Aseismic ridge  274, 275
Ash  90★, *91*
  andesitic  98★
  basaltic  98, 99★
  crystal  91
  dacitic  98★
  lithic  91
  rhyolitic  98, 99★
  vitric  91

Ash bed  337
Ash flow sheet  98, 375
Ash flow tuff (see pyroclastic flow deposits)
Ash L  170, 171★
Ash layers
  stratigraphic marker horizons  125
Ash layers, methods of dating  178
  $^{40}Ar/^{39}Ar$  178
  $^{14}C$  178
  distinctive fossils or faunal assemblages  178
  fission-track  178
  K/Ar  178
  paleomagnetic methods  178
  superhydration  178
  U/Th  178
Augite  19, 158★, 387, 388★
Authigenic feldspar  334, 335
Autoclastic  89
Autolith  237
Avalanche  227★

Back arc basin  385★, 391–393
Ballistic fragment  *128*, 152★
Ballistic trajectory  125
Basalt  17★–20
Basaltic glass, alteration (see palagonitization)
Basaltic hornblende  388★
Basaltic intraplate volcano  18★
Basaltic volcanoes  262★
Basanite  17★, 19, 20, 43
Base surge  247–257
Base surge deposits (see also hydroclastic deposits)  116★, 240★, 247, 249–257
  antidunes  107★, 250–252
  bed forms  249–255
  bedding sags  107★, 246★, 247★
  chute and pool structures  252–254
  convolute bedding  107★
  dunes  250–253
  facies  254–256
  flow regime and bed forms  250, 251
  imbrication of platy fragments  254
  layers  243★
  load casts  107★
  massive bed form  249, 254
  maximum radial distance  249★

449

Base surge deposits, mudcracks  107★
   planar bed form  243★, 249, 254
   phonolitic  257
   rills  107★
   ripples  250
   sandwave bed form  250
   size parameters  252★
   soft sediment deformation  *242*, 244★, 245
   structures  107★
   U-shaped channels  256, 257
   vesiculated tuff  242–245
Batholith evolution  375★
Bed  *107, 108*, 350
   thickness nomenclature  108★
Bed forms, experimental  112★
Bedding
   coset  108★, 111★
   set  108★, 111
   subset  108★
Benioff Zone  13★, 14, 20
Bentonite  168, 182, 312, *336–340*, 347
   formation  338
   illite(i)-bentonite  338
   kaolinite(k)-bentonite  338, 340
   smectite(s)-bentonite  338
Bimodal magmatic suite  374
Bingham plastic  *52*, 54
Biostratigraphic stages, California  366★
Biotite  21, 22, 29, 98★, 105, 197, 338, 355★, 397, 398★
Block  91, 95, 152
Block-and-ash flow  *189*, 190★, 193★, 211, 223, 226★
Bomb  90★–96, 237, 238, 243★
   armored  92
   breadcrust  92
   cauliflower  92
   cored  92, 95★
   cow-dung  92
   mud bomb  124
   phreatomagmatic  240★
   ribbon  92
   spindle  92
Breccia, pyroclastic  91, 92
Bubbles  55–58
   coalescence  55, 56
   disruption  55, 57
   growth  55, 56
   nucleation  55
   residual pressure  56
   surface tension  55
Burial diagenesis  340–345
Burial metamorphism  317, 340–345

Calc-alkalic arc terrane  397
Calc-alkalic suite  21
Calcite  341, 344★, 345, 388★
Caldera  15, 98★, 228★, 375
   area versus ash-flow eruption volume  23★

   cluster  372★
   collapse  *230*, 348, 360★
   diameter  63
   facies  359–361
   ring fracture  230
   size  23★
   subaqueous  293
Carbonate  18, 267, 293, 317, 320, 325, 343★, 385, 392
Carbonatite  44, 262
Carbonatite tuff  130
Celadonite subfacies  367
Cenozoic volcanic episodes  184, 185
Chabazite  319★, 320, 331★, 333★, 335
Chemical composition of tephra  6, 7, 24★, 25★
Chert  387, 388★, 392, 393★, 395
Chevkinite  355★
Chlorine  48
Chlorite  317, 318★, 338–344
   Al-rich  343
   Si-rich  343
Chute and pool, flume experiment  113★
Cinder cone  14, 19, 96, 260, 261, 262★
   stages of development  150★
Cinders  92, 96
Circum-Pacific sedimentary basins  392
Clastolava  228
Climate, volcanic control  182
Clinoptilolite  319, 320, 331, 333–335, 341–343
   biogenic siliceous skeletons  335
   dissolution of volcanic glass  335
   distribution in Tertiary-Mesozoic sediment  335★
   facies  367
Clinopyroxene  19, 21, 29, 34, 94★, 130★, 333★, 355★, 387, 388★
$CO_2$  40, *43–47*
   abundance in peridotite  44
   diffusion in basaltic magmas  50
   explosions  241
   solubility in magmas  44–47
   solubility mechanism  46, 47, 53
Co-ignimbrite airfall tephra  157
Co-ignimbrite ashfall  139
Co-ignimbrite lag-fall deposit  222
Composite lapilli  149★, 236★, *237*
Compositional gap  26
Conduit
   circular  70, 73★
   fissure  69
   magma rise speed  71–74
   mass eruption rate  73
   overpressure  71
   radius  72★
Conglomerate  91★, 380, 382
Continental margin  384, 385
   alluvial piedmont apron  385
   laharic piedmont apron  385
   subaerial stratovolcano  385

Convergent margin  11–14, *383–392*
Cordilleran geosyncline  392–396
    Cenozoic tectonism and volcanism, North America  398–401
    Paleozoic, Western North America  392–396
    Upper Mesozoic, South America  396–398
Correlation of ash layers, marine  176–179
    oxygen-isotope stratigraphy  178
    seismic reflection  178★
    tephrochronology  176, 352–356
Cristobalite  338
Cross bedding  107★, *110–113*
    backset beds  110
    coset  110, 111★
    foreset beds  110
    geometry  110
    in base surge deposits  250–254, 255★
    in pyroclastic surge deposits  111, 200, 201
    set  110, 111★
Crustal rock types  266★
Cummingtonite  355★
Cumulative grain size curves  120★, 122★
Curie point  124

Dacite  17★, *22*, 95
Debris flow (see also lahar)  227★, 270, 276★, 297, *298, 299*, 304, 306★, 308, 380, 381, 391★, 402, 406
    clast orientation  308
    continuous phase  298, 299
    dispersed phase  299
    matrix strength  308
    mudflow  298, 403★
    plug flow  298, *308*
    reverse grading  307, 308
    yield strength  298, 308
Deep sea fan  385, 386★, 396, 397★, 398
Deep sea pelagic deposits  392
Deep sea tephra (see fallout tephra, marine)
Delayed reworking  90
Density current  112, 115
Diagenesis  312–314
Diagenetic alteration  182
Diamictite  398
Diatreme  20, 261
Diffusion
    diffusion rate of $CO_2$  56
    diffusion rate of $H_2O$  56
    diffusional fractionation  56
    heat diffusion  54
    mass diffusion  54
Diopsidic clinopyroxene  21
Directed blast  125, 223★, 227
Distribution curves  118, 120★
Divergent margin  11–13
Dolomite  18
Dome collapse  8★, 95★, 190★, 192, 220★, 222, 223★, 227★
Domes  9, 80★, 98★, 99★

Drift pumice  *173*, 405

East Pacific Ridge  12, 13
Ejecta velocity  82, 83, 84★
    drag force  83
    maximum ejecta velocity  83★
    shock wave  83
Elutriation of particles  208
Energy of volcanic eruptions (see also eruption) 85–87
Eolian fractionation  156–162
Epiclastic  *89*, 91★
Epidote  343★, 344★, 388★, 389★
Erionite  331, 332, 333★, 335
Eruption  *347*
    destructive potential  87
    dispersive power  87
    hydroclastic  99★
    intensity  87
    magmatic  98★
    magnitude  87
    rheological regimes  65★
    shallow level  224
    size  87
    violence  87, 105
Eruption column  8★, 61–68, 125–127★, 188★, 189★, 223–225, 230
    beneath water  *83, 84*, 292★
    collapse  8★, 47, 67, 68★, 200, 204★, 221★, 223–225, 230
    convective thrust  *61, 62*, 125
    density  62
    gas-thrust  *61, 62*, 125
    heat exchange  62
    height and discharge rate  63★, 175
    initial velocity  62
    maximum height  67
    size and density sorting of particles  125
Eruption unit  348★, *349*, 371
Eruption velocity  65–68
Eruptive volcanic activity  59, *347–349*
    eruption  59, 347, 348★
    eruptive epoch  348★
    eruptive period  348★
    eruptive phase  59, 347, 348★
    eruptive pulse  348★
Eugeosynclinal assemblage  392, *393, 394*
Eugeosyncline  383, 392, 393
Evolution of sedimentary rocks  6★
Eutaxite  187
Explosive eruptions  51, 55, 61–68, 82–84
    initial phase  51
    viscosity of magma  51–56
Explosive volcanic activity
    episodicity  181
    global volcanic episodes  182
    influence on climate  181, 182
    rates  182
Explosivity index  88★

Fabric
  anisotropic  124
  apposition  123
  deformation  124
  dimensional  124
  isotropic  124
  magnetic  124
Fall unit  349
Fallout ejecta
  area of dispersal  60
  axis of dispersal  60
  cumulative curves  120
  degree of fragmentation  60
  maximum thickness  60
Fallout tephra, lacustrine  163, 166, *168*
Fallout tephra, subaerial  8 *, 15, 125
  accelerated erosion  147
  accretionary lapilli  127
  andesitic  133
  antitrade winds  406 *, 407
  area of tephra sheets  133–138, 140, 142
  areal distribution  132
  asymmetrical sheet  135
  axes of fans  133 *, 135–138
  basaltic  25 *, 129–131, 137
  bedding  141–147, 370
  bedding sag  152 *
  characteristics  126 *
  circular pattern  132, 137
  components  128–130
  composition  126 *
  crystals  128, 129 *, 153 *, 154 *
  curved fan axis  135, 136 *
  dacitic  133
  density of pumice  155
  displaced maxima  133–136
  distance-thickness graphs  139 *, 141 *
  distribution  126 *, 132–139, 155 *
  elliptical pattern  132, 138 *
  eolian fractionation  156–162
  eolian differentiation  157
  eruption column  127 *
  eruption diagram  144 *
  fabric  148
  facies  126 *, 358, 359
  fall velocity  128
  fans  132–137, 138 *, 160 *
  fan-shaped pattern  132, 152 *, 160 *
  fan-shaped sheet  135
  frequency size distribution  118, 120 *, 153 *, 154 *
  geometry of sheets  132–139
  glass shards  96, 101 *, 102 *, 128
  gradational bedding  145 *
  graded bedding  109 *, 148, 149 *
  heavy minerals versus distance  160 *
  isopach maps  135–138 *, 152 *, 160 *, 370 *
  isograde maps  151
  lateral variation of oxides  157*, 160, 161*, 162*
  lithic fragments  128, 129 *
  major element zonation  32 *
  mantle bedding  144–148
  maximum diameters  150–152 *
  median diameter  151–153, 158 *
  median diameter-sorting diagrams  121 *
  median diameter versus distance  155 *
  mineralogical zonation  32 *
  multiple maxima  133, 135 *
  near source deposits  141–143 *
  phonolitic  25 *, 117, 132 *
  phreatoplinian  61 *, 130
  Plinian and sub-Plinian pumice deposits  61 *, *130*, 132 *, 137, 143 *, 145 *, 148, 149 *, 201 *, 204 *, 225, 230
  premature fallout of very small particles  *127, 128*, 154
  pumice  93 *, 104 *, 128, 129 *, 131 *, 132 *, 143 *, 145, 155, 156
  rain flushing  *127*, 148
  reverse (inverse) grading  149 *
  rock associations  126 *
  sheets  135
  silicic ash  354 *, 355 *
  size parameters  118 *, 120 *, 150
  sorting  126 *, 127, *153–155*, 158 *
  sorting coefficient  153 *, 154 *
  structures  107 *, 126 *, 141–148
  ternary diagrams  129 *, 130
  textures  126 *, 150–155
  thickness  132–139
  time versus thickness curve  371 *
  topographic smoothing  147
  trachytic  25 *, 64, 131, 143 *
  units of different origins  136
  volume  *139–141*, 142
  widespread tephra layers  134 *
Fallout tephra, submarine  163–185
  ash layer reflections  178 *
  ash provinces  176 *, 178, 181 *
  attrition by floating pumice  10 *, 167, 173
  bioturbation  166–170
  characteristics  164 *
  chemical composition  163–166
  color  164, 165 *
  composition  164 *
  cores  183
  correlation  177–179
  cumulative size distribution  167
  dating  177–179
  deep sea  172
  depth and composition  165 *
  dish structures  167, 168
  distribution  164 *, *168–173*, 176 *, 177, 182 *
  downwind grain size patterns  173
  duration of large magnitude explosive eruptions  175
  facies  164 *
  frequency per million years  184 *, 185 *

grading  164★, *166–168*, 175
grain-size distribution  167★
isopach maps  169★, 171★, 172★, 407★
lobate sheets  15
maximum grain size and distance  174★
median grain size and distance  174★
ocean currents  174★
peralkalic  184★
post-depositional compaction  169, 170
pumice rafts  10★, 173, 174
redistribution by marine currents  170, 171★
rock association  164★
settling velocity  170★
size characteristics  173
sorting  173
source  176, 177
spreading rate and ash frequency  185★
structures  164★–168
superhydration curves  178, 180★
textures  164★
thickness  164★, 168–173
trace elements  179★
vertical distribution of shards  166
water-escape structures  167
Fallout velocity  128
Fayalite  22, 355★
Feldspar  104, 158★, 159
  adularia  343★, 344★, 345
  albite  340–345
  alkali feldspar  197
  andesine  355★
  anorthoclase  22, 197
  oligoclase  355★
  plagioclase  17–19, 21, 29, 32★, 42, 98★, 105, 157, 158★, 159, 197, 355★, 356, 387, 388, 397, 398★
  potassium feldspar  331, 333★, 387, 388, 397, 398★
  sanidine  22, 29, 32★, 98★, 105, 197, 355★, 356
  variation with distance  159
Feldspathoids  22
Fe-Mn oxides  317
Fe-montmorillonite  336
Fe-Ti oxides  197, 317
Fe-Ti oxide thermometry  30★, 162
Ferroaugite  22
Ferromagnesian minerals  98★, 158★
Ferromagnetic minerals  356
Fiamme  187★, 209, 215
Fibrous palagonite (see palagonite)
Fines-depleted ignimbrite  222
Fire fountain (see lava fountain)
Fissures  73
Flood basalts  14, 15
Flow regime  111–113, 250
  antidune  107★, 111–113
  bedforms  112
  chute-and-pool  113
  dunes  111★, 112

flow power  111★
high flow regime  112, 113
hydraulic jump  113
low flow regime  111★, 112
plane beds  111★, 112
ripples  111★, 112
transition  111★, 112
Fluid inclusions  18, 47
Fluidization  226
Fluvial deposits  309
Fluvial-deltaic sedimentation  389, 390★
Flyschlike graywacke  396
Foam lava  228
Fore-arc basin  391, 393
  andesitic  391
  epiclastic volcaniclastic materials  391
  graywacke  391
  reworked land-derived pyroclastics  391
  subaqueous fallout  391
  subaqueous pyroclastic flows  391
Fractionation
  crystal  27, 48
Fragmentation level  66★, 71★
Frequency curves  120★
Froth flow  228
Fuel-coolant interaction  80, 81
Fumaroles, in pyroclastic flow deposits  200, 202
Fused tuff  215

Garnet  388★
Garnet peridotite  17
Gas
  adiabatic expansion  75
  vesicles in submarine lava  36
  volume fraction  71
Gas bubbles  73
Gas escape structures  200–203
  fumaroles  200, 202
  fumarolic mounds  202
  gas pipes  203★
  vapor phase activity  202
Gel palagonite  *317*, 323
Geosyncline, reconstruction  394–401★
Geyser  79, 80
Glass  2, 75, 76
  alteration  312–345
  $CO_2$-concentration  44, 45
  crusts  105, 106
  hydration  312
  inclusions in phenocrysts  42
  interaction with sea water  322
  leaching of  327–329
  mobility of U  312
  rinds  105
  shards  *96*, 159
  surface pits  105
  transformation temperature  76
  transition in basaltic melts  76

Glauconite   336
Globule lava   228
Glowing avalanche   187, 225
Glowing cloud   187, 225
Gmelinite   335
Grade size terminology   119★
Graded bedding   107★, 109, 110, *307, 308*
  density grading   109★, 110
  multiple grading   109★, 110
  normal grading   109★, 110
  reverse grading   109★, 110, 148, 149★, 307, 308
  symmetric grading   109★, 110
Granodiorite   50
Granulation   77
  brittle failure   77
  decrepitation   77
Graptolite shales   392, 395, 396
Gravity segregation   219, 226★
Graywacke   7, 391, 393, 396
  volcaniclastic   5, 7
Ground layer   222
Groundwater
  behavior during intrusion of magma   78, 82, 83, 311
Group   352
Guyot   14, 394★
Gypsum   317

Harmotome   319★, 335
Hauyne   22
Hawaiian eruptions   69, 70★, 99★, 106
  fissures   69
  lava flows   69
  lava fountains   69, 73, 105★
  Pele's Hair   69, 101
  Pele's Tears   69, 102★
  pumice cones   69
  shield volcanoes   69
  spatter cones   69
Hawaiite   17★
HCl   48
Heat
  capacity   76
  transfer of   79–81
Heat flow   85
  global heat flow   85
  hot spring activity   85
Hedenbergite   22
Heulandite   *319*, 333, 340, 341★, 342★, 343★, 344★
HF   48
Hisingerite   325
Histograms   156★
$H_2O$
  boiling curve   78
  boiling point   79
  critical point   78, 79★
  density   79★

diffusion rate in magma   50
influx from wall rocks   50
mole fraction in magma   35
network-modifier   53
rate of solubility   49
saturation values   49★
transport by convection   50
vaporization   79
Hornblende   21, 157, 158★, 355★, 388★
Hyaloclastite   *233*, 268–272, 277★, 317, 336
  silicic   279–281
  spalling and granulation of pillow rinds   233, 270
  spalling shards   *270*, 272★
  subglacial   279★, 280★
Hyalotuff   234
Hydration   312, 327–330, 356
Hydraulic jump   113
Hydroclastic deposits (see also phreatomagmatic eruptions)   9★, 10, 99★, 231–245, 232★
  basanite bomb   238★
  bedding   240★, 242, 243★, 246★, 277★
  bedding sags (bomb sags)   245, 246
  blocky shards   99★, 234★, 235★, 236★, 272★
  breadcrust surface of bombs   238
  cauliflower surface of bombs   237, 238
  characteristics of essential components   233–238
  clasts   97★, 99★, 232★
  composite lapilli   236★, 237
  composition   232★
  convolute lamination   242
  decollement folds   242
  fragments   89
  fumarolic alteration   232★
  geometry   232★
  grain size distribution   232★, 234, 235
  granulation   233, 236, 270
  lithics   232★, 233, 239
  maximum fragment sizes   239–241
  mosaic cracks in glass shards   235★
  mudcracks   232★, 246
  origin   231
  platy fragments   242
  rock association   232★
  seamounts   237, 265, 268, 270★, *273, 274*, 276, 277★
  shards   97★, 234★, 235–237, 270, 272★, 314★, 318★
  shear ripples   242
  sideromelane grains   97, 99, 234★, 235, 236, 314★, 318★, 321
  sintering   232
  soft sediment deformation structures   242
  spalling of pillow rind   13, 236, 267–270
  structure   232★, 242
  tuff cone   99★, *258*, 259★, 260★
  tuff ring   99★, 236, 258, 259★, 260–262
  vesicles   242–245

Hydroclastic eruptions  9*, 10, 62*, *74*, 78, 99*, 219, *231*, 232*, 247, 248*
  granulation  77
  hydroexplosions  231
  interaction of magma and external water  74, 75, 231
  physical properties of essential clasts  78
  systems  61
Hydrogen  48
Hydrothermal explosions  233
Hydrothermal ore deposits  323
Hydrothermal systems  85
  overpressured hydrothermal systems  80
  sealed hydrothermal systems  80
Hyperconcentrated stream  298
Hypersthene  21, 22, 158*

Ignimbrite (see pyroclastic flow deposits)  192
Ignimbrite magma  228
Ignimbrite sheets
  facies  203, 204
  facies model  204*
  standard ignimbrite flow unit  204–206
  volume calculation  378*
Ignimbrite veneer deposit  *194*, 222
Ignimbrite vent  230
Ignispumite  228
Illite  5, 325, 337, 338, 342*, 343, 344*, 345
Illite(i)-bentonite  338
Illite-montmorillonite, mixed layer  325
Illmenite  20, 29, 162, 355*
Inclusions, ultramafic (see nodules)
Inman sorting coefficient  118*, 120, 153, 154*, 206*, 291*
Interarc basin  384, 392
Intra-arc basin  384, 385*
Intraplate volcanism  14
Island arc  11, 13, 281, 383–387, 400–407, 409

Jet stream  62

Kaersutite amphibole  19
Kaolinite  325, 336–338
Kaolinite(k)-bentonite  337–340
Keratophyre  387
Kimberlite  18, *20*, 44, 92, 237, 241
Komatiite  408
Krakatoan eruptions  59

Lacustrine tephra
  graded bedding  168
Lahar  8, 9*, 99*, 124, 211, *297–311*
  apron  188*
  areas of extensive accumulation  297*
  basal contact  302, 303
  breccia  115*, 305*
  cold rock avalanche  211
  comparison with non-volcanic debris flows  306
  components  303
  dacitic  305*
  dimensions  301*
  distribution  299
  emplacement temperature  302
  fabric  308
  fans  300*
  grading  307, 308
  grain size distribution  303–305
  grain size parameters  304*, 305*
  imbrication  308
  landslide-debris flow (rockslide-avalanche)  302
  lateral change to pyroclastic flow deposits  188*
  non-Newtonian fluid  298
  oriented fragments  308
  origin  311
  reverse grading  307, 308
  stratovolcanoes, association with  297
  surface  301, 302
  tephritic composition  300
  thermoremanent magnetism  303, 309
  thickness  299, 301*
  veneers  299
  vesicles  306
Lapilli  90*, 130*, 131*
Lapillistone  90*, 91
Lapilli-tuff  91
Lapillus  90*
Latite  17*
Laumontite  313, 319*, 335, 340–345
Lava fountains  69, 73, 75, 99*, 105*, 106
  submarine  270
Lava lake  98
Leucite  20, 22
Leucitite  19, 20
Leucoxene  322
Lithic fragments, sources  197
  conduit walls  197
Lithospheric plates  11–14
Littoral cone  263, 264, 275
  accretionary lapilli  263
  bedding in  263, 264
  deposits  263
  geometry  263
  lava tubes  264
  origin  264
Littoral explosions  233
Low aspect ratio ignimbrite  222, 227

Maar  99*, *257*, 260*
  areal extent  260
  composition of juvenile ejecta  262
  crater size versus spread of surge deposit  249*
  dimensions  260
  geometry  260
  list of 20th century eruptions  260*

455

Maar, origin   44, *258–260*
   volume of deposits   261
   volume versus diameter   262
Magma   16–22
   alkalic basaltic   17
   andesitic   20
   anhydrous   53★
   basaltic   16, 17
   basanite   ·17
   cooling rate   75
   crystallization temperature   75
   derivative   16
   differentiated   16, *21, 22*
   hierarchy of volumes   23
   mixing   33
   mixing with water   80, 81
   olivine nephelinite   17
   origin of   16
   parental   16
   phonolite   33
   primary   17
   production rate   13★, 86
   quenched   77
   rise speed (velocity)   69–74
   rhyolitic   21
   supercooled liquid   75
   tholeiitic   17
   undersaturated   18
   volume increase during melting   69
Magma chamber   *22*, 26★
   concentration gradient   31★
   mineralogical and chemical zonation   22–27
   near-surface silicic   22
   size   22
   thermal gradient   26
   volume   22, 27
   zoned   26★
Magma column   64
   exsolution surface   64
   fragmentation level   65
   fragmentation surface   64
Magma fracturing   69
Magmatic arc   383, 384, 386, 398
   rock assemblage   384, 385★
Magnesite   18
Magnetite   29, 159, 162, 355★
Mantle xenoliths (see nodules, ultramafic)
Marginal sea   392
Marine current   406
Marine sediments
   clay minerals   336
   pore fluids   336
Mass eruption rate   68★, 69, 73
Mass flows   10, 305
Massive bed   107★, *114*
Maximum pyroclast size   117
Median diameter-distance plot   155★
Median diameter-sorting diagram   121★
Melilite nephelinite   261, 262

Melting spots   14
Member   369
   bed   369
   eruptive cycle   369, 371★
Metamorphism   312, 340
Metasomatizing mantle fluids   51
Mica-montmorillonite, mixed layer   325
Mid-Atlantic Ridge   12, 13
Mid-Oceanic Ridge   12–17
   basalts (MORB)   *18*, 39, 316★
Miogeosyncline   392
Mn
   manganese crusts, nodules   322
Montmorillonite   5, 317, 326, 330, 334★, 336, 341, 342★
Montmorillonite subfacies   367
Montmorillonite-beidellite   338
Moon   101
Mordenite   319★, 333★, 335, 341, 342★, 343★
Mud flow   227★
Mud rain   245
Mudstone   91★, 397, 398
Mugearite   17★
Muscovite   398★

Natrolite   319★, 320, 335
Near-source facies   356, 357★, *358, 359*
Nepheline   20, 22
Nephelinite   17★, 18–20, 262, 316★
Nephelinite tuff   320
Noble gases   48
Nodules, ultramafic   17, 239, *241*
Nontronitic clays   317
Nosean   22
Nucleation rate of bubbles   55, 56
Nuée ardente   8★, 122★, 186, *187*, 188★, 189, 202, 221★, 224, 295
   boiling-over   189
   column collapse   189
   dome collapse   189
   lateral projection   189
   origin   189
Nuée ardente deposits (see pyroclastic flow deposits)

Obsidian   325, *327–329*
   hydration rate dating   329
Ocean margin basin   392
Oceanic crust   13, 396★
Oceanic formations   352
Oceanic island arcs   400
   volcaniclastic facies   400–405
Oceanic islands   14, 20, 33, 42, 265, 274
Olivine   19, 20, 32★, 34, 94★, 105, 388★
Olivine nephelinite   17
Opal   315★, 320
Opal-A   334, 341
Opal-CT   341, 342★
Opaque oxides   158★, 159

Ophiolite 398
Orientation bedding 107*, 115
Orthopyroxene 19, 29, 355*, 388*
Oxygen isotopes 20, 178, 356

Palagonite 314–327, 330*
  Archean hyaloclastites 317
  chemical changes 318*, 320–323
  crystallinity 317
  density 316*
  fibrous palagonite 314, 317, 323
  formation 323
  fumarolic palagonitization 315
  gel-palagonite 317, 323
  growth rate 325*
  immobile product layer 317
  microtubules 317
  mineralogical changes 317
  oxide gains and losses 322*
  physical properties 315–317
  potassium uptake 322, 324
  refractive index 316*, 317*
  sea floor weathering 323
  texture 315, 317
  tuff 315, 316*, 324*
  water 321, 325
Palagonitization 321–327
  Al-solubility 325
  diagenesis 323
  diffusion 327
  hydrothermal 323
  low temperature 326
  marine 317
  pore solution pH 325
  rates of 323, 324*, 326, 327
  redistribution of elements 321, 322*
  REE behavior 323, 325
  thermal control 324*, 325*
  trace element mobility 322
  water/rock ratio 325, 326
  weathering 323
Paleogeographic reconstructions 383
Paleomagnetism 356
Paleotectonic reconstructions 383
Paleovolcano model 379
Partial melting 17, 18
Particle
  roundness 122
  shape 122
  size 116–118
Pelagic limestones 392, 399
Peléean 59, 220*
Pele's Hair 69, 98*, 101
Pele's Tears 69, 98*, 101, 102*
Penecontemporaneous deformation 115, *242*
Penecontemporaneous volcanism 89
Pépérite *264*, 400, 404*
Peralkalic rhyolitic ashes 185
Peridotite 18
  garnet 17
  plagioclase 17
  spinel 17
Perlite *327*, 328
Phillipsite 317, 319, 320, 330*, 331, 333*, 334, 335*
  distribution in Tertiary-Mesozoic sediments 334
  in deep sea sediments 334
Phlogopite 18–20
Phonolite 15, 16*, 17*, 21, 22, 25, 30–33, 67, 100*, 104*, 117, 132*
  alkalic 21
  peralkalic 21, 22
Phosphate 339
Phreatic eruptions 82, *233*
Phreatomagmatic eruptions (see also hydroclastic deposits) 42, 64, 80*, 82, *233*
  base surges 83
  eruption columns 83
  hydraulic fracturing 82
  thermal expansion of water 82
Phreatoplinian 61, 130, 232
Pillow breccia 267–270, 271*, 272*, 273*
  debris flow 270
  in situ breccia 267
  isolated-pillow breccia 270
  mini pillows 270
  pillow-fragment breccia 267*, 268–270, 269*
  talus breccia 270
Pillow lava 267, 268, 270, 385
  glassy lava rinds 268, 271*
  vesicularity 39–41
Pillow volcano 13, 270
Piperno 229
Plagioclase (see feldspar)
Playa-lake complex 331
Plate, lithospheric 11–14
  constructive margin 11
  continental 14
  convergent margin 11, 13, 383–392
  destructive margin 11
  divergent margin 11
Plinian eruption 60–68, 99*, 106, 230, 380*
  control of eruptive processes 63
  energy 63
  eruption velocity 65, 66*, 68*
  gas content of magma 64
  mass discharge rate 63
  pyroclastic surge 8, 63, 219
  vent radius 63
  volume of fallout 63
  volume of magma erupted 63
  volume of pyroclastic flows 63
Plinian eruption column 63*, 67, 204*, 221*, 225, 230
  collapse 67, 204*
  column height 63*, 67
  gas velocity 67
Plinian fallout deposits (see also fallout tephra, subaerial) 64*, 117*, 201*, 230

457

Pore pressure 78
Potassium feldspar (see feldspar)
Pozzuolan (see puzzolan)
Pre-Cambrian 378, 379, 408, 409
Pre-Tertiary submarine tephra 168
Prehnite 341, 342*, 343*, 344*, 396
Pumice 92, 96–105, 117*, 131*, 132*, 143*, 146*, 167, 174*, 191*, 201*, 207
  basaltic 105
  bubble-wall shards 96, 97, 103
  fibrous fragments 103, 104*
  flow 191*, 221*
  lapillus 93, 104*, 117*, 130, 131*, 215*
  mafic 103
  mechanical strength 207
  morphology 96
  pulled apart 216*, 217
  reticulite 103
  scoria 103
  silicic 103
  spherical vesicles 103, *197*
  streched and imbricated 216*, 217
  swarms 115
  trains 115
  tubular vesicles 103, *197*
Pumpellyite 313, 341, 342*, 343*, 344*, 388*, 397
Puzzolan (see also Trass) 211
Pyrite 339, 343*
Pyroclastic avalanche 224
Pyroclastic bed (stratum) 107, 108
  bedding set 108
  co-set 108
  structures 107
  subset 108*
Pyroclastic breccia 90*, 91, 92*
Pyroclastic cone, development 379
Pyroclastic debris 89
  remobilization 406
  reworking 147
Pyroclastic/effusive ratio 4
Pyroclastic flow 15, 107*, 186, 192, 227*
  ash cloud 225, 226*
  ash-flow eruption volume versus caldera area 23
  basal avalanche 225
  block-and-ash flow 226*
  boiling over 8, 222, 223
  classification 8, 218
  collapse of a growing dome 222
  components 197, 198
  dense welding 196
  emplacement temperature 210, 211
  energy-line 227
  engulfed air 226
  exsolution of gas 226
  gravity segregation 226*
  laminar movement 200, 219
  mobility 225
  nomenclature 218–222
  origin 222–225
  particulate flow systems 15
  plateaus 15
  release of gases from particle breakage 226
  small-volume flow 192, 211
  topographic overtopping 227, 228*, 229*
  transport 225
  underflow 225
  velocity 225
Pyroclastic flow deposits 8, 9*, 104*, 188–226, 309
  altitude-velocity curves 229
  ash flow deposits 219
  ash size component 197
  ash size range 197
  basal layers 207
  block 198*
  block-and-ash flow deposits 190*, 193*, 205*, 219, 226*
  blocking temperature 218
  "boudinaged" feldspar tablets 197
  calc-alkalic 27, 29, 211
  chemical composition 209
  classification 220, 221*
  columnar jointing 214*, 215
  comenditic ignimbrite 210, 215, 216
  compaction 213
  compositional zonation 28*, *29, 30, 209*
  compound cooling unit 195, 196*
  concentration of pumice clasts at the top 208
  cooling unit 195, 196
  crumpled mica flakes 197
  crystal abundance 197
  crystal enrichment 209
  crystals 197, 206
  Curie point 218
  deflation of a pyroclastic flow 215
  directional structures 217
  distal deposits 201
  distribution curves 118
  enrichment in crystals 32*, 157
  enrichment, trace elements 31
  eruption unit 204*, 205*
  eutaxite 186*, 187
  facies 203, 204, 205*, 226*
  fiamme 187*, 209, 215
  fine-grained basal layer 207
  flattened and elongate gas cavities 217
  flattening ratio 217, 218
  flow banding 217
  flow directions 216
  flow folds 217
  flow foliation 217
  flow units 195
  flowage structures 216*
  fluidization 203
  folding 217
  folds 217

foliation plane  217
fumarolic mounds  202
gas pipes  202, 203, 209
glowing avalanche  225
glowing cloud  225
graded bedding  198, 199★
grading  198
grain size  206, 207, 208★
gravitational collapse of eruption column 222, 223★
height versus distance plot  227★
heterogeneous pumice lapilli  210
high temperature emplacement  218
high water marks  193
historic development of concepts  187
imbricated flat fragments  198, 200
imbricated pumice fragments  216★, 217
inclined blast  8, 222, 223★
intermediate- to large-volume ignimbrite sheets  209
intermediate-volume flow  192
inverse grading  191★, 198, 199★, 207
laminar flow  200, 217, *219*
lateral change to lahars  188★
lateral termination  146★
levees  191, 193★, 199★
lineation  217
lineation of gas pockets  217
lithics  *197, 198*, 206
lobate forms  191★, 193
magma mixing  210
magma zonation  210
magnetic orientation of fragments  218
massive  *114*, 192, 198, 212★
maximum diameter  207★, 208
mechanical compaction  213
median diameter  121★, 206
mineral zonation  29, 32★
mobility  225–227, 228★
multiple grading  198, 200
normal grading  198, 207
nuée ardente  78★, 122★, 186, *187*, 188, 189, 202, 221★, 295
orientation of elongate or platy particles  198–200
origin  222–225
paleovalley  194★, 217
pantelleritic ignimbrite  195★, 210, 212, 216
partly welded tuff  211, 213
peralkalic  209, 211, 215, 216★
peralkalic trachytic  230
phenocrysts, modal range  30★
phonolitic  100, 104★, 190★, 230
pierno  229
post-depositional compaction  217
pulled-apart pumice  215★, 216★, 217
pumice  100★, 104★, 191★, 206, 207, 215, 216★, 217
pumice flow deposits  191★, 221★
ramp structure  216★, 217
reddish tops  218, 309
relationship to topography  193–195
reversely graded  191★, 199★, 207
rhyolitic  104★, 187, 230
rounding of pumice  203★, 209
segregation structures  202
simple cooling unit  196
$SiO_2$ variation and volume  29★
size parameters  120★, 121★
small-volume flow  192, 198
sorting  121★, 206
stratification  201★
stretched pumice  215★, 216★, 217
strongly oriented flat fragments  200
structures  107★
St. Vincent type  122★
subaqueous (see subaqueous pyroclastic flow deposits)
symmetrical grading  198
temperature  210, 211
temperature profile  210★
tension cracks in matrix  216★, 217
texture  206–209
thermoremanent magnetism  218
trachytic  186, 199
trains of large fragments  198
tufolava  187★, 227, 228
unwelded base  196
unwelded deposits  198
U-shaped channel  217, 257★
vapor phase crystals  197
viscosity  215
viscous deformation of vitric fragments  213
vitroclastic texture  213
welded basal zone (vitrophyre)  210
welded shard  213★
welded tuff  214★, *219*, 222
welding and crystallization  211
welding compaction  213
welding threshold  213
zoned magma chamber  209
Pyroclastic fragments  89, 90★, 124
  anisotropic fabric  124
  apposition fabric  123
  Curie point  124
  deformation fabric  124
  dimensional orientation  124
  fabric  124
  isotropic fabric  124
  magnetic orientation  124
  origin  2, 8, *89*
  reworked  89
Pyroclastic stratigraphy  2, *346*, 368
Pyroclastic surge deposits  8, 107★, 146, 192, 194, *200*, 208, *219*
  antidune  107★
  ash cloud deposits  201★, 204★, 205★, *219*, 226
  ash hurricane  200

459

Pyroclastic surge deposits
  base surge deposits   107*, 219
  chute-and-pool   107*
  cross bedding   114*, 200, 201*
  crystal enrichment   208
  ground surge deposits   204*, 208, *219*
  lithic enrichment   209
  plane bed   107*
  size parameters   121*
  structures   107*
Pyroclastic turbidites (pyroturbidites)   282*, 292, 293, 385*
Pyroclasts   *89–91*, 116
  accidental   90
  angularity   123
  ash   90
  blocks   90, 91*
  bombs   90, 91*, *92*
  coarse ash   90*
  cognate (accessory)   90
  crystal   91
  fabric   116
  grain size   90, 116–118, 119*
  hydroclastic fragments   89
  juvenile (essential)   90
  lapilli   90
  lithic   90, 91
  maximum size   117
  mixture terms   92*
  pyrogenic crystals   90, *103*
  reworked   *89*, 91
  roundness   116, 122
  shape   116, *122*
  size distribution   116–122
  steam explosion   89
  vitric   91
Pyrogenic minerals   103–105
  glassy crusts   105
  impact pits   105, 106
  surface textures   105, 106
Pyroxene   22, 98, 105, 157, 159, 197, 398*

Quartz   22, 29, 98, 103, 104*, 197, 343*, 355*, 388*, 397, 398*
Quaternary tephra, dating   *352*, 356
  fission-track   356
  K-Ar   356
  radiocarbon   356

Radioactive waste   329
Rain flushing   127, 148
Rarefaction wave   80
REE pattern in Archean sedimentary rocks   7
Refractive index of glass   *163*, 165*, 354
Remanent magnetism   124, 309, 356
Resurgent cauldron   359–361
Reticulite   103
Retro-arc basin   384, 391
Retrograde boiling   51
  intrusive stocks   51

Reverse grading   67, 109*, 110, *148*, 149*, 289, *307, 308*
Rheological regimes   65*
Rheology   51–54
  elastic behavior   51
  fracture behavior   51
  plastic behavior   51
  viscous behavior   51
  yield stress   51, 298
Rhyodacite   17*, 22, 26*
Rhyolite   17*, 21
  alkalic   21
  calc-alkalic   22
  diffusional fractionation of $CO_2$   67
  diffusional fractionation of $H_2O$   67
  glass (obsidian)   43
  peralkalic   21, 22
  water content   43
Rift zones   14, 20, 330
Ring fractures   15, *230*
Rosin distribution   *118*, 120, 122
Rutile   338

Saline alkaline lake environment   329, *330–333,* 343
  alkaline hot springs   331
  alkalinity   331, 333
  alluvial fan   331
  closed hydrographic basins   330
  evaporation of groundwater   331
  groundwater circulation   331
  lateral zonation of authigenic minerals   331
  playa lake   331
  rift zones   330
  saline and alkaline brines   331, 332
Sandstone   91*, 399*
Sandwave beds   250–254
Sanidine (see feldspar)
Scoria   92–96
  cone   98*, 348
  flow   221*
Seamount   14, 19, 42, 265, *268–274*, 399
  alkali basalt   274
  lava-flow apron   273
  uplifted   274
Second boiling   50
Secondary base surge   248
Sediment gravity flows   276, 397, 398
Sedimentary basins, island arcs and magmatic arcs   384
  arc-trench gap   384, 389
  back-arc setting   384
  depositional facies   385*
  fore-arc basin   384
  intra-arc basin   384
  marginal basin   384
  remnant arc   384*
  retro-arc basin   384
  trench   384*
  trench-slope break   384*

Sedimentary rocks, volcanic contributions 3–6
Settling velocity 128
Shale, volcaniclastic 5, 6, 91★
Shard 96–106
  abrasion index 105, 106
  blister 106
  bubble 102★
  bubble junction 96, 101★
  bubble wall 96, 100★, 103
  bubble wall texture 105
  chemical etching 105, 106
  color 165
  cuspate 101★
  droplets 98★, 106
  fallout 106
  glass shards 96, 101★
  glassy crust 105
  impact pits 106
  intermediate composition 97, 98★
  mafic 97, 98★
  microcrater 98★, 106
  micropits 105, 106
  microprobe analysis 164, 165★
  morphology 96, 98★
  ovoids 98★, 101
  Pele's Hair 69, 98★, 101
  Pele's Tears 69, 98★, 101, 102
  peralkalic 97
  platy 94★, 96, 98★, 101★
  pumice 97, 101★
  pyroclastic flow 106
  pyroclastic surge 106
  refractive index 165★
  rhyolitic 98★, 100★
  shape 165★
  $SiO_2$ content 165★
  solution 332
  spheres 98★
  spheroidal bubbles 97, 102★
  surface texture 105, 106
  tear drops 98★, 101
  vesiculation 96
  welded tuff 96
Sheet lava flows 13, 19
Shield volcano 15, 348★, 379★
Shoaling submarine volcano 274, 275
Siderite 339
Sideromelane 76, 97–99★, 232★, 263, 279, 314, 315, 323, 334
Sideromelane tuff 97★, 235★, 314★, 344★
Silicic glass, alteration 327–330
  alkali ion exchange 328, 329
  alumino-silicate gel 332
  devitrification 329
  dissolution 330
  hydrated obsidian 326–329
  hydration 327
  hydration rate of obsidian 329
  hydrogen concentration profiles 328
  hydronium ion 328, 329
  ion exchange with groundwater 328
  leaching of U 329
  mobile alkali ions 328
  oxidation of Fe during hydration 328
  perlite 327, 328
Sillar 211
Siltstone 91★
Sintered tuff 219
$SiO_2$ variation with distance 156, 157, 160, 161★
Size distribution 116–120, 121★
  crystal fragments 207
  descriptive measures 118★, 120★
  distribution curves 117, 118
  Rosin's size-frequency distribution 118, 120
  size terminology 119★
Smectite 267, 315, 317, 323, 327, 334–338, 343, 344
  dioctahedral 317
  trioctahedral 317
Smectite(s)-bentonite 338
Sodalite foidite 262
Sorting-median diameter diagram 121★
Spalling of pillow rinds 13
Spalling shards 236, 270
Spatter 96
Spatter cone 96
Sphene 19, 22, 197, 338, 341–344★, 388★
Spilite 387
Spinel peridotite 17
Spot welding of shards 211
St. Vincent type eruption 59
Steam eruption 78–80
Steam explosion 231, 232
Stratigraphic belts 393★
Stratigraphic nomenclature 350–352
Stratovolcano 20, 98★, 297, 348, 368★, 381, 385, 399, 400
Strombolian eruption 59–74, 99★, 106, 243
Structural features of pyroclastic rocks 107★
Subaerial fallout (see fallout tephra, subaerial)
Subaerial deposits, major kinds 8, 9★
Subaqueous debris flow 289, 290★, 400–406, 407★
  mass transport, flow 400, 403★, 405
Subaqueous eruption column 83, 84, 292★
Subaqueous pyroclastic flow 281–296
  welding 295, 296
Subaqueous pyroclastic flow deposits 10, 283–296
  bedding 287, 288
  Bouma sequence 287, 291, 293
  components 285–287
  debris flow 289, 290
  depositional unit 288
  double grading 290–293
  emplacement temperature 295
Subaqueous pyroclastic flow deposits
  fabric 287
  facies, unwelded 288★

Subaqueous pyroclastic flow deposits,
   fiamme  293
   flame structure  294
   glass viscosity and water depth  295, 296
   grading  287–291
   grain size  287
   inverse grading  289
   load structure  294
   massive lower division  288*, 289
   median diameter  291
   normal grading  289
   occurrences  282*, 283*
   offshore  296
   pumice lapilli  286
   rip-ups  285
   size parameter  291*
   sorting  287, 291*
   underwater eruption  287, 292*
   upper division  290, 291
   viscosity of glass  295*, 296
   welded  293–296
Subaqueous pyroclastic processes  10
Subaqueous tephra  9, 10
Subduction zone  14, 387
Subglacial volcanoes  278, 280
Submarine ash layers (also see fallout tephra, submarine)
   correlation of  177–179
Submarine ash layers, Atlantic  182–185
   ice-rafted shards  183
Submarine ash layers, Pacific  176, 177, 179–182
   continental glaciation  181
   seismic reflections  178
   Worzel D Layer  177, 178
   Y-8 tephra  177
Submarine canyon  385, 386*
Submarine eruptions, silicic  279–281
Submarine explosive eruptions  83, 84, 232, 265, 292
   water depth  39, 265
Submarine fan  380, 385, 386, 391, 397
Submarine lava flow  77, 266–274, 279–281
Submarine lava fountaining  270
Submarine tuff (also see fallout tuff, submarine)  286
Submarine volcaniclastic rocks  163–185, 265–296
Submarine volcano  13, 14, 273, 274, 276, 400
Submarine weathering  323, 325
Subsonic to supersonic flow  65, 66
   geometry of reservoir-conduit-vent system  61
Sulfur  47, 48
Superhydration  178, 180*, 356
Surtseyan eruptions  82, 233

Tachylite  76, 96, 99*, 260, 263, 275, 314
Tectonic pattern, Cenozoic igneous rocks  400*
Temperature effects in pyroclastic flows  210, 211, 284, 295, 296
   adiabatic expansion  211
   atmospheric air  211
   charring of wood  211
   cooling rate  210*, 211
   degassing pipes  200–203, 211
   heat conserving mechanism  210
   height of eruption column  210
   liquidus temperature  210
   mineral geothermometers  211
   thermoremanent magnetism  211, 282*
   welding of glass shards  211, 295, 296
Tephra (also see fallout tephra)  89, 133*, 144
   chemical composition  6, 24*, 25*
   erosional processes  147
Tephrite  17*
   volatile content  43
Tephrochronology  352–356
Terminal settling velocity  128, 170
Texture  116–124
Thaumasite  335
Thermite-water systems  259
Thermodiffusion
   liquid-state  27
Thermoremanent magnetism  211, 218, 282*, 303, 309, 356
Tholeiite  17–19, 105*, 262, 316*
   Archean  409
   continental  19
   island arc  19
   ocean-island  19
   tuff  326
Thomsonite  335
Threshold settling velocity  125
Till  309, 310*
Titanaugite  19
Titanomagnetite  19, 22
Tonstein (kaolinite bentonite)  336–340
   chemistry of  339
   coal measures  338
   trace elements  339, 340
Tourmaline  388*
Trace fossils  397, 398
Trachyte  15, 17*, 408
   alkalic  21
   peralkalic  21, 22
Transform margin  14, 398
Transport velocity during eruption
   transition subsonic to supersonic  71
Trass (see also puzzolan)  189, 190*, 211
Tremolite-actinolite  388*
Trench  386, 392, 398
   accreted pelagic oceanic sediments  386
   andesitic volcanic terrane  387
   outer arc ridge  386
   stacked accretions  386
   subducting plate  386
Tridymite  338
Triple junction  398
   spreading center  398
   trench  398

Trona 331
Trondhjemite inclusions 33
Tropopause 62
Tuff 90, 91
  ash flow 91
  coarse 90★, 91
  fallout 91
  fine 90★, 91
  lacustrine 91
  subaerial 91
  submarine 91
Tuff cone 99★, 258, 259
Tuff ring 15, 99★, 258–262
  areal extent 260
  dimensions 260
  geometry 260
  melilite nephelinite 262
  volume 261
Tuffaceous breccia 91
Tuffaceous conglomerate 91
Tuffaceous mudstone 91
Tuffaceous sandstone 91
Tuffaceous siltstone 91
Tuffite 91★
Tufolava 187, 228
Turbidite 382, 385★, 391, 392, 402, 404★, 405, 406
Turbidity current 10, 113, 400, 403

Ugandite 53★
Ultramafic inclusions (see nodules)
Ultraplinian 61

Vent
  erosion 65, 67
  flaring 66
  maximum velocity in 66★
  pressures 66★, 67
  radius and gas content 67★
  radius and mass eruption rate 68★
  shape 65
Vent facies
  collapse breccia 358
  fumarolic alteration 358
  pyroclastic breccia 358
  radial dikes 358
Vent radius 67, 68★
Vesicles
  submarine lava 39, 45, 46
  vesiculation 55–58
  volatile content 39–42
Vesicular tuff 242–245
Viscosity 51–54
  alkali basalt magmas 53
  andesitic magma 53
  Bingham fluid (Bingham plastic) 52, 54
  calculated 53★
  density of melt 53
  Newtonian fluid 52, 54

  pressure 52, 53
  rhyolite 54★
  structural variations in melt 52
  tholeiite 54, 55★
Vitric particles 96–103
Vitric tuff 93★, 94★
Volatile fragmentation depth (VFD) 265
Volatiles 35–51
  Cl 48
  $CO_2$ 43–47
  composition in magma 37
  cupolas of magma chambers 27, 48
  F 48
  gradient in magma chambers 48
  $H_2O$ 38–43
  halogens 37
  inclusions in glass and crystals 36
  influence on chemical composition of magmas 18
  influence on degree of partial melting 18
  influence on eruptive behavior of magmas 36
  influence on melting temperature 18
  lava fountain 42, 44
  loss during subaerial explosive eruptions 42, 44
  methods of determination 36
  saturation 38, 43
  temperature-dependant equilibria 37
  viscosity of a magma 53★, 57, 58
Volcanic activity
  volcanoes, form 15
  volcanoes, tectonic setting 11–14
  world wide peaks 182
Volcanic activity units 347–349
  eruption 347, 348★
  eruptive epoch 348
  eruptive period 348
  eruptive phase 347, 348★
  eruptive pulse 347, 348★
Volcanic arc-basin region 392
Volcanic archipelago 14, 399
Volcanic blast 83
Volcanic center 368★
Volcanic chain 348, 368★
Volcanic district 368★
Volcanic dust veil 181
Volcanic energy 85–87
Volcanic eruption, sediment yield 389
Volcanic explosivity index (VEI) 87, 88★
Volcanic facies 356
  alluvial facies 357
  arc assemblage 357
  Archean caldera complex 358
  braided stream 357
  caldera outflow facies 356, *359–361*
  celadonite subfacies 367
  clinoptilolite facies 367
  compositional facies 362–365
  cone-complex facies 356
  detrital modes 398★, 399★

Volcanic, facies
  diagenetic rock facies   365–367
  distal facies   357
  distant-source facies   359
  facies diagram   367*
  fallout tephra   359
  fresh glass facies   367
  intermediate source facies   357, *359*, 374
  intracaldera facies   356, *359–361*
  lacustrine   361, 362*
  medial facies   357
  montmorillonite subfacies   367
  near-source facies   356, *358*, 374
  paleogeography   380*, 381*, 382
  petrofacies   365, 366*
  proximal facies   357
  reworked volcanic material   359
  stratigraphic succession   364
  stratovolcano   376*
  tectonic setting   385*, 390*
  vent facies   357, *358*
  volcanic arc   400–407
Volcanic formation   *350*, 362
  intertonguing   351*
  optimum thickness   350
Volcanic glass   312
Volcanic island chain   392
Volcanic production rate   4, 5
Volcanic province   368*
Volcanic region   368*
Volcanic rock stratigraphic units   348*, 349
Volcanic rocks
  classification   16
  composition   17*
  facies models   356
Volcanic sand   388*, 390*, 392, 406
Volcanic sediments, physical and chemical
  evolution   5
  volume   3–6
Volcanic stratigraphy   346
Volcanic turbidite   296, 402
Volcanic vent   368*
Volcaniclastic apron   278
  seismic reflectors   278
  submarine pyroclastic flows   278, 296
Volcaniclastic debris   357, 391
  depositional environments   391*, 403*
  types of deposits and position of magmatic arcs   385
Volcaniclastic flysch   397
  flame structure   397
  sole mark   397
  submarine fan   397
Volcaniclastic fragments   89
  alloclastic   89
  autoclastic   89
  epiclastic   89
  hydroclastic   89
  mixture   89, 92*
  pyroclastic   89
Volcaniclastic graywacke   7, 380, 382, 387, 391, 392–396
Volcaniclastic marine sequences   275
Volcaniclastic sandstone   404
Volcaniclastic siltstone   404
Volcaniclastic turbidite   387, 402
Volcanoes
  defined   11, 368*
  form   15
  growth of volcanic island   272*, 273, 274
  sheets   15
  shield volcanoes   15
  tectonic setting   11–14
Volcanotectonic depression   194
Volcanotectonic-sedimentary environment   389
Volume eruption rate   63*
Volume of pyroclastic rocks   3–6
Vulcanian eruption   60–62, *82, 83*, 233
  deposits   236–238, 243*
  ejecta velocity   82, 83
  mechanism   82

Wadati-Benioff subduction zones   14
Wairakite   344*
Waste disposal   312
Water (see $H_2O$)
Weathering
  pyroclastic rocks   313
Welded fallout tuff   124, 215, 219, 230
  degree of welding   215
Welded tuff   124, 186*, 187*, 189, 192, 212*, *213–218*
Welding in subaqueous environment   284, *293–296*
Worzel D layer   178–180

Xenoliths (see nodules, ultramafic)

Yield strength   52, 298
Yield stress   52, 298

Zeolites   313, 318–320
  chemical activity of pore fluid components   313
  facies   340
  formation of   325
  genesis   332
  saline alkaline lakes   319, 330–333
Zircon   29, 197, 338, 340, 356, 388

# Locality Index

For listings lacking page numbers, refer to localities in parentheses

Abitibi Volcanic Belt (Canada)
Absaroka Mountains (Wyoming)
Aegir Ridge   185
Africa   174, 408
   African Plate   12, 14
   African Rift Valley   260
   East Africa   15, 229, 262
   Karroo Basalt field   19
   Kilimanjaro (Tanzania)   12
   Mt. Suswa (Kenya)   100
   Nyiragongo (Zaire)   12, 44
   Olduvai Gorge (Kenya)   361
Aksitero Formation (Philippines)
Alaska (USA)   93, 95, 134, 176, 282, 368, 392, 394
   Adak Island   387
   Aleutian Arc   387
   Aleutian Basin   387
   Aleutian Islands   176
   Aleutian Ridge   182, 387, 388
   Aleutian Terrace   387
   Aleutian Trench   387
   Amlia Basin   387
   Andreanof Islands   387
   Aniakchak caldera   29, 227
   Atka Basin   387, 388
   Atka Island   387
   Augustine Volcano   27, 93, 95, 190, 192, 193, 198, 202, 210, 222, 245, 282, 296, 300
   Burr Point flows (Augustine Volcano)   296
   Fisher caldera (Unimak-Island)   227, 228
   Gulf of Alaska   177
   Hawley Ridge   387
   Juneau   134
   Katmai ash   177
   Katmai Volcano   12, 25, 97, 102, 177, 200, 210
   Kodiak Islands   386
   Mt. Bona   160
   Mt. Nathazha   160
   Nanwaksjiak Maar   239
   Novarupta   97, 102, 177, 200
   Tugamak Range   228
   Ukinrek Maars   241, 260, 261
   Valley of Ten Thousand Smokes   29, 189, 200, 211
   White River Ash   4, 134, 140, 160–162, 172

Aliamanu Crater (Hawaii)
America
   American Plate   373, 400
   Central America   179, 297, 352, 389
   North America   174, 337, 392–394, 396, 398, 399
   South America   174, 179, 182, 297, 368, 392, 396, 397, 407
Ammonia Tanks (Nevada)
Andes (South America)   386, 391
Andreanof Islands (Alaska)
Aniakchak (Alaska)
Antarctic Peninsula   397
Antarctic Plate   12
Antarctica   174, 260
Apache Spring (New Mexico)
Appalachian Basin   337
Arabian Plate   12, 14
Arizona (USA)   3
   Coronado Mesa   249
   Peridot Mesa   249
   Sugarloaf Mountain   249
Arkansas (USA)   282
Armenia (USSR)   187, 228, 229
Asama Volcano (Japan)
Ascension Island (Atlantic)   51
Ash layer L (eastern Pacific)   171
Asia   174
Askja (Iceland)
Aso caldera (Japan)
Atka Basin (Alaska)
Atlantic Ocean   46, 133, 172, 179, 266, 324, 334, 397, 406, 407
   Atlantic region   182, 336
   North Atlantic   180, 182–185, 266
   South Atlantic   185
Augustine Volcano (Alaska)
Australia   7, 408
   New South Wales   347
   Western Australia   408
Australian Plate   12
Azores (Portugal)   14, 153, 183, 185, 234
   Capelinhos (Fayal)   248
   Fayal   12
   Fogo A (São Miguel)   25, 29, 137, 142, 151, 153, 154, 156

465

Azores (Portugal)
  Fogo Volcano (São Miguel)   64, 131, 142
  Furnas tephra (São Miguel)   142
  Lake Fogo (São Miguel)   153, 154
  São Miguel   25, 51, 131, 153
  Setc Cidades Caldera (São Miguel)   143
  Terceira   137

Bali   133
Baltic Shield   6
Bandelier Tuff (New Mexico)
Basin and Range Province (USA)   401
Bay of Bengal   171
Bering Sea   228
Bestwig (Germany)
Bezymianny (Kamchatka)
Bikini Atoll (South Pacific)   247
Bishop Tuff (California)
Boos Volcano (Germany)
British Columbia (Canada)
Butte (Montana, USA)   135

California (USA)   287, 347, 392, 394–396
  Adobe (Bishop Tuff)   30
  Amargosa River   332
  Bishop Ash   134
  Bishop Tuff   4, 25, 29–31, 33, 104, 121, 134, 192, 201, 202, 214, 249
  Chicago Valley   332
  Chidago (Bishop Tuff)   30
  Coastal Belt Franciscan Province   401
  Coso Range   148
  Death Valley   116, 258
  Dublin Hills   332
  Gorges (Bishop Tuff)   30
  Great Valley Sequence   365, 366
  Greenwater Valley   332
  Klamath Mountains   365
  La Jolla-Navy-Coronado fan systems   397
  Lake Tecopa   331, 332
  Little Hebe crater   258
  Long Valley   23, 29, 33, 134, 192
  Mono (Bishop Tuff)   30
  Monterey fan   397
  Mt. Lassen   301
  Obispo Tuff   97, 289, 333
  Resting Spring Range   332
  Sacramento Valley   401
  San Andreas fault   401
  San Joaquin (Bishop Tuff)   30
  Santa Cruz Island   282
  Santa Maria   289
  Shoshone   332
  Sierra Nevada   282, 293, 297, 365, 366, 368, 396
  Tableland (Bishop Tuff)   30
  Ubehebe   116, 121, 194, 249, 251, 253, 255
  Wrightwood   302, 304, 306
Campanian tuff (Italy)

Canada   134, 135, 283, 357, 408
  Abitibi Volcanic Belt   408, 409
  Archean greenstone-belt volcanoes   367, 378, 408
  Birch-Uchi   408
  British Columbia   135, 160, 401
  Canadian Shield   408
  Dawson City   160
  Edmonton   134
  Lake of the Woods (Ontario)   379
  Lake of the Woods-Wabigoon   408
  Queen Charlotte Transform   401
  Rouyn – Noranda area (Quebec)   280, 282
  Superior Province   408
  Timmins-Kirkland Lake-Noranda (Ontario, Quebec)   408
  Wascana (Saskachawan)   180
  Western Peninsula of Lake of the Woods   379–382
  White Horse   134, 160
  White River   160
  White River ash   4, 134, 140, 160–162, 172
  Yellowknife Supergroup   7
  Yukon   160
  Yukon River   160
Canary Islands (Spain)   14, 25, 28, 184, 185, 276, 278, 307
  Agaete (Gran Canaria)   216
  Arguineguin (Gran Canaria)   216
  Arico (Tenerife)   186
  Duraznero Crater (La Palma)   240, 246
  Granadilla tephra (Tenerife)   142
  Gran Canaria   25, 28, 29, 33, 43, 115, 185, 195, 210, 212, 215, 216, 300, 330
  Ingenio (Gran Canaria)   216
  La Palma   12, 267, 269, 273, 274
  Las Palmas (Gran Canaria)   216
  Marteles Caldera (Gran Canaria)   236, 237, 246
  Medano (Tenerife)   199
  Mogan (Gran Canaria)   216
  Roque Nublo Formation (Gran Canaria)   115, 300, 330
  San Nicolas (Gran Canaria)   216
  Tejeda (Gran Canaria)   216
  Tenerife   142, 146, 186, 187, 199
Cape Verde Islands (Portugal)   14
Capel Curig Volcanic Formation (Wales)
Capelinhos (Azores)
Caribbean   180, 406
Caribbean Plate   12, 389
Caribbean Sea   172, 406, 407
Carlsberg Ridge (Indian Ocean)   12
Carnegie Ridge (eastern Pacific)   171
Carpenter Ridge (Colorado)   29
Cascade Range (USA)   297, 359, 368, 395, 399
  High Cascade Range   368
  Southern Cascades   297
  Western Cascade Range   368

Central America (see America)
Central Europe 338
Cerro Colorado (New Mexico)
Chile 340
    Corall Quemado 260
    Nilahue 260
    Quizapu 156, 157
Chuckanut – Swauk sequence (Cascades) 401
Coast Range Seamount Province (western USA) 401
Coatepeque Volcano (El Salvador)
Cocos Plate 12, 389
Cocos Ridge (eastern Pacific) 171
Columbia 171, 391
Columbia River Basalt field (USA) 19
Columbia River Plateau 401
Colorado (USA)
    Carpenter Ridge 29
    Conejos Formation 373
    Del Norte 372
    Fish Canyon (San Juan Mountains) 27, 29
    Gunnison River 372
    La Garita Caldera 192
    La Jara Canyon Member, Treasure Mountain Tuff 377
    Lake Fork Formation 373
    Platoro Caldera 372, 375, 377, 378
    Rio Grande 372
    Rio Grande Rift 401
    San Juan Basin 377
    San Juan Formation 373
    San Juan Mountains 23, 351, 367, 373, 375
    San Juan Volcanic Field 368, 371–375
    San Luis Valley 377
    Snowshoe Mountain 29
    Summer Coon Volcano 372, 374, 376, 377
    Treasure Mountain Tuff 375, 377, 378
    West Elk Formation 373
Conejos Formation (Colorado)
Corall Quemado (Chile)
Cordillera 347
Cordilleran geosyncline 392, 393, 401
Cordilleran Volcanic Region 368
Coronado Mesa (Arizona)
Coseguina (Nicaragua) 139
Coso Range (California)
Cotopaxi Volcano (Ecuador) 12, 187, 188, 222, 224, 301
Crater Lake (Oregon)
Crete 169
Cromwell Current (eastern Pacific) 171
Cudahy (Kansas, USA) 180
Cumberland Bay (South Georgia Island) 398, 399
Cyprus 266, 267, 271

Death Valley (California)
Deccan Basalt field (India) 19
Deception Island (Antarctica) 260

Deep Sea Drilling Project 386
    Site 186: 387, 388
    Site 192: 182
    Site 253: 275
    Site 369: 185
    Site 397: 185, 276
    Site 397a: 185
    Site 497: 278
Denmark 167
Devine Canyon (Oregon)
Dillenburg (Germany)
Dominica (Lesser Antilles)
Duraznero Crater (Canary Islands)

East Africa 15
East Pacific Rise (Pacific)
East Seamounts (Pacific)
Eastern Mediterranean 174
    Campanian layer 174
    Santorini layer 174
Ecuador 171
Eifel (Germany)
El Chichón (Mexico)
El Salvador 138
    Coatepeque Volcano 137, 138, 140
    Lake Coatepeque 138
    Santa Ana 138
    Santa Tecla 138
    Sonsonate 138
Elegante Crater (Mexico)
Ellensburg Formation (Washington)
England 336, 347
Eolian Islands (Italy)
Etna Volcano (Italy)
Eurasian Plate 12, 14
Europe 174, 340
Extinct Ridge axis 183

Faeroes Ridge (North Atlantic) 185
Falcon Island (Tonga Islands) 260
Famous Area (Atlantic) 46, 326
Fantale Tuff (Ethiopia) 29
Farallon Plate (Pacific) 400
Fayal (Azores)
Fiji 402, 405
Fiji Plateau 165, 169
Fisher Caldera (Aleutian Islands)
Fogo A (Azores)
Fogo Volcano (Azores)
Fort Benton (Wyoming)
Fraction Tuff (Nevada)
France 264
Fuego Volcano (Guatemala)

Galapagos Islands (Ecuador) 14, 171
Galapagos Mounds area (eastern Pacific) 25
Galungung (Java)
Germany 347
    Andernach 136

467

Germany
  Bestwig 286
  Boos Volcano (Eifel) 94
  Burgbrohl 136
  Dillenburg 268
  Eifel 34, 94, 95, 104, 114, 117, 129, 132, 143, 145, 146, 149, 152, 190, 194, 240, 244, 252, 253, 255, 257, 261, 262, 307
  Herchenberg Volcano (E-Eifel) 95, 149
  Laacher See tephra 25, 29, 31–33, 63, 104, 113, 114, 117, 132, 135, 136, 139, 140, 143, 145, 146, 149, 152, 172, 189, 190, 194, 205, 206, 242, 244, 245, 248, 251–253, 255–257, 361
  Lummerfeld Maar (E-Eifel) 238
  Madfeld 286
  Mayen 136
  Meerfelder Maar (W-Eifel) 262
  Moselle 136
  Neuwied 136
  Niedermendig 136
  Pulvermaar 262
  Rothenberg Volcano (E-Eifel) 34, 243
  Rhine River 136
  Ruhr 338, 339
  Ulmener Maar (W-Eifel) 262
  West Eifel 262
Gilliss Seamount 273
Gran Canaria (Canary Islands)
Great Valley Sequence (California)
Greater Antilles 14
Greenland 184, 185
Greenland Sea 184
Grenada Basin (Lesser Antilles)
Grouse Canyon (Nevada)
Guatemala 138, 357, 389
  Fore-arc area 357
  Fuego Volcano 25, 63, 222
  Guatemalan Highlands 180
  Lake Atitlan Caldera 177, 180
  Los Chocoyos Ash 178–182
  Santa Maria Volcano 63, 140, 301, 390
  Santiaguito Volcano (Santa Maria) 63, 223
  Tecum Uman Complex 179
Gulf of Mexico 180, 181

Halemaumau (Hawaii)
Hawaii (USA) 19, 39, 42, 44, 46, 48, 54, 55, 263, 275, 326, 327, 399
  Aliamanu Crater 258
  Halemaumau 249
  Hawaiian ash 139
  Hawaiian Islands 14, 176
  Honolulu Group (Oahu) 315
  Kilauea Iki 69
  Kilauea Volcano 12, 39, 42, 46, 48, 102, 105, 263, 272, 371
  Koko Crater (Oahu) 246, 258
  Makaopuhi Lava Lake 77

  Mauna Loa 12, 15, 263, 272
  Mauna Ulu 102, 105
  Oahu 315, 326
  Puu Hou 264
  Salt Lake Crater (Oahu) 258
Heimaey (Iceland)
Hekla (Iceland)
Herculaneum (Italy)
Hole-in-the-Ground Maar (Oregon)
Honduras 138
Honshu (Japan)
Hunt's Hole (New Mexico)
Hverfjall tuff ring (Iceland)

Iblean Mountains (Italy)
Iceland 14, 19, 33, 79, 82, 142, 182–185, 234, 262, 263, 274, 278, 311, 314, 320, 326, 352
  Askja 12, 27, 29, 33, 142
  Eastern Iceland 343
  Heimaey 12, 63
  Hekla 12, 27, 63, 133, 139, 140, 142, 160, 161, 172–175, 297
  Hverfjall tuff ring 246
  Reydarfjordur 318, 343, 344
  Reykjanes 279
  Surtsey 12, 82, 234, 276, 277, 323, 324
Iceland-Jan Mayen Ridge axis 183, 185
Idaho (USA) 135
Independence Volcano (Wyoming)
India 171, 408
Indian Ocean 133, 181
Indonesian Arc 181, 297
Iraklion (Crete) 169
Irazu Volcano (Costa Rica) 304, 306
Ireland 185, 282, 336
Italy 229
  Avellino 129, 142, 172
  Campanian tuff 174
  Eolian Islands 82
  Etna Volcano (Sicily) 12
  Herculaneum 63, 203
  Iblean Mountains (Sicily) 314
  Neapolitan volcanic province 177
  Palagonia (Sicily) 271, 314
  Pantelleria 213
  Phlegrean fields 229
  Pompeii 63, 129, 140, 142, 212
  Sardinia 214
  Sicily 271, 314
  Stromboli (Eolian Islands) 12, 69
  Vesuvius 12, 63, 86, 129, 131, 172, 203, 212, 297
  Vulcano (Eolian Islands) 82, 233, 238, 244
Ito pyroclastic flow (Japan)
Iwo Jima (Japan)
Izu volcanic chain (Japan)

Jan Mayen 182, 183
Jan Mayen Fracture Zone 183

Japan   23, 27, 120, 176, 282, 297, 304, 340, 341, 386
  Asama Volcano   127, 189, 364, 365
  Aso Caldera   209, 210
  Honshu   369
  Ito pyroclastic flow   227
  Iwo Jima   79, 260
  Izu volcanic chain   369
  Japan Trench   386
  Japanese Arc   392
  Komagatake pumice flow   199, 223, 224
  Mukaiyama Volcano   256
  Nomashi Formation (Oshima)   371
  Okinobe Caldera   285, 288, 293
  Ontake tephra   142
  Oshima Group   369, 370
  Oshima Volcanic Center   12, 367
  Oshima Volcano   86, 87, 350, 368–370
  Sashikiji Formation   371
  Shikotsu Caldera   27, 29
  Shikotsu tephra   142
  Tokachi-Dake Volcano   304, 306
  Tokiwa Formation   282, 283, 285
  Towada pumice flow   208
  Towada Volcano   27
  Usu Volcano   87
  Wadeira Tuff D   285, 291
  Yatsuga-Dake   301
  Yuba Formation (Oshima)   371
Java   133
  Galungung   301
  Java Sea   133
  Kelut ash   158, 159
  Kelut Volcano   12, 133, 159, 301
  Krakatau   12, 23, 27, 29, 156, 192, 282, 283
  Merapi   12
  Raung   301
John Day Formation (Oregon)
Juan de Fuca Ridge   39

Kamchatka (USSR)   166
  Bezymianny   63
  Kamchatka Peninsula   182
  Kliuchevskoi Volcano   12
  Stübel Crater   260
Kansas   170
Karpathos (Greece)   169
Karroo Basalt field (Africa)
Katmai Volcano (Alaska)
Kelut Volcano (Java)
Kenya   100, 332, 356
Kialineo (North Atlantic)   185
Kilauea Iki (Hawaii)
Kilauea Volcano (Hawaii)
Kilimanjaro (Africa)
Klamath Mountains (California)
Klirou (Cyprus)   271
Kliuchevskoi Volcano (Kamchatka)
Kodiak Islands (Alaska)

Koko Crater (Hawaii)
Kolbeinsey Ridge   185
Komagatake pumice flow (Japan)
Krakatau (Java)
Kuriles (USSR)   176, 260

La Garita Caldera (Colorado)
Laacher See (Germany)
La Palma (Canary Islands)
La Soufrière (Lesser Antilles)
Lake Atitlan Caldera (Guatemala)
Lake Fork Formation (Colorado)
Lake of the Woods (Canada)
Lake of the Woods-Wabigoon (Canada)
Lake Ruapehu (New Zealand)
Lake Magadi region (Kenya, Tanzania)   332
Lake Tecopa (California)
Lau Basin (South Pacific)   405
Lau Ridge (South Pacific)   405
Lava Creek Tuff (Wyoming)
Lesser Antilles   14, 283, 290, 406, 407
  Barbados   172, 407
  Dominica   172, 173, 282, 289, 290, 295, 407
  Grenada   407
  Grenada Basin   172, 282, 289, 290, 293, 406, 407
  Guadeloupe   172, 407
  La Soufrière (St. Vincent)   12, 27, 63, 80, 187, 223
  Lesser Antilles Arc   173, 405, 406
  Martinique   172, 189, 290, 407
  Micotrin Volcano (Dominica)   177
  Mt. Pelée (Martinique)   12, 23, 59, 90, 121, 186, 187, 192, 198, 202, 205, 206, 222, 224–227, 282, 295, 299, 303
  Rivière Blanche (Martinique)   225
  Roseau fallout tephra   172, 173, 177
  Roseau subaqueous pyroclastic flow   290, 291
  Roseau Valley   293
  St. Lucia   172, 290, 407
  St. Pierre (Martinique)   186, 225
  St. Vincent   172, 208, 290, 407
  Tobago   407
  Tobago Basin   290
Limagne (France)   264
Long Valley (California)
Los Chocoyos ash (Guatemala)
Lummerfeld Maar (Germany)

Madeira (Portugal)   14, 277
Madfeld (Germany)
Magdalena Lowland (Columbia)   391
Magdalena Valley (Columbia)   391
Makaopuhi lava lake (Hawaii)
Malaya   171
Malekula Island (New Hebrides)   400, 402–404
Mariana Arc   42, 386
Mariana Trench   386
Marteles Caldera (Canary Islands)

469

Martinique (Lesser Antilles)
Mascall Formation (Oregon)
Mauna Loa (Hawaii)
Mauna Ulu (Hawaii)
Mayon (Philippines)
Mazama ash (Oregon)
Mediterranean   169, 172, 174, 179
Meerfelder Maar (Germany)
Merapi (Java)
Mexico   42, 368
   El Chichón   192
   Elegante Crater (Sonora)   249, 256
   La Primavera   142
   Nevado de Toluca Volcano   140, 152, 299
   Paricutin   12, 42, 43, 137, 147
   Revillagigedo Islands   39
   Tala Tuff   29
   Toluca Pumice   129, 140, 142, 152, 172
Micotrin Volcano (Lesser Antilles)
Mid-Atlantic Ridge   12, 13, 182, 185, 272, 326, 327
Middle America Trench   389
Minoan tephra   169, 172
Missoula (Montana, USA)   135
Mohns Ridge axis   183, 185
Monotony Tuff (Nevada)
Montana (USA)   135, 357
Moon   101
Moscow (Idaho, USA)   135
Mt. Lamington (Papua)   222–225
Mt. Lassen (California)
Mt. Mayon (Philippines)
Mount Mazama (Crater Lake) (Oregon)
Mt. Pelée (Lesser Antilles)
Mt. Rainier (Washington)
Mount St. Helens (Washington)
Mukaiyama Volcano (Japan)

Nanwaksjiak Maar (Alaska)
Nazca Plate   12
Neapolitan volcanic province (Italy)
Nebraska (USA)   365
Nevada (USA)   3, 365, 394–396
   Ammonia Tanks   29
   Eureka   396
   Fraction Tuff   29
   Grouse Canyon Tuff   29
   Monotony Tuff   29
   Rainier Mesa   29
   Spearhead Tuff   29
   Timber Mountain Caldera   27
   Tiva Canyon Tuff   29
   Topopah Spring Member   29
   Topopanah ash flow   25
Nevado de Toluca Volcano (Mexico)
New Mexico (USA)   368
   Apache Spring   29
   Bandelier Tuff   67, 196, 201
   Cerro Colorado   246

Hunt's Hole   250
Rio Grande Rift   401
Tshirege Member   29, 201
Valles Caldera   23, 27, 134, 196, 201
Zuni Salt Lake   243
New South Wales (Australia)
New Zealand   43, 79, 80, 166, 189, 295, 297, 340, 350, 389, 405
   Auckland   249
   Hakonui Hills   341, 390
   Hapeotoroa   249
   Lake Ruapehu   247
   Lake Tarawera   249
   Mangaone tephra   142
   Moura   249
   Murihiku Supergroup   341
   Ngauruhoe Volcano   12, 63, 198, 221, 224
   North Island   195
   North Range Group   341
   Rotomahana   249
   Rotomakariri   249
   Rotorua   249
   Southland syncline   340
   Tarawera   12, 249
   Tarawera-Rotomahana-Waimangu area   249
   Taringatura Group   341
   Taupo   141, 142, 227, 350
   Te Anau   390
   Te Ariki   249
   Te Wairoa   249
Ngauruhoe Volcano (New Zealand)
Nilahue (Chile)
Ninetyeast Ridge (Indian Ocean)   274, 275
Nomashi Formation (Japan)
North America (America)
North American Plate   12
North Island (New Zealand)
North Sea tephra   185
Norwegian Sea   184
Novarupta (Alaska)
Nyiragongo (Africa)

Oahu (Hawaii)
Obispo Tuff (California)
Oklahoma (USA)   282
Olduvai Gorge (Africa)
Okinobe Caldera (Japan)
Onion Creek (Utah)   180
Oregon (USA)   28, 106, 123, 135, 262, 347, 362, 395, 399
   Basin Member (John Day Formation)   367
   Crater Lake   4, 23, 27, 28, 106, 123, 134, 142, 172, 192, 224, 358
   Devine Canyon   29
   Hole-in-the Ground Maar   239, 241, 262
   John Day Formation   147, 359, 366, 367
   Kimberly Member (John Day Formation)   367
   Mascall Formation   90

Mazama ash   29, 134, 135, 139, 159, 160
Mt. Mazama   4, 60, 106, 123, 134, 139–141, 157, 170, 192, 224, 358
Picture Gorge Ignimbrite   367
Portland   135
Prineville Tuff cone   246, 247, 259
Turtle Cove Member (John Day Formation)   367
Oshima (Japan)

Pacific   14, 46, 138, 168, 177, 179, 181, 185, 228, 273, 334, 336, 389
  East Pacific Rise   12, 13, 46, 273, 352
  East Seamount   273
  Eastern Pacific   25, 170, 178, 179, 181
  North Pacific   166, 169, 176, 182
  Northwest Pacific   180
  Pacific basin   182
  Pacific island arc   179, 368
  Pacific Ocean floor   352
  Pacific Ocean margin   14, 399
  Pacific Plate   12, 400
  South Pacific   247, 284
  West Seamount   273
Pahvant Butte (Utah, USA)   235
Palagonia (Italy)
Panama Basin (eastern Pacific)   171
Parana basalt field (South America)   19
Paricutin Volcano (Mexico)
Payson Canyon (Utah)   180
Pearlette ash (Kansas)   170, 180
Pelée (Mt. Pelée)
Pematang Bata (Sumatra)
Peridot Mesa (Arizona)
Peru   171
Philippine Plate   12
Philippines   62, 282
  Aksitero Formation   291
  Mt. Mayon   12, 23, 188, 192, 222–224
  Taal Volcano   12, 62, 74, 121, 238, 247, 251, 252, 260
Phlegrean fields (Italy)
Platoro Caldera (Colorado)
Pompeii (Italy)
Porto Santo (Madeira, Portugal)   277
Prineville Tuff cone (Oregon)
Puget Sound (Washington)
Pulvermaar (Germany)
Puu Hou (Hawaii)

Quizapu Volcano (Chile)

Radkevich (Tyatya Volcano, Kuriles, USSR)   260
Rainier Mesa (Nevada)
Ran Ridge   185
Raung (Java)
Revillagigedo Islands (Mexico)
Reykjanes Ridge   41, 46, 47, 185

Rhodos (Greece)   169, 282
Rio Grande Rise (South Atlantic)   275
Rivière Blanche (Lesser Antilles)
Rockall (North Atlantic)   185
Rocky Mountains (USA)   368, 372
Roque Nublo Formation (Canary Islands)
Roseau ash (Lesser Antilles)
Roseau Valley (Lesser Antilles)
Rothenberg Volcano (Germany)
Rotomahana (New Zealand)
Rotomakariri (New Zealand)
Rouyn-Noranda area (Canada)
Ruhr (Germany)

Sacramento Valley (California)
Salt Lake Crater (Hawaii)
Samoa   14, 278, 405
San Blas "Oceanic Formation"   352
San Juan Islands (Washington)
San Juan Mountains (Colorado)
San Juan Volcanic Field (Colorado)
Sandebugten sandstone (southern South America)   398
Santa Cruz Island (California)
Santa Maria (California)
Santa Maria (Guatemala)
Santiaguito Volcano (Guatemala)
Santorini Tuff (Greece)   174
Santorini Volcano (Greece)   4, 12, 169, 177
São Miguel (Azores)
Sardinia (Italy)
Sashikiji Formation (Japan)
Scotia Arc (South America)   397
Scotland   185, 347
Sentinel Gap (Washington)
Sicily (Italy)
Sierra Nevada (California)
Shikotsu Caldera (Japan)
Snake River Plain (Idaho, USA)   401
Snowdon (Wales)
Snowshoe Mountain (Colorado)
Soufrière (La Soufrière) (Lesser Antilles)
South America (America)
South American Plate   12
South Georgia Island (South Atlantic)   396–399
South Sandwich Islands (South Atlantic)   168, 185
Southland syncline (New Zealand)
Spearhead (Nevada)
Spirit Lake (Washington)
St. Pierre (Lesser Antilles)
St. Vincent (Lesser Antilles)
Stromboli (Italy)
Stübel Crater (Kamchatka)
Sugarloaf Mountain (Arizona)
Sumatra   171, 172
  Pematang Bata   260
  Toba   4, 23, 171, 172, 174–177, 192
Summer Coon Volcano (Colorado)

471

Superior Province (Canada)
Surtsey (Iceland)
Sweden   347

Taal Volcano (Philippines)
Tala Tuff (Mexico)
Tambora (Indonesia)   4
Tarawera (New Zealand)
Tarawera-Rotomahana-Waimangu area (New Zealand)
Taupo (New Zealand)
Tecum Uman Complex (Guatemala)
Tenerife (Canary Islands)
Terceira (Azores)
Tierra del Fuego (South America)   396–398
Timber Mountain Caldera (Nevada)
Tiva Canyon (Nevada)
Toba (Sumatra)
Tokachi-Dake Volcano (Japan)
Tokiwa Formation (Japan)
Tonga Arc   405
Tonga Ridge   405
Tonga Trench   405
Tonga-Kermadec   386
Topopah Spring Member (Nevada)
Topopanah ash flow (Nevada)
Tortuga Seamount   274
Toutle River (Washington)
Towada Volcano (Japan)
Treasure Mountain Tuff (Colorado)
Trinidad   407
Troodos ophiolite (Cyprus)   266, 267, 271
Turkey   169
Tutu Seamounts   274
Tyatya Volcano (Kuriles, USSR)   260

Ubehebe Maar (California)
Ukinrek Maars (Alaska)
Ulmener Maar (Germany)
United States   3, 347
   Western U.S.   3, 135, 354, 355, 367, 400
Utah (USA)   3, 394–396

Valles Caldera (New Mexico)
Valley of Ten Thousand Smokes (Alaska)
Vesuvius (Italy)
Viti Levu (Fiji Islands)   405
VTTS (Valley of Ten Thousand Smokes)
Vulcano (Italy)

Wadeira Tuff D (Japan)
Wagon Bed Formation (Wyoming)
Wales (United Kingdom)   168, 282, 294, 347
   Capel Curig Formation   293, 295, 296
   North Wales   189
   Snowdon   293, 294
   South Wales   282, 294

Walvis Ridge (South Atlantic)   275
Washington (USA)   93, 94, 100, 135, 399
   Cowlitz River   135
   Electron lahar   301
   Ellensburg Formation   93, 94, 100, 301, 303, 305
   Glacier Peak ash   134
   Lewis River   135
   Morton   135
   Mt. Adams   135
   Mt. Rainier   134, 282, 297, 304–306, 386
   Mount St. Helens   2, 4, 8, 9, 12, 23, 25, 60, 75, 83, 84, 125, 126, 132, 134–136, 142, 156, 189, 191–194, 202, 206, 210, 218, 222, 223, 225–227, 263, 297, 299, 301, 302, 304–306, 349
   Olympia   135
   Osccola lahar   301
   Packwood   135
   Paradise lahar (Mt. Rainier)   301
   Puget Sound lowland   303
   Randle   135
   San Juan Islands   395
   Seattle   177
   Sentinel Gap   93, 94
   Spirit Lake   75, 135, 136, 301
   Spokane   135
   Toutle River   135, 301
West Eifel (Germany)
West Elk Formation (Colorado)
West Seamount (Pacific)
White River ash (Alaska and Canada)
Worzel D layer (eastern Pacific)   175, 177–180, 181
Wrightwood (California)
Wyoming (USA)   168, 337
   Absaroka Mountains   297
   Absaroka Volcanics   357, 363, 365
   Fort Benton   336
   High Plain Sequence   365
   Huckleberry Ridge (Yellowstone)   29
   Independence Volcano   363–365
   Lava Creek Tuff (Yellowstone)   27, 29
   Wagon Bed Formation   333
   Yellowstone Park   23, 79, 80, 134, 192, 363, 365

Y-5 ash layer (Mediterranean)   177
Y-8 tephra (Caribbean)   177, 180, 181
Yahgan Formation (Tierra del Fuego)   398
Yatsuga-Dake (Japan)
Yellowknife Supergroup (Canada)
Yellowstone Park (Wyoming)
Yuba Formation (Japan)
Yukon basin (USA)   4

Zuni Salt Lake (New Mexico)

W. E. Galloway, D. K. Hobday

# Terrigenous Clastic Depositional Systems

Applications to Petroleum, Coal, and Uranium Exploration

1983. 237 figures. XV, 423 pages
ISBN 3-540-90827-7

**Contents:** The Fuel-Mineral Resource Base. – Approaches to Genetic Stratigraphic Analysis. – Alluvial-Fan Systems. – Fluvial Systems. – Delta Systems. – Clastic Shore-Zone Systems. – Terrigenous Shelf Systems. – Terrigenous Slope and Basin Systems. – Lacustrine Systems. – Eolian Systems. – Depositional Systems and Basin Hydrology. – Coal. – Sedimentary Uranium. – Petroleum. – References. – Index.

Here for the professional geologist and advanced student is a state-of-the-art account of clastic depositional environments and their associated mineral fuel deposits. Key concepts for the recognition and three dimensional sub-surface analysis of sedimentary basins are presented and their applications to resource development demonstrated. The authors, internationally recognized experts in fundamental and applied sedimentology and sedimentary economic geology, utilize a multi-disciplinary approach that integrates genetic stratigraphy, hydrogeology and elements of geochemistry and tectonics. They start by reviewing and integrating processes and genetic facies of clastic depositional systems, including fluvial, deltaic, shorezone, shelf, submarine slope-basin, eolian, lacustrine and alluvial fan systems.

Throughout, the emphasis is on the recognition, mapping and three dimensional reconstruction of clastic deposits, primarily from sub-surface data. An important and unique chapter reviews the hydrology of sedimentary basins. Final chapters discuss applications of genetic facies analysis to mineral fuel resource appraisal, exploration and development.

Profusely illustrated, this book is unique in its focus and emphasis on processes and interrelationships among stratigraphic, hydrologic and economic aspects of depositional systems. It will be welcomed by the explorationist, sedimentologist and student alike.

Springer-Verlag
Berlin
Heidelberg
New York
Tokyo

U. Förstner, G.T.W. Wittmann

# Metal Pollution in the Aquatic Environment

With contributions by F. Prosi, J. H. van Lierde
Foreword by E. D. Goldberg

Springer Study Edition

2nd printing of the 2nd revised edition. 1983.
102 figures, 94 tables. XVIII, 486 pages
ISBN 3-540-12856-5

**Contents:** Introduction. – Toxic Metals. – Metal Concentrations in River, Lake, and Ocean Waters. – Metal Pollution Assessment from Sediment Analysis. – Metal Transfer Between Solid and Aqueous Phases. – Heavy Metals in Aquatic Organisms. – Trace Metals in Water Purification Processes. – Concluding Remarks. – Appendix. – References. – Subject Index.

**From the reviews:** "Only two years after the publication of the first edition, a second reprint of this book became necessary. The authors made use of this opportunity to update their reviews of certain topics that recently received increasing interest in the field of research. ... The list of references, containing not less than 2500 entries, unequivocally stresses the fact that all efforts were made to give this study an extremely well documented background. At a time when increasing efforts are being made to understand, and resolve, but above all prevent ecological disasters, this book is a useful contribution that covers an important field of the environmental sciences. Whoever is concerned with anti-pollution programs will find it impossible to bypass this study." *Hydrobiologia*

"... This volume is a must for enironment protection agencies, health authorities, local authorities, manufacturing concerns and the waste disposal industry..."
*Waste Disposal and Water Management in Australia*

Springer-Verlag
Berlin
Heidelberg
New York
Tokyo

THE LIBRARY
ST. MARY'S COLLEGE OF MARYLAND
ST. MARY'S CITY, MARYLAND 20686